Study Guide for Blei and Odian's

GENERAL, ORGANIC, AND BIOCHEMISTRY
Connecting Chemistry to Your Life
Second Edition

Marcia L. Gilette
Indiana University Kokomo

Wendy Gloffke
Cedar Crest College

W. H. FREEMAN AND COMPANY
New York

ISBN: 0-7167-6167-X
EAN: 97807167-6167-9

© 2006 by W. H. Freeman and Company

All rights reserved.

Printed in the United States of America

First printing

W. H. Freeman and Company
41 Madison Avenue
New York, NY 10010
Houndmills, Basingstoke RG21 6XS England
www.whfreeman.com

Contents

Periodic Table of the Elements

Key: 1 — Atomic number; H — Symbol; 1.0079 — Atomic mass

Metals / Nonmetals / Metalloids

	1 I IA	2 II IIA	3 IIIB	4 IVB	5 VB	6 VIB	7 VIIB	8 ←	9 VIIIB	→ 10	11 IB	12 IIB	13 III IIIA	14 IV IVA	15 V VA	16 VI VIA	17 VII VIIA	18 VIII VIIIA
1	1 H 1.0079																	2 He 4.002
2	3 Li 6.941	4 Be 9.012											5 B 10.81	6 C 12.01	7 N 14.01	8 O 16.00	9 F 19.00	10 Ne 20.18
3	11 Na 22.99	12 Mg 24.31											13 Al 26.98	14 Si 28.09	15 P 30.97	16 S 32.07	17 Cl 35.45	18 Ar 39.95
4	19 K 39.10	20 Ca 40.08	21 Sc 44.96	22 Ti 47.87	23 V 50.94	24 Cr 52.00	25 Mn 54.94	26 Fe 55.85	27 Co 58.93	28 Ni 58.69	29 Cu 63.55	30 Zn 65.41	31 Ga 69.72	32 Ge 72.64	33 As 74.92	34 Se 78.96	35 Br 79.91	36 Kr 83.80
5	37 Rb 85.47	38 Sr 87.62	39 Y 88.91	40 Zr 91.22	41 Nb 92.91	42 Mo 95.94	43 Tc [98]	44 Ru 101.07	45 Rh 102.91	46 Pd 106.42	47 Ag 107.87	48 Cd 112.41	49 In 114.82	50 Sn 118.71	51 Sb 121.76	52 Te 127.60	53 I 126.90	54 Xe 131.29
6	55 Cs 132.91	56 Ba 137.33	71 Lu 174.97	72 Hf 178.49	73 Ta 180.95	74 W 183.84	75 Re 186.21	76 Os 190.23	77 Ir 192.22	78 Pt 195.08	79 Au 196.97	80 Hg 200.59	81 Tl 204.38	82 Pb 207.21	83 Bi 208.98	84 Po [209]	85 At [210]	86 Rn [222]
7	87 Fr [223]	88 Ra [226]	103 Lr [262]	104 Rf [261]	105 Db [262]	106 Sg [266]	107 Bh [264]	108 Hs [277]	109 Mt [268]	110 Ds [281]	111 Rg [272]	112 Uub		114 Uuq		116 Uuh		118 Uuo

Lanthanide series	57 La 138.91	58 Ce 140.12	59 Pr 140.91	60 Nd 144.24	61 Pm [145]	62 Sm 150.36	63 Eu 151.96	64 Gd 157.25	65 Tb 158.93	66 Dy 162.50	67 Ho 164.93	68 Er 167.26	69 Tm 168.93	70 Yb 173.04
Actinide series	89 Ac [227]	90 Th 232.04	91 Pa 231.04	92 U 238.03	93 Np [237]	94 Pu [244]	95 Am [243]	96 Cm [247]	97 Bk [247]	98 Cf [251]	99 Es [252]	100 Fm [257]	101 Md [258]	102 No [259]

Brackets [ ] indicate the most stable isotope.

INTRODUCTION: HOW TO USE THIS STUDY GUIDE

Congratulations! You are about to embark on a fascinating journey through the world of chemistry. Perhaps your biggest concern is that you won't understand some of the material and no one will be available to help you at the times you have free to study. In many ways, this Study Guide is intended to be your study partner. It is designed to help you focus on the important concepts in each chapter and to help you construct your own picture of how the information fits together. The format is as follows:

1. **Chapter Outline:** Each chapter is outlined to give you a succint summary of the essential material. The outline is keyed to tables and figures to help you use the text effectively. Each outline section also includes a description of the general type of example problem found there. Helpful notes are included where needed.
2. **Are You Able To...:** This section helps you to determine if you have gotten everything that you should out of the reading and problems. The objectives are keyed to examples and in-chapter problems as well as to end-of-chapter exercises.
3. **Worked Text Problems:** All in-chapter problems are worked out with commentary. When appropriate, you are referred to figures and tables in the text.
4. **Fill-ins:** This exercise helps you to assess your qualitative understanding of the material in each chapter. All answers are provided.
5. **Test Yourself:** You can evaluate your understanding of concepts through these exercises. Test Yourself exercises present the material in a slightly different manner from the text problems and often ask you to make generalizations based on your understanding of the material. All answers are provided.
6. **Making Connections:** The first eight chapters include concept maps to help you see the interconnectedness of the material in each chapter. Concept maps move you beyond simple definitions to look at the way in which ideas and terms are related to one another. Beginning with Chapter 9, you are asked to construct your own concept maps by using suggested groupings of terms. If you are not familiar with concept maps, you can write a paragraph describing the relations among the terms in a grouping.

Constructing a concept map: Concept maps resemble spider webs when they are complete. You construct them to see how ideas within a chapter are related to one another. Every tie line in a concept map depicts a relation between the two things that are linked. The relation is defined in writing on the line. Major concepts can be set apart from derivatives or minor topics by circles, squares, and so forth. Each concept map is unique, so do not be alarmed if yours looks different from someone else's. You can start by building small concept maps and eventually link them together to provide a larger system of connections.

Example: How are the following three terms related? orange juice, vitamin C, antioxidant

The relation among these three terms can be: orange juice contains vitamin C; vitamin C has antioxidant properties; orange juice has antioxidant properties. Put into a concept map, they may look like this:

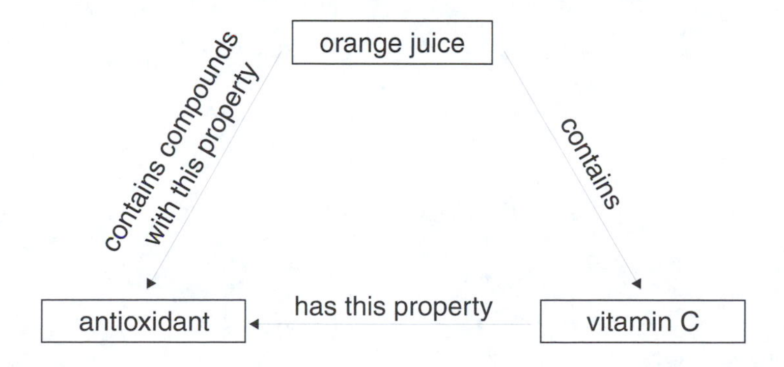

This Study Guide has been written to accompany *General, Organic, and Biochemistry,* Second Edition; *An Introduction to General Chemistry,* Second Edition; and *Organic Chemistry,* Second Edition. For Study Guide material to accompany the first two chapters of *Organic and Biochemistry,* Second Edition, please visit our Web site at www.whfreeman.com/bleiodian2e.

chapter 1

The Language of Chemistry

OUTLINE

1.1 The Composition of Matter

A. Applications of chemistry to medical breakthroughs
B. Matter can be categorized into
  1. pure substances
    a. obtained by separation of mixtures
    b. have consistent physical properties
      melting points, freezing points, color, densities
    c. have consistent chemical properties
      reactivity with other substances
    d. they are compounds if they can be decomposed chemically
    e. they are elements if they cannot be decomposed chemically
      elements have one- or two-letter symbols
      elements combine in fixed ratios to form compounds
  2. mixtures
    a. contain two or more pure substances mixed in varying proportions
    b. may be heterogeneous suspension; can see individual components
    c. may be homogeneous solution; cannot see individual components
    d. physical and chemical properties vary

Key Terms chemical property, compound, element, heterogeneous, homogeneous, mixture, scientific method, solution, physical property, pure substance

Key Figures
Figure 1.6 Physical changes of a pure substance: solid to liquid to gas
Figure 1.10 Categories of matter and their relationships

1.2 Measurement and the Metric System

A. Measurement compares a fundamental property to a standard unit.
B. Common standard metric units used in chemistry are

1. meter	m	length
2. kilogram	kg	mass
3. kelvin	K	temperature
4. second	s	time
5. mole	mol	amount of substance

C. Measurement of some properties is derived from fundamental units.
  1. area = m^2
  2. volume = m^3
  3. velocity = m/s
  4. density = kg/m^3

D. Standard metric units are divided or multiplied by units of ten.
   1. a prefix indicates whether unit has been divided or multiplied and by how much

Prefix	Standard unit multiplied by
giga	1,000,000,000
mega	1,000,000
kilo	1,000
hecto	100
deka	10
deci	0.1
centi	0.01
milli	0.001
micro	0.000001
nano	0.000000001
pico	0.000000000001

E. Prefix abbreviations are combined with standard unit abbreviations to define unit.

Example: mg ⇒ milli (m) and gram (g)
nm ⇒ nano (n) and meter (m)
dl ⇒ deci (d) and liter (l)

Key Terms fundamental unit, measure, metric system, SI system, unit
Key Tables
Table 1.2 Fundamental SI units
Table 1.3 Common non-SI units
Table 1.4 Metric prefixes, their values and symbols

Example 1.1 Using metric system prefixes to change between metric units
Problem 1.1 Reverse of Example 1.1

1.3 Measurement, Uncertainty, and Significant Figures

A. All measurements contain indeterminate error.
   1. degree of uncertainty expressed in last digit in measurement
   2. measuring device determines degree of uncertainty
   3. significant figures are used to identify degree of uncertainty in a measurement
   4. locations of zeros have to be analyzed (using rules) for significance

Key Terms error, significant figure, uncertainty, variability

1.4 Scientific Notation

A. Method of
   1. writing very large or very small numbers in shorthand notation
   2. avoiding ambiguity in reporting significant figures of measurements
B. Expresses numbers as coefficient multiplied by exponential factor
   1. coefficient must be between 1 and 10
   2. positive sign on exponential factor tells you the number is larger than one; the magnitude of the exponent is equal to the number of decimal places moved to the left when the actual number was written in scientific notation.
   3. negative sign on exponential factor tells you the number is smaller than one; the magnitude of the exponent is equal to the number of decimal places moved to the right when the actual number was written in scientific notation.

Key Terms coefficient, exponential factor, exponential notation, scientific notation

Example 1.2 Expressing a decimal number smaller than one in scientific notation
Problem 1.2 Expressing a decimal number smaller than one in scientific notation

1.5 The Use of Scientific Notation in Calculations

Example 1.3 Adding numbers written in scientific notation
Problem 1.3 Subtracting numbers written in scientific notation
Example 1.4 Multiplying numbers written in scientific notation
Problem 1.4 Multiplying numbers written in scientific notation
Example 1.5 Dividing numbers written in scientific notation
Problem 1.5 Dividing numbers written in scientific notation

1.6 Calculations and Significant Figures

A. Calculations involving numbers written in scientific notation
1. addition/subtraction of numbers written in scientific notation
   a. rewrite numbers so that both have same exponential factor
   b. add/subtract numbers, keep exponential factor same
   c. round to appropriate number of significant figures as determined by number with least number of decimal places
2. multiplication of numbers written in scientific notation
   a. multiply coefficients together; add exponential terms
   b. round to appropriate number of significant figures as determined by number with least number of significant figures
3. division of numbers written in scientific notation
   a. divide coefficients; subtract exponent in denominator from exponent in numerator
   b. round to appropriate number of significant figures as determined by number with least number of significant figures

Key Term rounding

Example 1.6 Dividing measured quantities and determining number of significant figures in the answer
Problem 1.6 Multiplying measured quantities and determining number of significant figures in the answer
Example 1.7 Adding measured quantities and determining number of significant figures in the answer
Problem 1.7 Adding measured quantities and determining number of significant figures in the answer
Example 1.8 Subtracting measured quantities and determining number of significant figures in the answer
Problem 1.8 Subtracting measured quantities and determining number of significant figures in the answer

1.7 The Use of Units in Calculations: The Unit-Conversion Method

A. Conversion factors relate two things
1. in the form of a ratio
2. relationship can be numerical between two measurements of the same property

   Example: 100 cm/1 m; both are lengths expressed in different units

3. relationship can be between two measurements of different properties that have a defined relationship

   Example: 1 g water/1 mL water; a mass to volume relationship

4. relationship can be between two things that have a defined relationship

   Example: 1 head/1 body

B. Conversion factors can be written in either of two ways

unit 1/unit 2 or unit 2/unit 1

1. units you want in the answer are in the numerator
2. units you want to get rid of are in the denominator

C. Choose the appropriate conversion factor as defined by algebraic relationship of units required in the answer and units given in the problem.

m × [conversion factor] = cm
(given unit) (wanted unit)
conversion factor = 100 cm/m

D. Conversion factors within the same measurement system are exact numbers and contain no uncertainty.
  1. are not used to determine significant figures
E. Conversion factors between different measurement systems may not be exact numbers.
  1. may be used to determine significant figures

Key Terms conversion factor, unit-conversion method
Key Table Table 1.5 Relations between English and metric units

Example 1.9 Converting liters to milliliters by using a conversion unit (metric to metric conversion)
Problem 1.9 Reverse of above; converting milliliters to liters by using a conversion unit (metric to metric conversion)
Example 1.10 Converting pounds to grams by using a conversion unit (English to metric conversion)
Problem 1.10 Reverse of above; converting grams to pounds by using a conversion unit (metric to English conversion)
Example 1.11 Converting gallons to milliliters by using more than one conversion unit (English to metric to metric conversion)
Problem 1.11 Converting days to seconds by using more than one conversion unit

1.8 Two Fundamental Properties of Matter: Mass and Volume

A. Mass
  1. measures quantity of matter
  2. differs from weight
  3. weight is measured by balances
  4. commonly measured in grams
B. Volume
  1. measures amount of space a sample occupies
  2. is measured by graduated cylinders, volumetric flasks, pipets, burets, and syringes
  3. commonly measured in milliliters
  4. 1 mL = 1 cm^3
  5. 1 L = 1000 cm^3

Key Terms mass, volume

Example 1.12 Converting between metric measures of mass; converting grams to kilograms
Problem 1.12 Reverse of above; converting kilograms to grams
Example 1.13 Converting between metric measures of volume; converting milliliters to liters
Problem 1.13 Reverse of above; converting liters to milliliters

1.9 Density

A. Density
  1. measures the ratio of a substance's weight to its volume
  2. is mathematically defined as
     density = mass (g)/ volume (cm^3)
  3. is dependent upon temperature because volume changes when temperature changes
  4. is used as a conversion factor to determine the mass equivalent of a volume or the volume equivalent of a mass
  5. is a physical property of a substance
B. Specific gravity
  1. is the ratio of the density of a liquid to the density of a reference liquid (often water)
  2. is mathematically defined as

specific gravity = density of test liquid (g/mL)/density of reference liquid (g/mL)

3. is a unitless measurement
4. is used clinically to determine densities of blood and urine samples

Key Terms density, hydrometer, specific gravity
Key Box Box 1.4 Density and the "fitness" of water

Example 1.14 Calculating the mass of a solution when given its density and its volume
Problem 1.14 Calculating the mass of a solution when given its density and its volume

1.10 Temperature

A. Temperature
1. measures how "hot" an object is
2. is measured with thermometers
3. is measured on three different scales
   a. Fahrenheit scale
      freezing point water = 32°F
      boiling point water = 212°F
   b. Celsius scale
      freezing point water = 0°C
      boiling point water = 100°C
   c. Kelvin scale
      freezing point water = 273 K
      boiling point water = 373 K
4. temperature scales are mathematically related
   a. K = °C + 273
   b. °F = °C × (9/5)(°F/°C)
   c. °C = °F × (5/9)(°C/°F)

Key Terms Celsius scale, Fahrenheit scale, Kelvin scale, temperature, thermometer
Key Figure Figure 1.14 Relationship between sizes of degrees and numerical values of freezing and boiling points in two systems

Example 1.15 Converting degrees Celsius to degrees Fahrenheit
Problem 1.15 Reverse of above; converting degrees Fahrenheit to degrees Celsius
Example 1.16 Converting degreess Celsius temperature to degrees Fahrenheit temperature
Problem 1.16 Converting degrees Celsius temperature to degrees Fahrenheit temperature
Example 1.17 Converting degrees Fahrenheit temperature to both degrees Celsius and degrees Kelvin temperatures
Problem 1.17 Converting degrees Fahrenheit temperature to both degrees Celsius and degrees Kelvin temperatures

1.11 Heat and Calorimetry

A. Heat
1. is related to temperature
2. is measured in units of joules (J) or calories (cal)

B. Specific heat
1. describes ability of one gram of a substance to absorb heat
2. is defined as being equal to the amount of absorbed per degree change in temperature per gram of substance
3. is mathematically defined as:

$$C_p = \text{joules/gram} \times \Delta°\text{C}$$

4. is a physical property of a substance
5. is different from heat capacity

C. Basal metabolic rate
  1. human metabolism produces heat
  2. BMR is minimal metabolic activity of human at rest with empty gastrointestinal tract
  3. can be calculated by using equations that differ for men and women

Key Terms basal metabolic rate, calorie, Calorie, calorimetry, heat, heat capacity, joule, specific heat
Key Table Table 1.6 Specific heats of common substances
Key Box Box 1.5 Specific heat of water and the "fitness" of water

Example 1.18 Calculating the specific heat of a substance when given its mass, the amount of heat added, and the rise in temperature of the sample
Problem 1.18 Same as above
Example 1.19 Using specific heat to determine how much heat must be added to a given mass of sample in order to cause a defined temperature increase
Problem 1.19 Same as above
Example 1.20 Estimating human caloric requirements of a male by using a given equation
Problem 1.20 Estimating human caloric requirements of a female by using a given equation

ARE YOU ABLE TO. . . AND WORKED TEXT PROBLEMS

Objectives Section 1.1 The Composition of Matter

Are you able to . . .
- explain the differences between elements and compounds (Fig. 1.10)
- explain the differences between pure substances and mixtures (Fig. 1.10, Exercise 1.4)
- explain the differences between homogeneous and heterogeneous mixtures
- explain how substances are characterized (Fig. 1.10)
- describe/identify/provide examples of physical properties (Figs. 1.5, 1.6, 1.9, Exercises 1.1, 1.2, 1.49)
- describe/identify/provide examples of physical changes (Exercises 1.1, 1.2, 1.49)
- describe/identify/provide examples of chemical properties (Fig. 1.7, Exercises 1.1, 1.2, 1.49)
- describe/identify/provide examples of chemical changes (Exercises 1.1, 1.2, 1.49)
- distinguish between physical and chemical properties (Exercises 1.1, 1.2, 1.49)
- relate sample purity to physical and chemical properties
- explain the symbolic language of elements (Table 1.1)
- outline the steps of the scientific method (Box 1.1, Fig. 1.1)
- explain the difference between a theory and a law (Expand Knowledge 1.64)

Problem: Opening Scenario You arrive for your shift at the skilled nursing facility and read on a patient's chart that the doctor has prescribed a 100-mg dose of Colace, a stool softener. The pharmacy sends up a bottle of the medication in syrup form, containing 20 mg of medicine in each 5 mL of syrup. How many milliliters of the syrup do you give to your patient?

Think unit conversions and draw a picture of what is given to you.

You are given: a bottle of syrup
every 5 mL of syrup contains 20 mg of Colace
You want to give your patient enough syrup so that he receives 100 mg of Colace
How many mL of syrup will contain 100 mg of Colace?

$$\underset{\text{(units you have)}}{100\ \cancel{\text{mg Colace}}} \times \frac{5\text{ mL syrup}}{20\ \cancel{\text{mg Colace}}} = \underset{\text{(units you want)}}{25\text{ mL syrup}}$$

Objectives Section 1.2 Measurement and the Metric System

Are you able to . . .
- explain how measurement is done (Figs. 1.11, 1.12)
- describe the metric system and its utility

- describe/identify the standard metric units of measurement (Table 1.2, Exercises 1.5, 1.6)
- assign numerical values to metric prefixes (Table 1.4)
- assign metric prefixes to numerical values (Table 1.4)
- assign abbreviations to metric units and prefixes (Table 1.4)
- use metric prefixes to generate conversion factors
- convert between metric units of measurement (Ex. 1.1, Prob. 1.1, Exercises 1.7, 1.8, 1.50)

Problem 1.1 Express (a) 2 ms in seconds; (b) 5 cm in meters; (c) 100 mL in liters.
(a) Recognize that "ms" is the abbreviation for millisecond (Tables 1.2 and 1.4). Examine the units given and the units wanted:

given: milliseconds want: seconds

What is a relationship between them and where can you find that information? To determine the value of the prefix milli- (m), use Table 1.4: 1000 ms = 1 s; this generates a conversion factor for seconds and milliseconds:

$$2\ \cancel{\text{ms}} \times \frac{1\ \text{s}}{1000\ \cancel{\text{ms}}} = 0.002\ \text{s}$$

(b) Ask the same questions as in (a): What units are given? What units are wanted? What is the relationship between them? Again use Table 1.4 to determine the numerical value of the prefix centi- (c): 100 cm = 1 m.

$$5\ \cancel{\text{cm}} \times \frac{1\ \text{m}}{100\ \cancel{\text{cm}}} = 0.05\ \text{m}$$

(c) Ask the same questions as in (a) and (b). Use Table 1.4 to determine the numerical value of the prefix milli- (m): 1000 mL = 1 L

$$100\ \cancel{\text{mL}} \times \frac{1\ \text{L}}{1000\ \cancel{\text{mL}}} = 0.100\ \text{L}$$

Objectives Section 1.3 Measurement, Uncertainty, and Significant Figures

Are you able to . . .

- explain indeterminate error
- explain significant figures
- count the number of significant figures in a number or in a measurement (Exercises 1.9, 1.10, 1.11,1.12, 1.52)
- use the location of a zero in a number to determine whether or not it is significant

Objectives Section 1.4 Scientific Notation

Are you able to . . .

- explain the meaning of coefficient
- explain the meaning of exponential factor
- write numbers in scientific notation (Ex. 1.2, Prob. 1.2, Exercises 1.13, 1.15, 1.16, 1.51)
- write out numbers that are written in scientific notation (Exercise 1.14)
- recognize that numbers are larger or smaller than one by examining their exponential factor

Problem 1.2 Express the number 0.0007068 in scientific notation.
A first question might be, "Is the number larger or smaller than 1?" The answer to this will determine the sign of the exponential factor. If the number is larger than 1, the sign will be positive; if it is smaller than 1, the sign will be negative. In this problem the number is smaller than 1, so the exponential factor's sign will be negative. Next ask what form the number must be in: a coefficient between 1 and 10 multiplied by a power of 10. To get the coefficient, the decimal point must be moved right 4 places:

$$0.0007068 \Rightarrow \underset{\text{(number between 1 and 10)}}{7.068} \times 10^{-4} \leftarrow \text{number corresponds to number of decimal places the point has been moved}$$

Note: Moving the decimal point three places will result in a coefficient smaller than 1: 0.7068×10^{-3}. Moving the decimal point five places will result in a coefficient larger than 10: 70.68×10^{-5}.

Objectives Section 1.5 The Use of Scientific Notation in Calculations

Are you able to . . .

- add, subtract, multiply, and divide numbers that are written in scientific notation (Exs. 1.3, 1.4, 1.5, Probs. 1.3, 1.4, 1.5, Exercises 1.17, 1.18,1.19, 1.20, 1.54, 1.55)

Problem 1.3 What is the result of subtracting 6.42×10^{-5} from 7.953×10^{-4}?

Note that the exponential factors are different. Do you have to rewrite the numbers to perform the subtraction? Not necessarily, but it would be easier if you did. 7.953×10^{-4} is the larger number, so you might change the smaller number to: 0.642×10^{-4}. So, the equation becomes:

$$(7.953 - 0.642) \times 10^{-4} = 7.311 \times 10^{-4}$$

Note that the answer is in the appropriate form for scientific notation.

Problem 1.4 Multiply 4.2×10^5 by 0.64×10^{-4}.

Refer to the text for the rules governing multiplication of numbers written in scientific notation: Multiply the coefficients and add the exponents:

$$4.2 \times 0.64 = 2.688$$
$$10^5 + 10^{-4} = 10^{5+(-4)} = 10^1$$
$$2.688 \times 10^1 \text{ or } 26.88$$

The answer should be reported to two significant figures: 27 or 2.7×10^1.

Problem 1.5 Divide 3.45×10^4 by 7.2×10^{-2}.

Refer to the text for the rules governing division of numbers written in scientific notation: Divide the coefficients and subtract the exponents:

$$3.45 \div 7.2 = 0.48 \text{ (report two significant figures)}$$
$$10^4 - 10^{-2} = 10^{4-(-2)} = 10^6$$
$$0.48 \times 10^6$$

The answer is not in the appropriate form: rewrite as 4.8×10^5.

Objectives Section 1.6 Calculations and Significant Figures

- use the rules governing significant figures to determine the correct number of significant figures in calculated numbers derived from measurements (Exs. 1.6, 1.7, 1.8, Probs. 1.6, 1.7, 1.8, Exercise 1.56)

Problem 1.6 Calculate the volume of a cube that is 8.5 cm on a side.

The note in the text states that the volume of a cube 1 cm on each side is 1 cm × 1 cm × 1 cm. This provides the form that your equation will take: 8.5 cm × 8.5 cm × 8.5 cm. Note that you are multiplying BOTH numbers and units and that each number has two significant figures:

$$8.5 \times 8.5 \times 8.5 = 614.125$$

Rewrite in scientific notation and report two significant figures: 6.1×10^2.
Don't forget to multiply the units: $\text{cm} \times \text{cm} \times \text{cm} = \text{cm}^3$.
Combine the two: $6.1 \times 10^2\ \text{cm}^3$

Problem 1.7 Add 1.9375, 34.23, and 4.184.

Note that each measurement has a different number of decimal places. Your answer will have two decimal places because the least well known measurement has two decimal places.

$$34.23 + 4.184 + 1.9375 = 40.3515, \text{ which is rounded to } 40.35$$

Problem 1.8 Subtract 7.6 m from 94.935 m.

Note that the least well known measurement has one decimal place so your answer will be rounded to one decimal place. Note also that units are not changed when adding or subtracting; meters will remain meters.

$$94.936\ \text{m} - 7.6\ \text{m} = 87.336\ \text{m}, \text{ which is rounded to } 87.3\ \text{m}$$

Objectives Section 1.7 The Use of Units in Calculations: The Unit-Conversion Method

Are you able to . . .

- explain how conversion factors "work"
- distinguish between exact numbers and measurements
- use conversion factors to change measurements from one unit to another (Ex. 1.9, Prob. 1.9, Exercises 1.7, 1.8, 1.9)
- use conversion factors to change metric measurements to English measurements and vice versa (Table 1.5, Exs. 1.10, 1.11, Probs.1.10, 1.11, Exercises 1.21, 1.22, 1.23, 1.24, 1.25, 1.26, 1.29, 1.30, 1.53)
- generate conversion factors

Problem 1.9 Convert 74.1 mL to liters.

This is exactly like Problem 1.1. Ask yourself what units you are given, what units you want, and what relationship exists between them. You are given milliliters and want liters. Refer back to Table. 1.4 to find the relationship

$$1000 \text{ mL} = 1 \text{ L}$$

This relationship gives you a way to convert between milliliters and can be used in one of two forms:

$$\frac{1000 \text{ mL}}{1 \text{ L}} \quad \text{or} \quad \frac{1 \text{ L}}{1000 \text{ mL}}$$

The form you need is determined by what the problem gives you and what it asks you to do. In this case, you are given milliliters and you want to convert to liters. You choose the conversion factor that allows you to cancel the unit you are given by placing it in the denominator and leaves you with the unit you want in the numerator: Which unit do you want in the numerator? Liters: so the form you use will be

$$\underset{\text{(unit given)}}{74.1 \cancel{\text{mL}}} \quad \times \quad \underset{\text{(conversion factor)}}{\frac{1 \text{ L}}{1000 \cancel{\text{mL}}}} \quad = \quad \underset{\text{(unit wanted)}}{0.0741 \text{ L}}$$

Note that this problem asks you to convert between two metric measurements; Table 1.4 can be used to generate conversions between the base unit and any other unit.

Problem 1.10 How many pounds are in 752.4 g?

The first thing you should note is that the conversion will be between a metric unit and an English unit. Table 1.5 has the relationships that are useful. What units are you given? Grams. What units do you want? Pounds. Does Table 1.5 provide any direct relationship between those two units? Yes; 1 lb = 453.59 g. Consider the form in which the conversion factor will be: You have grams; therefore, you want those to cancel—which means they must be in the denominator. You want pounds—which means they should be in the numerator:

$$\underset{\text{(unit given)}}{752.4 \cancel{\text{g}}} \quad \times \quad \underset{\text{(conversion factor)}}{\frac{1 \text{ lb}}{453.59 \cancel{\text{g}}}} \quad = \quad \underset{\text{(unit wanted)}}{1.659 \text{ lb}}$$

Note that the answer is reported to four significant figures because the measurement given had four significant figures.

Problem 1.11 How many seconds are in seven days?

The first thing to ask is, What units are given and what units are wanted? You are given days and asked to change them to seconds. Relying upon what you know about time and the hints given in the text.is there a single conversion factor that you can immediately think of that directly relates the two units? Not easily. That means that you will probably have to use more than one conversion factor:

$$\text{days} \quad \Rightarrow \ ? \Rightarrow \quad \text{seconds}$$

One thing you should note is that your first conversion factor must have days in the denominator and your last conversion factor must have seconds in the numerator:

$$\underset{\text{(unit given)}}{\text{days}} \times \underbrace{\frac{???}{\text{days}} \times \frac{\text{seconds}}{???}}_{\text{(conversion factors)}} = \underset{\text{(unit wanted)}}{\text{seconds}}$$

Notice that the numbers that go with the units haven't been filled in; if you can determine the placement of the units, the numbers follow. In this form, you can see that if you can think of a unit that can be related to both days and seconds, your job is made easy. In this case, however, it may be easier to think of first converting days to something that can be more easily related to seconds: minutes or hours, for example:

$$\cancel{\text{days}} \times \frac{?\ \cancel{\text{hour}}}{?\ \cancel{\text{day}}} \times \frac{?\ \cancel{\text{minutes}}}{?\ \cancel{\text{hour}}} \times \frac{?\ \text{seconds}}{?\ \cancel{\text{minutes}}} = \text{seconds}$$

Without filling in the numbers, you can see that the units cancel correctly. Now fill in the numbers: You know the relationships between hours and days; between minutes and hours; and between minutes and seconds:

$$7\ \cancel{\text{days}} \times \frac{24\ \cancel{\text{hours}}}{1\ \cancel{\text{day}}} \times \frac{60\ \cancel{\text{min}}}{1\ \cancel{\text{hour}}} \times \frac{60\ \text{s}}{1\ \cancel{\text{min}}} = 604800\ \text{s}$$

The answer makes sense: there are a lot of seconds in seven days! It can be rewritten in scientific notation as 6.04800×10^5 seconds. The text said the conversions were exact numbers, so all figures can be reported. Note that whenever you write a conversion factor, you are saying that the numerator and the denominator are the same thing in different units; therefore, you wouldn't write a conversion factor that had 10 hours in the numerator and 60 minutes in the denominator. Other conversion factors could be used, but these are the most familiar.

Objectives Section 1.8 Two Fundamental Properties of Matter: Mass and Volume

Are you able to . . .

- describe the fundamental properties of matter
- define mass
- describe how mass is measured (Fig. 1.11)
- identify/convert between the units in which mass is measured (Ex. 1.12, Prob. 1.12, Exercises 1.12, 1.27, 1.28, 1.29)
- define volume
- describe how volume is measured (Fig. 1.12)
- identify/convert between the units in which volume is measured (Ex. 1.13, Prob. 1.13)
- explain the relationship between units of length and units of volume (Table 1.3, Fig. 1.13, Exercises 1.31, 1.32)

Problem 1.12 Express a mass of 2.87 kg in grams.

This is exactly like Problem 1.1. Ask yourself what units you are given, what units you want, and what relationship exists between them. You are given kilograms and want grams. Refer back to Table. 1.4 to find the relationship. Note also that the metric units are being converted to other metric units.

$$\underset{\text{(unit given)}}{2.87\ \cancel{\text{kg}}} \times \underset{\text{(conversion factor)}}{\frac{1000\ \text{g}}{1\ \cancel{\text{kg}}}} = \underset{\text{(unit wanted)}}{2870\ \text{g}}$$

The answer can be rewritten in scientific notation as 2.87×10^3 g.

Problem 1.13 Express 3.97 L in milliliters.

This is exactly like Problem 1.1. Ask yourself what units you are given, what units you want, and what relationship exists between them. You are given liters and want milliliters. Refer back to Table. 1.4 to find the relationship. Note also that the metric units are being converted to other metric units.

$$3.97\ \cancel{\text{L}} \times \frac{1000\ \text{mL}}{1\ \cancel{\text{L}}} = 3970\ \text{mL}$$

The answer can be rewritten in scientific notation as 3.97×10^3 mL.

Objectives Section 1.9 Density

Are you able to . . .

- define density in words and mathematically
- solve problems by using the equation for density; given any two terms, solve for the third (Ex. 1.14, Prob. 1.14, Exercises 1.33, 1.34, 1.41, 1.42, 1.45, 1.46, 1.47, 1.48)
- describe how density is measured
- identify the units in which density is measured
- use density as a conversion factor to change mass measurements to volume equivalents and vice versa (Exercise 1.48)
- explain the relationship between density and temperature
- define specific gravity
- explain how specific gravity measurements are used clinically

Problem 1.14 Calculate the mass of 135.0 mL of a liquid whose density is 0.8758 g/mL at 20°C.

What's being asked? Find the mass of a volume of liquid, how much it weighs. What information are you given:

a volume of liquid, 135.0 mL the density of the liquid, 0.8758 g/mL a temperature, 20°C

Which piece(s) of information will be useful? Think about the relationships between each piece of information given and what you are looking for. You may see that the temperature is not very useful, but it is usually defined when given a density problem because densities are temperature-dependent properties.

What is the density? It is a measure of the weight of a known volume of a substance. In this case, it tells you that 1 mL of the liquid weighs 0.8758 grams. But you aren't given 1 mL; you are given 135.0 mL. The density can be used as a conversion factor to go from grams, or weight, of the liquid to milliliters, or volume, of the liquid:

$$\underset{\text{(unit given)}}{135.0\ \cancel{\text{mL}}} \times \underset{\text{(conversion factor)}}{\frac{0.8758\ \text{g}}{1\ \cancel{\text{mL}}}} = \underset{\text{(unit wanted)}}{118.2\ \text{g}}$$

Objectives Section 1.10 Temperature

Are you able to . . .

- define temperature
- describe how temperature is measured
- identify the units in which temperature is measured
- explain how the Fahrenheit, Celsius, and Kelvin scales are related (Fig. 1.14)
- use the mathematical relationship between temperature scales to convert degrees and measured temperatures from one scale to another (Exs. 1.16, 1.17, Probs. 1.16, 1.17, Exercises 1.35, 1.36, 1.37, 1.38)
- explain the relationship between heat and temperature

Problem 1.15 A temperature rise of 65 degrees on a Fahrenheit scale is equal to how many degrees on a Celsius scale?

The temperature you are given and the temperature you want determine the appropriate conversion factor to use: You are given degrees Fahrenheit and asked to convert to degrees Celsius:

$$\underset{\text{(unit given)}}{65\cancel{^\circ\text{F}}} \times \underset{\text{(conversion factor)}}{\frac{5^\circ\text{C}}{9\cancel{^\circ\text{F}}}} = \underset{\text{(unit wanted)}}{36^\circ\text{C}}$$

Problem 1.16 What Fahrenheit temperature is equivalent to 100°C?

You are given a Celsius temperature and must find the equivalent Fahrenheit temperature. The relationship between them comes from the equation given in the text:

$$^\circ\text{F} = \left(\cancel{^\circ\text{C}} \times \frac{9^\circ\text{F}}{5\cancel{^\circ\text{C}}}\right) + 32^\circ\text{F}$$

Fill in the degrees Celsius given:

$$°F = \left(100°\text{C} \times \frac{9°F}{5°\text{C}}\right) + 32°F$$

(unit given) (conversion factor)

°F = 212 When the temperature is 100°C, it is 212°F.

Note that the degrees Celsius and conversion factor are multiplied together and then that number is added to 32.

Problem 1.17 Convert 98.6°F (normal body temperature) to both Celsius and Kelvin temperatures. You are given a Farenheit temperature and must find the equivalent Celsius temperature and Kelvin temperature.

The text only shows a relationship between °C and Kelvin degrees, so you should first convert °F to °C:

$$(98.6°\text{F} - 32) \times \frac{5°C}{9°\text{F}} = 37.0°C$$ (*Note:* 32 is an exact number.)

The Celsius temperature just calculated is then converted to Kelvin degrees by using the relation:

$$K = °C + 273$$
$$K = 37°C + 273 = 310\ K$$

Objectives Section 1.11 Heat and Calorimetry

Are you able to . . .

- define specific heat in words and mathematically
- solve problems using the equation for specific heat; given any three terms, solve for the fourth (Exs. 1.18, 1.19, Probs. 1.18, 1.19, Exercises 1.39, 1.40, 1.58)
- describe how specific heat is measured
- identify the units in which specific heat is measured
- explain how heat is measured
- identify the units in which heat is measured
- use the mathematical relationship between calories and joules to convert heat measurements between the two units
- explain heat capacity
- explain how specific heat differs from heat capacity

Problem 1.18 What is the specific heat of a substance if 334 J of heat added to 52 g of it cause the temperature to rise from 16°C to 48°C?

What you are looking for? The specific heat of a substance. The specific heat is the temperature increase that can be expected when a defined amount of heat is added to a specified amount of a given substance; every substance has its own specific heat.

What are you given?

the amount of substance, 52 g (the substance is not identified)
the amount of heat added to the substance, 334 J
the rise in temperature caused by the added heat, 48°C − 16°C = 32°C

How do these pieces of information relate to the specific heat? The units of specific heat are:

$$\frac{\text{joules}}{\text{grams} \times \Delta°C}$$

All of these units are given in the problem; fill in the numbers:

$$\frac{334\ J}{52\ \text{grams}} \times 32°C = \frac{0.20\ J}{g \times °C}$$ or this can be written: 0.20 J/g· °C

Note that the specific heat of the substance now reads that 0.20 J of heat added/lost per 1 g of substance will cause a change of 1°C.

Problem 1.19 How much heat must be added to 37 g of a substance that has a specific heat of 0.12 J/g°C to cause its temperature to rise from 11°C to 22°C?

What is being asked? How much heat has to be added to a known amount of substance to get a specific temperature increase?

What are you given?

amount of substance, 37 g (substance is not identified)
specific heat of substance, 0.12 J/g·°C
the desired temperature increase, 22°C − 11°C = 11°C

The specific heat tells you that 1 g of the substance will change the temperature by 1°C when 0.12 J of heat is added/lost; you aren't given 1 g, you're given 37 g and your temperature change will be 11°C, not 1°C.

The units of specific heat are

$$\underset{\text{(given)}}{\text{specific heat}} = \frac{\text{joules}}{\underset{\text{(given)}}{\text{grams}} \times \underset{\text{(given)}}{\Delta°\text{C}}}$$

The problem gives you grams, temperature change, and specific heat and asks for the amount of heat required (J).

Rearrange the equation to get joules, what you want, on one side and everything else on the other:

$$\underset{\text{(given)}}{\text{specific heat}} \times \underset{\text{(given)}}{\text{grams}} \times \underset{\text{(given)}}{\Delta°\text{C}} = \underset{\text{(unknown)}}{\text{joules}}$$

$$\frac{0.12\ \text{J}}{\cancel{\text{g}}\cdot\cancel{°\text{C}}} \times 37\ \cancel{\text{g}} \times 11\cancel{°\text{C}} = 49\ \text{J}$$

Problem 1.20 Calculate the daily minimal Calorie requirement of a 22-year-old female who is 66 inches tall and weighs 132 lb, using the equation given:

BMR (female) = 655 + [4.36 ×weight in lb] + [4.32 × height in inches] − [4.7 × age in years]
= 655 + [4.36 ×132 lb] + [4.32 × 66 inches] − [4.7 × 22 years]
= 655 + 575 + 285 − 103 = 1412 Kcal per day

EQUATIONS

Temperature

$$\textbf{K} = \textbf{°C} + \textbf{273}$$

- used for converting temperature readings in Celsius degrees to kelvins
  Example: Convert 21°C to the equivalent Kelvin temperature.

$$\textbf{°C} = \textbf{5/9 (°F} - \textbf{32)}$$

- used for converting temperature readings in degrees Fahrenheit to Celsius degrees
  Example: Convert 68°F to degrees Celsius.

$$\textbf{°F} = \textbf{9/5 (°C} + \textbf{32)}$$

- used for converting temperature readings in Celsius degrees to Fahrenheit degrees
  Example: Convert 22°C to degrees Fahrenheit.
- can be combined:
  Example: Convert 67°F to its equivalent temperature in kelvins.

Hints

- The degrees you *have* are plugged into the equation. The degrees you *want* are alone on one side of the equation.

Density

$$\textbf{density} = \frac{\textbf{mass (g)}}{\textbf{volume (mL)}}$$

- calculated by dividing the mass of something by the volume it occupies

  Example: An object that weighs 15 g displaces 3.5 mL of water when placed in a graduated cylinder. What is the density of the object?

$$\textbf{volume (mL)} = \frac{\textbf{mass (g)}}{\textbf{density (g/mL)}}$$

- used to convert a mass measure into its equivalent volume

  Example: A piece of gold weighs 1.2 g. What volume does it occupy?

$$\textbf{mass} = \textbf{density (g/mL)} \times \textbf{volume (mL)}$$

- used to convert volume measurements to equivalent weights

  Example: What is the mass of 500 mL of a solution whose density is 1.02 g/mL?

- can be used to identify a substance

  Example: Identify the metal whose density is 19.3 g/mL.

Hints

- Density *tells* you something.

  Example: The density of gold is 19.3 g/mL. That says that 19.3 g of gold displace 1 mL of water. It also says that the equivalent of 1 mL of gold weighs 19.3 g.

- Each substance—solids, liquids, and gases—has a characteristic density.
- When densities are not given in a problem, refer back to tables in the book or look on a periodic table.
- The units of density are grams per milliliter; make sure your units are correct.
- Density varies with temperature; double check problems to determine whether you need to consider this.
- The density of water at 4°C is 1.000 g/mL; conversions between water volumes and masses are close to 1:1.

Specific Heat

$$\mathbf{C_p} = \frac{\textbf{joules}}{\mathbf{g \cdot \Delta ^\circ C}}$$

- calculated by dividing the amount of heat a sample absorbs by the temperature change times the mass of the sample

  Example: What is the specific heat of a substance if 10.0 J of heat added to 10 grams of the substance cause its temperature to rise from 20°C to 40°C?

$$\textbf{Joules} = \mathbf{C_p \times {}^\circ C \times g}$$

- used to determine the amount of heat a given mass of sample will absorb as it changes temperature within a defined temperature range

  Example: How much heat is absorbed when a 10-g sample of aluminum, with a specific heat of 0.89 J/g · °C, is heated from 21°C to 50°C?

$$\Delta°C = \frac{\text{joules}}{C_p \cdot g}$$

- used to determine the change in temperature that can be expected when a known amount of heat is added to a sample

  Example: What temperature change can be expected when 10 J of heat are applied to a 15-g sample of gold, with a specific heat of 0.13 J/g·°C?

$$°C_2 - °C_1 = \frac{\text{joules}}{C_p \cdot g}$$

- can determine final or initial temperatures by using the equation: put temperature term in the form ($T_{final} - T_{initial}$); rearrange to solve for either $°C_1$ or $°C_2$

  Example: What is the final temperature when 20 g of olive oil, with a specific heat of 1.97 J/g·°C, has 50 J of heat added at an initial temperature of 20°C?

$$°C_2 = \frac{50\text{ J}}{(1.97\text{ J/g}\cdot°\text{C})\cdot 20\text{ g}} + °C_1$$

$$\text{grams} = \frac{\text{joules}}{Cp \cdot °C}$$

- can determine the mass of a sample if the amount of heat added, temperature change and specific heat are known

  Example: What is the mass of a sample that has a specific heat of 1.26 J/g·°C and that shows a temperature increase of 20°C when 100 J of heat are absorbed.

Hints

- Specific heat *tells* you something. On a relative scale, a substance with a higher specific heat than another will absorb more heat before exhibiting a temperature change than with a substance with a lower specific heat.
- Each substance—solids, liquids, and gases—has a characteristic specific heat.
- When specific heat is not given in a problem, refer back to tables in the text.
- The units of specific heat are joules per degree Celsius per gram.
- The specific heat of water is 1 cal/ g·°C OR 4.184 J/ g·°C.

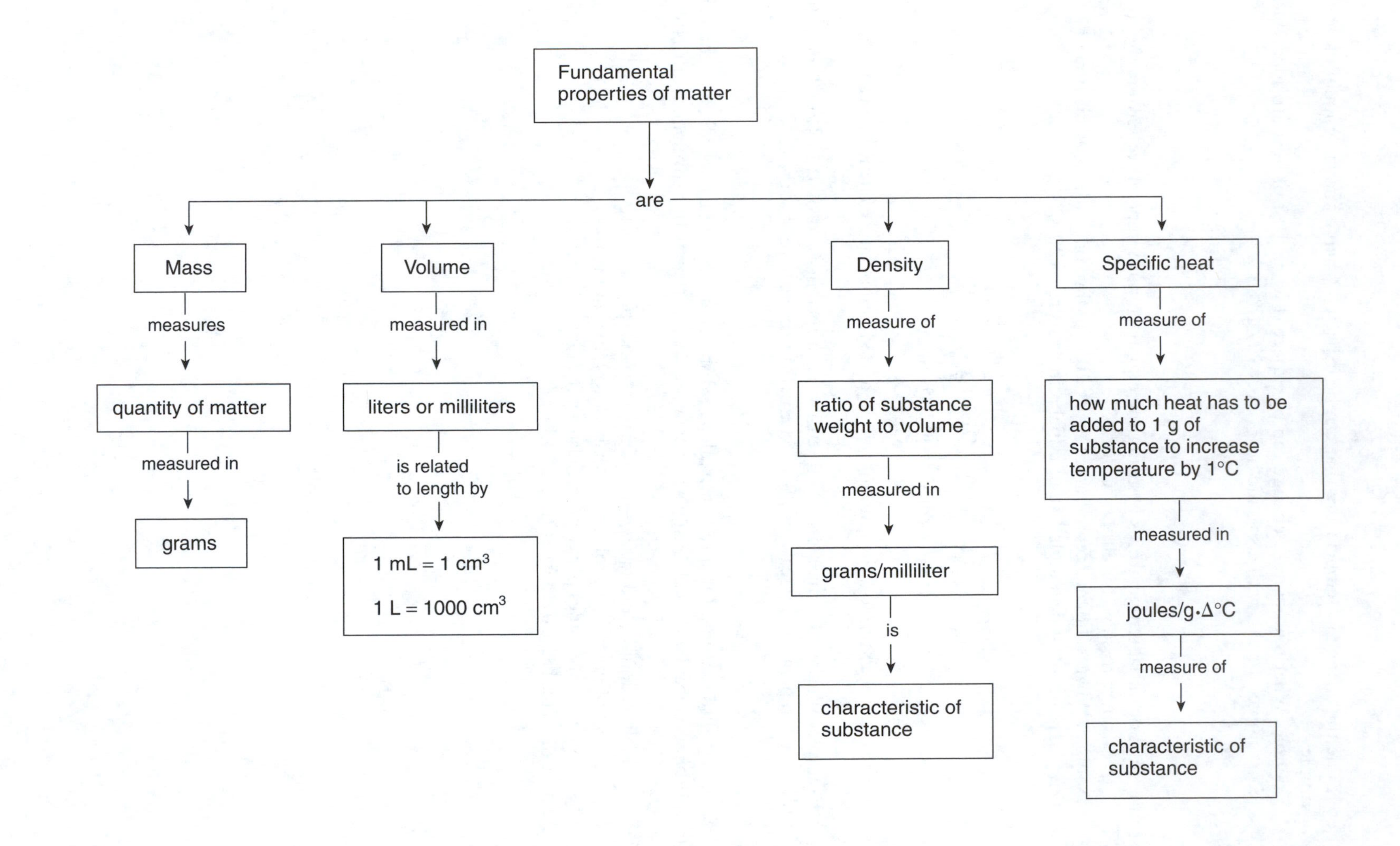
Fundamental properties of matter
are
Mass
measures
quantity of matter
measured in
grams
Volume
measured in
liters or milliliters
is related to length by
1 mL = 1 cm^3
1 L = 1000 cm^3
Density
measure of
ratio of substance weight to volume
measured in
grams/milliliter
is
characteristic of substance
Specific heat
measure of
how much heat has to be added to 1 g of substance to increase temperature by 1°C
measured in
joules/g·Δ°C
measure of
characteristic of substance

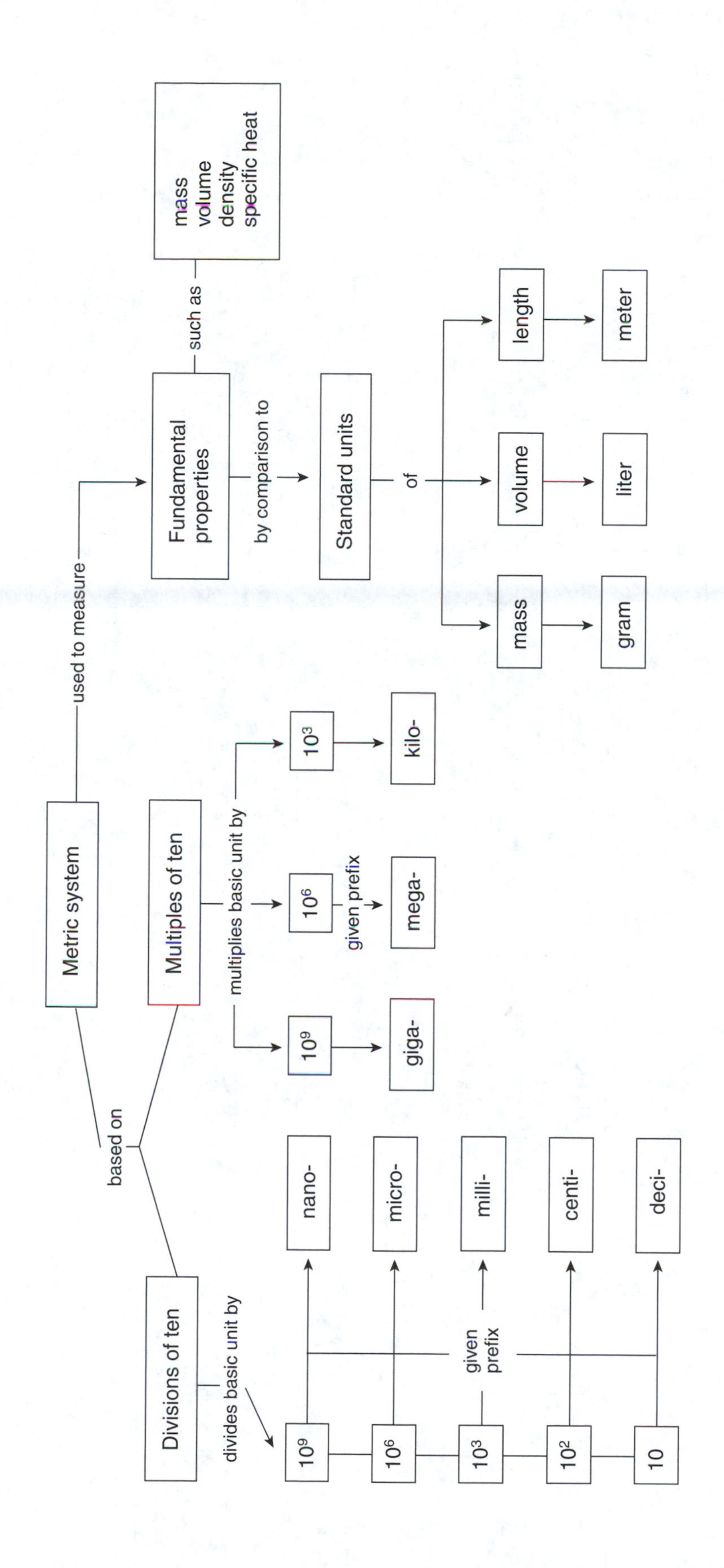

Metric system
used to measure
Fundamental properties
such as
mass
volume
density
specific heat
by comparison to
Standard units
of
mass
volume
length
gram
liter
meter
based on
Multiples of ten
multiplies basic unit by
10^9
10^6
10^3
given prefix
giga-
mega-
kilo-
Divisions of ten
divides basic unit by
10^9
10^6
10^3
10^2
10
given prefix
nano-
micro-
milli-
centi-
deci-

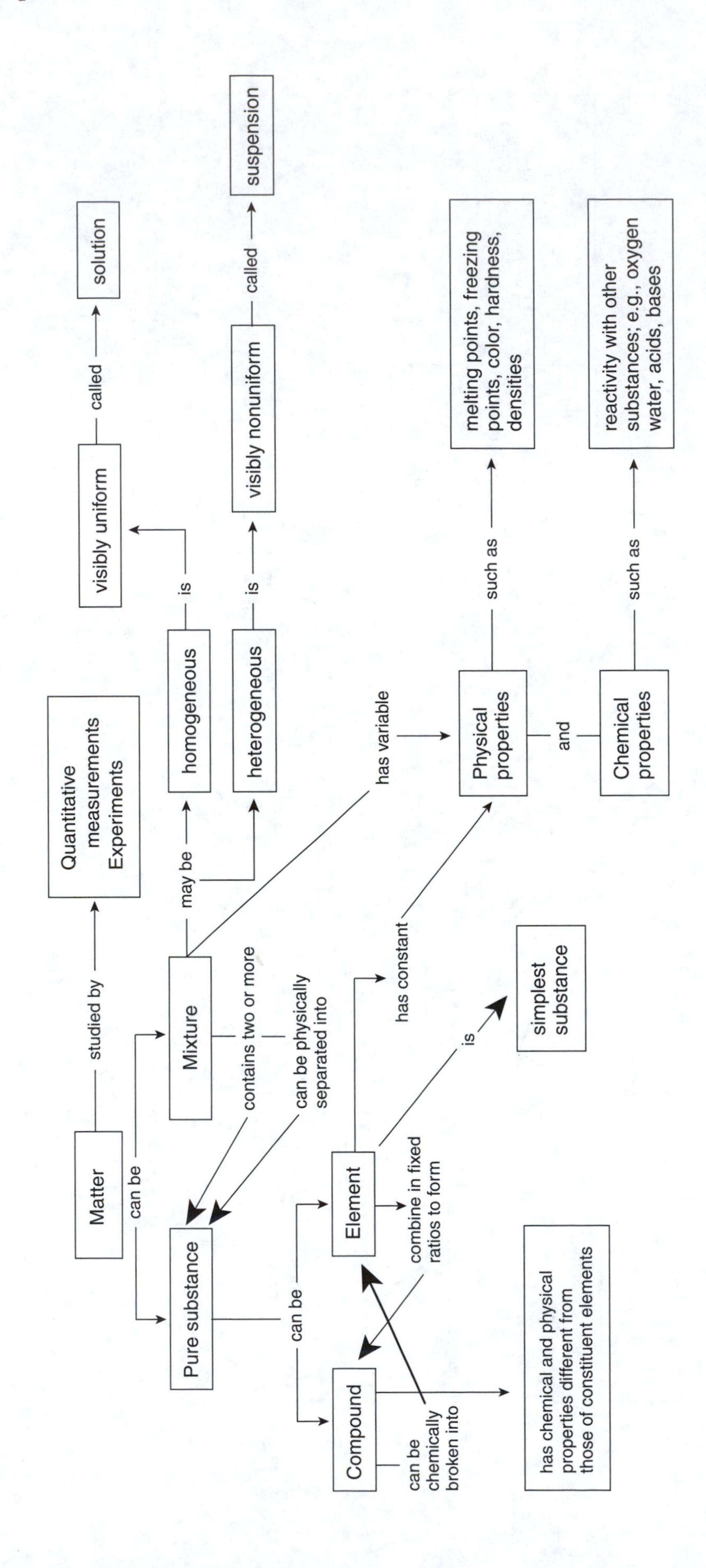

Matter
studied by
Quantitative measurements Experiments
can be
Mixture
Pure substance
may be
homogeneous
heterogeneous
is
visibly uniform
called
solution
is
visibly nonuniform
called
suspension
has variable
Physical properties
and
Chemical properties
such as
melting points, freezing points, color, hardness, densities
such as
reactivity with other substances; e.g., oxygen water, acids, bases
contains two or more
can be physically separated into
can be
Element
Compound
has constant
is
simplest substance
combine in fixed ratios to form
can be chemically broken into
has chemical and physical properties different from those of constituent elements

FILL-INS

Chemistry is the study of the ____ and ____ properties of matter. Matter can be broadly classified as being either a ____ or a ____, depending on the ____ of its physical and chemical properties. When a substance exhibits constant physical and chemical characteristics, it is called a____. When a substance exhibits variable chemical and physical characteristics, it is called a ____. These categories can be further subdivided. Mixtures, which can be separated by ____ methods, can be classified as visually ____ or ____. Uniform mixtures, such as sugar dissolved in water, are called ____. Nonuniform mixtures, such as a pizza, are called ____. Pure substances can be classified as ____, which cannot be separated chemically into simpler substances, or ____, which are combinations of elements in ____ ratios by weight. Compounds can be separated by ____ means into their component elements. Elements are symbolically represented by ____ derived from their English, Latin, German, or Arabic names.

The study of chemical and physical properties requires a ____ system of measurements. The system of measurement used all over the world is called the ____. There are five standard metric units of measurement that are commonly used to describe chemical and physical properties. Length is measured relative to the ____; mass (sometimes called weight) relative to the ____; time is reported in ____; the amount of a chemical substance is reported in ____; and temperature in degrees ____. However, the ____ and ____ temperature scales are also routinely used. Other units describing properties are derived from the five standard units; for example, the standard unit for volume is the ____. ____ is the ratio of the mass of one gram of a substance to the volume that the gram occupies. ____ is the amount of heat that one gram of a sample absorbs as it heats by one degree Celsius. The numerical values of these properties are ____ for each substance and can be used as a means of identification.

Numerically, the metric system is based on ____ or ____ of the standard units by powers of ____. A ____ corresponding to some power of ten is attached to the standard unit. For example, the prefix deci- denotes that the unit before which it appears is ____ of the standard unit. ____ means one-tenth of a gram. The prefix kilo- shows that the standard unit has been ____. ____ means one thousand meters. Conversion factors are ____ that relate two different units of measurement and can be used to ____ between different systems of measurement.

Measurements always contain an unavoidable uncertainty called ____. This uncertainty in a measurement is conveyed by the number of ____ that a measure contains. Theses significant figures are a result of the ____ used to make a measurement. ____ is a method of writing numerical measurements that avoids ambiguity and also allows very large and small numbers to be handled easily. Writing numbers in scientific notation requires that a number be rewritten as a ____, a number between ____ and ____, multiplied by the appropriate ____.

Answers

Chemistry is the study of the physical and chemical properties of matter. Matter can be broadly classified as being either a pure substance or a mixture, depending on the constancy of its physical and chemical properties. When a substance exhibits constant physical and chemical characteristics, it is called a pure substance. When a substance exhibits variable chemical and physical characteristics, it is called a mixture. These categories can be further subdivided. Mixtures, which can be separated by physical methods, can be classified as visually uniform or nonuniform. Uniform mixtures, such as sugar dissolved in water, are called homogeneous. Nonuniform mixtures, such as a pizza, are called heterogeneous. Pure substances can be classified as elements, which cannot be separated chemically into simpler substances, or pure compounds, which are combinations of elements in fixed ratios by weight. Compounds can be separated by chemical means into their component elements. Elements are symbolically represented by letters derived from their English, Latin, German, or Arabic names.

The study of chemical and physical properties requires a quantitative system of measurements. The system of measurement used all over the world is called the metric system. There are five standard metric units of measurement that are commonly used to describe chemical and physical properties. Length is measured relative to the meter; mass (sometimes called weight) relative to the kilogram; time is reported in seconds; the amount of a chemical substance is reported in moles; and temperature in kelvins. However, the Fahrenheit and Celsius temperature scales are also routinely used. Other units describing properties are derived from the five standard units; for example, the standard unit for volume is the cubic meter. Density is the ratio of the mass of one gram of a substance to the volume that the gram occupies. Specific heat is the amount of heat that one gram of a sample absorbs as it heats by one degree Celsius. The numerical values of these properties are unique for each substance and can be used as a means of identification.

Numerically, the metric system is based on multiples or divisions of the standard units by powers of ten. A prefix corresponding to some power of ten is attached to the standard unit. For example, the prefix deci- denotes that the unit before which it appears is one-tenth of the standard unit. Decigram means one-tenth of a gram. The prefix kilo- shows that the standard unit has been multiplied by one thousand. Kilometer means one thousand meters. Conversion factors are ratios that relate two different units of measurement and can be used to mathematically convert between different systems of measurement.

Measurements always contain an unavoidable uncertainty called indeterminate error. This uncertainty in a measurement is conveyed by the number of significant figures that a measure contains. Theses significant figures are a result of the calibration of the specific device used to make a measurement. Scientific notation is a method of writing numerical measurements that avoids ambiguity and also allows very large and small numbers to be handled easily. Writing numbers in scientific notation requires that a number be rewritten as a coefficient, a number between 1 and 10, multiplied by the appropriate power of ten.

TEST YOURSELF

1. Classify the following as a chemical or physical property and support your answer.
   (a) a gold ring does not rust
   (b) a gold ring melts at 1337 K
   (c) a "gold" ring turns your finger green
   (d) a piece of gold is hammered into a ring
   (e) a gold ring has a density of 19.3 g/mL

2. Indicate which of the following is a physical property and which is a chemical property.
   (a) color
   (b) souring of milk
   (c) flammability
   (d) conduction of heat
   (e) luster
   (f) a gas is produced when substance is heated
   (g) size
   (h) melting point
   (i) ripening of an apple

3. Assign significant or not significant to the numbers described below:
   (a) a zero at the end of a large number
   (b) a zero after the decimal place, but before the first digit of a small number
   (c) a zero after the last digit of a number with a decimal point
   (d) a zero between two digits in a number
   (e) a zero followed by a decimal point at the end of a large number

4. Perform the calculations below, making sure that the answer is reported with the proper number of significant figures.
   (a) 1.25×23.68
   (b) $1.25 + 23.68$
   (c) $1.25 \times (2.368 \times 10^2)$
   (d) $1.25 \div (2.368 \times 10^2)$

5. For each of the metric prefixes below, explain how it numerically changes a standard unit.
   (a) deci- (b) giga- (c) micro- (d) centi- (e) milli- (f) kilo-

6. Match the metric prefix with its numerical value.

(a) kilo-	(1) 10^{-6}
(b) milli-	(2) 10^{6}
(c) micro-	(3) 10^{3}
(d) nano-	(4) 10^{-1}
(e) centi-	(5) 10^{-9}
(f) mega-	(6) 10^{-3}
(g) deci-	(7) 10^{-2}

7. For each relationship below, generate a pair of conversion factors.
   (a) a car and tires
   (b) a person and shoes
   (c) the number of pounds in a kilogram
   (d) the number in a dozen
   (e) the number of seconds in a day
   (f) the number of deciliters in a liter
8. Calculate the number of kilograms in 5678 g.
9. Calculate the number of kilograms in 175 lb.
10. Calculate the number of liters in 2.5 qt.
11. Calculate the number of microliters in 12 mL.
12. Report the value for each unit of measurement below to the appropriate number of decimal places.

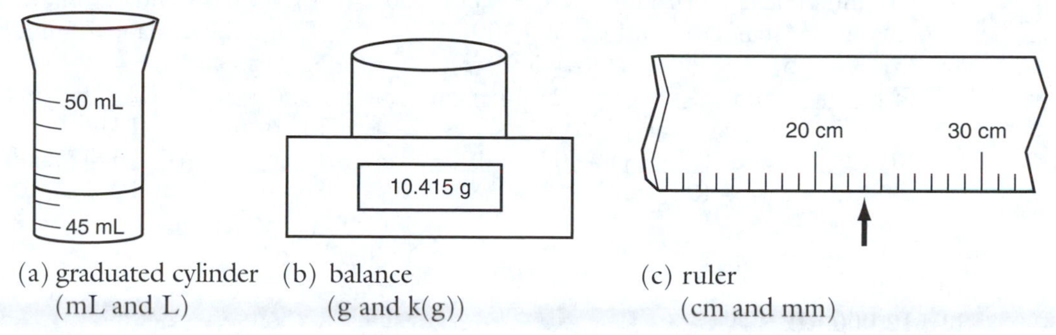

(a) graduated cylinder (mL and L) (b) balance (g and k(g)) (c) ruler (cm and mm)

13. Determine which of the numbers below are written in proper scientific notation. Change those that aren't.
   (a) 4.35×10^5 (b) $1.02 \times 10^2 \times 10^{-7}$ (c) 21.7×10^{12} (d) 0.090×10^2
14. A 21.7-lb child is prescribed medication that is given in a daily dosage of 5.00 mg per kilogram of body weight. The medication must be given for 14 days. Determine how many grams of medication the child will receive over the two-week period.
15. A metalsmith's guide lists the following information:

Substance	Specific gravity	Melting point, °C
lead	11.3	327
brass	8.5	954
gold	19.2	1063
silver	10.5	960.5

   Design two possible tests to determine the composition of an enameled ring.
16. Express the melting points of the metals above in °F and kelvins.
17. 5334 kJ of energy can theoretically be generated by eating a certain pastry. What mass of water could be heated from 20°C to 30°C with the energy?
18. A piece of jewelry that is supposed to be gold displaces 2.73 mL of water when placed in a graduated cylinder. The weight of the piece is 23.21 g. Is the piece real gold?
19. Match the terms on the right to the statements on the left.
   (a) mass
   (b) temperature
   (c) volume
   (d) specific heat
   (e) specific gravity
   (f) heat capacity
   (g) density

   (1) the weight of 1 cm^3 of a sample
   (2) the amount of heat a given sample can absorb
   (3) can be used to determine to what extent one sample will heat relative to another
   (4) the amount of matter in a sample
   (5) determines whether a sample is less dense than water
   (6) the amount of space an object occupies
   (7) a measure of how fast molecules in a sample are moving

20. How many joules of energy would be required to heat 200 g of aluminum from 20°C to 100°C?

Answers

1. (a) chemical (b) physical (c) chemical (d) physical (e) physical
2. (a) physical (b) chemical (c) chemical (d) physical (e) physical (f) chemical (g) physical (h) physical (i) chemical
3. (a) nonsignificant (b) nonsignificant (c) significant (d) significant (e) significant
4. (a) 29.6 (b) 24.93 (c) 296 (d) 5.28×10^{-3}
5. (a) divides standard unit by 10 (b) multiplies standard unit by 1,000,000,000 (c) divides standard unit by 1,000,000 (d) divides standard unit by 100 (e) divides standard unit by 1000 (f) multiplies standard unit by 1000
6. (a) 3 (b) 6 (c) 1 (d) 5 (e) 7 (f) 2 (g) 4
7. (a) $\frac{1\ \text{car}}{4\ \text{tires}}$ and $\frac{4\ \text{tires}}{1\ \text{car}}$ (b) $\frac{1\ \text{person}}{2\ \text{shoes}}$ and $\frac{2\ \text{shoes}}{1\ \text{person}}$ (c) $\frac{2.2\ \text{lb}}{1\ \text{kg}}$ and $\frac{1\ \text{kg}}{2.2\ \text{lb}}$ (d) $\frac{12}{1\ \text{dozen}}$ and $\frac{1\ \text{dozen}}{12}$ (e) $\frac{86{,}400\ \text{s}}{1\ \text{day}}$ and $\frac{1\ \text{day}}{86{,}400\ \text{s}}$ (f) $\frac{10\ \text{dl}}{1\ \text{L}}$ and $\frac{1\ \text{L}}{10\ \text{dL}}$
8. 5.678 kg
9. 79.5 kg
10. 2.3 L
11. 12,000 μL
12. (a) 46.5 mL or 0.0456 L (b) 10.415 g or 0.010415 kg (c) 23.2 cm or 232 mm
13. (a) correct (b) 1.02×10^{-5} (c) 2.17×10^{13} (d) 9.0×10^{0}
14. 0.690 g
15. Determine the densities of the metals by using the displacement method outlined in the chapter or melt them.
16. lead: 621°F; 600 K brass: 1750 °F; 1227 K gold: 1945 °F; 1336 K silver: 1761 °F; 1233.5 K
17. 1.2×10^{5} g
18. d = 8.50 g/mL; not gold; could be brass
19. (a) 4 (b) 7 (c) 6 (d) 3 (e) 5 (f) 2 (g) 1
20. 14000 J

chapter 2

Atomic Structure

OUTLINE

2.1 *Chemical Background for the Early Atomic Theory*

A. Law of conservation of mass: In chemical reactions mass is neither created nor destroyed

$$\begin{array}{ccc} A + B & \Rightarrow & A\text{-}B \\ \text{mass 1} & = & \text{mass 2} \end{array}$$

B. Law of constant composition: Elements combine in fixed ratios by mass when they form a compound; H_2O is always 2 H atoms: 1 O atom

Key Terms law of conservation of mass, law of constant composition, percent composition

Example 2.1 Calculating the mass percent of a compound
Problem 2.1 Calculating the mass percent of a compound

2.2 *Dalton's Atomic Theory*

A. Dalton's model
1. matter is composed of atoms
2. atoms of an element are chemically identical
3. atoms of different elements have different masses
4. elements are combinations of atoms of that element
5. compounds are combinations of atoms of different elements
6. compounds have properties different from those of their constituent elements
7. in chemical reactions, starting compounds exchange atoms to form new compounds

B. Dalton's model explains the law of constant composition.

Key Terms atoms, compound particle, element, elementary particle
Key Figure Figure 2.1 Dalton's picture of a chemical reaction

2.3 *Atomic Masses*

A. Relative mass scale developed because individual atoms can't be weighed.
1. atomic mass is a relative mass
2. carbon is assigned 12 atomic mass units
3. all elements have atomic masses relative to carbon

Key Terms amu, atomic mass, atomic mass unit, atomic weight
Key Figure Figure 2.2 A mass spectrometer

Example 2.2 Relating mass to numbers of objects
Problem 2.2 Relating numbers of objects to mass

2.4 *The Structure of Atoms*

A. Atoms are made up of subatomic particles.
  1. mass of the atom is concentrated in nucleus
    a. the mass of the nucleus is the combined masses of its protons and neutrons
    b. protons and neutrons have equal masses: 1 amu
  2. electrons define volume of the atom outside the nucleus
  3. protons (+) and electrons (−) have equal but opposite charges
  4. number of protons is the atomic number of the element
  5. in a neutral atom, number of protons equals number of electrons
  6. charged atoms are called ions
    a. in cations (+), the number of protons is greater than the number of electrons
    b. in anions (−), the number of protons is less than the number of electrons

Key Terms anion, atomic number, cation, electron, ion, neutron, nucleus, proton
Key Table Table 2.1 Properties of subatomic particles
End Paper Table (inside front cover) Atomic numbers and atomic masses of the elements

Example 2.3 Calculating the charge of an atom
Problem 2.3 Calculating the charge of an atom
Example 2.4 Calculating the charge of an ion
Problem 2.4 Calculating the charge of an ion
Example 2.5 Computing the mass of an atom by adding masses of subatomic particles
Problem 2.5 Identifying an element by its atomic mass

2.5 *Isotopes*

A. Isotopes are atoms of the same element that have different numbers of neutrons.

number of protons + number of neutrons = mass number

B. Isotopes can be designated in two ways.
  1. protons + neutrons → $^{16}_{8}O$ ← elemental symbol
     protons →

  2. O-17 ← protons + neutrons
     ↑
     elemental symbol

Key Terms isotope, mass number, natural abundance

Example 2.6 Using isotopic notation
Problem 2.6 Using isotopic notation
Example 2.7 Using isotopic notation to calculate the number of protons and neutrons in an isotope
Problem 2.7 Using isotopic notation to calculate the number of protons and neutrons in an isotope
Example 2.8 Calculating the average atomic mass of an element from its natural abundances
Problem 2.8 Calculating the average atomic mass of an element from its natural abundances

2.6 *The Periodic Table*

A. Elements are given letter symbols in the periodic table.
B. Properties of elements repeat periodically when elements are arranged in order of increasing atomic number (number of protons).
C. As atomic number increases, the number of electrons increases accordingly.
D. Each box contains the elemental symbol, atomic number, and mass number.

E. Elements having similar properties are arranged in vertical columns.
   1. Roman numerals above the main-group columns
   2. correlated to number of valence electrons
   3. noble gases are the last column on the right side of the periodic table
   4. halogens are next to noble gases
   5. alkali metals are the first column on the left side of the periodic table
   6. alkaline earth metals are next to them.
F. The periodic table can be divided into
   1. metals
      a. shiny, malleable, good conductors, able to be melted and cast into shapes
      b. tend to lose electrons in chemical reactions
   2. metalloids
      a. properties between metals and nonmetals
      b. used to be called semimetals
      c. located diagonally right of center in periodic table
   3. nonmetals
      a. don't conduct electricity, can't be cast into shapes
      b. tend to gain electrons in chemical reactions
G. Horizontal rows are called periods.
   1. represent major energy shells
   2. transition elements first appear in fourth row
H. Biologically important elements; the first four periods contain the elements of most importance to life processes.

Key Terms alkali metal, alkaline earth metal, group, halogen, inner transition elements, main-group elements, metal, metalloid, noble gas, nonmetal, period, Periodic Law, periodic table, semimetal, transition element

Key Tables and Figures

Table 2.2	Essential elements in human body
Table 2.3	Elements essential to human nutrition and consequences of deficiencies of essential elements
Figure 2.3	Abbreviated periodic table
Figure 2.4	Violent reaction of potassium and water
Figure 2.5	Drano and Easy-Off, caustic products
Figure 2.6	The halogens chlorine, bromine, and iodine at room temperature

2.7 *Electron Organization Within the Atom*

A. Atomic spectra
   1. Emission spectrum

solid at high temperature ∿∿∿∿ white light ∿∿∿∿ spectrometer ⇒ component colors

atom in flame ∿∿∿∿ light ∿∿∿∿ spectrometer ⇒ colored lines: emission spectrum, with pattern, colors characteristic of element

B. Electromagnetic radiation and energy
   1. electromagnetic spectrum
   2. unit is given in terms of frequency: the number of times per second a vibration is completed

hertz (Hz) = cycles / second

3. energy of electromagnetic radiation is directly related to frequency

higher energy = higher frequency

C. Atomic energy states
   1. excited state
      ↑ ↓ ⇝ emitted energy
      ground state
   2. atoms absorb energy in quantized packets
   3. emission of energy after absorption of energy results in line spectra that have specific frequencies; frequencies can be correlated to energy of the state
D. Bohr's Atomic Model
   1. electrons orbit nucleus in shells
   2. shells are at fixed distances from nucleus
   3. looks like planets circling sun

Key Terms absorption spectrum, atomic emission spectrum, electromagnetic radiation, excited state, frequency, ground state, photon, quanta, radiation, visible spectrum

Key Tables, Figures, and Box

Table 2.4	Frequencies and energies of electromagnetic radiation
Figure 2.7	White light resolved into component colors
Figure 2.8	Colors of light emitted by Group I metals in flame tests
Figure 2.9	Spectrometer resolves light into its component colors
Figure 2.10	Atomic emission spectra of various elements
Box 2.1	Absorption spectra and chemical analysis

2.8 *The Quantum Mechanical Atom*

A. Mathematical theory that successfully predicts a variety of atomic properties
B. Answers are solutions to mathematical equations; hard to describe physically.
C. Replaced Bohr's atomic model
D. Quantum mechanics specifies four features of electronic organization in the atom.
   1. shells are energy levels and are specified by principal quantum numbers: 1, 2, 3, 4. . .
   2. subshells are contained within shells and are identified by lowercase letters: *s, p, d, f*
   3. orbitals exist within subshells
      a. region of space where there is a high probability of finding the electron
      b. have characteristic shapes

s: spherical
p: dumbbell

   4. Electrons have spin.
      a. 2 electrons in orbitals must be spin paired ↑ ↓

Key Terms atomic orbital, lone electron, principal quantum number, quantum mechanics, shell, spin, spin-paired electrons, subshell, unpaired electron

Key Tables and Figures

Table 2.5	Relationship between principal quantum number, number of subshells, and orbital types
Table 2.6	Number of orbitals within each subshell type
Table 2.7	Number of orbitals and electrons in each type of subshell
Figure 2.11	*s* and *p* atomic orbitals
Figure 2.12	1*s*, 2*s*, and 3*s* atomic orbitals

2.9 *Atomic Structure and Periodicity*

A. Aufbau principle: build the elements by adding electrons one at a time
B. Electron configuration describes the organization of electrons in an atom.
C. Designating electron configurations:
  1. energy shells are designated by the principal quantum number; as the principal quantum number increases, the number of subshells increases

     principal quantum number of shell ⇒2 *s*⇐ subshell

  2. subshells have different numbers of orbitals
  3. a maximum of two electrons can occupy an orbital
     a. Pauli exclusion principle
     b. if two electrons occupy the same orbital, they must be spin-paired
  4. orbitals have shapes
     a. *s* orbitals are spherical
     b. *p* orbitals are dumbbell shaped; oriented along *x*-, *y*-, *z*-axes perpendicular to each other
  5. each orbital in a subshell gets one electron before any gets two; this principle is called Hund's rule
  6. electrons enter available shells in a specific order

Key Terms Aufbau procedure, core configuration, electron configuration, Hund's rule, Pauli exclusion principal, quantized

Key Tables and Figure

Table 2.8 Electron configurations of first 11 elements in periodic table (Method 1; quantum numbers)
Table 2.9 Electron configurations (shorthand notation) of selected Group I, VII, and 0 elements
Table 2.10 Electron configurations (shorthand notation) of main-group elements in periods 2 and 3
Figure 2.13 Box representation of quantized energy levels of atomic subshells

Example 2.9 Determining the type and number of subshells associated with a principal quantum number
Problem 2.9 Determining the type and number of subshells associated with a principal quantum number
Example 2.10 Using quantum numbers (Method 1) to diagram electron configuration
Problem 2.10 Using quantum numbers (Method 1) to diagram electron configuration
Example 2.11 Using orbital boxes (Method 2) to diagram electron configuration
Problem 2.11 Using orbital boxes (Method 2) to diagram electron configuration
Example 2.12 Using orbital boxes (Method 2) to diagram electron configuration; emphasizes Hund's rule
Problem 2.12 Using orbital boxes (Method 2) to diagram electron configuration; emphasizes Hund's rule
Example 2.13 Using shorthand notation to diagram electron configuration
Problem 2.13 Using shorthand notation to diagram electron configuration

2.10 *Atomic Structure, Periodicity, and Chemical Reactivity*

A. All members of the same group have the same valence electron configuration.
  1. elements with the same valence electron configuration have similar chemical properties
  2. all noble gases except He have eight electrons in outer shell
  3. all alkali metals have one electron in outer shell
  4. all halogens have seven electrons in outer shell
  5. main group element's group number = number of electrons in valence shell
  6. eight electrons in a valence shell is the most stable configuration for most main group elements
B. Electron-dot diagrams can be used to depict valence shell electrons of atoms and ions.

Key Terms Lewis symbols, octet rule, valence electrons, valence shell

Example 2.14 Using Lewis symbols to illustrate a chemical reaction
Problem 2.14 Using Lewis symbols to illustrate a chemical reaction

ARE YOU ABLE TO ... AND WORKED TEXT PROBLEMS

Objectives Section 2.1 Chemical Background for the Early Atomic Theory

Are you able to . . .

- define the law of conservation of mass (Exercise 2.1)
- show how mass is conserved in a chemical reaction (Fig. 2.1)
- define the law of constant composition (Exercise 2.2)
- explain what is meant by "percent composition"
- determine mass percentages of individual elements in compounds (Ex. 2.1, Prob. 2.1, Exercises 2.3, 2.4, Expand Your Knowledge 2.62)

Problem 2.1 Analysis of 4.800 g of niacin (nicotinic acid), one of the B complex vitamins, yields 2.810 g of carbon, 0.1954 g of hydrogen, 0.5462 g of nitrogen, and 1.249 g of oxygen. Calculate the mass percentages of each element in the compound.
What does mass % mean?

$$\frac{\text{mass individual element}}{\text{mass compound}} \times 100$$

Calculating mass %'s requires that you know or can calculate the total mass of the compound. The total mass of the compound, niacin, is given as 4.800 g; alternatively, the total mass could be calculated by adding the individual masses.

Individual elemental masses		Mass % element in compound
Carbon	2.810 g	C = 2.810 g / 4.800 g × 100 = 58.54%
Hydrogen	0.1954 g	H = 0.1954 g / 4.800 g × 100 = 4.071%
Nitrogen	0.5462 g	N = 0.5462 g / 4.800 g × 100 = 11.38%
Oxygen	1.249 g	O = 1.249 g / 4.800 g × 100 = 26.02%

The %'s of individual contributions should add up to 100%.

Objectives Section 2.2 Dalton's Atomic Theory

Are you able to . . .

- describe Dalton's atomic theory
- define atom
- describe atomic features
- explain how a molecule is different from an atom
- explain what happens to atoms in chemical reactions (Fig. 2.1)
- define Dalton's elementary particle
- define Dalton's compound particle
- explain how Dalton's model supported the law of constant composition

Objectives Section 2.3 Atomic Masses

Are you able to . . .

- define atomic mass
- define atomic weight
- define atomic mass unit (amu)
- explain the relationship between the mass of carbon and the relative atomic masses of other elements (Exercises 2.5, 2.8)

- calculate the relative masses of elements in a compound (Exercises 2.6, 2.7)
- calculate the relative masses of individual items in a group (Ex. 2.2)
- calculate numbers of items on the basis of their relative masses (Prob. 2.2)

Problem 2.2 Suppose, in one crate, there were 4500 g of oranges weighing 150 g each and, in another crate, 4500 g of cantaloupes weighing 450 g each. Calculate the numbers of oranges and cantaloupes in each crate.

The first suggestion would be to draw a picture to get an idea of the situation described in the problem.

What information is given?	What is wanted?
1 orange weighs 150 g	number of oranges in the crate
all oranges in the crate weigh a total of 4500 g	number of cantaloupes in the crate
1 cantaloupe weighs 450 g	
All cantaloupes in the crate weigh a total of 4500 g	

You know the total weight of all the fruit in each crate and you know the individual weight of one piece of fruit. The problem can be set up as a unit conversion: the question asks for the number of oranges per crate; you want oranges in the numerator and crate in the denominator:

Again, using unit conversions, you want cantaloupes in the numerator and crate in the denominator:

$$\frac{4500\ \cancel{g}}{1\ \text{crate}} \times \frac{1\ \text{orange}}{150\ \cancel{g}} = \frac{30\ \text{oranges}}{\text{crate}}$$

$$\frac{4500\ \cancel{g}}{1\ \text{crate}} \times \frac{1\ \text{cantaloupe}}{450\ \cancel{g}} = \frac{10\ \text{cantaloupes}}{\text{crate}}$$

Objectives Section 2.4 The Structure of Atoms

Are you able to . . .

- describe subatomic particles in terms of location, mass, and charge (Table 2.1, Exercise 2.9)
- use information about subatomic particles to determine the magnitude and sign of a charge on an atom or ion (Table 2.1, Exs. 2.3, 2.4, Probs. 2.3, 2.4, Exercises 2.12, 2.13, 2.14)
- define cation (Exercise 2.13)
- define anion (Exercise 2.14)
- define atomic number (Exercises 2.10, 2.47)
- calculate mass contributions made by subatomic particles to total atomic mass (Ex. 2.5, Prob. 2.5, Exercises 2.11, 2.57)
- identify elements by using information about protons, neutrons, and/or electrons (Exercises 2.10, 2.12)

Problem 2.3 What are the sign and magnitude of the charge of an atom containing 9 protons, 10 neutrons, and 9 electrons?

Set up a table showing the charge distribution:

$$\text{protons } (+1) \times 9 = 9 \text{ positive charges}$$
$$\text{electrons } (-1) \times 9 = 9 \text{ negative charges}$$
$$\text{neutrons } (0) \times 10 = 0 \text{ charge contribution}$$

There are 9 positive charges and 9 negative charges; therefore, the charge on the atom is 0.

Problem 2.4 What are the sign and magnitude of the charge of an atom containing 13 protons, 14 neutrons, and 10 electrons?

Set up a table showing the charge distribution:

$$\text{protons } (+1) \times 13 = 13 \text{ positive charges}$$
$$\text{electrons } (-1) \times 10 = 10 \text{ negative charges}$$
$$\text{neutrons } (0) \times 14 = 0 \text{ charge contribution}$$

There are 13 positive charges and 10 negative charges; therefore, the charge on the atom is 3+; it is a cation.

Problem 2.5 What are the name and approximate mass of an element that has 22 protons and 26 neutrons?

The number of protons is the identifying factor for an element: 22 protons corresponds to titanium, Ti. The mass number is the sum of the masses of the protons and neutrons: 22 protons = 22 amu plus 26 neutrons = 26 amu; therefore, the mass is 48 amu.

Objectives Section 2.5 Isotopes

Are you able to . . .

- define isotope (Exercise 2.58)
- define mass number (Expand Your Knowlege 3)
- describe isotopes, using both types of symbolic notation (Exs 2.6, Prob. 2.6, Exercises 2.17, 2.18, 2.55)
- determine the number of protons and neutrons in an atom from its isotopic notation (Ex. 2.7, Prob. 2.7)
- use mass numbers and atomic numbers to calculate the number of neutrons in an atom (Ex. 2.12)
- use % of naturally occurring isotopes to calculate the average atomic mass of an element (Ex. 2.8, Prob. 2.8, Exercise 2.15, Expand Your Knowlege 2.63)
- use the average atomic mass of an element and isotopic information to calculate the % of each naturally occurring isotope (Exercise 2.16)

Problem 2.6 What is the symbolic notation for the two isotopes of nitrogen that contain 7 and 8 neutrons, respectively?

How is an isotope written in symbolic notation? There are two ways to do this.

(1) The element symbol followed by its mass number (number of protons plus number of neutrons). The elemental symbol can be found on the periodic table. The number of protons is the same for each isotope: Nitrogen has 7 protons.

N-14 is the isotope with 7 protons and 7 neutrons
N-15 is the isotope with 7 protons and 8 neutrons

(2) The elemental symbol can have subscripts and superscripts. How are superscript/subscript notations used to designate isotopes? The elemental symbol has a superscript corresponding to the atomic mass (number of protons plus number of neutrons) and a subscript corresponding to the atomic number (number of protons)

$$\begin{array}{r} \text{number of protons + number of neutrons} \rightarrow \\ \text{number of protons} \rightarrow \end{array} {}^{14}_{7}\text{N} \leftarrow \text{elemental symbol}$$

${}^{14}_{7}N$ is the isotope with 7 protons and 7 neutrons. ${}^{15}_{7}N$ is the isotope with 7 protons and 8 neutrons.

Problem 2.7 Calculate the number of protons and neutrons in the two isotopes of nitrogen. (a) ${}^{14}_{7}N$; (b) ${}^{15}_{7}N$.

The number of protons is the same for each isotope: Nitrogen has 7 protons. The subscript of the isotopic designation is the number of protons. The superscript of the isotopic designation is the mass number, the sum of the number of protons and the number of neutrons in the atom:

$$\text{number of neutrons} = \text{mass number of} - \text{number of protons}$$
$$\text{(a) number of neutrons} = 14 - 7 = 7 \text{ neutrons}$$
$$\text{(b) number of neutrons} = 15 - 7 = 8 \text{ neutrons}$$

N-14 has 7 protons and 7 neutrons, and N-15 has 7 protons and 8 neutrons.

Problem 2.8 Magnesium consists of three isotopes of masses 24.0 amu, 25.0 amu, and 26.0 amu (to the closest amu) with abundances of 78.70%, 10.13%, and 11.17%, respectively. Calculate the average atomic mass of magnesium.

How do the natural abundances relate to the atomic mass of the element? The atomic mass is the weighted average of the isotopic masses. A weighted average is calculated by first multiplying the % natural abundance by the atomic mass of that isotope then the individual contributions are added together to get the final average:

% isotope (in decimal form)	×	atomic mass (amu)	=	contribution to average (amu)
0.7870	×	24.0	=	18.9
0.1013	×	25.0	=	2.5
0.1117	×	26.0	=	2.9
			average atomic mass of Mg	24.3

The average atomic mass should be closest to the mass of the isotope that makes the greatest contribution, i.e., the one that exists in the greatest natural abundance. In this case, 78.79% of magnesium exists as the isotope that weighs 24.0 amu . . .very close to the calculated average.

Objectives Section 2.6 The Periodic Table

Are you able to . . .

- describe the organization of the periodic table (Fig. 2.3)
- interpret and use information contained in the periodic table (Exercises 2.19, 2.20)
- define periodic law
- define period and row (Exercises 2.21, 2.22)
- describe reasoning behind the horizontal arrangement of atoms (Exercise 2.25)
- describe the reasoning behind the vertical arrangement of atoms (Exercise 2.51)
- identify noble gases, halogens, alkali metals, alkaline earth metals, transition elements, main group elements, metals, nonmetals, metalloids (Exercises 2.23, 2.24, 2.25, 2.26, 2.27, 2.28, 2.52, 2.53, 2.54)
- state the elements that are important to biological processes (Tables 2.2, 2.3)

Objectives Section 2.7 Electron Organization Within the Atom

Atomic Spectra

Are you able to . . .

- explain how emission spectra arise (Figs. 2.7, 2.8, 2.9, 2.10, Box. 2.1, Exercises 2.48, 2.49, 2.60)

Electromagnetic Radiation

Are you able to . . .

- define electromagnetic radiation (Exercise 2.52)
- define frequency and its units
- explain the relationship between frequency and energy (Table 2.4, Exercise 2.59)

Atomic Energy States and Quanta

Are you able to . . .

- define ground state
- define excited state
- define quanta
- relate emission spectra and energy states (Exercises 2.29, 2.30)

Bohr's Atomic Model

Are you able to . . .

- describe Bohr's atomic model (Exercise 2.33)
- define shell
- define principal quantum number

Objectives Section 2.8 The Quantum Mechanical Atom

Are you able to . . .

- define quantum mechanics
- explain the four features of electronic organization that are specified by quantum mechanics (Exercise 2.31)

- describe how shells are designated
- explain the relationship between principal quantum number, shell energy, and distance from the atomic nucleus (Fig. 2.12, Exercise 2.49)
- define subshell
- describe subshell designation (Table 2.5, Exercise 2.32)
- relate subshell designation and number of orbitals (Tables 2.6, 2.7, Exercise 2.35)
- define orbital (Exercise 2.33)
- discuss relationship between type of subshell and number of electrons in each subshell (Table 2.7, Exercises 2.34, 2.36, 2.50)
- describe the characteristic orbital shapes and spatial orientations (Fig. 2.11, Exercises 2.37, 2.38)
- explain spin-pairing of electrons

Objectives Section 2.9 Atomic Structure and Periodicity

Are you able to . . .

- define the aufbau principle
- explain what is meant by electron configuration
- describe the Pauli exclusion principle
- define Hund's rule
- diagram the electron configuration of an atom, using quantum number designations (Table 2.8, Exs. 2.9, 2.10, Probs. 2.9, 2.10, Exercises 2.43, 2.44, 2.45, 2.46)
- diagram the electronic configuration of an atom, using energy level designation (boxes) (Fig. 2.13, Exs. 2.11, 2.12, Probs. 2.11, 2.12, Exercise 2.39)
- use shorthand notation to represent electron configurations (Tables 2.9, 2.10)

Problem 2.9 How many and what kind of subshells are allowed in principal quantum state 4?

The principal quantum number 4 has four subshells; *s, p, d,* and *f.* Refer to Table 2.5.

Problem 2.10 Diagram the electron configuration of element 16 by using Method 1.

How do you know what element 16 is? Look at the periodic table. Element 16 is sulfur; it has 16 protons and 16 electrons; it is in period 3. Use Tables 2.5, 2.6, and 2.7 to determine the principal quantum number and available subshells. The 16 electrons must be distributed according to rules outlined on p. 53 in your textbook.

Shell	Subshell	Number of electrons	Electron configurations
1	*s*	2	$1s^2$
2	*s*	2	$2s^2$
2	*p*	6	$2p^6$
3	*s*	2	$3s^2$
3	*p*	4	$3p^4$ $(3p_x^{\,2}, 3p_y^{\,1}, 3p_z^{\,1})$

The electron configuration of sulfur is $1s^2 2s^2 2p^6 3s^2 3p_x^{\,2} 3p_y^{\,1} 3p_z^{\,1}$.

Problem 2.11 Diagram the electron configuration of element 5 by using method 2.

Element 5 is boron; it has 5 protons and 5 electrons and is in period 2. The box method is described in your text on pp. 55–56. Each orbital is represented by a box that can contain only 2 electrons; electrons fill boxes singly before pairing up.

Shell	Subshell	Number of electrons	Box configuration
1	*s*	2	1*s* [↑↓]
2	*s*	2	2*s* [↑↓]
2	*p*	1	[↑] $2p_x$ [ ] $2p_y$ [ ] $2p_z$

Problem 2.12 Diagram the electron configuration of element 7 by using method 2.

Element 7 is nitrogen; it has 7 protons and 7 electrons and is in period 2. The box method is described in your text on pp. 55–56. Each orbital is represented by a box that can contain only 2 electrons.

Shell	Subshell	Number of electrons	Box configuration		
1	*s*	2	1*s*	↑↓	
2	*s*	2	2*s*	↑↓	
2	*p*	3	↑	↑	↑
			$2p_x$	$2p_y$	$2p_z$

Problem 2.13 Write the electron configuration of element 16, sulfur, by using the compact method.

The text describes the method for writing electron configurations in shorthand form on pp. 56–57. Referring back to Problem 2.10, note that the electron configuration of sulfur is $\mathbf{1s^22s^22p^63s^23p^4}$. Which noble gas is represented in sulfur's electron configuration? Period 2 is filled and the noble gas represented here is neon, $\mathbf{1s^22s^22p^6}$. Sulfur's configuration can be rewritten as **[Ne]** $\mathbf{3s^23p^4}$.

Objectives Section 2.10 Atomic Structure, Periodicity, and Chemical Reactivity

Are you able to . . .

- the octet rule (Exercises 2.56, 2.61)
- define valence shell
- define valence electron
- diagram electron configurations of elements and ions by using energy-level notation (Exercises 2.40, 2.41, 2.42, 2.43, 2.44, 2.45, 2.46)
- diagram valence-shell electrons of elements by using electron-dot (Lewis symbols) notation
- diagram the valence-shell electrons of ions by using electron-dot notation
- use electron-dot diagrams to predict whether an element will lose or gain electrons
- explain the stability of a valence-shell octet (Exercise 2.61)
- relate valence electron configurations to chemical reactivity (Ex. 2.14, Prob. 2.14)

Problem 2.14 Use Lewis symbols to illustrate the formation of sodium and chloride ions.

The first question to ask is, What group is each element in? The answer to this will tell you how many valence electrons each has and how many electrons will be lost or gained when the element ionizes. Sodium is in Group I and has 1 valence electron; chlorine is in Group VII and has 7 valence electrons.

Lewis symbols show the valence electrons as

$$\text{Na}\cdot \qquad :\dot{\underset{\cdot\cdot}{\text{Cl}}}:$$

Na will lose 1 electron to form Na^+

$$\text{Na}\cdot \rightarrow \text{Na}^+ + 1\text{e}^-$$

Cl will gain 1 electron to form Cl^-

$$:\dot{\underset{\cdot\cdot}{\text{Cl}}}: + 1\text{e}^- \rightarrow :\ddot{\underset{\cdot\cdot}{\text{Cl}}}:^-$$

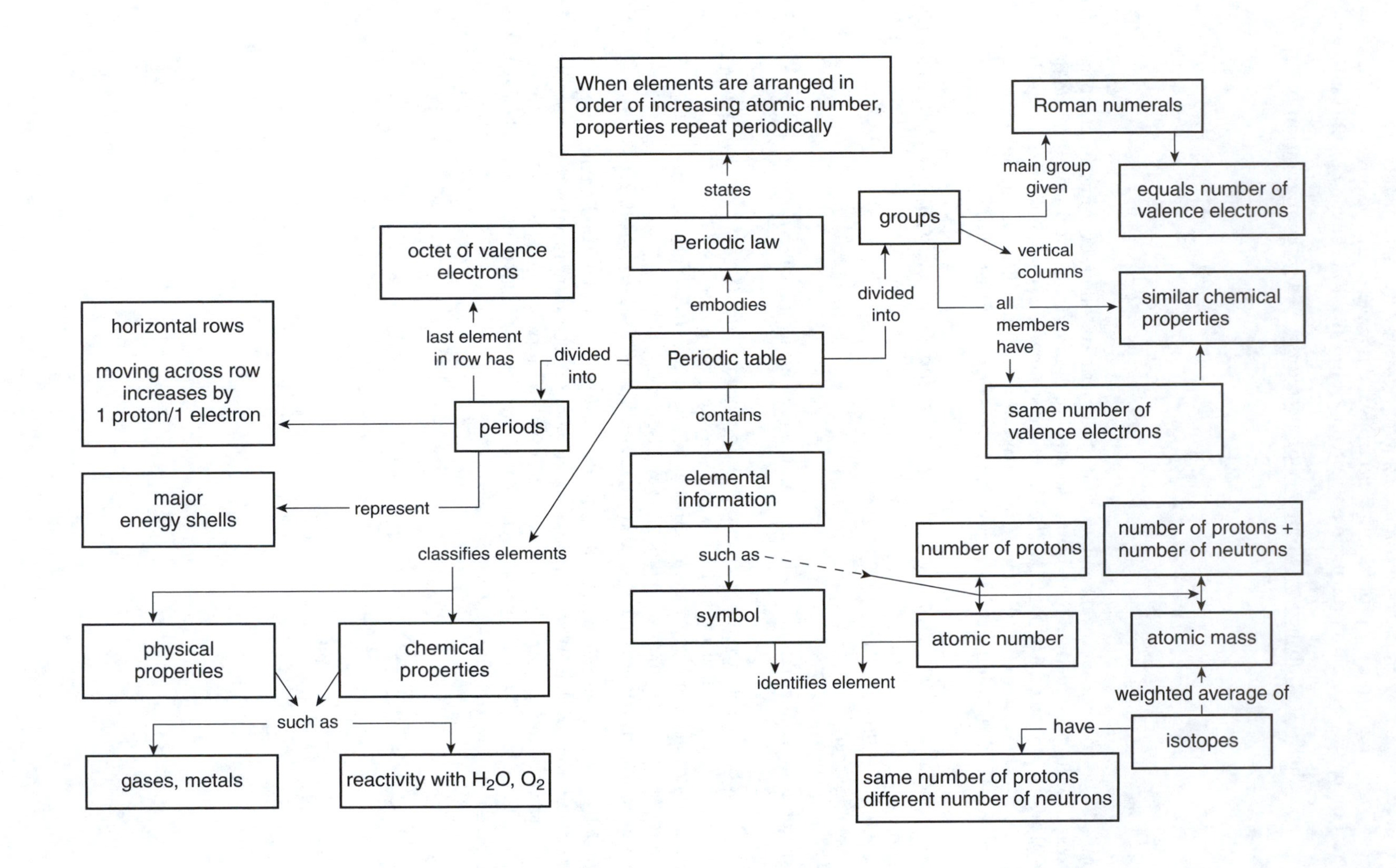
When elements are arranged in order of increasing atomic number, properties repeat periodically
states
Periodic law
embodies
Periodic table
divided into
groups
main group given
Roman numerals
equals number of valence electrons
vertical columns
all members have
similar chemical properties
same number of valence electrons
contains
elemental information
such as
symbol
identifies element
number of protons
number of protons + number of neutrons
atomic number
atomic mass
weighted average of
isotopes
have
same number of protons different number of neutrons
divided into
periods
last element in row has
octet of valence electrons
horizontal rows
moving across row increases by 1 proton/1 electron
represent
major energy shells
classifies elements
physical properties
chemical properties
such as
gases, metals
reactivity with H_2O, O_2

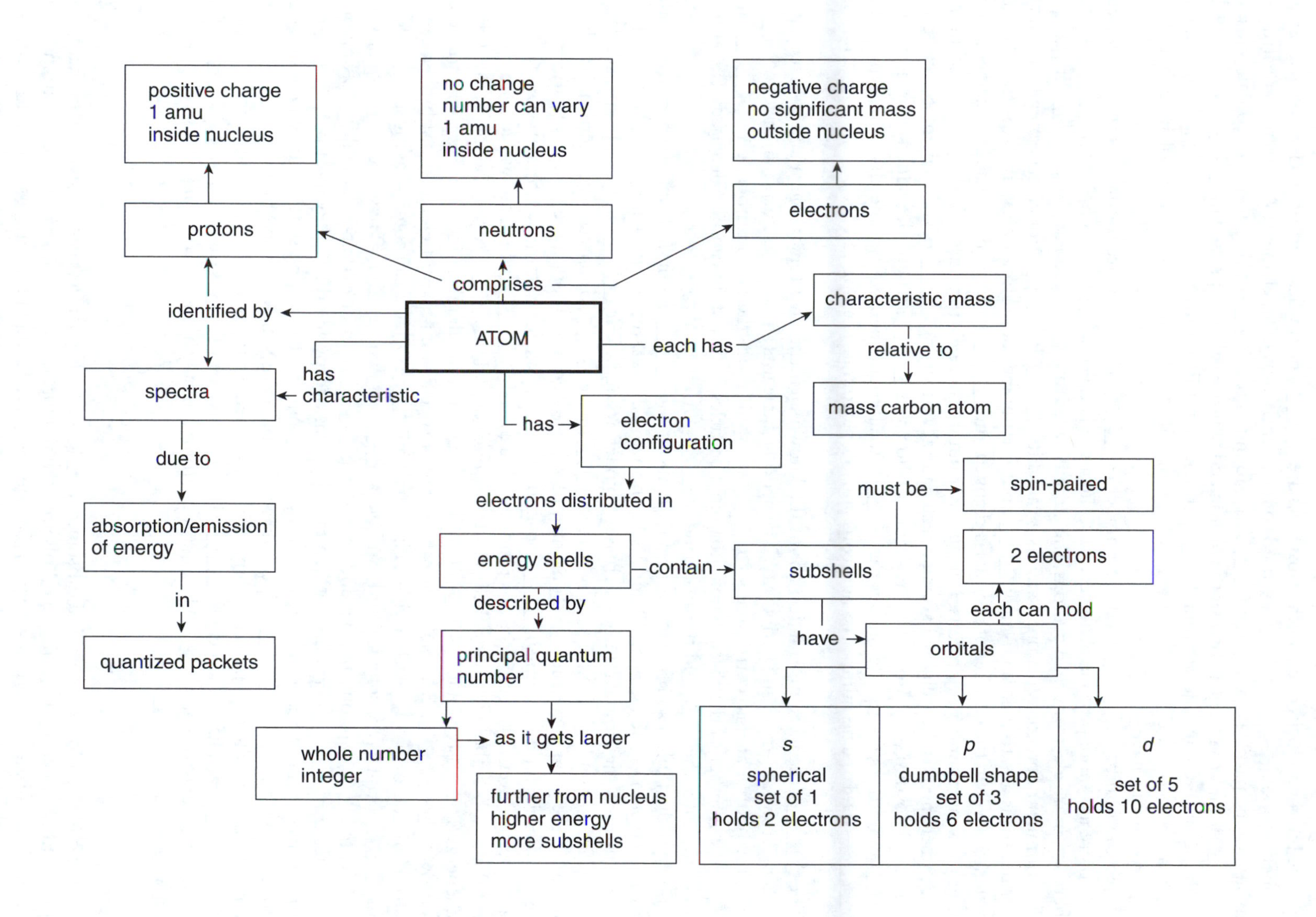

positive charge
1 amu
inside nucleus
no change
number can vary
1 amu
inside nucleus
negative charge
no significant mass
outside nucleus
protons
neutrons
electrons
comprises
identified by
ATOM
each has
characteristic mass
relative to
mass carbon atom
has
characteristic
spectra
has
electron configuration
due to
absorption/emission of energy
in
quantized packets
electrons distributed in
energy shells
contain
subshells
must be
spin-paired
2 electrons
each can hold
have
orbitals
described by
principal quantum number
whole number
integer
as it gets larger
further from nucleus
higher energy
more subshells
s
spherical
set of 1
holds 2 electrons
p
dumbbell shape
set of 3
holds 6 electrons
d
set of 5
holds 10 electrons

FILL-INS

The basic building blocks of chemical substances are ______.The smallest part of an element that retains its properties is an ______. ______ are used to represent atoms of an element. An atom is constructed from subatomic particles. ______ and ______ are found in the ______ of the atom and contain most of the ______ of the atom. The masses of the proton and neutron can be expressed in ______. ______ define the volume of the atom and are found ______ the nucleus. Electrons are ______ charged, but contain ______ mass. Protons are ______ charged and the magnitude of their charge exactly balances that of the ______. Neutrons have ______.

When the number of ______ and the number of ______ in an atom are the same, the atom is neutral. If there are more protons than electrons, the atom is called a ______ and has a net ______ charge. If there are more electrons than protons, the atom is called an ______ and has a net ______ charge.

The number of protons an atom contains is called its ______. It is ______ for each element. The number of neutrons an atom contains may ______, but the number of protons always remains the same for that element. When the number of protons is changed, the ______ of the element changes. Elements containing the same number of protons, but different numbers of neutrons, are called ______. The atomic mass of an element is the weighted average of all naturally occurring ______ of the element. Special notation is used to designate isotopes; they can be shown as ______, where the superscript designates the______ and the subscript designates the ______. An alternate method uses the elemental symbol and the ______ : O-16. The number of neutrons an isotope contains can be calculated by subtracting the ______ from the ______.

The ______ organizes information about the known elements in a logical form and is based upon increasing atomic number and periodicity. The periodic table lists all elements known in order of ______, which, for neutral atoms, corresponds to ______ as well. From the two numbers given, ______ and ______, the number of the three major subatomic particles in that atom can be calculated. The atomic number indicates the ______ in the atom and the ______ in a neutral atom.

The periodic table also provides information about the ______ configuration (arrangement of the atom's electrons) of the elements. The ______, horizontal rows, represent major ______ that are filled with electrons as the atomic number increases from left to right. The ______, vertical groups, are designated by group numbers that correspond to the______ in the ______, or outermost, electronic shells. All elements in the group have the ______ number of valence electrons, though the ______ number of electrons is different for each. Similar ______ result from similar valence electron configurations.

The energy levels in which electrons reside are designated by ______, which are whole numbers. The larger the number, the ______ the shell is from the nucleus and the ______ the energy of the shell. Principal energy shells also contain ______ subshells as their quantum numbers increase. Subshells are designated by ____________. Each subshell can contain a specified number of electrons. Subshells can be resolved into sets of______, each having a characteristic ______. Orbitals can contain, at most, ______electrons. ______ orbitals are spherical and can hold a ______ of two electrons. There are ______ *p* orbitals, each oriented ______ to one another and shaped like ______. A set of *p* orbitals can hold a maximum of ______ electrons, ______ per orbital. When two electrons occupy an orbital, they must be ______.

Electrons can move between energy levels by ______ energy. The resting state, or lowest energy state, of an electron is called its ______. Upon absorption of energy, the electron enters an ______. As the electron returns to the ground state, some of the absorbed energy is ______. The light can be resolved into its ______, each having a specific ______. The resultant set of lines is called the atom's ______. This spectrum can be used as an ______. The energy of atoms can only be increased by absorbing discrete packets of energy called ______.

Answers

The basic building blocks of chemical substances are elements. The smallest part of an element that retains its properties is an atom. Chemical symbols are used to represent atoms of an element. An atom is constructed from subatomic particles. Protons and neutrons are found in the nucleus of the atom and contain most of the mass of the atom. The masses of the proton and neutron can be expressed in atomic mass units (amu). Electrons define the volume of the atom and are found outside the nucleus. Electrons are negatively charged, but contain virtually no mass. Protons are positively charged and the magnitude of their charge exactly balances that of the electrons. Neutrons have no charge.

When the number of protons and the number of electrons in an atom are the same, the atom is neutral. If there are more protons than electrons, the atom is called a cation and has a net positive charge. If there are more electrons than protons, the atom is called an anion and has a net negative charge.

The number of protons an atom contains is called its atomic number. It is unique for each element. The number of neutrons an atom contains may vary, but the number of protons always remains the same for that element. When the number of protons is changed, the identity of the element changes. Elements containing the same number of protons, but different numbers of neutrons, are called isotopes. The atomic mass of an element is the weighted average of all naturally occurring isotopes of the element. Special notation is used to designate isotopes; they can be shown as $^{A}_{Z}E$, where the superscript designates the mass number and the subscript designates the atomic number. An alternate method uses the elemental symbol and the mass number: O-16. The number of neutrons an isotope contains can be calculated by subtracting the atomic number from the mass number.

The periodic table organizes information about the known elements in a logical form and is based upon increasing atomic number and periodicity. The periodic table lists all elements known in order of increasing atomic number, which, for neutral atoms, corresponds to increasing number of electrons as well. From the two numbers given, atomic number and atomic mass, the number of the three major subatomic particles in that atom can be calculated. The atomic number indicates the number of protons in the atom and the number of electrons in a neutral atom.

The periodic table also provides information about the electron configuration (arrangement of the atom's electrons) of the elements. The periods, horizontal rows, represent major energy shells that are filled with electrons as the atomic number increases from left to right. The columns, vertical groups, are designated by group numbers that correspond to the number of electrons in the atoms' valence, or outermost, electronic shells. All elements in the group have the same number of valence electrons, though the total number of electrons is different for each. Similar chemical properties result from similar valence electron configurations.

The energy levels in which electrons reside are designated by principal quantum numbers, which are whole numbers. The larger the number, the further the shell is from the nucleus and the higher the energy of the shell. Principal energy shells also contain more subshells as their quantum numbers increase. Subshells are designated by lower-case letters, *s, p, d, f*. Each subshell can contain a specified number of electrons. Subshells can be resolved into sets of orbitals, each having a characteristic shape. Orbitals can contain, at most, two electrons. *s* orbitals are spherical and can hold a maximum of two electrons. There are three *p* orbitals, each oriented perpendicular to one another and shaped like dumbbells. A set of *p* orbitals can hold a maximum of six electrons, two per orbital. When two electrons occupy an orbital, they must be spin-paired.

Electrons can move between energy levels by absorbing or releasing energy. The resting state, or lowest energy state, of an electron is called its ground state. Upon absorption of energy, the electron enters an excited state. As the electron returns to the ground state, some of the absorbed energy is emitted as visible light. The light can be resolved into its component colors, each having a specific frequency. The resultant set of lines is called the atom's emission spectrum. This spectrum can be used as an elemental fingerprint. The energy of atoms can only be increased by absorbing discrete packets of energy called quanta.

TEST YOURSELF

1. An Alka-Seltzer tablet dissolving seems to violate the law of conservation of mass. Explain why it doesn't.
2. Explain how you could determine the number of jellybeans in a bowl without counting them.
3. Explain how the volumes of a cation and of an anion differ from that of a neutral atom.
4. Match the periodic table labels on the left with the information on the right:

(a) main group number	____	1 can tell number of neutrons in most abundant isotope
(b) period number	____	2 identity of element
(c) position on the table	____	3 number of valence electrons
(d) number above elemental symbol	____	4 energy shell occupied by valence electrons
(e) number below elemental symbol	____	5 whether it has metallic properties

5. Match the information on the left with the elements on the right; there may be more than one right answer.

(a) alkali metal	____	1 iron, Fe
(b) inert gas	____	2 calcium, Ca
(c) transition metal	____	3 potassium, K
(d) metal	____	4 oxygen, O
(e) nonmetal	____	5 neon, Ne

6. Which of the following would you expect to be most like sodium in its chemical activity and why: calcium, iron, chlorine, potassium?
7. Draw the orbital shapes associated with $2p$ and $3p$ subshells.
8. Match the terms on the left with the descriptions on the right.

(a) Lewis structure	____	1 a volume in space where an electron may be found
(b) principal quantum number	____	2 the tendency for electrons to occupy orbitals singly before pairing up
(c) orbital	____	3 the sequential addition of electrons as atomic nuclei increase in number of protons
(d) Hund's rule	____	4 specifies energy level of an electron
(e) Aufbau	____	5 diagrams the valence shell electrons of an atom

9. Identify the element with the electron configuration: $1s^2 2s^2 2p^2$.
10. Write the electron configuration of a Mg^{2+} ion.

Answers

1. The tablet dissolves and produces a fizz, which is gas and has mass. If the mass of the gas produced were added to the mass of the solids that dissolve, the sum would equal the mass of the original tablet.
2. Weigh one jellybean; weigh the entire bowl; weigh the empty bowl. Subtract the mass of the empty bowl from the mass of the bowl with the jellybeans in it. This is the weight of all the jellybeans. Divide this by the weight of a single jellybean.
3. A cation has lost at least one electron and will have a smaller volume than the neutral atom. An anion has gained one electron and will have a larger volume than the neutral atom.
4. (a) 3 (b) 4 (c) 5 (d) 1 (e) 2
5. (a) 3 (b) 5 (c) 1 (d) 1, 2, 3 (e) 4, 5
6. potassium; it is in the same group as sodium and has the same number of valence electrons.
7. three sets of dumbbell shapes oriented along x-, y-, z-axes; the $3p$ set is larger than the $2p$ set.
8. (a) 5 (b) 4 (c) 1 (d) 2 (e) 3
9. carbon
10. $1s^2 2s^2 2p^6$; the 2 electrons in the $3s$ orbital are lost.

chapter 3

Molecules and Chemical Bonds

OUTLINE

3.1 Ionic versus Covalent Bonds

A. Bond formation results as most atoms strive for an octet of valence electrons.
   1. ionic bonds form as a result of electron transfer from/to valence shells
   2. covalent bonds form as a result of electron sharing between valence shells

B. Ionization energy
   1. measures how easily valence electrons can be removed from outer shell
   2. depends on strength of nuclear attraction for electron
   3. is a periodic property
      a. increases across period
      b. decreases down group

C. Chemical reactivity of elements depends on tendency to lose, gain, or share valence electrons.

Key Terms covalent bond, ionic bond, ionization energy

Key Figures and Box

Figure 3.1 Ionization energies and trends of elements through first four periods.

Box 3.1 On the intensity of electrical fields

3.2 Ionic Bonds

A. Ionic bond
   1. results from electron transfer between valence shells; one atom can attract electrons much more strongly than the other
   2. usually forms between metal and nonmetal
      a. both attain noble-gas electron configuration
      b. metal forms cation; loss of electron is oxidation
      c. nonmetal forms anion; gain of electron is reduction
   3. combining power determines number of electrons lost/gained
      a. metals lose number of electrons equal to group number if I, II, III
      b. nonmetals gain number of electrons equal to (8 − group number)
   4. Lewis structures can be used to show valences before and after ionic bond forms

B. Ionic formula; identifies types (elemental symbol) and numbers (subscripts) of ions in compound

$$\text{1 sodium ion} \rightarrow NaCl \leftarrow \text{1 chloride ion}$$

$$\text{1 magnesium ion} \rightarrow MgF_2 \leftarrow \text{2 fluoride ions}$$

C. Binary compound
   1. contains two elements
   2. combining ratios determine formula
   3. combining ratios can be determined as follows: net-charge approach (Ex. 3.2)
      a. determine number of electrons lost by metal
      b. determine number of electrons gained by nonmetal
      c. calculate what ratio will produce equal numbers of positive and negative charges; a net charge of zero
      d. multiplication factors become subscripts

Group I: loses 1 electron → $NaCl$ ← Group VI: gains 1 electron

$$1(1+) + 1(1-) = 0$$

Group II: loses 2 electrons → MgF_2 ← Group VI: gains 1 electron

$$1(2+) + 2(1-) = 0$$

   4. combining ratios can be determined as follows: cross-over approach (Ex. 3.3)
      a. determine number of electrons lost by metal: charge becomes anion subscript
      b. determine number of electrons gained by nonmetal: charge becomes cation subscript

Al^{3+} I^{-} ←3 goes here

1 goes here ↑

   5. Figure 3.2 shows combining ratios for nonmetal anions and metal cations

Key Terms binary compound, combining power, cross-over approach, electron transfer, formula, ionic compound, net-charge approach

Key Figure Figure 3.2 Table of combining ratios for ionic compounds

Example 3.1 Predicting the electrical charges of ions by using periodic table
Problem 3.1 Predicting the electrical charges of ions by using periodic table
Example 3.2 Predicting the formula of an ionic compound by using ionic charges and net charge
Problem 3.2 Predicting the formula of an ionic compound by using net charge
Example 3.3 Predicting the formula of an ionic compound by using ionic charges and cross-over approach
Problem 3.3 Predicting the formula of an ionic compound by using ionic charges and cross-over approach

3.3 Naming Binary Ionic Compounds

A. Binary ionic compounds are named by using the Stock system.
   1. metal keeps its name; placed first
   2. metal cation charge, if variable, is noted
      a. Roman numeral equal to charge is enclosed in parentheses
      b. parenthetical unit is placed next to metal
   3. nonmetal ending is replaced by –ide; follows metal

$FeCl_3$ ← 3 Cl^- indicate Fe^{3+}; number of positives = number of negatives
iron(III) chloride

   4. Table 3.1 lists common and Stock system names of cations with variable charges

Key Terms Stock system
Key Table Table 3.1 Names and formulas of ions with variable charge

Example 3.4 Writing names of ionic compounds by using Stock system
Problem 3.4 Writing common names of ionic compounds by using Table 3.1

3.4 Polyatomic Ions

A. Polyatomic ions
   1. two or more nonmetal atoms combined covalently
   2. unit carries a charge

B. Table 3.2 lists names, formulas, and charges of common polyatomic ions
C. Ionic compounds containing polyatomic ions can have variety of forms
   1. element + polyatomic ion

$$NaCN$$

   2. element + more than one polyatomic ion

$$Mg(CN)_2$$

   3. two polyatomic ions

$$(NH_4)_2SO_4$$

D. Combining ratios determine formula
   1. combining ratios can be determined as follows: net-charge approach (Ex. 3.2)
      a. determine charge on cation
      b. determine charge on anion
      c. calculate what ratio will produce equal numbers of positive and negative charges; a net charge of zero
      d. multiplication factors become subscripts
      e. if two or more of the same polyatomic ion are in formula, enclose in parentheses and subscript with appropriate number

$$NH_4^+ \rightarrow (NH_4)_2SO_4 \leftarrow SO_4^{2-}$$
$$2(1+) + 1(2-) = 0$$

   2. combining ratios can be determined as follows: cross-over approach (Ex. 3.3)
      a. charge on cation becomes anion subscript
      b. charge on anion becomes cation subscript

$$\mathbf{(NH_4)_2SO_4} \leftarrow \text{1 goes here}$$
2 goes here ↑

Key Term polyatomic ion
Key Table Table 3.2 Names, formulas, and charges of common polyatomic ions

Example 3.5 Writing formulas of compounds containing polyatomic ions by using Table 3.2
Problem 3.5 Writing names of compounds containing polyatomic ions by using Table 3.2

3.5 Does the Formula of an Ionic Compound Describe Its Structure?

A. Structure of ionic compounds
   1. ions locked into positions in a three-dimensional lattice
      a. lattice has no structural unit; cations surround anions, which surround cations
      b. ratio of ions in structure represented by ionic formula
   2. solid is neutral
   3. if melted; ions are free to move around
   4. if dissolved in water, ions are free to move around
   5. chemical properties of ionic solids may be different from the gas or liquid phase
B. Electrolytes
   1. ionic compounds that conduct electricity when melted or dissolved in water
   2. moving current carries charge that can be measured
   3. nonelectrolytes do not conduct electricity when dissolved
C. Electrolytes in blood
   1. Table 3.3 lists cations in blood
   2. can be measured as independent ions

Key Terms electrolyte, nonelectrolyte

Key Figures and Table

Figure 3.3	Electrical conductivity cell, showing movement of electrical charges and current flow
Figure 3.4	Crystal structure of the ionic compound sodium chloride
Table 3.3	Cations found in blood and tissue

3.6 *Covalent Compounds and Their Nomenclature*

A. Covalent bond
   1. results from electrons being shared between atoms; both atoms attract electrons about equally

B. Naming covalent compounds that contain two different elements
   1. many covalent compounds have common names, e.g., water, ammonia
   2. systematic naming of covalent compounds:
      a. examine formula; note order, identity, and number of each atom

$$\text{2 nitrogens are first} \rightarrow N_2O \leftarrow \text{1 oxygen is last}$$

      b. first element retains its elemental name
         i. if more than one atom is in compound, you must use a prefix that tells you the number of atoms present (Table 3.4)

$$\text{dinitrogen} \rightarrow N_2O$$

         ii. don't use prefix mono- if only one atom is present; use elemental name only
      c. second element is given suffix -ide
         i. if more than one atom is in the compound, you must use a prefix that tells you the number of atoms present (Table 3.4 and marginal note on p. 74)

$$N_2O \leftarrow \text{oxide}$$

         dinitrogen monoxide; common name is nitrous oxide; used as an anesthetic (laughing gas)

Common "-ide's"

Group VA	Group VIA	Group VIIA
*nitr*ide	*ox*ide	*fluor*ide
*phosph*ide	*sulf*ide	*chlor*ide
		*brom*ide
		*iod*ide

Key Term	binary molecular compound
Key Table	Table 3.4 Systemic names of covalent compounds

Example 3.6	Naming covalent compounds by using Table 3.4
Problem 3.6	Naming covalent compounds by using Table 3.4

3.7 *Representation of Covalent Bonds*

A. Various methods used to represent bonding arrangements in molecules
   1. Lewis structures

B. Electron-dot notation
   1. based on number of electrons in valence shells of atoms
   2. uses octet rule to fill valence shells
      a. most elements (except hydrogen) need eight valence electrons to achieve stability
      b. hydrogen only needs two valence electrons to achieve stability
   3. produces Lewis structures
      a. tells how atoms are connected to one another within molecule
      b. tells whether bonds connecting atoms are single, double, triple
      c. does not describe three-dimensional structure of molecule

C. See "Drawing Lewis Structures" and Table 3.5
   1. Lewis structures show how atoms are connected in a molecule
   2. Lewis structures show single, double, and triple bonds between atoms
   3. Lewis structures show nonbonding electrons
   4. Lewis structures can be simplified by leaving out nonbonding electrons to produce structural formula

Drawing Lewis Structures

1. Count the number of **valence** electrons in each atom in the molecule; add up all valence electrons
   a. the group number will tell you this information
   b. if the species is a negative ion, add one additional electron for each negative charge
   c. if the species is a positive ion, subtract one electron for each positive charge
2. Choose the most likely atom to be central and see how the other atoms might be connected to it; the most symmetric arrangement has the highest probability of being correct
   a. **Look at Table 3.5;** the atom with the least number of valence electrons can form the most bonds to fill its octet
      - the atom that can form the most bonds is a good candidate
      - the least electronegative atom is a good candidate
      - hydrogen can never be a central atom because it only forms one bond
      - Group VIA atoms will usually not be central atoms
   b. molecules with two central atoms are usually symmetric
   c. oxygen atoms do not usually bond with other oxygen atoms except in the cases of O_2^{2-} and O_2^{-}
3. Draw a single bond between each pair of atoms; subtract two electrons for each bond drawn from total number of valence electrons
4. Use remaining electrons to complete the octets of the atoms bonded to the central atom; these will be lone pairs
   a. if there are not enough to go around, then multiple bonds will be needed
   b. if all octets are satisfied, no multiple bonds are needed
   c. **Hydrogen is complete with only two electrons**
5. If any atom still has less than an octet, form multiple bonds so that each atom has an octet
6. Exceptions to the octet rule are
   a. some elements form compounds where there are fewer than eight electrons around them; e.g., boron forms compounds and only has six electrons around B
   b. some elements form compounds in which there are more then eight electrons around the element; e.g., phosphorus forms compounds in which there are ten electrons around P
   c. Some molecules have an odd number of electrons, resulting in unpaired electrons.

Key Terms alkane, combining power, double bond, duet, hydrocarbon, Lewis structure, lone pair, molecular formula, multiple bond, nonbonded electron, triple bond

Key Table Table 3.5 Bonding requirements for octet formation of some nonmetals

Example 3.7 Predicting formulas for simple molecular compounds by using bonding requirements for an octet
Problem 3.7 Predicting formulas for simple molecular compounds by using bonding requirements for an octet
Example 3.8 Drawing Lewis structures for simple molecules with one lone pair
Problem 3.8 Drawing Lewis structures for simple molecules with one lone pair
Example 3.9 Drawing Lewis structures for simple molecules with two lone pairs

Problem 3.9 Drawing Lewis structures for simple molecules with two lone pairs
Example 3.10 Drawing Lewis structures for complex molecules
Problem 3.10 Drawing Lewis structures for complex molecules
Example 3.11 Drawing Lewis structures for molecules with double bonds
Problem 3.11 Drawing Lewis structures for molecules with double bonds
Example 3.12 Drawing Lewis structures for molecules with triple bonds
Problem 3.12 Drawing Lewis structures for molecules with triple bonds

3.8 Lewis Structures of Polyatomic Ions

A. Charged species have more or fewer electrons than neutral species.
   1. atoms are covalently bonded
   2. the group carries a charge
B. Polyatomic ions can be formed as a result of
   1. reactions occurring with water in aqueous solutions
   2. reaction with another substance in a solution
C. Coordinate covalent bond
   1. results when two atoms share electron pair donated by one of the atoms
   2. used only to keep track of electrons; is not different from regular covalent bond
D. See "Drawing Lewis Structures."
   1. if negatively charged, extra electrons equal to charge are added
   2. if positively charged, number of electrons equal to charge are subtracted

Example: SO_4^{2-}

Element	(Number of atoms)	×	(Number of valence electrons)	=	(Total valence electron contribution)
S	1	×	6	=	6
O	4	×	6	=	24
					2 extra electrons
total number of valence electrons					32

O
|
O — S — O
|
O

1. Sulfur is most likely as the central atom; there's only 1
2. 4 oxygens can be symmetrically arranged around the S atom
3. This arrangement takes care of 8 electrons: 32 − 8 = 24
4. 24 remaining electrons can be distributed around the oxygens to fill their octets: 4 × 6 = 24

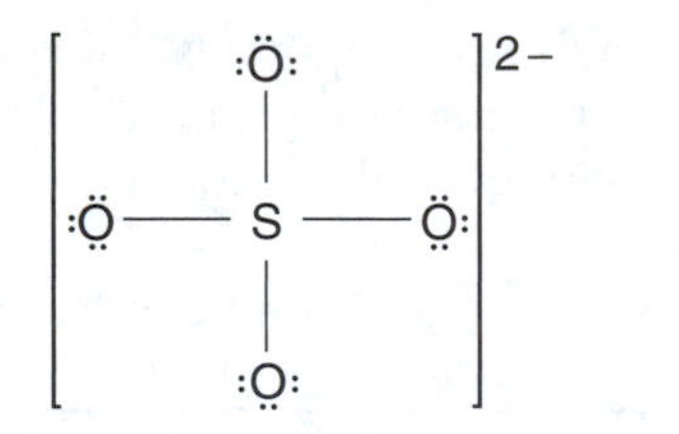

Key Term coordinate covalent bond

3.9 Polar and Nonpolar Covalent Bonds

A. Electrons shared between two atoms are
   1. equally shared if both nuclei have same attraction for electrons; nonpolar covalent bond results
   2. unequally shared if one nucleus has greater attraction for electrons; polar covalent bond results

B. Electronegativity is measure of ability of element to attract electrons to itself.
   1. it is a periodic trend; generally increases up a group and across a period
   2. fluorine is the most electronegative element (4.0); cesium is the least electronegative element (0.79)
   3. when you look at the periodic table, the most electronegative elements are at the upper right and the least electronegative are at the lower left
   4. electrons in a covalent bond to stay closer to the nucleus of the more electronegative atom
   5. Figure 3.7 shows a spectrum of bond types based on differences in electronegativity between bonded atoms

      general rule: if electronegativity difference between two bonded atoms is

more than 2.0	less than 1.5
ionic bond	covalent bond

   6. generally, metals have electronegativities less than 2 and nonmetals have electronegativities greater than 2; hydrogen has an electronegativity of 2.1

C. Electronegativity differences between bonded atoms
   1. produce asymmetric distribution of charge
      a. relatively more negative charge on more electronegative atom: show as partial charge, delta negative
      b. relatively less negative charge on less electronegative atom: show as partial charge, delta positive
   2. bonds in this case are polar
   3. number of total electrons *still equals* number of total protons, therefore molecule is neutral overall
   4. shape of molecule will determine whether polar bonds produce a polar molecule
      a. dipoles of individual bonds have direction and magnitude

      $$+\!\!\longrightarrow$$

      less electronegative end ↑ ↑ more electronegative end

      b. dipoles can be additive or subtractive, or can cancel each other
      c. shape of molecule determines whether they cancel each other
         linear molecules may not have dipoles
         tetrahedral molecules may not have dipoles
         bent molecules may have dipoles

D. Polarity of molecules determines how they interact with other substances
   1. polar molecules can interact with other polar molecules

      $$(\delta^- \; \delta^+) \Leftrightarrow (\delta^- \; \delta^+)$$

   2. nonpolar molecules will not interact with polar molecules
   3. molecular polarity affects physical and chemical properties because it determines
      a. which substances the compound can interact with
      b. to what extent the compound will interact with other substances
      c. to what extent the compound interacts with molecules of its own kind

Key Terms dipole moment, electronegativity, nonpolar covalent bond, polar covalent bond, polar molecule

Key Figures

Figure 3.5 Table of electronegativities of some elements
Figure 3.6 Electronegativities of main-group elements
Figure 3.7 Effect of electronegativity difference on bond type
Figure 3.8 Response of polar and nonpolar substances to electrically charged rod

3.10 Three-Dimensional Molecular Structures

A. Three-dimensional arrangement of atoms describes how atoms are connected and aligned in space; called stereochemistry of molecule (Chapter 17)
  1. cause unique characteristics of molecule
  2. stereochemistry of biological molecules influences biological functions: structure-function relationship

B. VSEPR theory: can describe three-dimensional structure of molecule
  1. based on number of electron pairs around central atoms
  2. pairs of electrons stay as far away from each other as possible
  3. mutual repulsion of all electron pairs in a molecule forces the atoms in a molecule to occupy positions in space that minimize the repulsions

C. VSEPR structures
  1. produces geometric figures (see Summary of Molecular Shapes)
  2. based on electron-dot structures
  3. considers pairs of electrons around a central atom
  4. nonbonding electrons occupy larger volume in space and distort geometric figures

Key Terms ball-and-stick model, equilateral triangle, linear, molecular absorption spectra, space-filling model, stereochemistry, tetrahedron, trigonal pyramidal, valence-shell electron-pair repulsion theory, VSEPR theory

Key Figures and Table

Table 3.6 Arrangement of electron pairs around a central atom
Figure 3.9 Tetrahedral shape of methane
Figure 3.10 X-ray diffraction pattern of a crystal of NaCl
Figure 3.11 Ball-and-stick and space-filling models of methanol
Figure 3.12 Space-filling model of cholesterol

Example 3.13 Predicting molecular structure by using VSEPR theory
Problem 3.13 Predicting molecular structure by using VSEPR theory
Example 3.14 Predicting molecular structure of molecule with lone pair by using VSEPR theory
Problem 3.14 Predicting molecular structure of molecule with lone pair by using VSEPR theory
Example 3.15 Predicting molecular structure of molecule with two lone pairs by using VSEPR theory
Problem 3.15 Predicting molecular structure of molecule with two lone pairs by using VSEPR theory

ARE YOU ABLE TO. . . AND WORKED TEXT PROBLEMS

Objectives Section 3.1 Ionic Versus Covalent Bonds

Are you able to . . .

- explain the driving force behind bond formation (Exercises 3.1, 3.2, 3.3, 3.4)
- describe an ionic bond (Exercise 3.34)
- describe a covalent bond
- define ionization energy
- explain the relationship between ionization energy and removal of valence electrons from outer shell
- explain what determines the magnitude of ionization energy
- recognize the periodicity of ionization energy
- explain the periodicity of ionization energy
- comment on chemical reactivity and its relationship to valence-shell configurations

Objectives Section 3.2 Ionic Bonds

Are you able to . . .

- define ionic bond
- discuss how ionic bonds are formed and the driving force behind their formation
- recognize the distinctive features of ionic bonds

- recognize ionic compounds by their constituent elements
- determine whether an element is likely to engage in ionic bonding (Exercise 3.45)
- determine an ion's electrical charge from its position in the periodic table (Exs. 3.1, 3.2, 3.3, Probs. 3.1, 3.2, 3.3)
- explain the relationship between an element's combining power and its position on the periodic table (Exercises 3.13, 3.14, 3.15, 3.16)
- use Lewis structures to illustrate the formation of an ionic compound
- determine the combining ratio for various elements and ions (Exs. 3.1, 3.2, 3.3, Probs. 3.1, 3.2, 3.3)
- explain the information contained in a chemical formula
- determine the formula for a binary ionic compound (Exs. 3.2, 3.3, Probs. 3.2, 3.3, Exercises 3.6, 3.9, 3.11, 3.17, 3.19, 3.21, 3.22, 3.36, 3.38, 3.40, 3.44, 3.47, 3.50)

Problem 3.1 Predict the charge of ions formed from the elements (a) potassium; (b) calcium; (c) indium; (d) phosphorus; (e) sulfur; (f) bromine.

The electrical charges on elemental ions can be predicted from their position in the periodic table.

- Groups IA and IIA: metals will form cations; the number of electrons lost will be equal to the group number
- Groups IIIA, IVA, VA, VIA, and VIIA: nonmetals will form anions; the number of electrons gained will be equal to the Group number subtracted from 8
- Groups IIIA, IVA, VA, and VIA metals: will form cations; the number of electrons lost may vary; generally, the number of electrons lost for Groups IIIA and IVA will equal the group number
- transition metals: will form cations; the number of electrons lost may vary
- metalloids may form both anions and cations

Element	Group number	Metal/nonmetal	Valence	Ion charge	Symbol
(a) potassium (K)	I	metal ⇒ cation	1	1+	K^{+}
(b) calcium (Ca)	II	metal ⇒ cation	2	2+	Ca^{2+}
(c) indium (In)	III	metal ⇒ cation	3	3+	lu^{3+}
(d) phosphorus (P)	V	nonmetal ⇒ anion	5	3−	P^{3-}
(e) sulfur (S)	VI	nonmetal ⇒ anion	6	2−	S^{2-}
(f) bromine (Br)	VII	nonmetal ⇒ anion	7	1−	Br^{-}

Problem 3.2 Use the net-charge approach to predict the formulas for the ionic compounds formed from (a) K and Br; (b) Ga and F; (c) Ca and P.

Determine which group each element is in and predict the charge each ion will have.

(a) Element	Group number	Metal/nonmetal	Valence	Ion charge
K	I	metal ⇒ cation	1	1+
Br	VII	nonmetal ⇒ anion	7	1−

Total the charges to determine net charge on a 1:1 binary compound: (1+) + (1−) = 0. If the net charge for a 1:1 binary compound is zero, the formula is 1 cation and 1 anion: KBr.

(b) Element	Group number	Metal/nonmetal	Valence	Ion charge	Multiplier
Ga	III	metal ⇒ cation	3	3+	(3+) × 1 = 3+
F	VII	nonmetal ⇒ anion	7	1−	(1−) × 3 = 3−
			net charge	2+	0

If the net charge is not zero, multiply the positive charge by the number of the negative charge and multiply the negative charge by the number of the positive charge; the multiplier of the ion is the subscript of that ion; if multiplier is 1, no subscript is needed: GaF_3.

(c) Element	Group number	Metal/nonmetal	Valence	Ion charge	Multiplier
Ca	II	metal ⇒ cation	2	2+	$(2+) \times 3 = 6+$
P	V	nonmetal ⇒ anion	5	3−	$(3-) \times 2 = 6-$
			net charge	1−	0

The multiplier of the ion is the subscript of that ion; Ca_3P_2.

Problem 3.3 Use the cross-over approach to predict the formulas of the ionic compounds formed from (a) K and Cl; (b) Mg and P; (c) Ga and O

Determine which group each element is in and predict the charge each ion will have. The cation charge becomes the anion subscript; the anion charge becomes the cation subscript.

	Element	Group number	Metal/nonmetal	Valence	Ion charge	Formula
(a)	K	I	metal ⇒ cation	1	1+	KCl
	Cl	VII	nonmetal ⇒ anion	7	1−	
(b)	Mg	II	metal ⇒ cation	2	2+	Mg_3P_2
	P	V	nonmetal ⇒ anion	5	3−	
(c)	Ga	III	metal ⇒ cation	3	3+	Ga_2O_3
	O	VI	nonmetal ⇒ anion	6	2−	

Objectives Section 3.3 Naming Binary Ionic Compounds

Are you able to . . .

- give the definition of binary ionic compound
- identify binary ionic compounds
- name binary compounds by using the Stock system (Exercises 3.5, 3.6, 3.10, 3.12, 3.18, 3.20, 3.22, 3.35, 3.37, 3.39, 3.41, 3.48)
- name common anions and cations
- identify elements that commonly form ions of variable charge
- write the chemical formulas of ionic compounds from their names

Problem 3.4 Use Table 3.1 to write the common names for (a) $TiCl_3$; (b) $HgCl_2$; (c) $FeCl_2$; (d) PbO_2.

Determine the charge on the cation by looking at the subscript and charge on the anion; multiply the anion subscript by the anion charge to determine the total negative charge; the cation positive charge must equal this.

Formula	Anion charge × anion subscript	Net (−) charge	Formula name
(a) $TiCl_3$	$(1-) \times 3$	3−	titanium (III) ⇒ titanous chloride
(b) $HgCl_2$	$(1-) \times 2$	2−	mercury (II) ⇒ mercuric chloride
(c) $FeCl_2$	$(1-) \times 2$	2−	iron (II) ⇒ ferrous chloride
(d) PbO_2	$(2-) \times 2$	4−	lead (IV) ⇒ plumbic oxide

Objectives Section 3.4 Polyatomic Ions

Are you able to . . .

- define polyatomic ion
- name common polyatomic ions (Exercise 3.7)
- write the formulas and charges of common polyatomic ions (Exercise 3.8)

- describe the structure of a polyatomic ion
- identify ionic compounds that contain polyatomic ions
- determine the combining ratio of polyatomic ions
- write the formulas for ionic compounds containing polyatomic ions
- name ionic compounds containing polyatomic ions

Problem 3.5 Using Table 3.2, write the names of the following compounds: (a) NH_4HSO_3; (b) $CaCO_3$; (c) $Mg(CN)_2$; (d) $KHCO_3$; (e) $(NH_4)_2SO_4$.

Determine the cation and anion in each compound; if the cation is elemental, the name is the elemental name. If the cation or anion is a polyatomic ion, refer to Table 3.2 to find the appropriate name.

Formula	Cation	Name	Anion	Name	Formula name
(a) NH_4HSO_3	NH_4^+	ammonium	HSO_3^-	hydrogen sulfite	ammonium hydrogen sulfite
(b) $CaCO_3$	Ca^{2+}	calcium	CO_3^{2-}	carbonate	calcium carbonate
(c) $Mg(CN)_2$	Mg^{2+}	magnesium	CN^-	cyanide	magnesium cyanide
(d) K	K^+	potassium	HCO_3^-	hydrogen carbonate	potassium hydrogen carbonate
(e) $(NH_4)_2SO_4$	NH_4^+	ammonium	SO_4^{2-}	sulfate	ammonium sulfate

Objectives Section 3.5 Does the Formula of an Ionic Compound Describe its Structure?

Are you able to . . .

- describe the solid structure of ionic compounds
- explain what happens to ions in a solid upon melting or dissolution
- explain why the chemical properties of ionic solids may be different from their gas or liquid phases
- define electrolyte
- define nonelectrolyte
- explain what the chemical formula for an ionic compound represents
- explain how the presence of electrolytes can be detected
- identify important biological electrolytes and what they do

Objectives Section 3.6 Covalent Compounds and Their Nomenclature

Are you able to . . .

- define covalent bond (Exercise 3.33)
- explain how covalent bonds differ from ionic bonds
- discuss how covalent bonds are formed and the driving force behind their formation
- recognize the distinctive features of covalent bonds
- recognize covalent compounds by their constituent elements
- determine whether an element is likely to engage in covalent bonding (Exercise 3.45)
- use Lewis structures to illustrate the formation of a covalent compound
- determine the combining ratio for various nonmetals
- identify the Greek prefixes used in chemical nomenclature and correlate them to numbers
- explain how electronegativity is used to determine the order of elements in a binary compound's formula
- write the formula for a binary covalent compound
- name binary covalent compounds
- explain what the chemical formula for a covalent compound represents

Problem 3.6 Name the following compounds: (a) NI_3; (b) P_2O_5; (c) S_2Cl_2; (d) SO_3.

Note that all of the compounds contain nonmetals only; this tells you that they are molecular compounds and should be named by referring to Table. 3.4. The first element retains its name; if there are more than one atom of the element, the appropriate prefix is used (see Greek prefix list); the second element is given the suffix –ide, the appropriate prefix is used (see Greek prefix list).

Formula	First element/ Number		Prefix	Second element/ Number		Prefix	Formula name
(a) NI_3	nitrogen	1	none	iodine	3	tri-	nitrogen triiodide
(b) P_2O_5	phosphorus	2	di-	oxygen	5	penta-	diphosphorus pentoxide
(c) S_2Cl_2	sulfur	2	di-	chlorine	2	di-	disulfur dichloride
(d) SO_3	sulfur	1	none	oxygen	3	tri-	sulfur trioxide

Objectives Section 3.7 Representation of Covalent Bonds

Are you able to . . .

- identify the various methods used to represent molecular structures
- explain how Lewis structures are used to represent molecules
- define the octet rule (Exercise 3.2)
- recognize exceptions to the octet rule (Exercise 3.49)
- explain the relationship between an element's combining power and its position in the periodic table
- develop Lewis structures for covalent molecules with one central atom; more than one central atom; multiple bond; electron deficits or excesses (Exercises 3.23, 3.24, 3.25, 3.27, 3.28, 3.42, 3.46, Expand Knowledge 3.53, 3.55)
- explain the what information can be gained from Lewis structures
- explain the limitations of Lewis structures
- draw structural formulas based on Lewis structures

Problem 3.7 For each of the following pairs of elements, predict the formula for the product of the reaction between the two elements: (a) carbon and hydrogen; (b) phosphorus and chlorine; (c) carbon and oxygen.

Note these are all nonmetals bonding to nonmetals and the bonds between them will be covalent. The group number subtracted from 8 tells the number of covalent bonds each element is able to form. Table 3.5 provides information about the bonding requirements for these nonmetals.

First element	Number of bonds formed	Second element	Number of bonds formed	Product formula
carbon	4	hydrogen	1	CH_4
phosphorus	3	chlorine	1	PCl_3
carbon	4	oxygen	2	CO_2

Problem 3.8 Construct the Lewis structure for phosphorus trichloride, PCl_3

First construct a table of all the elements in the compound and the number of valence electrons each element has; the group number tells you the number of valence electrons.

Element	Number of valence electrons	Number of atoms in compound	Total valence electrons
phosphorus	5	1	5
chlorine	7	3	21
		total number of valence electrons in compound	26

Twenty-six valence electrons have to be distributed around the atoms so that each atom has an octet. Choose a structure that seems reasonable; the atom that forms the most bonds should be a first choice for the central atom: phosphorus forms three bonds and chlorine forms one bond (Table 3.5), so P is a reasonable central atom. Arrange the chlorines as symmetrically as possible around the P and give each bond two valence electrons:

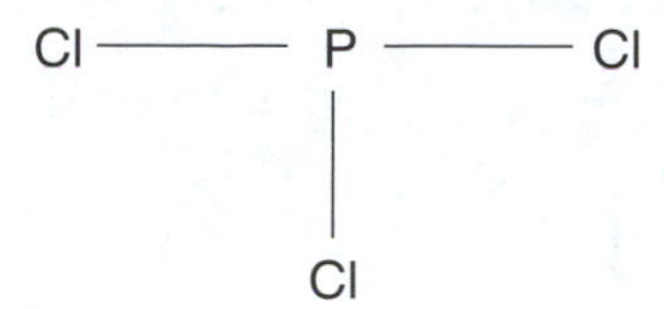

Each line represents two electrons; subtract this number from the total number of valence electrons: 26 − 6 = 20. Next, distribute the 20 remaining electrons around each atom to complete its octet:

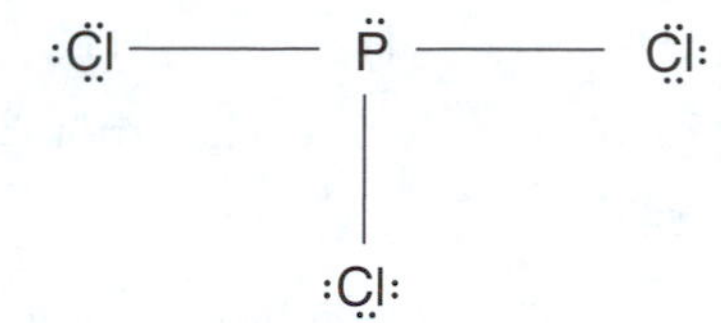

Each atom has 8 electrons and there are left over: 20 − 20 = 0. Notice that P has a lone pair of electrons.

Problem 3.9 Construct the Lewis structure for sulfur dichloride, SCl_2.

First construct a table of all the elements in the compound and the number of valence electrons each element has; the group number tells you the number of valence electrons.

Element	Number of valence electrons	Number of atoms in compound	Total valence electrons
sulfur	6	1	6
chlorine	7	2	14
		total number of valence electrons in compound	20

Twenty valence electrons have to be distributed around the atoms so that each atom has an octet. Choose a structure that seems reasonable; the atom that forms the most bonds should be a first choice for the central atom: sulfur forms two bonds and chlorine forms one bond (Table 3.5), so S is a reasonable central atom. Arrange the chlorines as symmetrically as possible around the sulfur and give each bond two valence electrons:

Cl —— S —— Cl

Each line represents two electrons; subtract this number from the total number of valence electrons: 20 − 4 = 16. Next, distribute the 16 remaining electrons around each atom to complete its octet:

:C̤̈l —— S̤̈ —— C̤̈l:

Each atom has an octet, and there are no electrons left over: 16 − 16 = 0. Note that S has two lone pairs of electrons.

Problem 3.10 Write the Lewis structure for C_3H_8.

First construct a table of all the elements in the compound and the number of valence electrons each element has; the Group number tells you the number of valence electrons.

Element	Number of valence electrons	Number of atoms in compound	Total valence electrons
carbon	4	3	12
hydrogen	1	8	8
		total number of valence electrons in compound	20

Twenty valence electrons have to be distributed around the atoms so that each atom has an octet. Choose a structure that seems reasonable; the atom that forms the most bonds should be a first choice for the central atom; remember that hydrogen can only form one bond, so it can only bond to one carbon.

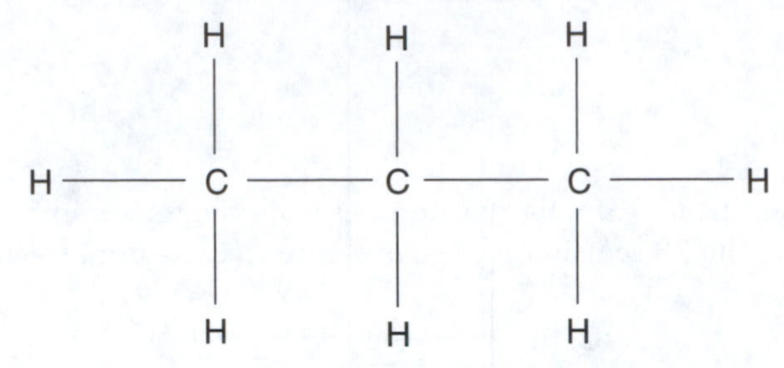

H requires only two electrons to fill its valence shell. Each carbon atom has an octet, and there are no electrons left over: 20 − 20 = 0

Problem 3.11 Write the Lewis structure for carbon dioxide, CO_2.

First construct a table of all the elements in the compound and the number of valence electrons each element has; the group number tells you the number of valence electrons.

Element	Number of valence electrons	Number of atoms in compound	Total valence electrons
carbon	4	1	4
oxygen	6	2	12
		total number of valence electrons in compound	16

Sixteen valence electrons have to be distributed around the atoms so that each atom has an octet. Choose a structure that seems reasonable; the atom that forms the most bonds should be a first choice for the central atom: carbon forms four bonds and oxygen forms two bonds (Table 3.5), so C is a reasonable central atom. Arrange the oxygens symmetrically around the C and give each bond two valence electrons:

$$\mathrm{O{-}C{-}O}$$

Each line represents two electrons; subtract this number from the total number of valence electrons: 16 − 4 = 12. If you try to distribute the 12 remaining electrons around each atom to complete its octet, you will find there are not enough electrons to give each atom an octet; try adding a double bond:

$$\mathrm{O{=}C{-}O}$$

This leaves (16 − 6 = 10) 10 electrons to distribute among three atoms; again not enough to give each atom an octet. Add another double bond:

$$\mathrm{O{=}C{=}O}$$

This leaves (16 − 8 = 8) 8 electrons to distribute over two oxygen atoms; carbon has an octet in this arrangement. If each oxygen is given 4 electrons, each octet is filled and no electrons are left over.

$$\ddot{\underset{..}{\mathrm{O}}}\mathrm{{=}C{=}}\ddot{\underset{..}{\mathrm{O}}}$$

Problem 3.12 Write the Lewis structure for the rocket fuel hydrazine, N_2H_4.

First construct a table of all the elements in the compound and the number of valence electrons each element has; the group number tells you the number of valence electrons.

Element	Number of valence electrons	Number of atoms in compound	Total valence electrons
nitrogen	5	2	10
hydrogen	1	4	4
		total number of valence electrons in compound	14

Fourteen valence electrons have to be distributed around the atoms so that each nitrogen atom has an octet and each hydrogen atom has a duet. Choose a structure that seems reasonable; hydrogen forms only one bond, so it can only be at the ends of a molecule; this leaves nitrogen as a central atom:

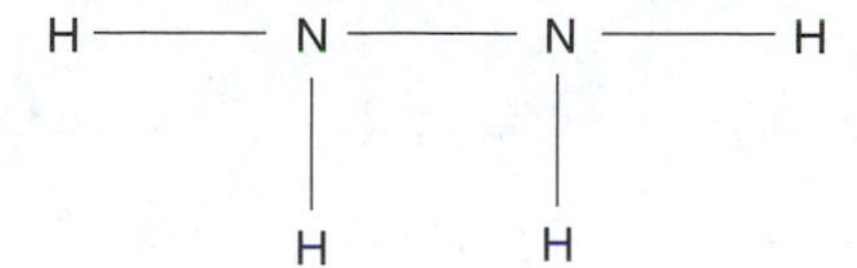

Each line represents two electrons; subtract this number from the total number of valence electrons: 14 − 10 = 4. Both hydrogen atoms have duets. If you try to distribute the 4 remaining electrons around each nitrogen atom to complete its octet, you will be able to give each nitrogen a lone pair:

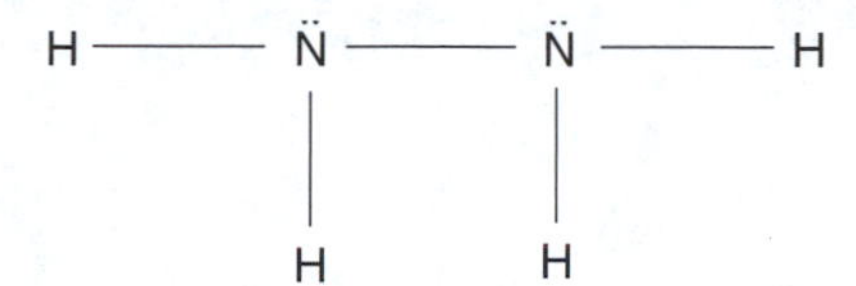

Both nitrogen atoms have octets; both hydrogen atoms have duets and there are no electrons left over.

Objectives Section 3.8 Lewis Structures of Polyatomic Ions

Are you able to . . .
- identify polyatomic ions
- describe the structure of polyatomic ions
- explain how polyatomic ions can be formed
- define coordinate covalent bond
- draw Lewis structures of polyatomic ions

Objectives Section 3.9 Polar and Nonpolar Covalent Bonds

Are you able to . . .
- describe how electrons can be shared between two atoms
- define electronegativity
- describe the electronegativities of various elements
- explain the significance of electronegativity as it relates to covalent bonds
- define nonpolar covalent bond
- define polar covalent bond
- discuss the various bonds that can form between atoms in terms of the relative electronegativities of the atoms sharing the bond
- recognize the periodicity of electronegativity
- explain the periodicity of electronegativity
- assign partial negative and positive charges to atoms sharing a bond
- describe the polarity of a bond
- define dipole moment
- assign a dipole moment to a bond by using proper notation
- explain the significance of an unequal charge distribution in a bond and in a molecule
- explain the effects that molecular shape can have on molecular polarity (Exercises 3.31, 3.32)

Objectives Section 3.10 Three-Dimensional Molecular Structures

Are you able to . . .
- explain VSEPR theory
- draw molecular structures by using VSEPR theory (Exs. 3.13, 3.14, 3.15, Probs. 3.13, 3.14, 3.15, Exercises 3.29, 3.30, 3.43, Expand Your Knowledge 3.54)
- explain how lone pairs of electrons affect the shape of a molecule
- discuss the advantages and limitations of VSEPR structures and compare them to Lewis structures
- identify other models that can be used to illustrate molecular structure

- explain how molecular structure can be clarified by analytical techniques such as X-ray diffraction and absorption of UV and visible light
- describe how ball-and-stick and space-filling models are used to represent three-dimensional molecular structures

Problem 3.13 Use the VSEPR theory to predict the three-dimensional structure of carbon tetrachloride, CCl_4.

From Table 3.5, you can determine that C will form four bonds and Cl will form one bond. The Lewis structure of carbon tetrachloride can be determined as illustrated in Problems 3.8–3.12. The result is

Lewis structure VSEPR structure

There are four electron pairs arranged around a central carbon atom; there are no lone pairs. Use the table of VSEPR structures to determine how four pairs of electrons will be arranged around a central atom. The tetrahedral structure is the most likely one.

Problem 3.14 Predict the molecular structure of nitrogen triiodide, NI_3, by using VSEPR theory.

From Table 3.5, you can determine that N will form three bonds and I will form one bond. The Lewis structure of nitrogen triiodide can be determined as illustrated in Problems 3.8–3.12. The result is

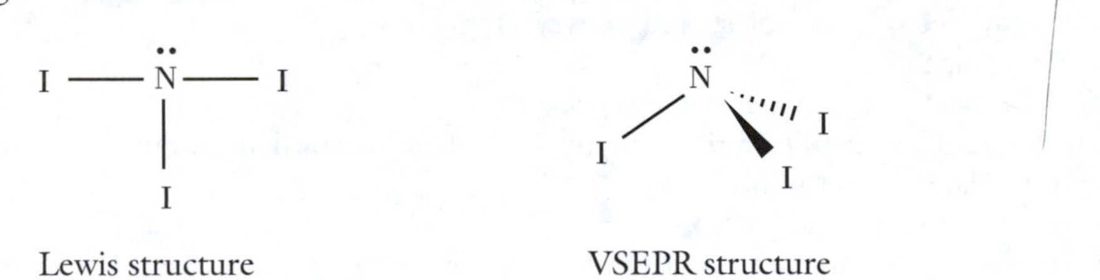

Lewis structure VSEPR structure

There are four electron pairs arranged around a central nitrogen atom; there is one lone pair. Use the table of VSEPR structures to determine how three pairs of bonding electrons will be arranged around a central atom when one lone pair is present. Trigonal pyramidal structure is the most likely one.

Problem 3.15 Use VSEPR theory to predict the three-dimensional structure of oxygen difluoride, OF_2.

From Table 3.5, you can determine that O will form two bonds and F will form one bond. The Lewis structure of oxygen difluoride can be determined as illustrated in Problems 3.8–3.12. The result is

Lewis structure VSEPR structure

There are four electron pairs arranged around a central nitrogen atom; there are two bonding pairs and two lone pairs. Use the table of VSEPR structures to determine how two pairs of bonding electrons will be arranged around a central atom when two lone pairs are present. A bent structure is the most likely one.

Molecular Geometries Predicted by VSEPR

Structure	Number of electron pairs	Geometric arrangement
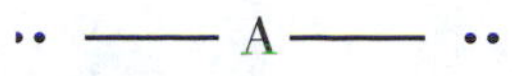 ·· —— A —— ··	2	Linear
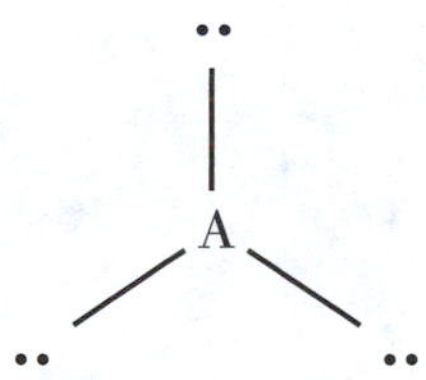	3	Trigonal planar
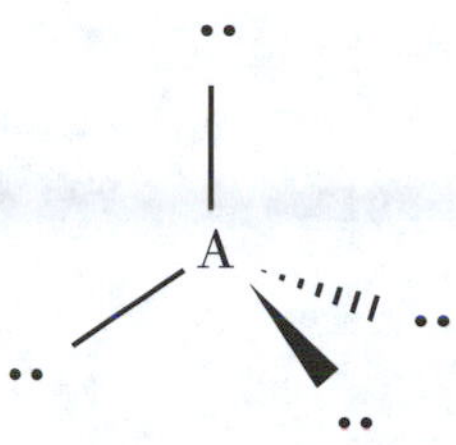	4	Tetrahedral

Molecular Shapes with Lone Pair Electrons Predicted by VSEPR Theory

Number of electron pairs	Number of lone pairs	Geometry	
3	1		Angular
4	1	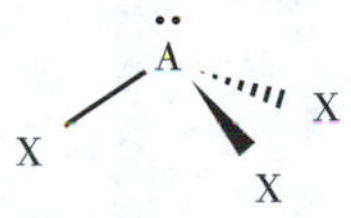	Trigonal pyramidal
4	2	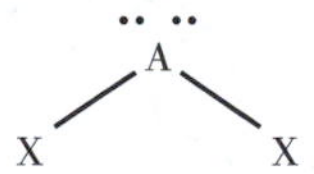	Bent

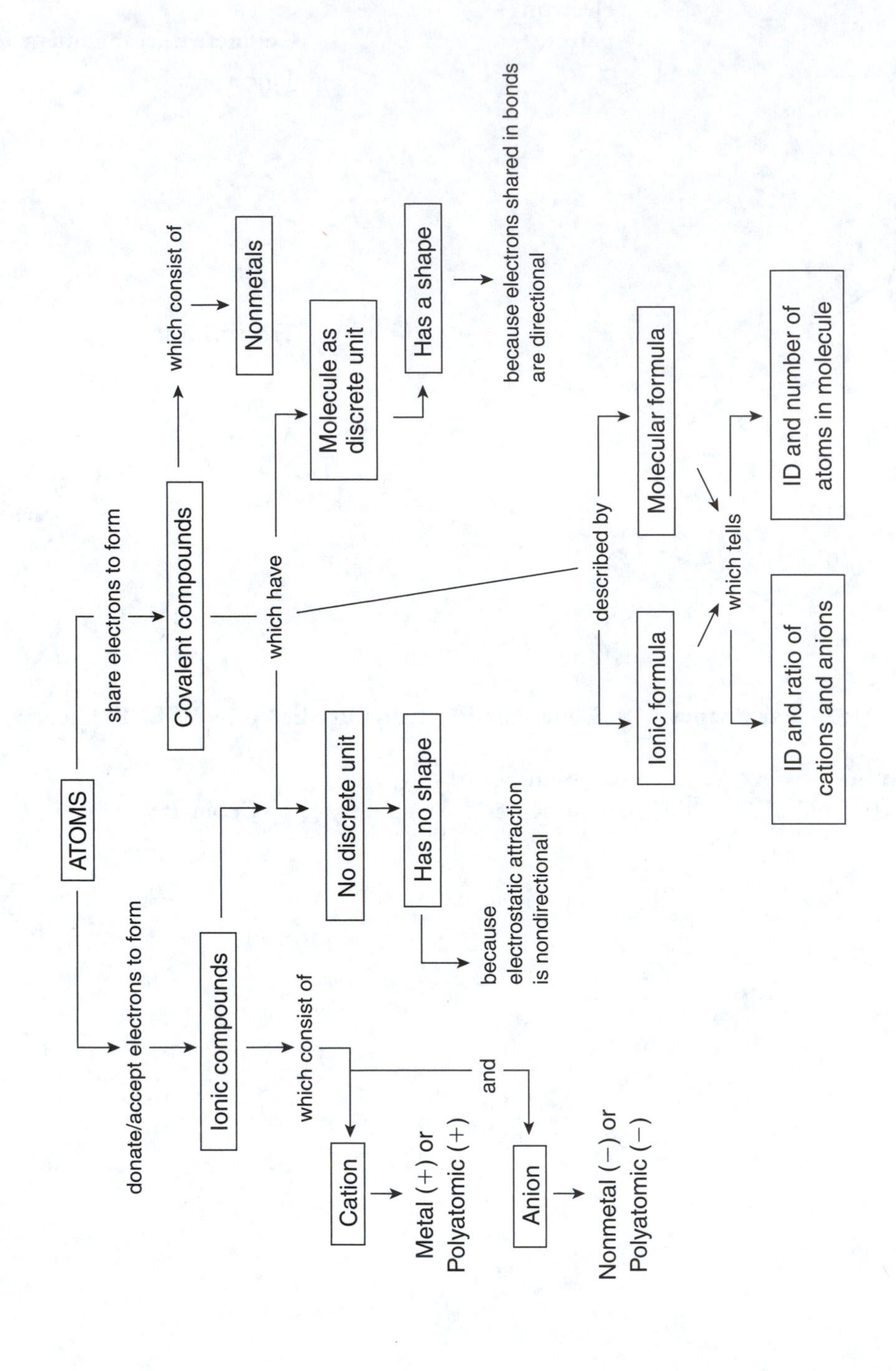

ATOMS
donate/accept electrons to form
share electrons to form
Ionic compounds
Covalent compounds
which consist of
Nonmetals
which have
Molecule as discrete unit
Has a shape
because electrons shared in bonds are directional
No discrete unit
Has no shape
because electrostatic attraction is nondirectional
which consist of
Cation
Metal (+) or Polyatomic (+)
and
Anion
Nonmetal (−) or Polyatomic (−)
described by
Ionic formula
Molecular formula
which tells
ID and ratio of cations and anions
ID and number of atoms in molecule

FILL-INS

Atoms combine in two different ways. They can share valence electrons to form _____ or they can donate or accept valence electrons to form_____ that are attracted to each other due to their charge differences. Both combinations are driven by the tendency of atoms to fill their _____ shells to form a stable_____ arrangement. This arrangement mimics the electronic arrangement of the _____.

The _____ of the attraction between the positively charged nucleus and negatively charged valence electrons determines how easily an element can lose a valence electron. _____ have relatively low ionization energies and _____ have relatively high ionization energies. Ionization energy is a _____ property that generally _____ across a period and _____ down a group. An element's _____ depends on its tendency to lose, gain, or share valence electrons.

_____ is a measure of an element's ability to attract bonding electrons to itself. If two bonding elements have very different electronegativities, an _____ will form. If, on the other hand, they have similar electronegativities, a _____ will form.

The _____ of ionic and covalent compounds result from the way in which atoms are joined together. In ionic compounds, the _____ between positively charged cations and negatively charged anions produces a superstructure in which ions are arranged around each other in ratios determined by their _____. There is no _____ in an ionic compound, but the ionic formula indicates the _____ of ions present. Covalent bonds are _____ because the electrons being shared exist in specific regions in space. This produces _____, which are discrete units. The chemical formula of a covalent compound indicates the _____ present in a molecule of the substance.

Ionic formulas are determined by examining the _____ on the ions involved in bonding. The net charge must always be _____, meaning that the number of negative charges must _____ the number of positive charges. Ions may be _____, such as Na^+ and Cl^-, or _____, such as NH_4^+ or PO_4^{3-}. Ionic formulas are named according to rules that include identifying the charge on the cation and giving the suffix _____ to the anion.

Covalent formulas are determined by looking at the _____ in order to fill its octet. The least electronegative element is _____ in the formula. In binary covalent compounds, both elements are quantified by using a Greek prefix; and the most electronegative element, the _____ element, is given the suffix _____.

The structures of molecules and individual ions can be represented by using _____. These models depict the _____ electrons around an atom or ion and can be rearranged to show how covalent or ionic bonding will occur between two species. _____ can show how atoms are connected in a molecule, whether multiple bonds exist between atoms, and whether nonbonding electrons are present. They cannot show the _____. _____ provides a basis for determining the three-dimensional arrangement of atoms in a molecule. The theory is based on the idea that electron pairs will try to _____ the distance between them in order to repulsive forces. The geometries that result from various bonding schemes can be predicted if one knows how many pairs of _____ electrons are arranged around a central atom.

Answers

Atoms combine in two different ways. They can share valence electrons to form covalent bonds or they can donate or accept valence electrons to form ions that are attracted to each other due to their charge differences. Both combinations are driven by the tendency of atoms to fill their valence shells to form a stable octet arrangement. This arrangement mimics the electronic arrangement of the noble gases.

The magnitude of the attraction between the positively charged nucleus and negatively charged valence electrons determines how easily an element can lose a valence electron. Metals have relatively low ionization energies and nonmetals have relatively high ionization energies. Ionization energy is a periodic property that generally increases across a period and decreases down a group. An element's chemical reactivity depends on its tendency to lose, gain, or share valence electrons.

Electronegativity is a measure of an element's ability to attract bonding electrons to itself. If two bonding elements have very different electronegativities, an ionic bond will form. If, on the other hand, they have similar electronegativities, a covalent bond will form.

The properties of ionic and covalent compounds result from the way in which atoms are joined together. In ionic compounds, the electrostatic attraction between positively charged cations and negatively charged anions produces a superstructure in which ions are arranged around each other in ratios determined by their charges. There is no discrete unit in an ionic compound, but the ionic formula indicates the types and ratios of ions present. Covalent bonds are directional because the electrons being shared exist in specific regions in space. This produces molecules, which are discrete units. The chemical formula of a covalent compound indicates the types and numbers of elements present in a molecule of the substance.

Ionic formulas are determined by examining the charges on the ions involved in bonding. The net charge must always be zero, meaning that the number of negative charges must equal the number of positive charges. Ions may be simple elemental ions, such as Na^+ and Cl^-, or polyatomic ions, such as NH_4^+ or PO_4^{3-}. Ionic formulas are named according to rules that include identifying the charge on the cation and giving the suffix -ide to the anion.

Covalent formulas are determined by looking at the number of bonds each atom needs to form in order to fill its octet. The least electronegative element is first in the formula. In binary covalent compounds, both elements are quantified by using a Greek prefix; and the most electronegative element, the second element, is given the suffix -ide.

The structures of molecules and individual ions can be represented by using Lewis structures. These models depict the valence electrons around an atom or ion and can be rearranged to show how covalent or ionic bonding will occur between two species. Lewis structures can show how atoms are connected in a molecule, whether multiple bonds exist between atoms, and whether nonbonding electrons are present. They cannot show the three-dimensional arrangement of atoms in the molecule. VSEPR theory provides a basis for determining the three-dimensional arrangement of atoms in a molecule. The theory is based on the idea that electron pairs will try to maximize the distance between them in order to minimize repulsive forces. The geometries that result from various bonding schemes can be predicted if one knows how many pairs of bonding and nonbonding electrons are arranged around a central atom.

TEST YOURSELF

1. Determine whether the formulas below are right or wrong; correct if wrong and explain your answer.
   (a) CCl_6
   (b) AlO
   (c) H_2O
   (d) NaF_3
   (e) $Ca(NH_4)_2$
2. Determine whether the names below are right or wrong; correct if wrong and explain your answer.
   (a) carbon chloride
   (b) aluminum trioxide
   (c) oxygen dihydride
   (d) sodium(I) fluoride
   (e) calcium diammonium
3. Draw the atomic level pictures of sodium chloride, NaCl, and water, H_2O. Identify each as an ionic or a covalent substance and list the ways in which their structures differ.
4. Determine how the relative electronegativities of the following pairs of atoms compare and predict whether they will form ionic or covalent bonds. Refer to the periodic table and Figure 3.5.
   (a) Mg and Cl
   (b) Mg and O
   (c) S and O
   (d) S and Cl

5. Determine the relative ionization energies of the pairs of atoms in question 4. What is the relationship between relative ionization energy and the type of bond formed?
6. Explain each of the following:
   (a) Na loses one valence electron when it ionizes
   (b) C forms four covalent bonds
   (c) F gains one electron when it forms an ion
   (d) Fe can combine with either two or three chloride ions
   (e) H never has an octet of valence electrons
7. Identify the statements below as being about ionic or covalent compounds:
   (a) produce electrolytes when melted
   (b) do not conduct electricity when dissolved
   (c) chemical properties may be different in different phases (i.e. solid, gas, liquid)
   (d) exist as molecules that have shapes
   (e) exist because of electrostatic attractions
8. List at least five ionic and five covalent compounds that are found in the human body.
9. Fill in the table below by using the words "nonpolar" or "polar" to indicate the type of covalent bond formed between the atoms in each pair:

	Carbon	**Nitrogen**	**Oxygen**
Carbon			
Nitrogen			
Oxygen			

Answers

1. (a) CCl_6; wrong. This formula indicates that one carbon atom (a nonmetal) is covalently bonded to six chlorine atoms (nonmetals). Covalent compound ratios are determined by the number of bonds the atoms form. Carbon is in Group IV and only forms four bonds. CCl_4 is the correct formula.
   (b) AlO; wrong. Aluminum is a metal in Group III and oxygen is a nonmetal in Group VI. Aluminum will lose three electrons when it forms a cation and oxygen will accept two electrons when it forms an anion. This would mean Al (3+) and O (2−) produce a compound with a net charge of 1+. Because ionic compounds have to be neutral, the charges of the ions indicate the ratios in which they combine: Al_2O_3 .
   (c) H_2O; right. Oxygen (a nonmetal) in Group VI will form two bonds covalently. Hydrogen (a nonmetal) in Group I only forms one covalent bond.
   (d) NaF_3; wrong. Na is a metal in Group I and will lose one electron to form Na^+. F is a nonmetal in Group VII and will accept one electron to form F^-. They would combine in a 1:1 ratio: NaF. The formula here indicates that Na is a 3^+ ion, which is wrong.
   (e) $Ca(NH_4)_2$; wrong. This formula contains two cations: Ca^{2+} and NH_4^+. Two cations do not combine to form an ionic compound; there is no attractive force between them.
2. (a) carbon chloride; wrong. See (a) above. Carbon will combine with four chlorine atoms, not one carbon tetrachloride is correct.
   (b) aluminum trioxide; wrong. See (b) above. The ionic compound Al_2O_3 would be named aluminum oxide.
   (c) oxygen dihydride; wrong. See (c) above. The least electronegative element in a covalent compound is first and the most electronegative is second. Dihydrogen oxide would be the correct name for water.
   (d) sodium(I) fluoride; wrong. See (d) above. When the metal in an ionic compound only forms one ion, the Roman numeral is not used. NaF is correct.
   (e) calcium diammonium; wrong. See (e) above.

3. (Cl^-) (Na^+) (Cl^-) (Na^+)
   (Na^+) (Cl^-) (Na^+) (Cl^-)
   (Cl^-) (Na^+) (Cl^-) (Na^+)
   Ionic compound; metal cations and nonmetal anions
   No discrete units; repeating lattice based on attractions of opposite charges

   O
   H H
   Covalent compound; nonmetals
   Molecular unit with distinct shape and size
4. (a) Mg (metal, 1.2) and Cl (nonmetal, 3.0); ionic
   (b) Mg (metal, 1.2) and O (nonmetal, 3.5); ionic
   (c) S (nonmetal, 2.5) and O (nonmetal, 3.5); covalent
   (d) S (nonmetal, 2.5) and Cl (nonmetal, 3.0); covalent
5. Relative ionization energies from Figure 3.1: Mg very much less than Cl; Mg very much less than O; S slightly less than O; S slightly less than Cl. When there is a large difference in ionization energies, ionic compounds are favored; when there is a small difference between ionization energies, covalent compounds are favored.
6. (a) Na, a metal in Group I, loses one valence electron when it ionizes because it can achieve an octet (Ne) easily by losing one electron.
   (b) C, nonmetal in Group IV, forms four covalent bonds because it is the most energetically favorable way for carbon to gain a valence octet. Losing four electrons or accepting four electrons are energetically unfavorable situations.
   (c) F, a nonmetal in Group VII, gains one electron when it forms an ion because it can achieve an octet (Ne) easily by accepting one electron
   (d) Fe, a transition metal, can combine with either two or three chloride ions because it is able to lose either two or three electrons to form Fe^{2+} or Fe^{3+} in ionic compounds.
   (e) H never has an octet of valence electrons because its valence shell can only hold two electrons.
7. (a) ionic
   (b) covalent
   (c) ionic
   (d) covalent
   (e) ionic
8. Ions in the human body: Na^+ (sodium ion), K^+ (potassium ion), Ca^{2+} (calcium ion), Cl^-(chloride ion) HCO_3^- (bicarbonate ion). Molecules in the human body: H_2O (water), CO_2 (carbon dioxide), O_2 (oxygen), NO (nitrogen monoxide), cholesterol.
9.

	Carbon	**Nitrogen**	**Oxygen**
Carbon	nonpolar	polar	polar
Nitrogen	polar	nonpolar	polar
Oxygen	polar	polar	nonpolar

chapter 4

Chemical Calculations

OUTLINE

4.1 Chemical Formulas and Formula Masses

A. The chemical formula of a compound
  1. represents the formula unit
     a if a molecule, the chemical formula represents one molecule
     b. if an ionic compound, the chemical formula represents the unit containing ions in correct ratios for the compound
  2. identifies elements that occur in a compound; elements are identified by their symbols
  3. indicates number of atoms of each element in formula unit
     a. numbers are indicated by subscripts
     b. polyatomic ions may be enclosed in brackets and subscripted if more than one is present in the formula unit
  4. indicates proportion of each element in compound
  5. does not convey information about structure
  6. is needed to calculate the formula mass

B. The formula mass of a compound
  1. is determined by adding masses, in atomic mass units, of all elements in formula

Key Terms formula, formula mass, formula unit, molecular mass
Key Figure Figure 4.1 Masses of compounds in which there are equal numbers of formula units

Example 4.1 Calculating the formula mass of a compound
Problem 4.1 Calculating the formula mass of a compound
Example 4.2 Converting formula units to masses in grams
Problem 4.2 Calculating masses in grams of equivalent formula units

4.2 The Mole / 4.3 Avogadro's Number

A. The mole
  1. is 6.022×10^{23} of anything (Avogadro's number). Avogadro's number is defined as the number of carbon-12 atoms in exactly 12 g of carbon-12.
  2. is the number of atoms contained in the atomic mass (in grams) of an element
  3. is the number of formula units contained in the formula mass (in grams) of a compound
  4. can be used as a conversion factor:

$$\frac{1 \text{ mol}}{6.022 \times 10^{23} \text{ atoms}} \qquad \frac{1 \text{ mol substance}}{\text{formula mass in grams}} \qquad \frac{1 \text{ mol element}}{\text{atomic mass in grams}} \leftarrow \text{from periodic table}$$

B. Mass, in grams, of 1 mol of substance is called molar mass.
C. Molar masses of anything contain 6.022×10^{23} particles but are different according to their identities.

Key Terms Avogadro's number, mol, molar mass, mole
Key Table Table 4.1 Relationships between formula mass (amu), molar mass (mol), particles/mole, and number of moles for hydrogen atoms, hydrogen molecules, and water

Example 4.3 Converting mass into moles
Problem 4.3 Converting mass into moles
Example 4.4 Converting moles into mass
Problem 4.4 Converting mass into moles
Example 4.5 Calculating the number of moles of elements in a compound
Problem 4.5 Calculating the number of moles of atoms in 1 mol of a substance
Example 4.6 Calculating atomic masses in grams
Problem 4.6 Calculating the masses of atoms in grams
Example 4.7 Relating number of atoms to mass of sample
Problem 4.7 Calculating the mass, in grams, of a sample of atoms

4.4 Empirical Formulas

Empirical formula:
1. is based on the percent composition of a substance
2. provides the simplest whole-number ratio of the elements in a compound
3. may not tell the molecular formula of a substance
4. can be determined from percent composition data

Determining Empirical Formulas

1. Empirical formulas are based on the ratio of moles of elements in a compound; when given grams, these must be converted to moles.
2. Percent composition data are usually given for the constituent elements; these numbers can be changed to "per 100 g sample" because percent means "per 100."

   39.3% Na ⇒ 39.3 g Na/100 g sample   11.2% H ⇒ 11.2 g H
   60.7% Cl ⇒ 60.7 g Cl/100 g sample   88.8% O ⇒ 88.8 g O

3. Convert all gram masses to mols (refer to Sections 4.2 and 4.3):

   39.3 g Na ⇒ 1.71 mol Na   11.2 g H ⇒ 11.1 mol H
   60.7 g Cl ⇒ 1.71 mol Cl   88.8 g O ⇒ 5.55 mol O

4. If moles of each element are about equal, the ratios are 1:1; the empirical formula is NaCl.
5. If moles of each element are not equal, divide each number by the smallest number of moles:

   11.1 mol H/5.55 mol O = 2.00 H/1 O   5.55 mol O/5.55 mol O = 1

The ratio is 2 H for each O; the empirical formula is H_2O.

Key Terms empirical formula, percent composition

Example 4.8 Finding the empirical formula of a sample
Problem 4.8 Determining the empirical formula of a compound if given percent composition data
Example 4.9 Finding the empirical formula of a sample
Problem 4.9 Determining the empirical formula of a compound if given percent composition data
Example 4.10 Finding the empirical formula of a sample
Problem 4.10 Determining the empirical formula of a compound if given percent composition data

4.5 *Molecular Formulas*

A. Molecular formula
   1. elemental symbols identify atoms in molecule; subscripts identify number of each type of atom
   2. can be calculated using percent composition and molar mass data

B. The molecular weight of a compound is the sum of the atomic weights of the atoms in the molecule.

Determining Molecular Formulas

1. Start with empirical formula data; empirical formulas provide the smallest whole-number ratios of atoms in an element but do not give exact numbers of each type of atom, only proportions.
2. The molar mass of a sample combined with empirical formula information can be used to determine molecular formula:

   Example: Suppose the empirical formula of a compound is CH_2O; its molar mass is 60.0 g

3. The molar mass is a multiple of the empirical formula weight:

   Example: 30 g is empirical formula weight of CH_2O

4. Divide the molar mass by the empirical formula weight:

   Example: $\frac{60\text{ g}}{30\text{ g}} = 2$

The molecular formula is twice the empirical formula: $2\,(CH_2O) \Rightarrow C_2H_4O_2$

Key Term molecular formula

4.6 *Balancing Chemical Equations*

A. A chemical reaction is written as a chemical equation:

$$\text{Reactants} \rightarrow \text{Products}$$

$\uparrow$ indicates a chemical reaction has occurred; reactants have their chemical identities changed

B. The number and type of atoms on the reactant side always equals the number and type of atoms on the product side (chemical equations balance).
   1. mass is conserved in a chemical reaction
   2. atoms do not change their identity (number of protons) in a chemical reaction; they are only rearranged

C. Coefficients in front of each substance indicate the number of moles of substance that react or are produced in the chemical reaction. Coefficients are the only numbers changed when balancing a chemical equation. *Do not change subscripts* because this changes the identity of the substance.

Balancing Chemical Reactions

1. Make sure all formulas are correct; you cannot properly balance an equation without the correct chemical formulas.
2. Check to see whether the reaction is balanced.
3. A table may help to keep information organized (see worked Problems 4.11–4.13).

(continued on next page)

Balancing Chemical Reactions *(cont.)*

4. If applicable, start balancing by choosing a metal with a relatively high charge on the left side of the equation. Count the number of those atoms on the left and right sides; if they do not balance, use a coefficient in front of the formula containing the metal to balance both sides. If there is no metal in the reaction, choose a nonmetal in a high oxidation state (e.g., carbon). It is generally best to leave H and O until last.
5. Work from element to element, changing coefficients as needed to balance each side.
   a. The coefficient becomes the multiplier of the entire formula:

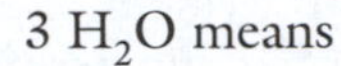

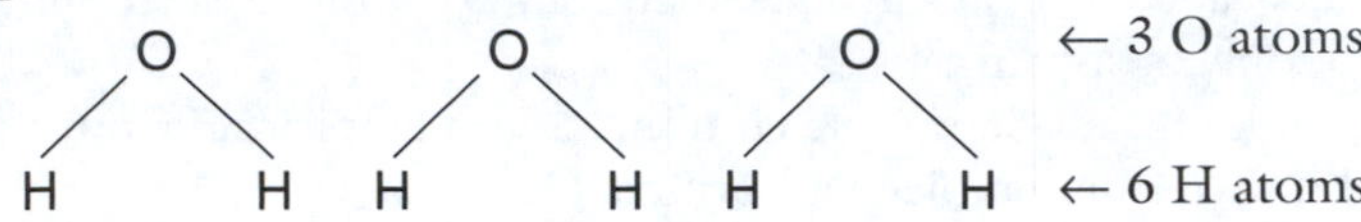

3 $Ba_3(PO_4)_2$ means:
← don't forget the subscripts; this means there are 2 P atoms and 8 O atoms in a unit

3×3 Ba = 9 Ba
3×2 P = 6 P
3×8 O = 24 O

   b. Coefficients should be whole numbers; multiply fractional coefficients to determine smallest whole number.
6. Once you think all atoms are balanced, write the equation and double check the number of atoms on either side of the equation.

Key Terms balanced chemical equation, coefficient, Law of Conservation of Atoms, products, reactants, unbalanced chemical equation

Example 4.1 Balancing a chemical equation
Problem 4.11 Balancing a chemical equation

4.7 Oxidation-Reduction Reactions

A. Chemical reactions can be classified into types; one type is the oxidation-reduction reaction.
B. Oxidation-reduction reactions can be subdivided into distinct categories:
   1. reactions in which oxygen is gained or lost
      a. if oxygen is gained, an oxidation has occurred
      b. if oxygen is lost, a reduction has occurred
   2. reactions in which electrons are gained and lost (transferred from one element to another)
      a. the element losing electrons is oxidized
      b. the element gaining electrons is reduced
   3. reaction in which the oxidation states of elements change
      a. atoms whose oxidation states become more positive are oxidized
      b. atoms whose oxidation states become more negative are reduced
C. Oxidation reactions include
   1. direct reaction of metals with oxygen to form oxides
      a. rusting is an example: metal is oxidized; oxygen is reduced

$$4\,Fe + 3\,O_2 \rightarrow 2\,Fe_2O_3$$

   2. direct reaction of compound or element with oxygen accompanied by burning is called combustion
      a. carbon-containing compounds react with oxygen to produce carbon dioxide and water; carbon is oxidized; oxygen is reduced

$$CH_4 + 2\,O_2 \rightarrow CO_2 + 2\,H_2O$$

b. if insufficient oxygen is available, carbon-containing compounds react with oxygen to produce carbon monoxide
c. metabolic reactions that yield energy are slow, controlled combustion reactions
3. removal of hydrogen from a molecule is called dehydrogenation; energy can be extracted from small biological molecules by removing hydrogen

$$H_3CCH_2OH + \text{oxidizing agent} \rightarrow H_3CCOH + \text{reduced oxidizing agent (gains H)}$$

4. loss of electrons is an oxidation; metals lose electrons to form cations

$$Mg \rightarrow Mg^{2+} + 2\ e^-$$

D. Reduction reactions:
1. loss of oxygen from a metal oxide is a reduction of the metal; metal oxide is reacted with a chemical that is easily oxidized (reducing agent) to separate the metal from the oxide; metal is reduced, reducing agent is oxidized

$$2\ Fe_2O_3 + 3\ C \rightarrow 4\ Fe + 3\ CO_2$$

2. gain of electrons is a reduction; nonmetals gain electrons to form anions

$$Cl + 1\ e^- \rightarrow Cl^-$$

E. Changes in oxidation numbers of oxidant and reductant used to balance redox reaction equations.
1. Rules for assigning oxidation numbers:
a. Oxidation number for free elements = 0
b. Sum of oxidation numbers of all atoms in species = net charge on species
c. In compunds, oxidation numbers of Group I metals (Na, etc.) = +1
d. In its compounds, oxidation number of F = −1
e. In compounds, oxidation numbers of Group II metals (Ca, etc.) = +2; oxidation numbers of main Group III metals (Al, etc.) = +3
f. H, in compounds, has oxidation number = +1
g. O, in compounds, has oxidation number = −2

F. Oxidation-reduction reactions can be balanced by using the ion-electron method (Box 4.2).

Key Terms combustion reaction, dehydrogenation, ion-electron method, oxidation, oxidation number, oxidation state, redox, reduction

Key Box Box 4.2 Balancing oxidation-reduction reactions by the ion-electron method

Example 4.12 Determining oxidation states
Problem 4.12 Balancing an oxidation-reduction equation
Example 4.13 Assigning oxidation states
Problem 4.13 Assigning oxidation states
Example 4.14 Balancing a combostion equation
Problem 4.14 Writing and balancing a combustion equation
Example 4.15 Balancing a combustion equation
Problem 4.15 Writing and balancing a combustion equation

4.7 Stoichiometry

A. Stoichiometry refers to the quantitative aspects of chemical reactions.
1. atoms react to form molecules in simple whole-number ratios.
2. the ratio in which atoms react is the same as the ratio in which moles of those atoms react
3. coefficients in a balanced chemical equation describe the ratios of all participants in the reaction.
4. if the mole ratios are known for a given reaction, the gram or mass ratios are also known because the mass of 1 mol of an element or a compound can be calculated from the periodic table

B. Stoichiometric relationships can be used to
   1. predict how much product will form from a given amount of reactant
   2. determine how much reactant will be required to produce a given amount of product

C. The coefficients in a balanced chemical equation can be used to generate conversion factors that relate any two participants to each other

$$2\text{ A} + 3\text{ B} \rightarrow 4\text{ C} + \text{D}$$

The relationships are

$$\frac{2 \text{ mol A}}{3 \text{ mol B}} \quad \frac{2 \text{ mol A}}{4 \text{ mol C}} \quad \frac{2 \text{ mol A}}{1 \text{ mol D}} \quad \frac{3 \text{ mol B}}{4 \text{ mol C}} \quad \frac{3 \text{ mol B}}{1 \text{ mol D}} \quad \frac{4 \text{ mol C}}{1 \text{ mol D}}$$

Each conversion factor can be used as its reciprocal, so there are a total of 12 conversion factors generated from the coefficients in the balanced reaction.

D. The mass relationships can be generated by converting each mole amount to its equivalent in grams.

Stoichiometric Relationships

$2\ H_2$	+	O_2	→	$2\ H_2O$	
2 mol H		1 mol O_2		2 mol H_2O	← coefficients give mole relationships
2 (2.02 g)		1 (32.00 g)		2 (18.02 g)	← change 1 mol to equivalent weight in grams
4.04 g H_2		32.00 g O_2		36.04 g H_2O	← multiply by coefficients to get gram ratios

These relationships can generate conversion factors in either moles or grams:

$$\frac{2 \text{ mol } H_2}{2 \text{ mol } H_2O} \quad \frac{4.04 \text{ g } H_2}{36.04 \text{ g } H_2O}$$

Stoichiometric Problems

There are several types of stoichiometric problems that are encountered in chemistry; all require that you start with a balanced equation.

$$2\text{ A} + \text{B} \rightarrow 3\text{ C}$$

1. Given the amount of a reactant, in either grams or moles, predict the amount of a product, in either grams or moles, that will be produced.
   (a) Given 2.5 mol of A, how many moles of C will be produced?

$$2.5 \text{ mol A} \times \frac{3 \text{ mol C}}{2 \text{ mol A}} = 3.8 \text{ mol C}$$

↑ from balanced equation

   (b) Given 10 g of A, how many grams of C will be produced?

$$10 \text{ g A} \times \frac{1 \text{ mol A}}{\text{molar mass of A, in grams}} \times \frac{3 \text{ mol C}}{2 \text{ mol A}} \times \frac{\text{molar mass of C, in grams}}{1 \text{ mol C}} = \text{number of grams of C produced}$$

↑ from chemical formula and periodic table; ↑ from balanced equation; ↑ from chemical formula and periodic table

(continued on next page)

Stoichiometric Problems *(cont.)*

2. Given the amount of a reactant, in either grams or moles, predict the amount of a second reactant needed to react completely with the given reactant.
   (a) Given 2.5 mol A, how many moles of B would be needed to react completely with this amount of A?

$$2.5 \text{ mol A} \times \frac{1 \text{ mol B}}{2 \text{ mol A}} = 1.3 \text{ mol B}$$

↑ from balanced equation

   (b) Given 10 g of A, how many grams of B would be required to react completely with this amount of A?

$$10 \text{ g A} \times \frac{1 \text{ mol A}}{\text{molar mass of A, in grams}} \times \frac{3 \text{ mol B}}{2 \text{ mol A}} \times \frac{\text{molar mass of B, in grams}}{1 \text{ mol B}} = \text{number of grams of B produced}$$

↑ from chemical formula and periodic table ↑ from balanced equation ↑ from chemical formula and periodic table

3. Given the amount, in either grams or moles, of a product, determine how much of a reactant was required to produce the amount of product given.
   (a) 3.75 mol of C were produced when A and B reacted. How many moles of A were required to produce 3.75 mol C?

$$3.75 \text{ mol C} \times \frac{2 \text{ mol A}}{3 \text{ mol C}} = 2.50 \text{ mol A}$$

↑ from balanced equation

   (b) 10 g of C were produced when A and B reacted. How many grams of A were required to produce 10 g of C?

$$10 \text{ g C} \times \frac{1 \text{ mol C}}{\text{molar mass of C, in grams}} \times \frac{2 \text{ mol A}}{3 \text{ mol C}} \times \frac{\text{molar mass of A, in grams}}{1 \text{ mol A}} = \text{number of grams of A produced}$$

↑ from chemical formula and periodic table ↑ from balanced equation ↑ from chemical formula and periodic table

4. Given the amounts, in either grams or moles, of more than one reactant, determine which reactant will be the limiting factor (that is, which one will be used up first) and then determine how much product is produced.
   Given 1.0 mol A and 2.0 mol B, which of these will determine how much C will be produced?
   The balanced equation requires 1 mol of B for every 2 mol of A:

$$1.0 \text{ mol A} \times \frac{1 \text{ mol B}}{2 \text{ mol A}} = 0.50 \text{ mol B}$$

↑ from balanced equation ← amount required to react completely with 1 mol of A

OR

$$2.0 \text{ mol B} \times \frac{2 \text{ mol A}}{1 \text{ mol B}} = 4.0 \text{ mol A}$$

↑ from balanced equation ← amount required to react completely with 2 mol B

There is not enough A to use up all the B; there is enough B to use up all the A; A will be the limiting factor and will determine how much C is produced.

Key Term stoichiometry

Key Figures

Figure 4.2	Flow chart for using balanced reaction equations to calculate mass or moles
Figure 4.3	Summary of stoichiometric relationships

Example 4.16	Creating conversion factors by using stoichiometric relationships
Problem 4.16	Deriving conversion factors from a balanced equation
Example 4.17	Determining the amount of product produced in a reaction by using stoichiometric relationships
Problem 4.17	Determining the amount of reactant required in a reaction by using stoichiometric relationships
Example 4.18	Using stoichiometric relationships to create conversion factors
Problem 4.18	Determining moles of product formed by using stoichiometric relationships
Example 4.19	Using stoichiometric relationships to predict amounts of reactants required and amounts of products produced
Problem 4.19	Using stoichiometric relationships to predict amounts of reactants required and amounts of products produced

ARE YOU ABLE TO ... AND WORKED TEXT PROBLEMS

Objectives Section 4.1 Chemical Formulas and Formula Masses

Are you able to . . .

- define formula unit
- explain what information is contained in a formula unit
- define formula mass
- determine the formula mass, in amu, of a substance (Ex. 4.1, Prob. 4.1, Exercise 4.48)
- determine the formula mass, in grams, of a substance (Ex. 4.2, Prob. 4.2, Exercises 4.1, 4.2, 4.3, 4.4, 4.48)

Problem 4.1 Calculate the formula masses of the following compounds.

The formula mass is the sum of all the constituent atomic masses, in amu. The mass, in amu, of any element is given on the periodic table.

(a) $Zn(HPO_4)_2$

Atom	Atomic mass (amu)	Number of atoms in formula unit	Total contribution (amu)
Zn	65.38	1	65.41
H	1.01	2	2.02
P	30.97	2	61.94
O	16.00	8	128.00
		total formula mass	257.37amu = 257.4 amu

(b) Fe_3O_4

Atom	Atomic mass (amu)	Number of atoms in formula unit	Total contribution (amu)
Fe	55.85	3	167.55
O	16.00	4	64.00
		total formula mass	231.55 amu = 231.6 amu

(c) H_2O

Atom	Atomic mass (amu)	Number of atoms in formula unit	Total contribution (amu)
H	1.01	2	2.02
O	16.00	1	16.00
		total formula mass	18.02 amu

(d) H_2SO_4

Atom	Atomic mass (amu)	Number of atoms in formula unit	Total contribution (amu)
H	1.01	2	2.02
S	32.07	1	32.07
O	16.00	4	64.00
		total formula mass	98.09 amu

(e) $C_3H_6O_2N$

Atom	Atomic mass (amu)	Number of atoms in formula unit	Total contribution (amu)
C	12.01	3	36.03
H	1.0079	6	6.05
O	16.00	2	32.00
N	14.01	1	14.01
		total formula mass	88.09 amu

Problem 4.2 Calculate the masses, in grams, that will contain equal numbers of formula units of each compound.

If you wanted the same number of feathers and bricks in a sample, 100, for example, would they weigh the same? No. This question is like that. Each sample has the same number of formula units, but they all have different masses. The mass of each formula unit can be scaled up from atomic mass units to grams while retaining the same proportion of each element. This mass, expressed in grams, represents the same number of formula units for any compound: 6.02×10^{23} or 1 mol.

(a) Na_2SO_4

Atom	Atomic mass (g)	Number of atoms in formula unit	Total contribution (g)
Na	22.99	2	45.98
S	32.07	1	32.07
O	16.00	4	64.00
		total formula mass	142.05 g = 142.1 g

(b) KBr

Atom	Atomic mass (g)	Number of atoms in formula unit	Total contribution (g)
K	39.10	1	39.10
Br	79.90	1	79.90
		total formula mass	119.00 g = 119.0

(c) MgO

Atom	Atomic mass (g)	Number of atoms in formula unit	Total contribution (g)
Mg	24.31	1	24.31
O	16.00	1	16.00
		total formula mass	40.31 g

(d) C_6H_{14}

Atom	Atomic mass (g)	Number of atoms in formula unit	Total contribution (g)
C	12.01	6	72.06
H	1.0079	14	14.11
		total formula mass	86.17 g

Objectives Section 4.2 The Mole

Are you able to . . .

- define mole
- define molar mass
- calculate the molar mass of an element or compound (Exs. 4.3, 4.4, Probs. 4.3, 4.4)
- describe the relationships between formula masses and molar masses
- compare weights of equal molar amounts of elements or compounds
- use the mole as a conversion factor to relate number of particles and masses (Exercises 4.39, 4.40, 4.41, 4.42)
- convert moles to grams and grams to moles (Exs. 4.3, 4.4, Prob. 4.3, 4.4, Exercises 4.5, 4.6, 4.7, 4.8, 4.36, 4.37, 4.38, 4.49)

Problem 4.3 Calculate the number of moles of $CaCl_2$ in 138.8 g of $CaCl_2$.

Use the periodic table to determine the mass of one mole.

Atom	Atomic mass (g)	Number of atoms in formula unit	Total contribution (g)
Ca	40.08	1	40.08
Cl	35.45	2	70.90
		mass of 1 mol $CaCl_2$	110.98 g

$$\text{Conversion factor: } 1 \text{ mol } CaCl_2 = 110.98 \text{ g } CaCl_2$$

$$\frac{1 \text{ mol } CaCl_2}{110.98 \text{ g } CaCl_2} \quad \text{or} \quad \frac{110.98 \text{ g } CaCl_2}{1 \text{ mol } CaCl_2}$$

You are given 138.8 g of $CaCl_2$ and your calculations tell you that 1 mol of $CaCl_2$ weighs 110.98 g. The first question to ask yourself is: Do you have more or less than 1 mol in what you are given? A little bit more. This is a good way to check your answer; it should be a little larger than 1 mol.

$$138.8 \, \cancel{\text{g } CaCl_2} \times \frac{1 \text{ mol } CaCl_2}{110.98 \, \cancel{\text{g } CaCl_2}} = 1.251 \text{ mol } CaCl_2$$

Problem 4.4 How many moles are in 0.3600 g of glucose?

Refer to Example 4.1 for the calculation of the formula mass of glucose, $C_6H_{12}O_6$. Use the periodic table to determine the mass of one mole.

Atom	Atomic mass (g)	Number of atoms in formula unit	Total contribution (g)
C	12.01	6	72.06
H	1.01	12	12.12
O	16.00	6	96.00
		mass of 1 mol $C_6H_{12}O_6$	180.18 g

$$\text{Conversion factor: } 1 \text{ mol } C_6H_{12}O_6 = 180.18 \text{ g } C_6H_{12}O_6$$

$$\frac{1 \text{ mol } C_6H_{12}O_6}{180.18 \text{ g } C_6H_{12}O_6} \quad \text{or} \quad \frac{180.18 \text{ g } C_6H_{12}O_6}{1 \text{ mol } C_6H_{12}O_6}$$

You are given 0.3600 g of glucose and your calculations tell you that 1 mol of glucose weighs 180.18 g. The first question to ask yourself is, Do you have more or less than 1 mol in what you are given? A lot less! This is a good way to check your answer; it should be a lot smaller than 1 mol.

$$0.3600 \, \cancel{g\, C_6H_{12}O_6} \times \frac{1 \text{ mol } C_6H_{12}O_6}{180.18 \, \cancel{g\, C_6H_{12}O_6}} = 1.998 \times 10^{-3} \text{ mol } C_6H_{12}O_6$$

Objectives Section 4.3 Avogadro's Number

Are you able to . . .

- define Avogadro's number
- use Avogadro's number as a conversion factor to relate number of particles to moles and grams (Exercises 4.9, 4.10, 4.11, 4.12, 4.39, 4.40, 4.41, 4.42)
- determine the number of moles of constituent elements or ions in 1 mol of a substance (Ex. 4.5, Prob. 4.5)
- calculate the masses of individual atoms by using Avogadro's number (Ex. 4.6, Prob. 4.6)
- relate the number of atoms in a sample to its actual weight, in grams (Ex. 4.7, Prob. 4.7)

Problem 4.5 How many moles of atoms of each kind are there in 1 mol of each of the following substances?

The subscripts tell you how many moles of each atom are in 1 mol of the substance.

(a) $ZnHPO_4$

Zn	H	P	O
1 mol	1 mol	1 mol	4 mol

(b) H_2SO_4

H	S	O
2 mol	1 mol	4 mol

(c) Al_2O_3

Al	O
2 mol	3 mol

(d) Ca_3P_2

Ca	P
3 mol	2 mol

Problem 4.6 Calculate the mass, in grams, of the following atoms:

From the periodic table, the mass in grams of 6.022×10^{23} atoms (1 mol) of the element can be obtained. If you know the mass of 6.022×10^{23} atoms, then you can determine the mass of a single atom:

(a) lithium

$$\frac{6.944 \text{ g}}{6.022 \times 10^{23} \text{ atoms}} = \frac{1.153 \times 10^{-23} \text{ g}}{\text{atom}} \quad \text{each atom weighs } 1.153 \times 10^{-23} \text{ g}$$

(b) nitrogen

$$\frac{14.01 \text{ g}}{6.022 \times 10^{23} \text{ atoms}} = \frac{2.326 \times 10^{-23} \text{ g}}{\text{atom}} \quad \text{each atom weighs } 2.326 \times 10^{-23} \text{ g}$$

(c) fluorine

$$\frac{19.00 \text{ g}}{6.022 \times 10^{23} \text{ atoms}} = \frac{3.155 \times 10^{-23} \text{ g}}{\text{atom}} \quad \text{each atom weighs } 3.155 \times 10^{-23} \text{ g}$$

(d) calcium

$$\frac{40.08 \text{ g}}{6.022 \times 10^{23} \text{ atoms}} = \frac{6.656 \times 10^{-23} \text{ g}}{\text{atom}} \quad \text{each atom weighs } 6.656 \times 10^{-23} \text{ g}$$

(e) carbon

$$\frac{12.01 \text{ g}}{6.022 \times 10^{23} \text{ atoms}} = \frac{1.994 \times 10^{-23} \text{ g}}{\text{atom}} \quad \text{each atom weighs } 1.994 \times 10^{-23} \text{ g}$$

Problem 4.7 Using the numbers of atoms in Example. 4.7, calculate the corresponding masses, in grams, of calcium.

In Problem 4.6, you calculated the mass of a single calcium atom: 6.656×10^{-23} g

Number of atoms of calcium	Mass (g)	
10	6.656×10^{-22}	
100	6.656×10^{-21}	
100000	6.656×10^{-18}	
1.0×10^{15}	6.656×10^{-8}	
6.022×10^{20}	0.04008	
6.022×10^{23}	40.08	← Found on periodic table; Mass, in grams, of 1 mol of Ca is also mass, in amu, of 1 atom of Ca

Objectives Section 4.4 Empirical Formulas

Are you able to . . .

- define empirical formula
- calculate percent composition of a compound (refer to Section 2.1, Expand Your Knowledge 4.51)
- determine the empirical formula of a compound from percent composition data (Exs. 4.8, 4.9, 4.10, Probs. 4.8, 4.9, 4.10, Exercises 4.13, 4.14, 4.43, 4.44, 4.45, 4.46, Expand Your Knowledge 4.54, 4.55)
- describe the information contained in the empirical formula
- explain what information is *not* contained in the empirical formula

Problem 4.8 Quantitative analysis of magnesium oxide reveals that the composition of the compound is 60.31% magnesium and 39.69% oxygen. From those data, determine its empirical formula.

Any time you see "%" you can use 100 as the sample basis. For this example, in 100 g of sample, 60.31 g will be magnesium and 39.69 g will be oxygen. (Refer back to Section 2.1 to review the calculation of percent composition of a compound.)

The empirical formula is based on the simplest whole number ratio of atoms of each type in the formula unit; scaled up, it represents the simplest ratio of each element in a given sample. Converting the mass of each element to the corresponding number of moles will provide the basis for determining the ratio:

$$60.31\ \cancel{\text{g Mg}} \times \frac{1\ \text{mol Mg}}{24.31\ \cancel{\text{g Mg}}} = 2.481\ \text{mol Mg}$$

$$39.69\ \cancel{\text{g O}} \times \frac{1\ \text{mol O}}{16.00\ \cancel{\text{g O}}} = 2.481\ \text{mol O}$$

Divide the moles of Mg by the moles of O to determine the ratio:

$$\frac{2.481\ \text{mol Mg}}{2.481\ \text{mol O}} = \frac{1\ \text{Mg}}{1\ \text{O}}$$

The empirical formula is MgO.

Problem 4.9 Chemical analysis shows the composition of magnesium sulfate to be 20.20% Mg, 26.60% S, and 53.20% O. Determine its empirical formula.

Arrange the data in a table based on a 100-g sample:

Element	Number of grams in 100-g sample	Number of moles in 100-g sample
Mg	20.20	0.8309
S	26.60	0.8297
O	53.20	3.325

Calculate the number of moles of each element:

$$20.20 \text{ \sout{g Mg}} \times \frac{1 \text{ mol Mg}}{24.31 \text{ \sout{g Mg}}} = 0.8309 \text{ mol Mg}$$

$$26.60 \text{ \sout{g S}} \times \frac{1 \text{ mol S}}{32.06 \text{ \sout{g S}}} = 0.8297 \text{ mol S}$$

$$53.20 \text{ \sout{g O}} \times \frac{1 \text{ mol O}}{16.00 \text{ \sout{g O}}} = 3.325 \text{ mol O}$$

Mg and S are very close to each other; if rounded to two decimal places, they would both be 0.83 mol; these are also the smallest values. Divide each number of moles by the smallest, which is S in this case:

$$\frac{0.8309 \text{ mol Mg}}{0.8297} = 1.001 \text{ mol Mg} \quad \frac{0.8297 \text{ mol S}}{0.8297} = 1.000 \text{ mol S} \quad \frac{3.325 \text{ mol O}}{0.8297} = 4.007 \text{ mol O}$$

The empirical formula is $MgSO_4$.

Problem 4.10 By chemical analysis, the composition of an oxide of phosphorus was 43.66% P and 56.34% O. Determine its empirical formula.

Arrange the data in a table, based on a 100-g sample.

Element	Number of grams in 100-g sample	Number of moles in 100-g sample
P	43.66	1.410 mol P
O	56.34	3.521 mol O

Calculate the number of moles of each element:

$$43.66 \text{ \sout{g P}} \times \frac{1 \text{ mol P}}{30.97 \text{ \sout{g P}}} = 1.410 \text{ mol P}$$

$$56.34 \text{ \sout{g O}} \times \frac{1 \text{ mol O}}{16.00 \text{ \sout{g O}}} = 3.521 \text{ mol O}$$

P is the smallest value; divide each number of moles by the number of moles of P:

$$\frac{1.410 \text{ mol P}}{1.410} = 1.000 \text{ mol P} \qquad \frac{3.521 \text{ mol O}}{1.410} = 2.497 \text{ mol O}$$

This results in a ratio:

$$\frac{1 \text{ mol P}}{2.5 \text{ mol O}}$$

To get whole numbers, multiply each number of moles by 2

$$\frac{2 \text{ mol P}}{5 \text{ mol O}}$$

The empirical formula is P_2O_5.

Objectives Section 4.5 Molecular Formulas

Are you able to . . .

- define molecular formula
- explain the information contained in the molecular formula
- determine the molecular formula of a compound from its empirical formula and mass data (Expand Your Knowledge 4.55)

Objectives Section 4.6 Balancing Chemical Equations

Are you able to . . .

- define chemical equation
- explain what information is contained in a chemical equation
- define reactants
- identify reactants in a chemical equation
- define products
- identify products in a chemical equation (Exercise 4.47)
- explain how a coefficient affects the substance it precedes
- explain the differences between subscripts and coefficients
- recognize balanced and unbalanced chemical equations
- balance chemical equations (Ex. 4.11, Prob. 4.11, Exercises 4.15, 4.16, 4.17, 4.18, 4.35, 4.47)

Problem 4.11 Balancing the following equation:

$$Ca(OH)_2 + H_3PO_4 \rightarrow Ca_3(PO_4)_2 + H_2O$$

A good place to start is with either Ca because it is a metal ion.

	$Ca(OH)_2$ +	H_3PO_4 →	$Ca_3(PO_4)_2$ +	H_2O	
Number of Ca atoms	1		3		not balanced; multiply $Ca(OH)_2$ by 3

	3 $Ca(OH)_2$ +	H_3PO_4 →	$Ca_3(PO_4)_2$ +	H_2O	
Number of Ca atoms	3		3		Ca balances; next try P
Number of P atoms		1	2		not balanced; multiply H_3PO_4 by 2

	3 $Ca(OH)_2$ +	2 H_3PO_4 →	$Ca_3(PO_4)_2$ +	H_2O	
Number of P atoms		2	2		P and Ca both balance; next try H

	3 $Ca(OH)_2$ +	2 H_3PO_4 →	$Ca_3(PO_4)_2$ +	H_2O	
Number of H atoms	2	6		2	H does not balance; multiply H_2O by 6

	3 $Ca(OH)_2$ +	2 H_3PO_4 →	$Ca_3(PO_4)_2$ +	6 H_2O	
Number of H atoms	6	6		12	H balances; O is last
Number of O atoms	6	8	8	6	O balances

Check balancing:

	3 $Ca(OH)_2$ +	2 H_3PO_4 →	$Ca_3(PO_4)_2$ +	6 H_2O
Number of Ca atoms	3		3	
Number of P atoms		2	2	
Number of H atoms	6	6		12
Number of O atoms	6	8	8	6

All atoms are balanced; the final equation is

$$3\ Ca(OH)_2 + 2\ H_3PO_4 \rightarrow Ca_3(PO_4)_2 + 6\ H_2O$$

Objectives Section 4.7 Oxidation-Reduction Reactions

Are you able to . . .

- explain what an oxidation-reduction reaction is
- recognize oxidation-reduction reactions
- assign oxidation numbers to elements in reactant and product species
- explain what a combustion reaction is
- recognize combustion reactions
- write balanced combustion reactions (Exs. 4.12, 4.13, Probs. 4.12, 4.13, Exercises 4.19, 4.20, 4.21, 4.22, 4.23, 4.24)

- explain what a dehydrogenation reaction is
- recognize a dehydrogenation reaction
- write balanced dehydrogenation reactions
- recognize when electrons are transferred in a redox reaction
- balance an electron transfer reaction by using the ion-electron method

Problem 4.12 Complete and balance the following equation. The reaction takes place in acidic solution. $N_2H_4 + I^- \rightarrow NH_4^+ + I_2$

To recognize this as a redox reaction first assign oxidation numbers to the atoms in each species:

N_2H_4 Net charge = 0 so sum of oxidation numbers = 0.
Oxidation number of H = +1 (rule f.)
so contribution by four H atoms = (4)(+1) = (+4).
Oxidation number of N must be such that (2)(oxidation number of N) = −4.
Oxidation number of N = −2.

I^- Oxidation number of I in I^- must be −1 (rule b.).

NH_4^+ Net charge = +1. Oxidation number of H = +1 (rule f.).
(4)(+1) = +4. Oxidation number of N must be −3.
(−3) + (+4) = +1

I_2 Oxidation number of I in I_2 is zero (rule a.).

Separate the oxidation half reaction from the reduction half reaction.

I in I^- was oxidized because its oxidation number becomes more positive. $(-1 \rightarrow 0)$
N in N_2H_4 was reduced because its oxidation number becomes more negative $(-2 \rightarrow -3)$

Balance oxidation half reaction:	$2\,I^- \rightarrow I_2 + 2e^-$
Balance reduction half reaction: adding elements of H+ and/or H_2O as needed	$N_2H_4 + 2e^- + 4H^+ \rightarrow 2NH_4^+$
Multiply one and/or the other half reaction (as needed) so that the number of electrons gained = number of electrons lost.	No multiplication needed for this case
Add oxidation and reduction half reactions to get balanced overall reaction.	$2I^- + N_2H_4 + 4H^+ \rightarrow I_2 + 2\,NH_4^+$
Check to be certain atoms and net charge balance on both sides of equation.	Balanced.

Problem 4.13 Assign an oxidation state to chlorine in each of the following compounds: (a) Cl_2O; (b) Cl_2O_3; (c) ClO_2; (d) Cl_2O_6.

(a) Cl_2O According to Rule b, the sum of the oxidation numbers for Cl_2O is zero.
According to Rule g, the O atom in Cl_2O has an oxidation number of −2.
The oxidation number of Cl must be +1. (+1)(2) + (−2) = 0

(b) Cl_2O_3 According to Rule b, the sum of the oxidation numbers for Cl_2O_3 = 0.
According to Rule g, each O atom has an oxidation number of −2.
The oxidation number of Cl in Cl_2O_3 must be +3. (+3)(2) + (−2)(3) = 0

(c) ClO_2 According to Rule b, the sum of the oxidation numbers for ClO_2 is zero.
According to Rule g, each O atom in ClO_2 has an oxidation number of −2.
The oxidation number of Cl must be +4. (+4) + (−2)(2) = 0

(d) Cl_2O_6 According to Rule b, the sum of the oxidation numbers for Cl_2O_6 = 0.
According to Rule g, each O atom has an oxidation number of −2.
The oxidation number of Cl in Cl_2O_6 must be +6. (+6)(2) + (−2)(6) = 0

Problem 4.14 Write the balanced equation for the combustion of the alkane, C_4H_{10}.

A combustion reaction means that the hydrocarbon is reacted with oxygen to produce carbon dioxide and water:

$$C_4H_{10} + O_2 \rightarrow CO_2 + H_2O$$

Start with the carbon atoms; all carbon from the hydrocarbon wind up in the carbon dioxide

	C_4H_{10}	$+ O_2 \rightarrow$	CO_2	$+ H_2O$	
Number of C atoms	4		1		C doesn't balance; multiply CO_2 by 4

	C_4H_{10}	$+ O_2 \rightarrow$	$4\ CO_2$	$+ H_2O$	
Number of C atoms	4		4		C balances; next try H
Number of H atoms	10			2	H doesn't balance; multiply H_2O by 5

	C_4H_{10}	$+ O_2 \rightarrow$	$4\ CO_2$	$+ 5\ H_2O$	
Number of H atoms	10			10	H balances; O is last
Number of O atoms		2	8	5	O doesn't balance; coefficient will be a fraction $13/2\ O_2$ will balance with 13 O on product side; multiply everything by 2 to get rid of the fraction

Check balancing:

	$2\ C_4H_{10}$	$+ 13\ O_2 \rightarrow$	$8\ CO_2$	$+ 10\ H_2O$
Number of C atoms	8		8	
Number of H atoms	20			20
Number of O atoms		26	16	10

All atoms are balanced; the final equation is

$$2\ C_4H_{10} + 13\ O_2 \rightarrow 8\ CO_2 + 10\ H_2O$$

Problem 4.15 Write the balanced equation for the combustion of lactic acid, $C_3H_6O_3$, a product of carbohydrate metabolism.

A combustion reaction means that the reactant molecule reacts with oxygen to produce carbon dioxide and water:

$$C_3H_6O_3 + O_2 \rightarrow CO_2 + H_2O$$

Start with the carbon atoms; all carbon from the hydrocarbon winds up in the carbon dioxide

	$C_3H_6O_3$	$+ O_2 \rightarrow$	CO_2	$+ H_2O$	
Number of C atoms	3		1		C doesn't balance; multiply CO_2 by 3

	$C_3H_6O_3$	$+ O_2 \rightarrow$	$3\ CO_2$	$+ H_2O$	
Number of C atoms	3		3		C balances; next try H
Number of H atoms	6			2	H doesn't balance; multiply H_2O by 3

	$C_3H_6O_3$	$+ O_2 \rightarrow$	$3\ CO_2$	$+ 3\ H_2O$	
Number of H atoms	6			6	H balances; O is last
Number of O atoms	3	2	6	3	O doesn't balance; multiply O_2 by 3

	$C_3H_6O_3$	$+ 3\ O_2 \rightarrow$	$3\ CO_2$	$+ 3\ H_2O$	
Number of O atoms	3	6	6	3	O balances

Check balancing:

	$C_3H_6O_3$	$+ 3\ O_2 \rightarrow$	$3\ CO_2$	$+ 3\ H_2O$
Number of C atoms	3		3	
Number of H atoms	6			6
Number of O atoms	3	6	6	3

All atoms are balanced; the final equation is

$$C_3H_6O_3 + 3\ O_2 \rightarrow 3\ CO_2 + 3\ H_2O$$

Objectives Section 4.7 Stoichiometry

Are you able to . . .

- define stoichiometry
- interpret the stoichiometry of a balanced chemical equation
- determine the conversion factors inherent in a balanced chemical equation (Ex. 4.14, Prob. 4.14, Exercises 4.25, 4.26)
- use conversion factors derived from balanced chemical equations to relate moles of substances (Exs. 4.15, 4.16, Probs. 4.15, 4.16, Exercises 4.27, 4.28, 4.29, 4.33, 4.34)
- use conversion factors derived from balanced chemical equations to relate grams of substances (Ex. 4.17, Prob. 4.17, Exercises 4.31, 4.32, 4.33. 4.34)
- determine moles or grams of product(s) produced from a given amount of reactant(s) and vice versa (Exs. 4.15, 4.16, Probs. 4.15, 4.16, Exercises 4.27, 4.28, 4.29, 4.33, 4.34)
- determine moles or grams of reactant(s) required from a given amount of reactant(s) or products(s) and vice versa (Ex. 4.17, Prob. 4.17, Exercises 4.31, 4.32, 4.33, 4.34)

Problem 4.16 Derive all possible unit conversion factors from the following balanced equation:

$$3\ Mg(OH)_2 + 2\ FeCl_3 \rightarrow 2\ Fe(OH)_3 + 3\ MgCl_2$$

Each coefficient can be related to any other coefficient. These can be used as their reciprocals, so each of the following relationships generates two conversion factors:

$$\frac{3\text{ mol }Mg(OH)_2}{2\text{ mol }FeCl_3} \qquad \frac{3\text{ mol }Mg(OH)_2}{2\text{ mol }Fe(OH)_3} \qquad \frac{3\text{ mol }Mg(OH)_2}{3\text{ mol }MgCl_2} \qquad \frac{2\text{ mol }FeCl_3}{2\text{ mol }Fe(OH)_3}$$

$$\frac{2\text{ mol }FeCl_3}{3\text{ mol }MgCl_2} \qquad \frac{2\text{ mol }Fe(OH)_3}{3\text{ mol }MgCl_2}$$

Problem 4.17 In the reaction of aluminum hydroxide with sulfuric acid, how many moles of aluminum hydroxide are required to produce 3.200 mol of $Al_2(SO_4)_3$?

Refer to the conversion factors derived in Example 4.16 and choose the one that relates the two quantities in the question: $Al(OH)_3$ and $Al_2(SO_4)_3$:

$$\frac{2\text{ mol }Al(OH)_3}{1\text{ mol }Al_2(SO_4)_3}$$

The way in which you use the conversion factor depends on what you want: you are interested in the number of moles of aluminum hydroxide; aluminum hydroxide will go in the numerator:

$$3.200\ \cancel{\text{mol }Al_2(SO_4)_3} \times \frac{2\text{ mol }Al(OH)_3}{1\ \cancel{\text{mol }Al_2(SO_4)_3}} = 6.400\text{ mol }Al(OH)_3$$

Problem 4.18 How many moles of ammonia will be formed from 0.5000 mol of N_2 and 1.500 mol of H_2?

Refer to Example 4.18 to determine the balanced equation for the formation of ammonia from N_2 and H_2:

	N_2	+ $3\ H_2$	$\rightarrow$ $2NH_3$	
	1 mol	3 mol	2 mol	
given →	0.5000 mol	1.500 mol	→ ? mol	←wanted

The question you must ask is, Which of the reactants will determine how much ammonia will be formed? The reactant that is available in the least amount will determine how much ammonia will be formed; the reactant that is available in excess can only be used as long as the other reactant is around to combine with it. We know that 0.5000 mol of N_2 will require 1.500 mol of H_2 to react completely. These are the amounts available, so all the reactants will form products. Choose a conversion factor that relates either reactant to moles of ammonia.

$$\frac{1\text{ mol }N_2}{2\text{ mol }NH_3} \quad \text{or} \quad \frac{3\text{ mol }H_2}{2\text{ mol }NH_3}$$

These need to be in a form that puts what is wanted, ammonia, in the numerator.

$$0.5000 \text{ mol } N_2 \times \frac{2 \text{ mol } NH_3}{1 \text{ mol } N_2} = 1.000 \text{ mol } NH_3$$

$$1.500 \text{ mol } H_2 \times \frac{2 \text{ mol } NH_3}{3 \text{ mol } H_2} = 1.000 \text{ mol } NH_3$$

When 0.5000 mol of N_2 react with 1.500 mol of H_2, 1.000 mol of NH_3 is formed.

Problem 4.19 Given the reaction

$$K_3PO_4 + BaCl_2 \rightarrow Ba_3(PO_4)_2 + KCl$$

calculate the number of grams of $BaCl_2$ needed to react with 42.4 g of K_3PO_4, and the number of grams of $Ba_3(PO_4)_2$ produced.

First, balance the equation to determine the stoichiometric relationships:

$$2\ K_3PO_4 + 3\ BaCl_2 \rightarrow Ba_3(PO_4)_2 + 6\ KCl$$

Make up a table to show the stoichiometric relationships from the balanced equation in moles and in grams:

Reactant or product	**Moles**	**Grams/mole**	**Grams in balanced equation**
K_3PO_4	2	212.27	424.54
$BaCl_2$	3	208.23	624.69
$Ba_3(PO_4)_2$	1	601.93	601.93
KCl	6	74.55	447.30

The mass relationships can be used to develop conversion factors between grams of reactants and grams of products. (Refer to Example 4.2 and Problem 4.2 for calculations to convert formula units to mass.)

$$2\ K_3PO_4 + 3\ BaCl_2 \rightarrow Ba_3(PO_4)_2 + 6\ KCl$$

given → 42.4 g ? g ? g ← wanted

Develop conversion factors for K_3PO_4 and $BaCl_2$ from the balanced relationship and the table you generated:

$$\frac{2 \text{ mol } K_3PO_4}{3 \text{ mol } BaCl_2} \text{ is equivalent to } \frac{424.54 \text{ g } K_3PO_4}{624.69 \text{ g } BaCl_2}$$

$$\frac{2 \text{ mol } K_3PO_4}{1 \text{ mol } Ba_3(PO_4)_2} \text{ is equivalent to } \frac{424.54 \text{ g } K_3PO_4}{601.93 \text{ g } Ba_3(PO_4)_2}$$

$$42.4 \cancel{\text{ g } K_3PO_4} \times \frac{624.69 \text{ g } BaCl_2}{424.54 \cancel{\text{ g } K_3PO_4}} = 62.4 \text{ g } BaCl_2$$

$$42.4 \cancel{\text{ g } K_3PO_4} \times \frac{601.93 \text{ g } Ba_3(PO_4)_2}{\cancel{424.54 \text{ g } K_3PO_4}} = 60.1 \text{ g } Ba_3(PO_4)_2$$

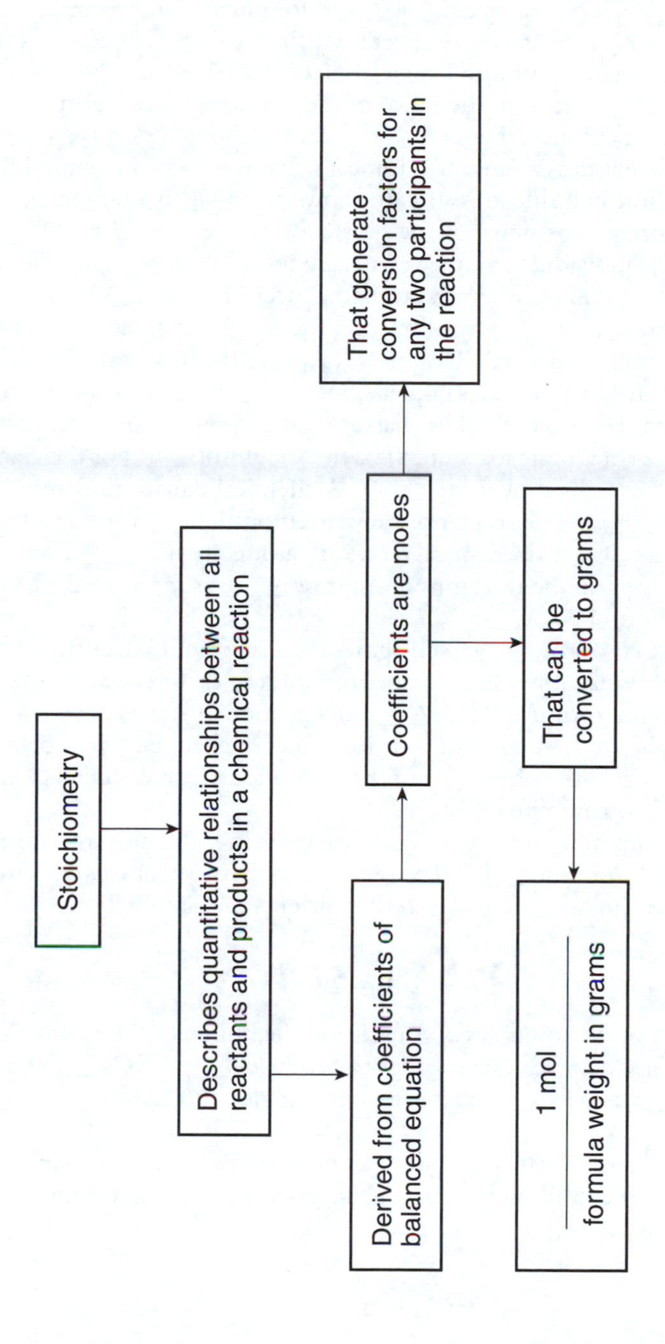
Stoichiometry
Describes quantitative relationships between all reactants and products in a chemical reaction
Derived from coefficients of balanced equation
Coefficients are moles
That generate conversion factors for any two participants in the reaction
That can be converted to grams
1 mol
formula weight in grams

FILL-INS

The chemical formula of a substance provides information about the _____ and _____ of atoms that make up the substance. The _____ indicates the type of atom and _____ to the right of the symbol tell how many atoms of the element are in a unit of the substance. Chemical formulas for molecules represent _____ of the substance. $C_6H_{12}O_6$, for example, is the _____ for glucose and represents one molecule of glucose. Chemical formulas for ionic substances represent one formula unit that contains the correct _____ of anions and cations in the substance. A formula unit of NaCl, for example, contains _____ sodium ion and _____ chloride, and is representative of the ratios of the two ions in the substance. $AlCl_3$ indicates that in the substance aluminum chloride there are _____ chloride ions for every _____ aluminum ion. This does not mean that $AlCl_3$ exists as a molecule; it means that the correct chemical formula will contain one aluminum ion and three chloride ions so as to show the _____ in a sample of the substance.

The formula weight, in atomic mass units, of any substance can be found by adding the _____ of its constituent atoms or ions. The individual atomic weights are found on the _____. These masses are not very useful when weighing out samples. Chemical reactions occur between _____ or _____, and in the lab samples are weighed out in _____; so a way of relating the weight of a sample in grams to the number of atoms or ions in the sample is needed. The _____ provides the link between a sample weight, in grams, and the number of atoms or ions in a sample. A mole is _____ atoms (or anything else!) and is equal to the _____ of a substance, in grams. The masses reported on the periodic table are either the weight of _____ of an element (in atomic mass units) or the weight of _____ of the element in grams.

Chemical reactions can be described by using _____. A chemical equation identifies the _____ that combine to form a specified _____. The reactants are written on the _____ of a right-pointing arrow and products are written on the _____. In chemical reactions, atoms are not _____, only rearranged. The number and type of atoms or ions on the reactant side must be _____ as those on the product side. When this is true, the equation is said to be _____.

Chemical reactions occur between _____,which allows the use of mole relationships to describe the quantitative aspects of chemical reactions. The number of moles of each reactant and product is identified by a in front of each chemical formula. When _____ an equation, it is the coefficients that are manipulated to achieve equality. The _____ of an atom or ion is never changed when balancing a reaction because a change in the subscript alters the _____ of the substance. Correct chemical formulas must be known in order to balance an equation properly.

The quantitative relationships among products and reactants in a chemical reaction is called _____. These relationships are derived from the _____. The relationship between any two participants in a reaction can be summarized by writing a _____ using the coefficients of the two. For example, if a chemical reaction is written

$$3\ A + 2\ B \rightarrow 2\ C$$

the relationship between the number of moles of A used and the number of moles of C produced is that _____ of A produces _____ of C; conversely, in order to produce _____ of C, _____ of A are required. The relationship between _____ and _____ is that every three moles of A require two moles of B in order to react completely.

The _____ can be used to determine the actual number of grams of reactants required or products produced in a reaction. The conversion of molar amounts to gram amounts is accomplished by using the relationship between _____, in grams.

Answers

The chemical formula of a substance provides information about the identity and number of atoms that make up the substance. The elemental symbol indicates the type of atom and subscripts to the right of the symbol tell how many atoms of the element are in a unit of the substance. Chemical formulas for molecules represent one discrete molecule of the substance. $C_6H_{12}O_6$, for example, is the chemical formula for glucose and represents one molecule of glucose. Chemical formulas for ionic substances represent one formula unit that contains the correct ratio of anions and cations in the substance. A formula unit of NaCl, for example, contains one sodium ion and one chloride, and is representative of the ratios of the two ions in the substance. $AlCl_3$ indicates that in the substance aluminum chloride there are three chloride ions for every one aluminum ion. This does not mean that $AlCl_3$ exists as a molecule; it means that the correct chemical formula will contain one aluminum ion and three chloride ions so as to show the ratio of cations to anions in a sample of the substance.

The formula weight, in atomic mass units, of any substance can be found by adding the individual atomic weights of its constituent atoms or ions. The individual atomic weights are found on the periodic table. These masses are not very useful when weighing out samples. Chemical reactions occur between atoms or ions, and in the lab samples are weighed out in grams, so a way of relating the weight of a sample in grams to the number of atoms or ions in the sample is needed. The mole provides the link between a sample weight, in grams, and the number of atoms or ions in a sample. A mole is 6.022×10^{23} atoms (or anything else!) and is equal to the formula weight of a substance, in grams. The masses reported on the periodic table are either the weight of one atom of an element (in atomic mass units) or the weight of 1 mol of the element in grams.

Chemical reactions can be described by using chemical equations. A chemical equation identifies the reactants that combine to form a specified product. The reactants are written on the left side of a right-pointing arrow and products are written on the right side. In chemical reactions, atoms are not created or destroyed, only rearranged. The number and type of atoms or ions on the reactant side must be the same as those on the product side. When this is true, the equation is said to be balanced.

Chemical reactions occur between whole atoms or ions, which allows the use of mole relationships to describe the quantitative aspects of chemical reactions. The number of moles of each reactant and product is identified by a coefficient in front of each chemical formula. When balancing an equation, it is the coefficients that are manipulated to achieve equality. The subscript of an atom or ion is never changed when balancing a reaction because a change in the subscript alters the identity of the substance. Correct chemical formulas must be known in order to balance an equation properly.

The quantitative relationships among products and reactants in a chemical reaction is called stoichiometry. These relationships are derived from the coefficients in the balanced equation. The relationship between any two participants in a reaction can be summarized by writing a conversion factor using the coefficients of the two. For example, if a chemical reaction is written

$$3\text{ A} + 2\text{ B} \rightarrow 2\text{ C}$$

the relationship between the number of moles of A used and the number of moles of C produced is that three moles of A produces two moles of C; conversely, in order to produce two moles of C, three moles of A are required. The relationship between the number of moles of A used and the number of moles of B used is that every three moles of A require two moles of B in order to react completely.

The mole relationships can be used to determine the actual number of grams of reactants required or products produced in a reaction. The conversion of molar amounts to gram amounts is accomplished by using the relationship between one mole of the substance and its molar mass, in grams.

TEST YOURSELF

1. Identify the types and numbers of each element in the list below. Write the empirical formula for each.
   (a) 2 mol of $Al(OH)_3$ (aluminum hydroxide, an antacid)
   (b) 3 mol of CH_2O (formaldehyde)
   (c) 1 mol of H_3CCH_2OH (ethanol, grain alcohol)
   (d) 2 mol of $(NH_4)_2CO_3$ (ammonium carbonate, an expectorant)
   (e) 2 mol of $C_6H_{12}O_6$ (glucose)
2. Balance the following equation:

$$C_6H_{12}O_6 + O_2 \rightarrow CO_2 + H_2O$$

   Based on the balanced equation:
   (a) Write a conversion factor for the number of moles of glucose reacted and the number of moles of carbon dioxide produced.
   (b) Write a conversion factor for the number of moles of oxygen reacted and the number of moles of glucose reacted.
   (c) Write a conversion factor for the grams of glucose reacted and the mass of water produced.

(d) How much oxygen would be required to react completely with 3.60×10^2 g of glucose?
(e) What type of reaction is this?

3. Indicate whether the following statements are correct or incorrect. If incorrect, explain why.
   (a) The subscripts in a chemical formula indicate the number of grams of the element in a substance.
   (b) The empirical formula of a substance is the same as the chemical formula of that substance.
   (c) The chemical formula describes a molecule's shape as well as its composition.
   (d) One mole of carbon weighs more than one mole of hydrogen.
   (e) One mole of carbon contains more atoms than one mole of hydrogen.
   (f) Oxidation reactions always involve oxygen.
   (g) Reduction reactions are always coupled with oxidation reactions.
4. Write the reactions described below and identify the element being oxidized. Make sure they are balanced.
   (a) methane (CH_4) burns in air to produce carbon dioxide and water
   (b) lead metal reacts with oxygen in air to produce lead oxide (refer to Table 3.1)
   (c) copper sulfate, $CuSO_4$, reacts with Zn metal in solution to form copper metal and zinc sulfate, $ZnSO_4$

Answers

1. (a) 2 mol of $Al(OH)_3$; 2 mol aluminum atoms, 6 mol oxygen atoms, 6 mol hydrogen atoms; empirical formula: $Al(OH)_3$
   (b) 3 mol of CH_2O; 3 mol carbon atoms, 6 mol hydrogen atoms, 3 mol oxygen atoms; empirical formula: CH_2O
   (c) 1 mol of H_3CCH_2OH; 6 mol hydrogen atoms, 2 mol carbon atoms, 1 mol oxygen atoms; empirical formula: C_2H_6O
   (d) 2 mol of $(NH_4)_2CO_3$; 4 mol nitrogen atoms, 16 mol hydrogen atoms, 2 mol carbon atoms, 6 mol oxygen atoms; empirical formula: $(NH_4)_2CO_3$
   (e) 2 mol of $C_6H_{12}O_6$; 12 mol carbon atoms, 24 mol hydrogen atoms, 12 mol oxygen atoms; empirical formula: CH_2O
2. $C_6H_{12}O_6 + 6\ O_2 \rightarrow 6\ CO_2 + 6\ H_2O$
   (a) 1 mol $C_6H_{12}O_6$ /6 mol CO_2
   (b) 6 mol O_2 /1 mol $C_6H_{12}O_6$
   (c) 180.18 g $C_6H_{12}O_6$/108.12 g H_2O
   (d)

$$3.60 \times 10^2\ \cancel{\text{g } C_6H_{12}O_6} \times \frac{192.00 \text{ g } O_2}{180.18\ \cancel{\text{g } C_6H_{12}O_6}} = 384 \text{ g } O_2$$

   (e) combustion
3. (a) Incorrect. The subscripts indicate the number of atoms in a molecule or the number of ions in a formula unit or the number of moles of that element in one mole of the substance.
   (b) Sometimes, but not always (see question 1).
   (c) Incorrect. The chemical formula does not provide any information about the shape of a molecule.
   (d) Correct. One mole of carbon weighs 12.01 g and one mole of hydrogen weighs 1.01 grams.
   (e) Incorrect. One mole of carbon and one mole of hydrogen both contain the same number of atoms: 6.022×10^{23}.
   (f) Incorrect. Sometimes they involve a loss of electrons or a loss of hydrogen.
   (g) Correct. Oxidation and reduction reactions are always coupled.
4. (a) $CH_4 + 2\ O_2 \rightarrow CO_2 + 2\ H_2O$, carbon is oxidized
   (b) $2\ Pb + O_2 \rightarrow 2\ PbO$, lead is oxidized
   (c) $CuSO_4 + Zn \rightarrow Cu + ZnSO_4$, zinc is oxidized

chapter 5

The Physical Properties of Gases

OUTLINE

5.1 Gas Pressure

A. Physical properties of gases include
   1. volume
   2. temperature
   3. pressure

B. Pressure
   1. is defined as a force applied over a unit area; mathematically,

$$\text{Pressure} = \frac{\text{force}}{\text{area}}$$

   2. results from collisions of gas molecules with a surface; anything that affects rate or number of collisions affects gas pressure
      a. increase number of molecules in fixed volume container
      b. increase temperature
   3. is measured in a variety of units
      a. atmospheres (atm)
      b. torr
      c. millimeters of mercury (mm Hg)
      d. pounds per square inch (psi)
      e. units can be converted by using relationships:

$$1 \text{ atm} = 760 \text{ mm Hg} = 760 \text{ torr}$$

   4. differences cause gas movement down pressure gradients

C. Atmospheric pressure
   1. weight of atmospheric gases pressing against earth's surface
   2. due mostly to nitrogen and oxygen gases: 79% nitrogen, 20% oxygen, 1% other
   3. varies by geographic location and temperature
      a. higher elevations have lower atmospheric pressure than low areas
      b. colder air is heavier and produces higher atmospheric pressure than warm air
   4. is standardized as the pressure that will support a column of Hg to 760 mm at 0°C: defined as 1 atmosphere (1 atm)

D. Pressure is measured using
   1. barometer: measures atmospheric pressure
   2. manometer: measures pressure in a closed container
   3. gas gauges: measures pressure in a gas cylinder

Key Terms atm, atmosphere, atmospheric pressure, barometer, manometer, mm Hg, pressure, torr

Key Figures, Boxes, and Table

Figure 5.1 Mercury barometer measuring atmospheric pressure
Figure 5.2 Differential manometer measuring pressures of two gases
Figure 5.3 Gas gauge on oxygen cylinder
Box 5.1 Diving Time and Gas Pressure
Box 5.2 Manometry and Blood Pressure
Table 5.1 Illustrates universal behavior of gases
Table 5.2 Summarizes Boyle's, Charle's, Gay-Lussac's, and Avogadro's laws

Example 5.1 Converting between atmospheres, mm of mercury, and torr
Problem 5.1 Converting between atmospheres, mm of mercury, and torr

5.2 *The Gas Laws*

A. Physical properties of gases depend on
   1. pressure (P)
   2. volume (V)
   3. Kelvin temperature (T):

$$\text{Kelvin } T = 273 + {}^{\circ}\text{C}$$

   4. number of moles (n)

B. Gas laws examine changes in gas properties when these are varied.
   1. relationship between pressure and volume when T and n are constant: Boyle's law
   2. relationship between volume and temperature when P and n are constant: Charles's law
   3. relationship between temperature and pressure when V and n are constant: Gay-Lussac's law
   4. relationship between volume and number of moles when T and P are constant: Avogadro's law

Key Terms kelvin, Kelvin temperature, mol, mole, pressure, volume

Key Figures

Figure 5.4 Changes in the volume of a gas as pressure is changed
Figure 5.5 Change in volume when temperature is changed
Figure 5.6 Effect of changing both temperature and pressure of a fixed sample of gas
Figure 5.7 Change in volume when more gas molecules are added to sample of gas

5.3 *Boyle's Law*

A. Boyle's law examines the relationship between pressure and volume when T and n are constant.
   1. as the volume of a sample decreases, the pressure increases because the gas molecules are crowded into smaller space and have more collisions with the container surface
   2. as the volume of a sample increases, the pressure decreases because the gas molecules spread out over a larger area and have fewer collisions with the container wall
   3. the product of a gas's pressure multiplied by its volume is a constant (Table 5.2).
      a. mathematically,

$$P \times V = \text{constant}$$

      b. inverse relationship
         i. as pressure decreases, volume increases
         ii. as pressure increases, volume decreases
      c. can be written as

$$P_1 \times V_1 = P_2 \times V_2$$

B. Boyle's law can be used to represent two experimental situations.
   1. what happens to the volume when pressure is varied (Ex. 5.2, Prob. 5.2)

$$V_2 = V_1\,(P_1/P_2)$$

2. what happens to the pressure when volume is varied (Ex. 5.3, Prob. 5.3)

$$P_2 = P_1 (V_1/V_2)$$

Key Terms Boyle's law, inverse relationship

Key Figure and Table

Figure 5.4 Changes in the volume of a gas as pressure is changed

Table 5.3 Relationship between volume and pressure at constant T and n

Example 5.2 Using Boyle's law to calculate the final volume of a gas sample as a result of a pressure change at constant temperature

Problem 5.2 Using Boyle's law to calculate the final volume of a gas sample as a result of a pressure change at constant temperature

Example 5.3 Using Boyle's law to calculate the final pressure of a gas sample as a result of a volume change at constant temperature

Problem 5.3 Using Boyle's law to calculate the final pressure of a gas sample as a result of a volume change at constant temperature

5.4 *Charles's Law*

A. Charles's law examines the relationship between volume and temperature when P and n are constant.
   1. as the temperature of a sample increases, the gas molecules move faster and the volume increases
      a. causes more collisions with container surface per unit time
      b. think of a balloon being heated
   2. as the temperature of a sample decreases, the gas molecules move more slowly and the volume decreases
      a. results in fewer collisions with container wall
      b. think of a balloon being chilled
   3. the ratio of a gas's volume to its Kelvin temperature is a constant (Table 5.4)
      a. mathematically

$$V/T = \text{constant can be rewritten} \quad \text{or} \quad V = \text{constant} \times T$$

      b. direct relationship
         i. as temperature increases, volume increases
         ii. as temperature decreases, volume decreases
      c. can be written as

$$V_1/T_1 = V_2/T_2$$

B. Charles's law can be used to represent two experimental situations.
   1. what happens to the volume when temperature is varied (Ex. 5.4, Prob. 5.4)

$$V_2 = V_1 (T_2/T_1)$$

   2. what happens to the temperature when volume is varied (Ex. 5.5, Prob. 5.5)

$$T_2 = T_1 (V_2/V_1)$$

Key Terms Charles's law, direct relationship, Kelvin temperature,

Key Figure and Table

Figure 5.5 Change in volume when temperature is changed

Table 5.4 Variation of gas volume with temperature change at constant P and n

Example 5.4 Using Charles's law to calculate the final volume of a gas sample when the temperature is changed at a constant pressure

Problem 5.4 Using Charles's law to calculate the final volume of a gas sample when the temperature is changed at a constant pressure

Example 5.5 Using Charles's law to calculate the final temperature of a gas sample when the volume is changed at a constant pressure

Problem 5.5 Using Charles's law to calculate the final temperature of a gas sample when the volume is changed at a constant pressure

5.5 *Gay-Lussac's Law*

A. Gay-Lussac's law examines the relationship between temperature and pressure when V and n are constant
   1. as the temperature of a sample increases, the pressure increases and the gas molecules move faster and have more collisions with container surface per unit time
   2. as the temperature of a sample decreases, the pressure decreases and the gas molecules move more slowly and have fewer collisions with container wall
   3. the ratio of a gas's pressure to its Kelvin temperature is a constant (Table 5.5)
      a. mathematically

$$P/T = \text{constant} \quad \text{or} \quad P = \text{constant} \times T$$

      b. direct relationship
         i. as temperature increases, pressure increases
         ii. as temperature decreases, pressure decreases
      c. can be written as

$$P_1/T_1 = P_2/T_2$$

B. Gay-Lussac's law can be used to represent two experimental situations.
   1. what happens to the pressure when temperature is varied (Ex. 5.6, Prob. 5.6)

$$P_2 = P_1\,(T_2/T_1)$$

   2. what happens to the temperature when pressure is varied (Ex. 5.7, Prob. 5.7)

$$T_2 = T_1\,(P_2/P_1)$$

Key Terms direct relationship, Gay-Lussac's law, Kelvin temperature

Key Figure, Table, and Box

Figure 5.6 To keep volume of a fixed amount of gas constant when temperature is changed, external pressure must be changed

Table 5.5 Variation of pressure and temperature at constant V and n

Box 5.4 Sterilization and the Autoclave

Example 5.6 Using Gay-Lussac's law to calculate the pressure of a gas when the temperature is changed at constant volume

Problem 5.6 Using Gay-Lussac's law to calculate the pressure of a gas when the temperature is changed at constant volume

Example 5.7 Using Gay-Lussac's law to calculate the temperature of a gas when the pressure is changed at constant volume

Problem 5.7 Using Gay-Lussac's law to calculate the temperature of a gas when the pressure is changed at constant volume

5.6 *Avogadro's Law*

A Avogadro's law examines the relationship between equal volumes of gas and number of molecules of gas when both samples are at the same temperature and pressure.
   1. at the same temperature and pressure, equal volumes of gas contain the same number of molecules
   a. can be any type of gas molecules
   b. basis for determining molar mass from a measured gas mass

2. the ratio of a gas's volume to the number of moles is a constant
   a. mathematically

$$V/n = \text{constant} \quad \text{or} \quad V = \text{constant} \times n$$

   b. direct relationship
      i. as number of moles increases, volume increases
      ii. as number of moles decreases, volume decreases
   c. can be written as

$$V_1/n_1 = V_2/n_2$$

B. Avogadro's law can be used to represent two experimental situations.
   1. what happens to the volume when the number of moles is varied

$$V_2 = V_1\,(n_2/n_1)$$

   2. what number of moles would be contained in a different volume

$$n_2 = n_1\,(V_2/V_1)$$

Key Term Avogadro's law
Key Figure Figure 5.7 Effect on volume when amount of gas is increased and temperature and pressure remain the same

5.7 *The Combined Gas Law*

A. Combined gas law examines the volume changes that occur when a fixed amount of gas is subjected to both a pressure and a temperature change.
   1. pressure and volume are inversely related: as one increases, the other decreases
   2. temperature and volume are directly related: as one increases, the other increases
   3. the ratio of a gas's volume multiplied by its pressure to the temperature is a constant
      a. mathematically

$$PV/T = \text{constant can be rewritten} \quad \text{or} \quad PV = \text{constant} \times T \quad \text{or} \quad T = PV/\text{constant}$$

      b. can be written as

$$P_1V_1/T_1 = P_2V_2/T_2$$

B. Combined gas law can be used to examine three experimental situations.
   1. calculating the new pressure of a gas when both the temperature and volume are simultaneously changed

$$P_2 = P_1V_1T_2/T_1V_2$$

   2. calculating the new temperature of a gas when both the pressure and volume are simultaneously changed

$$T_2 = P_2V_2T_1/P_1V_1$$

   3. calculating the new volume of a gas when both the pressure and temperature are simultaneously changed

$$V_2 = P_1V_1T_2/P_2T_1$$

Key Term combined gas law

Example 5.8 Using the combined gas law to calculate the volume of a fixed amount of gas when both the temperature and pressure are changed

Problem 5.8 Using the combined gas law to calculate the volume of a fixed amount of gas when both the temperature and pressure are changed

5.8 *The Ideal Gas Law*

A. Ideal gas law relates the four variables that describe gas properties.
   1. states that the ratio of gas pressure multiplied by its volume to amount of gas multiplied by its Kelvin temperature is a constant; mathematically,

$$PV/nT = \text{constant}$$

   2. usually written as

$$PV = nRT$$

   3. called the ideal gas law

B. Constant is given symbol R, which
   1. has units of liter × atm / mole × K
      a. rewritten as $\text{L·atm·K}^{-1}\text{·mol}^{-1}$
      b. units of R depend on units used to measure pressure and volume
   2. is determined at standard temperature and pressure (STP)
      a. standard temperature is 273 K
      b. standard pressure is 1 atm
      c. at STP, R has value of 0.0821 $\text{L·atm·K}^{-1}\text{·mol}^{-1}$
      d. at STP, 1.00 mol of any gas occupies 22.4 L

C. Ideal gas law can be used to determine
   1. the pressure of a gas when all other variables are known

$$P = nRT/V$$

   2. the volume of a gas when all other variables are known

$$V = nRT/P$$

   3. the temperature of a gas when all other variables are known

$$T = PV/nR$$

   4. the number of moles of gas when all other variables are known

$$n = PV/RT$$

   5. the mass of a gas sample from the number of moles (Ex. 5.9, Prob. 5.9)

Key Terms ideal gas law, gas constant, standard pressure, standard temperature, STP, universal gas constant

Example 5.9 Using the ideal gas law to calculate the mass of a gas sample
Problem 5.9 Using the ideal gas law to calculate the mass of a gas sample

5.9 *The Ideal Gas Law and Molar Mass*

A. Molar mass can be determined by using the ideal gas law and substituting for moles.
   1. molar mass (M) is grams/mole

$$\text{M} = \text{g/mol}$$

   2. relationship between moles and molar mass is

$$\text{mol} = \text{g/M}$$

   3. substitute g/M for n in the ideal gas equation

$$PV = (\text{g/M})\ RT$$

   4. used to solve for either grams or molar mass

B. Density can be related to molar mass through the ideal gas law.
   1. when n = g/M is substituted in ideal gas law

$$PV = (\text{g/M})\ RT$$

2. rewrite equation to get a density term: grams/volume

$$M = gRT/PV$$
$$M = (g/V)\ RT/P$$

3. used to determine the molar mass of a gas sample when the density and other variables are known (Ex. 5.10, Prob. 5.10)

$$M = (g/V)\ RT/P$$

Key Term molar mass

Example 5.10 Using the ideal gas law to calculate the molar mass of a gas sample when its density, temperature, and pressure are given
Problem 5.10 Using the ideal gas law to calculate the molar mass of a gas sample when its density, temperature, and pressure are given
Example 5.11 Using Avogadro's law to calculate the molar mass of a gas sample when its volume, temperature, mass, and pressure are given
Problem 5.11 Calculating the molar mass of a given amount of gas sample at STP

5.10 Dalton's Law of Partial Pressures

A. Dalton's law states that the total pressure of a gas sample is the result of the contributions of the individual pressures of the gases in the sample.
   1. individual gases exert partial pressures, P_{gas}
   2. partial pressure is the pressure that the gas would exert if it were the only gas in sample
   3. mathematically,

$$P_{total} = P_{gas\ a} + P_{gas\ b} + P_{gas\ c} + \ldots$$

B. Dalton's law can be used to
   1. determine total pressure of a system when partial pressures are known

$$P_{total} = P_{gas\ a} + P_{gas\ b} + P_{gas\ c}$$

   2. determine the partial pressure of a gas when the total pressure and other partial pressures are known

$$P_{total} - P_{gas\ a} = P_{gas\ b}$$

   3. determine partial pressures when total pressure and percentages of individual gases are known

$$P_{total}\ (\%\ \text{gas A}) = P_{gas\ a}$$

Key Terms Dalton's law, partial pressure, total pressure
Key Figure Figure 5.8 Collection of gas over water

Example 5.12 Using Dalton's law to calculate the partial pressure of a gas in a sample when the total pressure and other partial pressures are given
Problem 5.12 Calculating the partial pressures of gases in a sample from their relative ratios and the total pressure
Example 5.13 Calculating the partial pressure of a gas collected over water when water vapor contributes to the total pressure of the sample
Problem 5.13 Using Dalton's law to calculate the partial pressure of a gas sample collected over water and using the ideal gas equation to determine its volume

5.11 Gases Dissolve in Liquids

A. Henry's law states that the amount of gas that will dissolve in a liquid depends strongly on the pressure of the gas above the liquid.
   1. as gas pressure increases, more gas will dissolve

2. for gas mixtures, as the partial pressure of a gas in the mixture increases, more of that gas will dissolve in a liquid
3. it is assumed that there is no chemical reaction between the gas and the solvent

B. The solubility of a gas in a liquid depends on temperature.
1. as the temperature of a liquid increases, the solubility of gas in the liquid decreases
2. as the temperature of a liquid decreases, the solubility of gas in the liquid increases

C. Henry's law constant provides a measure of the solubility of a particular gas at a defined temperature.
1. mathematically,

$$C_H = \text{amount of gas dissolved/unit volume of solvent}$$

2. can be used to determine the amount of gas dissolved in a liquid at a particular temperature by using the equation:

$$\text{mL gas/mL solvent} = C_H \times P_{gas}$$

D. Clinically, gas solubility in blood is called gas tension.

Key Terms gas tension, Henry's law, Henry's law constant, solubility, solvent

Key Figure and Box

Figure 5.9 Solubility of gases in water as temperature is varied

Box 5.5 Gas Solubility and Caisson Disease (the Bends)

Example 5.14 Using Henry's law constant to calculate the amount of gas dissolved in a liquid at a given temperature and pressure

Problem 5.14 Using Henry's law constant to calculate the amount of gas dissolved in a liquid at a given temperature and pressure

Example 5.15 Using Henry's law for a mixture of gases

Problem 5.15 Using Henry's law for a mixture of gases

ARE YOU ABLE TO... AND WORKED TEXT PROBLEMS

Objectives Section 5.1 Gas Pressure

Are you able to. . .

- identify the variables that describe the physical properties of gases
- explain what pressure is and the factors that influence it
- provide the mathematical definition of pressure
- identify the common units of pressure measurement
- convert between atmospheres, torr, and mm Hg (Exercises 5.1, 5.2, 5.3, 5.4, 5.5, 5.6, 5.51)
- define atmospheric pressure
- explain how and why atmospheric pressure varies
- describe the composition of atmospheric gases
- describe the standard unit of measurement of atmospheric pressure
- explain how pressure is measured by a barometer and a manometer (Figs. 5.1, 5.2)

Problem 5.1 (a) Calculate the pressure, in atmospheres, that a gas exerts if it supports a 563-mm column of Hg.

Choose the relationship that includes both atm and mm Hg: 1 atm = 760 mm Hg

$$563 \, \cancel{\text{mm Hg}} \times \frac{1 \text{ atm}}{760 \, \cancel{\text{mm Hg}}} = 0.741 \text{ atm}$$

(b) Calculate the equivalent of 0.930 atm in mm Hg and torr.

Both mm Hg and torr have the same numerical value when related to 1 atm: 760 mm Hg and 760 torr

$$0.930 \, \cancel{\text{atm}} \times \frac{760 \text{ mm Hg}}{1 \, \cancel{\text{atm}}} = 707 \text{ mm Hg}$$

$$0.930 \, \cancel{\text{atm}} \times \frac{760 \text{ torr}}{1 \, \cancel{\text{atm}}} = 707 \text{ torr}$$

Objectives Section 5.2 The Gas Laws

Are you able to. . .

- describe the four variables that define the physical properties of a gas
- provide the appropriate abbreviations for these variables
- convert degrees Celsius or Fahrenheit to kelvins
- explain Boyle's law in qualitative terms (Fig. 5.4)
- describe the relationships that can be examined by using Boyle's law (Exercise 5.7)
- explain Charles's law in qualitative terms (Fig. 5.5)
- describe the relationships that can be examined by using Charles's law (Exercise 5.8)
- explain Gay-Lussac's law in qualitative terms (Fig. 5.6)
- describe the relationships that can be examined by using Gay-Lussac's law (Exercise 5.9)
- explain Avogadro's law in qualitative terms (Fig. 5.7)
- describe the relationships that can be examined by using Avogadro's law (Exercise 5.10)

Objectives Section 5.3 Boyle's Law

Are you able to. . .

- explain the relationship between pressure and volume as stated in Boyle's law (Table 5.3)
- predict qualitatively how a volume will respond to a pressure change (Fig. 5.4)
- predict qualitatively how gas pressure will respond to a volume change
- explain what happens at the molecular level in each of the cases above
- explain what is meant by an inverse relationship
- define the mathematical relationship between pressure and volume
- calculate what happens to the volume when pressure is varied (Ex. 5.2, Prob. 5.2, Exercise 5.13)
- calculate what happens to the pressure when volume is varied (Ex. 5.3, Prob. 5.3, Exercise 5.11)

Problem 5.2 An 862-mL sample of gas at 425 torr is compressed at constant temperature until its final pressure is 901 torr. What is its final volume?

First, imagine what will happen qualitatively as the gas is compressed. Will the final volume be larger or smaller than the starting volume? It should be smaller. The final pressure is a bit more than double the initial pressure, so the final volume should be less than half of the initial volume.

List the information given:

initial pressure: 425 torr	final pressure: 901 torr
initial volume: 862 mL	final volume: ???

Set up the Boyle's law relationship to solve for the second volume:

$$P_1 \times V_1 = P_2 \times V_2$$

$$V_2 = \frac{P_1 \times V_1}{P_2}$$

Plug in given values:

$$V_2 = \frac{(425\ \cancel{\text{torr}})\ (862\ \text{mL})}{901\ \cancel{\text{torr}}} = 407\ \text{mL}$$

This answer is in the range that was predicted.

Problem 5.3 A 1.80-L sample of gas at 0.739 atm must be compressed to 1.40 L at constant temperature. What pressure in atmospheres must be exerted to bring it to that volume?

First, imagine what will happen qualitatively as the gas is compressed. A pressure has to be applied to decrease the volume by 0.40 L. The final pressure needed will be less than twice the original pressure.

List the information given:

initial pressure: 0.739 atm	final pressure: ???
initial volume: 1.80 L	final volume: 1.40 L

Set up the Boyle's law relationship to solve for the second pressure:

$$P_1 \times V_1 = P_2 \times V_2$$

$$P_2 = \frac{P_1 \times V_1}{V_2}$$

Plug in given values:

$$P_2 = \frac{(0.739 \text{ atm})(1.80 \not{L})}{1.40 \not{L}} = 0.950 \text{ atm}$$

This answer is in the predicted range.

Objectives Section 5.4 Charles's Law

Are you able to. . . .

- explain the relationship between temperature and volume as stated in Charles's law (Table 5.4)
- predict qualitatively how a volume will respond to a temperature change (Fig. 5.5)
- predict qualitatively how gas temperature will respond to a volume change
- explain what happens at the molecular level in each of the cases above
- explain what is meant by an direct relationship
- define the mathematical relationship between temperature and volume
- calculate what happens to the volume when temperature is varied (Ex. 5.4, Prob. 5.4, Exercise 5.12)
- calculate what happens to the temperature when volume is varied (Ex. 5.5, Prob. 5.5, Exercise 5.16)

Problem 5.4 A 755-mL sample of a gas, in a cylinder with a movable piston, at 3.00°C is heated at a constant pressure of 0.800 atm at 142°C. What is its final volume?

First, imagine what will happen qualitatively as the gas is heated. The volume will be able to change because there is a movable piston. Heating results in more activity at the molecular level and more collisions per unit area. If the pressure is kept constant, the gas responds by expanding. Remember to convert °C to kelvins. The Kelvin temperature is not quite doubled, so the volume should increase, but not quite double.

List the information given:

initial temperature: 3.00°C	final temperature: 142°C
Kelvin temp: 273 + 3.00 = 276 K	Kelvin temp: 273 + 142 = 415 K
initial volume: 755 mL	final volume: ???

Set up the Charles's law relationship to solve for the second volume:

$$V_1/T_1 = V_2/T_2$$

$$V_2 = \frac{T_2 \times V_1}{T_1}$$

Substitute given values:

$$V_2 = \frac{(415 \not{K})(755 \text{ mL})}{276 \not{K}} = 1135 \text{ mL} = 1.14 \times 10^3 \text{ mL or } 1.14 \text{ L}$$

This answer is in the predicted range.

Problem 5.5 A 0.630-L sample of gas at 16.0°C is heated so that it expands at constant pressure to a final volume of 1.35 L. Calculate its final temperature.

First, imagine what will happen qualitatively as the gas is heated; it responds by expanding. The final volume is a little more than double the initial volume, so the final temperature should be a little more than double the initial temperature. Remember to convert °C to kelvins.

List the information given:

initial temperature: 16.0°C	final temperature: ???
Kelvin temp: 273 + 16.0 = 289 K	
initial volume: 0.630 L	final volume: 1.35 L

Set up the Charles's law relationship to solve for the second temperature:

$$V_1/T_1 = V_2/T_2$$

$$T_2 = \frac{T_1 \times V_2}{V_1}$$

Plug in given values:

$$T_2 = \frac{(289\ \text{K})\ (1.35\ \cancel{\text{L}})}{0.630\ \cancel{\text{L}}} = 619\ \text{K or } 619\ \text{K} - 273 = 346°\text{C}$$

This answer is in the predicted range.

Objectives Section 5.5 Gay-Lussac's Law

Are you able to. . . .

- explain the relationship between temperature and pressure as stated in Gay-Lussac's law (Table 5.5)
- predict qualitatively how a pressure will respond to a temperature change (Fig. 5.6)
- predict qualitatively how gas temperature will respond to a pressure change
- explain what happens at the molecular level in each of the cases above
- define the mathematical relationship between temperature and pressure
- calculate what happens to the pressure when temperature is varied (Ex. 5.6, Prob. 5.6, Exercise 5.54)
- calculate what happens to the temperature when pressure is varied (Ex. 5.7, Prob. 5.7)

Problem 5.6 Calculate the final pressure of a sample of gas in a fixed-volume cylinder at 1.25 atm and 14.0°C when the gas is heated to 125°C.

First, imagine what will happen qualitatively as the gas is heated. The gas cannot expand, so there will be more collisions against the container surface as gas molecules become more active. This increases the pressure. Remember to convert °C to kelvins. The temperature increases by about a third, so the pressure should increase by about a third.

List the information given:

initial temperature: 14.0°C
Kelvin temp: 273 + 14.0 = 287 K
initial pressure: 1.25 atm

final temperature: 125°C
Kelvin temp: 273 + 125 = 398 K
final pressure: ???

Set up the Gay-Lussac's law relationship to solve for the second pressure:

$$P_1/T_1 = P_2/T_2$$

$$P_2 = \frac{P_1 \times T_2}{T_1}$$

Plug in given values:

$$P_2 = \frac{(1.25\ \text{atm})\ (398\ \cancel{\text{K}})}{287\ \cancel{\text{K}}} = 1.73\ \text{atm}$$

This answer is within the predicted range.

Problem 5.7 A constant-volume sample of gas at 3.00°C and 0.650 atm is heated so that its pressure increases to a final value of 2.15 atm. What is its final temperature?

First, imagine what will happen qualitatively as the gas is heated. The gas cannot expand, so there will be more collisions against the container surface as gas molecules become more active. This increases the pressure. Remember to convert °C to kelvins. The pressure increases by a little more than three times the initial pressure, so the temperature should increase by about the same amount.

List the information given:

initial temperature: 3.00°C
Kelvin temp: 273 + 3.00 = 276 K
initial pressure: 0.650 atm

final temperature: ???
final pressure: 2.15 atm

Set up the Gay-Lussac's law relationship to solve for the second temperature:

$$P_1/T_1 = P_2/T_2$$

$$T_2 = \frac{P_2 \times T_1}{P_1}$$

Plug in given values:

$$T_2 = \frac{(2.15\ \cancel{\text{atm}})\ (276\ \text{K})}{0.650\ \cancel{\text{atm}}} = 913\ \text{K or } 913\text{K} - 273 = 640°\text{C}$$

This answer is within the predicted range.

Objectives Section 5.6 Avogadro's Law

Are you able to. . . .

- explain the relationship between gas volumes and number of moles of gas as stated in Avogadro's law (Table 5.2, Fig. 5.7)
- define the mathematical relationship between equal volumes of gas and number of moles
- calculate what happens to the volume when the number of moles is varied (Exercise 5.25)
- calculate the number of moles that would be contained in a different volume than the one given

Objectives Section 5.7 The Combined Gas Law

Are you able to. . . .

- explain the relationship between temperature, pressure, and volume as stated in the combined gas law
- predict qualitatively how a volume will respond to a temperature change and pressure change
- predict qualitatively how gas pressure will respond to a pressure change and a volume change
- predict qualitatively how a gas temperature will respond to a pressure change and a volume change
- explain what happens at the molecular level in each of the cases above
- define the mathematical relationship between temperature, pressure, and volume
- calculate what happens to the volume when pressure and temperature are varied simultaneously (Ex. 5.8, Prob. 5.8, Exercises 5.14, 5.19, 5.20, 5.21, 5.22, 5.43, 5.45)
- calculate what happens to the temperature when pressure and volume are varied simultaneously (Exercises 5.15, 5.40, 5.41)
- calculate what happens to pressure when volume and temperature are varied simultaneously (Exercises 5.17, 5.18, 5.42, 5.44)

Problem 5.8 A 1.76-L sample of gas at 6.00°C and 0.920 atm pressure is heated to 254°C and 2.75 atm final pressure. What is its final volume?

First, imagine what will happen qualitatively as the gas is both heated and the pressure is increased. An increase in temperature will cause an expansion of the gas. An increase in pressure will cause a compression of the gas. The question is, Which effect will dominate and what will the final effect be on the volume of the gas? The final pressure is about three times the initial pressure, which would compress the gas to about one-eighth of its original volume. The final temperature is about twice the initial temperature, which would approximately double the volume. Overall, the gas volume should decrease. Remember to convert °C to kelvins.

List the information given:

initial temperature: 6.00°C	final temperature: 254°C
Kelvin temp: 273 + 6.00 = 279 K	Kelvin temp: 273 + 254 = 527 K
initial pressure: 0.920 atm	final pressure: 2.75 atm
initial volume: 1.76 L	final volume: ???

Set up the combined gas relationship to solve for the second volume:

$$\frac{P_1 V_1}{T_1} = \frac{P_2 V_2}{T_2}$$

$$V_2 = \frac{(P_1)\ (V_1)\ (T_2)}{(P_2)\ (T_1)}$$

Plug in given values:

$$V_2 = \frac{(0.920\ \cancel{\text{atm}})\ (1.76\ \text{L})\ (527\ \cancel{\text{K}})}{(2.75\ \cancel{\text{atm}})\ (279\ \cancel{\text{K}})} = 1.11\ \text{L}$$

Objectives Section 5.8 The Ideal Gas Law

Are you able to. . .

- identify the four variables that describe gas properties
- identify the units for each variable
- explain the mathematical relationship between the four variables and the gas constant, R (Exercises 5.23, 5.24)
- give the quantity and units of standard temperature (Exercise 5.52)
- give the quantity and units of standard pressure (Exercise 5.52)
- calculate the value and units of R at STP
- calculate the volume occupied by 1 mol of any gas at STP
- calculate the pressure of a gas when all other variables are known (Exercise 5.53)
- calculate the volume of a gas when all other variables are known (Exercise 5.53)
- calculate the temperature of a gas when all other variables are known (Exercise 5.53)
- calculate the number of moles of gas when all other variables are known (Exercises 5.26, 5.29, 5.28, 5.46, 5.53)
- calculate the mass of a gas sample from the number of moles (Ex. 5.9, Prob. 5.9, Exercise 5.29)
- use density and the ideal gas law to calculate the molar mass of a sample of gas (Exercise 5.30)

Problem 5.9 Exhaled air contains CO_2 at a pressure of 28.0 torr. The capacity of a human lung is about 3.00 L. Body temperature is 37.0°C. Calculate the mass of CO_2 in a lung before expiration.

Draw a picture: a volume of air from the lung is at body temperature (37.0°C) and lung volume (3.00 L); carbon dioxide partial pressure in the volume is 28.0 torr. How much does this amount of gas weigh?

The mass of CO_2 can be calculated if the number of moles of exhaled gas can be calculated. From the information given, the ideal gas law can be used to calculate the number of moles of CO_2 in a sample. Remember to convert torr to atm and °C to kelvins in order to have agreement with the units of R: L, atm, K, mol.

List the information given:

body temperature: 37.0°C
Kelvin temp: 273 + 37.0 = 310 K
pressure: 28.0 torr (1 atm = 760 torr)
pressure in atm: 0.0368 atm
lung volume: 3.00 L
R: 0.0821 L·atm·K^{-1}·mol^{-1}

Set up the universal gas relationship to solve for the number of moles:

$$PV = nRT$$

$$n = \frac{PV}{RT}$$

Plug in given values:

$$n = \frac{(0.0368\ \cancel{\text{atm}})\ (3.00\ \cancel{\text{L}})}{(310\ \cancel{\text{K}})\ (0.0821\ \cancel{\text{L}}\cdot\cancel{\text{atm}}\cdot\cancel{\text{K}^{-1}}\cdot\text{mol}^{-1})} = 0.00434\ \text{mol}\ CO_2$$

The molar mass and the number of moles of CO_2 are used to calculate the mass of CO_2:

$$0.00434\ \cancel{\text{mol}\ CO_2} \times \frac{44.01\ \text{g}\ CO_2}{1\ \cancel{\text{mol}\ CO_2}} = 0.191\ \text{g}\ CO_2$$

Objectives Section 5.9 The Ideal Gas Law and Molar Mass

Are you able to. . .

- explain how molar mass of a gas is related to the ideal gas law
- calculate the molar mass of a gas sample by using the ideal gas law (Exercises 5.29, 5.30)
- calculate the number of grams in a gas sample by using its molar mass and the ideal gas law
- explain how density is related to molar mass through the ideal gas law
- calculate the molar mass or density of a gas sample when the density or molar mass and other variables are known (Ex. 5.10, Prob. 5.10, Exercises 5.48, 5.49)

Problem 5.10 The density of O_2 at 15.0°C and 1.00 atm is 1.36 g/L. Calculate its molar mass. First, go to the periodic table and calculate the molar mass of O_2: 32.00 g/mol. You know that your answer should be close to this value.

List information you are given:

density O_2: 1.36 g/L
temperature: 15.0°C
Kelvin temperature: 273 + 15.0 = 288 K
pressure: 1 atm
R: 0.0821 L·atm·K^{-1}·mol^{-1}

Example 5.10 uses the density factor to get to molecular mass, so follow that example if you have trouble.

$$M = (g/V)\ RT/P$$

↑ density term

Make sure the units you are using are the same as the units of *R*: L, atm, K, mol. Plug in the values:

$$M = \frac{(g)\ RT}{(V)\ P}$$

$$M = \frac{(1.36\text{ g})\ (0.0821\ \cancel{\text{L}}\cdot\cancel{\text{atm}}\cdot\cancel{\text{K}^{-1}}\cdot\text{mol}^{-1})\ (288\ \cancel{\text{K}})}{(1\cancel{\text{L}})\ (1\ \cancel{\text{atm}})} = 32.2\text{ g/mol}$$

The calculated answer is very close to the calculated value.

Problem 5.11 Determine the molar mass of argon, 14.10 g of which occupies 7.890 L at STP.

First, go to the periodic table and determine the molar mass of argon: 39.95 g/mol. You know that your answer should be close to this value.

List information you are given:

grams argon: 14.10 g
volume: 7.890 L
temperature: standard temperature
Kelvin temperature: 273 K
pressure: standard pressure, 1 atm
R: 0.0821 L·atm·K^{-1}·mol^{-1}

Example 5.11 uses STP values and relationships to get a molar mass; follow that example if you have trouble. Make sure the units you are using are the same as the units of *R*: L, atm, K, mol. Plug in the values:

$$M = \frac{(g)\ RT}{(V)\ P}$$

$$M = \frac{(14.10\text{ g})\ (0.0821\ \cancel{\text{L}}\cdot\cancel{\text{atm}}\cdot\cancel{\text{K}^{-1}}\cdot\text{mol}^{-1})\ (273\ \cancel{\text{K}})}{(7.890\ \cancel{\text{L}})\ (1\ \cancel{\text{atm}})} = 40.1\text{ g/mol}$$

The calculated answer is very close to the value given on the periodic table.

There is an alternate way to approach this problem. If you know how many liters 1 mol of gas occupies at STP, you can use that to determine the number of moles in the volume you are given:

$$7.890\ \cancel{L} \times \frac{1\ \text{mol}}{22.4\ \cancel{L}} = 0.352\ \text{mol argon}$$

We know that 0.352 mol of argon weighs 14.10 g, so the molar mass can be determined:

$$\frac{14.10\ \text{g argon}}{0.352\ \text{mol argon}} = \frac{40.1\ \text{g}}{\text{mol}}$$

Objectives Section 5.10 Dalton's Law of Partial Pressures

Are you able to...

- explain what is meant by partial pressure of a gas
- explain the relationship between the total pressure of a gas sample and the partial pressures of its constitutents
- define the mathematical relationship between the partial pressures of gases in a sample and the total pressure of the sample
- calculate the total pressure of a sample from partial pressures (Exercises 5.37, 5.38)
- determine the partial pressure of a gas when the total pressure and other partial pressures are known (Ex. 5.12, Prob. 5.12, Exercises 5.31, 5.32, 5.33, 5.34)

Problem 5.12 Water can be decomposed by an electric current into oxygen and hydrogen, whose collected volumes are in the ratio of 1 to 2, respectively. The total pressure of these two dry gases is 741 torr. What are their partial pressures?

$$P_{total} = P_{H_2} + P_{O_2}$$
$$741\ \text{torr} = 2P + 1P \leftarrow \text{set this up like an algebraic equation}$$
$$741\ \text{torr} = 3P$$
$$P = 247\ \text{torr (oxygen partial pressure)}$$
$$2P = 494\ \text{torr (hydrogen partial pressure)}$$

Alternately, if you know the percentages of each gas, you can determine each partial pressure contribution to the total pressure. A ratio of 1:2 says there are three parts total; one-third is oxygen and two-thirds are hydrogen. Changing to percentages: about 66.6% of the sample pressure comes from hydrogen pressure and about 33.3% comes from oxygen pressure.

$$741\ \text{torr}\ (0.666) = 494\ \text{torr}$$
$$741\ \text{torr}\ (0.333) = 247\ \text{torr}$$

Problem 5.13 Hydrogen was collected over water at 25.0 °C and a total pressure of 755 torr. The vapor pressure of water at that temperature is 23.8 torr. The amount of hydrogen collected was 0.320 mol. What was the volume of the gas?

First, determine what part of the total pressure comes from the hydrogen:

$$P_{total} = P_{H_2} + P_{water}$$
$$755\ \text{torr} = P_{H_2} + 23.8\ \text{torr}$$
$$P_{H_2} = 755\ \text{torr} - 23.8\ \text{torr} = 731\ \text{torr (partial pressure of hydrogen)}$$

Next, list the information given about the hydrogen:

pressure: 731 torr (1 atm = 760 torr)
pressure in atm: 0.962 atm
amount: 0.320 mol
temperature: 25.0°C
Kelvin temp: 273 + 25.0 = 298 K
volume: ???
↑this is what you want

You are asked to find the volume; the ideal gas law relates pressure, volume, temperature, and number of moles. Remember to convert pressure to atm and °C to kelvins because R is 0.0821 $L{\cdot}atm{\cdot}K^{-1}{\cdot}mol^{-1}$

$$PV = nRT$$

$$V = \frac{nRT}{P} = \frac{(0.320\ \cancel{mol})\ (0.0821\ L{\cdot}\cancel{atm}{\cdot}\cancel{K^{-1}}{\cdot}\cancel{mol^{-1}})\ (298\ \cancel{K})}{0.962\ \cancel{atm}} = 8.14\ L$$

Objectives Section 5.11 Gases Dissolve in Liquids

Are you able to. . .

- explain the relationship between the amount of gas that will dissolve in a liquid and the pressure of the gas above the liquid
- define the mathematical relationship between the amount of gas that will dissolve in a liquid and the pressure of the gas above the liquid
- explain how temperature affects solubility of a gas (Fig. 5.9)
- explain the qualitative meaning of Henry's law constant
- give the mathematical definition of Henry's law constant
- use Henry's law constant to determine the amount of gas dissolved in a liquid at a particular temperature by using the equation (Exs. 5.14, 5.15, Probs. 5.14, 5.15, Exercise 5.50, Expand Your Knowledge 5.57)
- define gas tension

Problem 5.14 The Henry's law constant for pure nitrogen in water at 20.0°C and 1.00 atm pressure of N_2 is 0.0152 mL N_2/mL H_2O. Calculate its solubility in water at 20.0°C and 1 atm pressure.

You can use the relationship:

$$\frac{\text{mL gas}}{\text{mL solvent}} = C_H \times P_{gas}$$

$$1\ \cancel{atm} \times \frac{0.0152\ \text{mL } N_2}{\text{mL } H_2O \times 1\ \cancel{atm}} = \frac{0.0152\ \text{mL } N_2}{\text{mL } H_2O}$$ ← this is the solubility at 1 atm

↑ this term says that 0.0152 mL nitrogen dissolves in 1 mL of water when the pressure of nitrogen is 1 atm

Problem 5.15 The solubility of pure nitrogen in water at 20.0°C and 1.00 atm pressure is 0.0152 mL N_2/mL H_2O. It constitutes about 80% of air. Calculate its solubility when air is in contact with water at 20°C and a pressure of 1.00 atm.

Nitrogen has a partial pressure of 0.800 atm in a sample of air.

$$0.800\ \cancel{atm} \times \frac{0.0152\ \text{mL } N_2}{\text{mL } H_2O \times 1.00} = \frac{0.0122\ \text{mL } N_2}{\text{mL } H_2O}$$ ← this is the amount of nitrogen that dissolves 1 mL of water when the nitrogen pressure is 0.800 atm

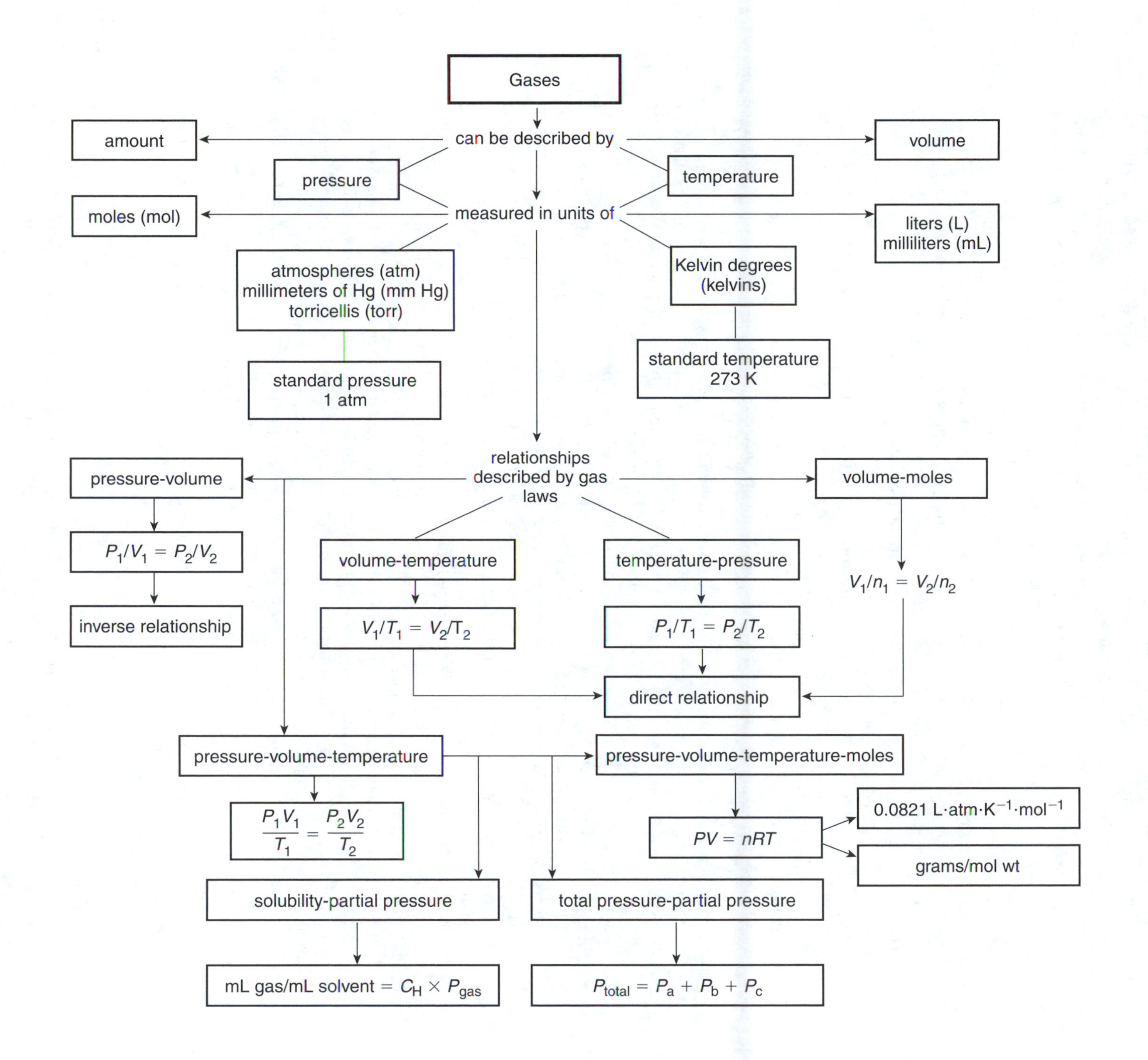
Gases
can be described by
amount
volume
pressure
temperature
measured in units of
moles (mol)
liters (L)
milliliters (mL)
atmospheres (atm)
millimeters of Hg (mm Hg)
torricellis (torr)
Kelvin degrees
(kelvins)
standard pressure
1 atm
standard temperature
273 K
relationships
described by gas
laws
pressure-volume
volume-moles
$P_1/V_1 = P_2/V_2$
volume-temperature
temperature-pressure
$V_1/n_1 = V_2/n_2$
inverse relationship
$V_1/T_1 = V_2/T_2$
$P_1/T_1 = P_2/T_2$
direct relationship
pressure-volume-temperature
pressure-volume-temperature-moles
$\frac{P_1 V_1}{T_1} = \frac{P_2 V_2}{T_2}$
$PV = nRT$
$0.0821\ L{\cdot}atm{\cdot}K^{-1}{\cdot}mol^{-1}$
grams/mol wt
solubility-partial pressure
total pressure-partial pressure
mL gas/mL solvent = $C_H \times P_{gas}$
$P_{total} = P_a + P_b + P_c$

FILL-INS

The four variables that describe a gas are _____. The pressure of a gas is the _____with which gas molecules hit a _____. Pressure is commonly measured in units of _____. Atmospheric pressure is defined as the pressure that will support a column of mercury _____ tall at _____ degrees C. Atmospheric pressure varies with to _____. The temperature of a gas is always reported in _____ when doing gas law calculations. Degrees Celsius can be converted to degrees Kelvin by _____ to the Celsius temperature. The volume of a gas is defined by _____ and is commonly measured in units of _____.

The properties of gases are related to each other by a variety of _____. The use of these gas laws allows for the examination of changes in gas properties when any of the four variables _____ are altered. For most gas law problems, one or two of the variables are held _____ when the others are changed. The relationship between pressure and volume is examined with _____ constant. _____ relationships can be determined when pressure and numer of moles are constant. Temperature and pressure relationships can be analyzed when _____ are constant. When _____ are held constant, the relationship between volume and number of moles can be examined.

Boyle's law states that there is an _____ relationship between gas pressure and volume at constant _____. When the volume of a sample is decreased, the pressure of the sample _____ and when the volume of a sample is increased, the pressure of the sample _____. _____ result from gas molecules hitting a container surface with an increased or decreased frequency or energy. Changing a volume either _____ the number of collisions molecules have with the container surface.

Charles's law states that there is an _____ relationship between the temperature of a gas sample and its volume when P and n are constant. When the temperature of a gas is increased, the volume will _____; and when the temperature is decreased, the volume will _____. There is also a _____ relationship between gas pressure and temperature when V and n are constant. As the temperature of a sample increases, pressure _____; and as the temperature of a sample decreases, pressure _____.

At the same temperature and pressure, equal volumes of gas contain the _____ of molecules. This is _____ law. As the number of moles of gas increases, volume _____. At STP, 1 mol of any gas occupies _____.

When the relationships among temperature, pressure, and volume are combined and the amount of gas is kept constant, the effects of varying two variables can be examined. This is the _____ and can be used to determine the new _____ of a gas when both temperature and volume change; the new temperature of a gas when both _____ change; or the new _____ of a gas when both temperature and pressure change.

The ideal gas law mathematically relates all four variable and is written _____. R is a _____ that has the units of _____ in the numerator and _____ in the denominator. When pressure is measured in atm and volume in L, at standard temperature, _____, and standard pressure, _____, R has the value of _____ L·atm/K·mol. Any one of the four variables can be calculated when the other three are known, using this equation. The _____ term in the equation can be substituted by sample mass/molar mass in order to determine the _____ of a gas. The equation can also be rearranged so that density (sample mass/volume) can be used to determine _____.

The total pressure of any sample of gas is the _____ of each partial pressure contribution of each of its constituents; this is _____ law. The partial pressure of a gas is the pressure the _____. When gases dissolve in a liquid, the partial pressure of any gas determines _____ will dissolve in the liquid. This is called _____ law and can be used to determine the _____ in a liquid at a defined temperature. Gas solubility in a liquid is dependent on _____: the warmer the liquid, the _____ gas will dissolve, and the colder the liquid, the _____ gas will dissolve.

Answers

The four variables that describe a gas are pressure, volume, temperature, and amount (number of moles). The pressure of a gas is the force with which gas molecules hit a defined surface area. Pressure is commonly measured in units of atmospheres, millimeters of mercury, or torr. Atmospheric pressure is defined as the pressure that will support a column of mercury 760 mm tall at 0 degrees C. Atmospheric pressure varies with elevation and temperature. The temperature of a gas is always reported in kelvins when doing gas law calculations. Degrees Celsius can be converted to degrees Kelvin by adding 273 to the Celsius temperature. The volume of a gas is defined by the container in which the gas is held and is commonly measured in units of liters or milliliters.

The properties of gases are related to each other by a variety of gas laws. The use of these gas laws allows for the examination of changes in gas properties when any of the four variables pressure, temperature, volume, or number of moles are altered. For most gas law problems, one or two of the variables are held constant when the others are changed. The relationship between pressure and volume is examined with temperature and number of moles kept constant. Volume and temperature relationships can be determined when pressure and number of moles are constant. Temperature and pressure relationships can be analyzed when volume and the number of moles are constant. When temperature and pressure are held constant, the relationship between volume and number of moles can be examined.

Boyle's law states that there is an inverse relationship between gas pressure and volume at constant T and n. When the volume of a sample is decreased, the pressure of the sample increases and when the volume of a sample is increased, the pressure of the sample decreases. Pressure changes result from gas molecules hitting a container surface with an increased or decreased frequency or energy. Changing a volume either increases or decreases the number of collisions molecules have with the container surface.

Charles's law states that there is a direct relationship between the temperature of a gas sample and its volume when P and n are constant. When the temperature of a gas is increased, the volume will increase; and when the temperature is decreased, the volume will decrease. There is also a direct relationship between gas pressure and temperature when V and n are constant. As the temperature of a sample increases, pressure increases, and as the temperature of a sample decreases, pressure decreases.

At the same temperature and pressure, equal volumes of gas contain the same number of molecules. This is Avogadro's law. As the number of moles of gas increases, volume also increases. At STP, 1 mol of any gas occupies 22.4 L.

When the relationships among temperature, pressure, and volume are combined and the amount of gas is kept constant, the effects of varying two variables can be examined. This is the combined gas law and can be used to determine the new pressure of a gas when both temperature and volume change; the new temperature of a gas when both volume and pressure change; or the new volume of a gas when both temperature and pressure change.

The ideal gas law mathematically relates all four variables and is written $PV = nRT$. R is a constant that has the units of pressure and volume in the numerator and Kelvin temperature and number of moles in the denominator. When pressure is measured in atm and volume in L, at standard temperature, 273 K, and standard pressure, 1 atm, R has the value of 0.0821 L·atm/K·mol. Any one of the four variables can be calculated when the other three are known by using this equation. The mole term in the equation can be substituted by sample mass/molar mass in order to determine the molar mass of a gas. The equation can also be rearranged so that density (sample mass/volume) can be used to determine molar mass.

The total pressure of any sample of gas is the sum of each partial pressure contribution of each of its constituents; this is Dalton's law. The partial pressure of a gas is the pressure the gas would exert if it were the only gas in a sample. When gases dissolve in a liquid, the partial pressure of any gas determines how much of it will dissolve in the liquid. This is called Henry's law and can be used to determine the solubility of a gas in a liquid at a defined temperature. Gas solubility in a liquid is dependent on temperature: the warmer the liquid, the less gas will dissolve, and the colder the liquid, the more gas will dissolve.

TEST YOURSELF

1. Explain how a gas can be described by using pressure, volume, temperature, and moles.
2. Mathematically, pressure is described as force per unit area: $P =$ force/area. Show how the magnitude of pressure is influenced by both force and area.
3. Describe the composition of a column of air at sea level and at the top of a mountain. In which case is the atmospheric pressure greater?
4. Why do pressure and amount of gas need to be held constant in order to examine the relationship between the temperature and volume of a gas? What would happen if they weren't held constant?

5. Explain what happens at the molecular level to a fixed amount of gas when its volume is compressed to one-half its original volume at constant temperature. By what factor will the gas pressure change?
6. Describe a situation that illustrates Charles's law. Describe the molecular-level activity of the gas molecules in your example.
7. Describe a situation that illustrates Boyle's law. Describe the molecular-level activity of the gas molecules in your example.
8. Explain why increasing the temperature of a gas increases its pressure.
9. Explain why decreasing the volume of a gas increases its pressure.
10. Explain why increasing the amount of a gas increases its pressure.
11. Why must the temperature and pressure of two gas samples be the same in order to apply Avogadro's law?
12. Explain the effect of changing both the pressure and temperature on a volume of gas.
13. From the list of equations, choose the one that would be used in the situation on the described:

    (1) $P_1V_1/T_1 = P_2V_2/T_2$ (2) $V_1/T_1 = V_2/T_2$ (3) $P_1/T_1 = P_2/T_2$ (4) $V_1/n_1 = V_2/n_2$
    (5) $P_1V_1 = P_2V_2$ (6) $PV = nRT$ (7) $P_{total} = P_1 + P_2$ (8) $C_H \times P_{gas}$ = mL gas/mL solvent

    (a) Determine what happens to a volume of gas when the temperature is decreased and the pressure is increased.
    (b) Determine the pressure of a gas mixture from its components.
    (c) Determine the volume of 1 mol of gas at room temperature and atmospheric pressure.
    (d) Determine how much oxygen will dissolve in blood at body temperature and atmospheric pressure.
    (e) Determine what happens to the pressure of a gas when the temperature is decreased.
    (f) Determine what happens to the volume of gas when the number of moles is decreased.
    (g) Determine what happens to the volume of a gas when the pressure is increased.
    (h) Determine what happens to the temperature of a gas when the volume is decreased.
14. Match the correct equations from Problem 13 with the problems below and solve them.
    (a) The volume of 2.0 mol of gas is changed at constant temperature from 1.5 L to 2.0 L. If the pressure of the gas was 10.0 atm, what will the pressure be at the new volume?
    (b) 1 mol of gas is heated from 100°C to 200°C at constant pressure. If the original volume of the gas was 30 L, what volume will it occupy after the temperature is raised?
    (c) The gas in a steel cylinder is heated from 25°C to 100°C. What will the pressure of the gas be at 100°C if its original pressure was 1.0 atm?
    (d) A 10-L volume contains 0.5 mol of gas. If 0.5 mol of gas is added to the sample and the temperature and pressure remain constant, what volume will the sample occupy?
    (e) At 20°C and 1.0 atm pressure, an amount of gas occupies 15 L. How many moles of gas are in the sample?
15. Show how molar mass is related to density through the ideal gas law.
16. Explain how the temperature of water affects the solubility of oxygen. What impact does this have on aquatic life?
17. Would you expect the partial pressure of oxygen at the top of a mountain to be more or less than it is at sea level? How does this affect the amount of oxygen dissolved in blood?

Answers

1. Pressure, volume, temperature, and moles are related through the gas constant, R. The ideal gas equation describes the relationship: $PV = nRT$
2. If force is increased, the number in the numerator increases and pressure increases. If force is decreased, the number in the numerator decreases and the pressure decreases. If the area is increased, the number in the denominator increases and the pressure decreases. If the area is decreased, the number in the numerator is decreased and pressure increases. There is a direct relationship between force and pressure and an inverse relationship between pressure and area.

3. Both air at sea level and air at the top of a mountain have the same composition, and the components are present in the same percentages. There are more particles of air, however, in a sample at sea level than in a sample at the top of a mountain, so the pressure is greater at sea level than at the top of a mountain.
4. If pressure and amount of gas are not held constant, their influences will combine with that of temperature to influence the volume and it won't be possible to isolate temperature's impact alone.
5. When a volume of gas is compressed, the space in which the molecules move is decreased. (This is analogous to decreasing the area in the Pressure = force/area equation.) The molecules have to travel less distance to hit the walls and will hit them more often, thus increasing the pressure. The gas pressure will double when the volume is halved.
6. Charles's law describes the effects of temperature on a volume of gas. If you examine a volume of gas in a balloon, the molecules are moving at a certain speed. If the balloon is heated, the volume of the balloon increases as the speed (kinetic energy) of the molecule increases. If the balloon is chilled, for example, in a freezer, the volume shrinks as the molecules slow down (less kinetic energy).
7. Boyle's law describes the relationship between the pressure of a gas and its volume. When breathing, the movement of the diaphragm causes small volume changes in the lungs. This in turn increases or decreases the pressure in the lungs. Gas molecules in a volume of gas have to travel farther before hitting a surface if the volume is increased (analogous to increasing the area in Pressure = force/area equation). The pressure is decreased as the number of times the molecule hits the wall is decreased.
8. When the temperature of gas is increased, the speed of the molecules increases, which, in turn, increases the frequency of collisions with the walls of the container.
9. When the volume of a gas is decreased, the same number of molecules is moving in a smaller volume, and the frequency of collisions with the container wall increases.
10. Increasing the amount of gas increases the number of particles hitting the container wall. There are more molecules in the same volume.
11. Volume is proportional to number of moles only at constant temperature and pressure.
12. Volume is directly proportional to temperature and inversely proportional to pressure. Therefore, changing both temperature and pressure simultaneously can result in an increase, decrease, or no change in volume.
13. a) 1 b) 7 c) 6 d) 8 e) 3 f) 4 g) 5 h) 2
14. a) $P_1V_1 = P_2V_2$ (10 atm) (1.5 L) = (x atm) (2.0 L) $x = 7.5$ atm
    b) $V_1/T_1 = V_2/T_2$ 30 L/ 373 K = x/ 573 K $x = 46$ L
    c) $P_1/T_1 = P_2/T_2$ 1.0 atm/298 K = x atm/373 K $x = 1.3$ atm
    d) $V_1/n_1 = V_2/n_2$ 10 L/ 0.5 mol = x/ 1.0 mol $x = 20$ L
    e) $PV = nRT$ $P = 1$ atm $V = 15$ L $T = 293$ K 0.0821 L atm/K mol
    $n = PV/RT$ = (1.0 atm) (15 L) / (0.0821 L·atm/K·mol) (293 K) = 0.62 mol
15. $PV = nRT$ n = sample mass/molar mass density = mass/volume
    put both n and density in terms of mass: mass = (n) (M) mass = (density) (volume)
    then: (n) (M) = (density) (volume) n = (density) (volume)/ M
    substitute for n in $PV = nRT$: $P\cancel{V}$ = (density) ~~(volume)~~ (RT)/M ← volumes cancel
    then: P = density (RT)/ M
    in terms of density: density = M (P)/RT
    in terms of M: M = density (RT)/P
16. As the temperature is increased the solubility of oxygen decreases. As temperature increases due to thermal pollution, the amount of oxygen dissolved is decreased and the amount of oxygen available to aquatic life is decreased.
17. The partial pressure of oxygen is less at mountaintop because the total pressure is less. The amount dissolved is less because the pressure is less.

chapter 6
Interactions Between Molecules

OUTLINE

6.1 The Three States of Matter and Transitions Between Them

A. Matter can exist as a solid, a liquid, or a gas.

B. Solids
   1. have a fixed volume; volume changes little with pressure
   2. have a fixed shape
      a. solids cannot be compressed
      b. molecules are strongly held in fixed positions
      c. molecules have low kinetic energy
   3. can be melted to form liquids; requires addition of heat, which increases kinetic energy of molecules

C. Liquids
   1. have a fixed volume; volume changes little with pressure
   2. do not have a fixed shape
      a. shape depends on container
      b. attractive forces between molecules are weaker than in solids and molecules move easily around one another
   3. can be frozen to form solids; requires removal of heat, which decreases kinetic energy of molecules
   4. can be vaporized to form gases
      a. requires the addition of heat, which increases kinetic energy of molecules
      b. results in a large increase in volume

D. Gases
   1. do not have a fixed volume
      a. volume of gas depends on volume of container
      b. can easily be compressed to smaller volume
   2. molecules have more kinetic energy than as liquid or solid
      a. molecules are very far apart and do not interact with each one another
      b. attractive forces between gas molecules are negligible
   3. can be condensed to form liquids
      a. requires the removal of heat, which decreases the kinetic energy of molecules
      b. results in a large decrease in volume

E. Changes between physical states are called phase transitions.
   1. phase transitions are reversible
   2. phase transitions occur as heat is added or removed
      a. addition of heat increases kinetic energy of molecules; opposes attractive forces between molecules
      b. removal of heat decreases kinetic energy of molecules

3. phase transitions include
   a. melting: solid changes to liquid as heat is added
   b. freezing: liquid changes to solid as heat is removed
   c. vaporization: liquid changes to gas as heat is added
   d. evaporation: continuous vaporization from the surface of a liquid
   e. condensation: gas changes to liquid as heat is removed

F. Amount of heat required to bring about changes in phases is a unique physical property.
1. melting point of pure solid can be used to help identify it
   a. melting point is a unique temperature at which a pure solid sample begins to melt
   b. the melting point of a pure solid equals the freezing point of the pure liquid
   c. molar heat of fusion is the amount of heat required to melt 1 mol of the solid
2. boiling point of pure liquid can be used to help identify it
   a. normal boiling point is a unique temperature at which gas bubbles form throughout a sample of the liquid that is open to the atmosphere
   b. molar heat of vaporization is the amount of heat required to vaporize 1 mol of the liquid
3. for a given pure substance, it takes more energy to vaporize liquid molecules than it does to melt the solid
   a. attractive forces in solid are loosened to cause melting
   b. attractive forces between liquid molecules must be completely overcome to cause vaporization

Key Terms condensation, freezing, freezing point, kinetic energy, melting, melting point, molar heat of fusion, molar heat of vaporization, normal boiling point, phase transition, reversible, vaporization

Key Figure Figure 6.1 A molecular view of the three states of matter

6.2 *Attractive Forces Between Molecules*

A. Attractive forces between molecules are called secondary forces.
1. secondary forces are weaker than bonds between atoms (chemical bonds)
2. secondary forces determine physical properties of solids and liquids
3. secondary forces are the basis of many physiological phenomena
4. attractive forces are electrical in nature

B. Secondary forces can be categorized on the basis of the polarity of the molecules involved.
1. dipole-dipole interactions
   a. occur between polar molecules
   b. result in relatively stronger intermolecular attractions
   c. cause substances to have relatively higher melting and boiling points than those of substances with only nonpolar interactions
2. London forces
   a. occur between nonpolar molecules
   b. result from transient unequal distributions of electronic charge due to electron movements
   c. can make significant contribution to physical properties if they are the only intermolecular forces present
   d. cause substances to have relatively lower melting and boiling points than polar interactions
   e. increase with increasing number of electrons (increasing molecular mass)

Key Terms dipole-dipole interaction, intermolecular force, London force, nonpolar molecule, polar molecule, secondary force, van der Waals force

Key Figure and Table
Figure 6.2 Interaction of polar molecules in the liquid state
Table 6.1 Boiling points of polar and nonpolar molecules

Example 6.1 Using structural formulas to predict polarity (refer to Chapter 3 for Lewis structures)
Problem 6.1 Using structural formulas to predict polarity

6.3 The Hydrogen Bond

A. Covalent bonds between hydrogen and either oxygen, nitrogen, or fluorine are very polar. Electronegativity differences between the bonded atoms produce a large dipole moment.
B. Intermolecular attractions between molecules containing these bonds are stronger than other dipole-dipole attractions.
   1. special case of dipole-dipole interactions
   2. produce higher boiling points in substances than would be predicted by molecular mass
C. Hydrogen bonds in water
   1. produce distinct arrangements of molecules in ice; results in a less dense structure than in liquid water
   2. are responsible for the properties of water

Key Terms hydrides, hydrogen bond, intermolecular, intramolecular
Key Figures, Table
Figure 6.3 Intermolecular attraction between hydrogen and oxygen
Figure 6.4 Hydrogen bonds between water molecules in ice
Figure 6.5 Boiling points of binary hydrides as a function of increasing mass
Table 6.2 Hydrides arranged by group and row of the periodic table

Example 6.2 Recognizing molecules that can form hydrogen bonds
Problem 6.2 Recognizing molecules that can form hydrogen bonds

6.4 Secondary Forces and Physical Properties

A. Secondary forces between molecules determine physical properties such as
   1. boiling points
   2. melting points
   3. surface tension
   4. solubility
B. Boiling points and melting points provide a measure of the relative strength of attractive forces between molecules (Table 6.1)
   1. London forces $<$ dipole-dipole interactions $<$ hydrogen bonds
      increasing strength $\Rightarrow$
      a. molecules may interact through a combination of forces
      b. magnitude of energy required for phase change is a measure of the strengths of the attractions
C. Surface tension is determined by the attractions among molecules at the surface of a liquid.
   1. molecules at the surface experience attractions only on the "underside" of the surface
   2. force of surface tension creates an elastic skin that can prevent some penetration of surface (Figs. 6.7 and 6.8)
   3. surfactants can reduce surface tension
      a. surfactants contain both polar and nonpolar areas
      b. surfactants are amphipathic
         i. polar areas dissolve in water; are hydrophilic
         ii. nonpolar areas dissolve in fats; are hydrophobic
D. Solutions are mixtures in which all molecules or ions are uniformly distributed.
E. Secondary forces between particles must be the same in order for them to form solutions: like dissolves like.
   1. polar molecules will mix with polar molecules or ions
   2. nonpolar molecules will not mix with polar molecules or ions
   3. nonpolar molecules will mix with nonpolar molecules

Key Terms amphipathic, hydrophilic, hydrophobic, solution, surface tension, surfactants

Key Figures, Boxes, and Table

Figure 6.6 Surface tension
Figure 6.7 Illustration of surface tension of water
Figure 6.8 Insect walking on water
Figure 6.9 Surfactant molecule
Figure 6.10 Formation of a solution
Box 6.1 Surfactants and the digestion of fats
Box 6.2 Surfactants in the lungs
Table 6.3 Normal boiling points of substances and related secondary forces

Example 6.3 Using molecular structure to predict physical properties
Problem 6.3 Predicting relative boiling points
Example 6.4 Predicting solution formation by using molecular structures
Problem 6.4 Predicting solution formation by using molecular structures
Example 6.5 Predicting solution formation by using molecular structures
Problem 6.5 Predicting solution formation by using molecular structures

6.5 The Vaporization of Liquids

A Vaporization is the process of changing a liquid into a gas.
1. vaporization occurs when molecules have enough kinetic energy to escape the attractive forces of the liquid
   a. as molecule acquires energy, it moves to the surface
   b. when it has enough energy to overcome attractive forces at the surface, it moves away as a gas
2. Evaporation is a case of continuous vaporization from the surface of a liquid; it requires that liquid be continuously absorbing energy from surroundings
   a. absorption of energy from surroundings produces cooling
   b. can be used to cool body temperatures (Box 6.3)

B. Vapor pressure occurs when a liquid and a gas are present simultaneously.
1. gas is called vapor under these circumstances
2. vapor exerts gas pressure
3. when vapor and liquid are at equilibrium, the pressure exerted by the gas is called the vapor pressure

Key Terms vapor, vapor pressure
Key Box Box 6.3 Topical anesthesia

6.6 Vapor Pressure and Dynamic Equilibrium

A. Vaporization and condensation are reverse processes.

gas molecules leave liquid phase ⇌ gas molecules enter liquid phase
(rate depends on temperature) (rate depends on number of vapor molecules and temperature)

B. In a closed system, equilibrium develops between vaporization and condensation.
1. rate at which liquid becomes gas equals rate at which gas becomes liquid
2. dynamic system: equilibrium indicated by two arrows

$$\text{liquid} \underset{\text{condensation}}{\overset{\text{vaporization}}{\rightleftharpoons}} \text{vapor}$$

3. equilibrium vapor pressure is vapor pressure of a liquid in a closed system

C. In an open system, equilibrium does not develop.
1. gas molecules vaporize from liquid, but do not condense back
2. evaporation is the case of continuous vaporization from an open system

Key Terms closed system, dynamic equilibrium, equilibrium state, equilibrium vapor pressure, evaporation, open system

Key Figure Figure 6.11 Equilibrium between liquid and vapor in a closed system

6.7 *The Influence of Secondary Forces on Vapor Pressure*

A. Vapor pressure is a measure of attractive forces between molecules in the liquid phase.
   1. small vapor pressure means that molecules don't leave the liquid easily and the attractive forces are very strong; polar interactions result in lower vapor pressure, e.g., water
   2. large vapor pressure means that molecules can easily leave the liquid phase and indicates weak attractive forces between molecules; nonpolar interactions result in higher vapor pressure, e.g., gasoline

B. Boiling results when vapor bubbles form in a liquid sample.
   1. atmospheric pressure presses against liquid surfaces in open containers
   2. vapor bubbles cannot form unless vapor pressure is higher than atmospheric pressure
   3. formation of stable vapor bubbles in liquid is called boiling
   4. atmospheric pressure decreases with increasing altitude

C. Normal boiling point is defined as temperature at which boiling occurs when atmospheric pressure is 1 atm.
   1. normal boiling point indicates strength of attractive forces between molecules (Fig. 6.12, Table 6.2)
   2. higher boiling points mean stronger secondary forces are operating; liquids with low vapor pressures have higher boiling points
   3. lower boiling points mean weaker secondary forces are operating; liquids with high vapor pressures have lower boiling points

Key Terms boil, escaping tendency, normal boiling point

Key Figure Figure 6.12 Relationship between vapor pressures and temperature

Example 6.6 Interpreting vapor pressure versus temperature curves using Fig. 6.12
Problem 6.6 Interpreting vapor pressure versus temperature curves using Fig. 6.12
Example 6.7 Relating vapor pressure to melting and boiling points
Problem 6.7 Explaining differences in vapor pressures from a molecular standpoint
Example 6.8 Relating boiling points to secondary forces (refer to Chapter 3 for Lewis structures)
Problem 6.8 Using molecular structures and secondary forces to predict boiling points

6.8 *Vaporization and the Regulation of Body Temperature*

A. Body regulates temperature through vaporization.
   1. vaporization requires heat and can be used to rid body of excess heat
   2. water is vaporized from skin as sweat and from lungs, mouth, and nasal passages; heat from body is carried away as kinetic energy of gas molecules
   3. water requires a lot of heat to overcome hydrogen bonds between liquid molecules; water has a high heat of vaporization as a result of hydrogen bonding
   4. rate of vaporization of water from body can be affected by relative humidity

B. Relative humidity is a measure of the partial pressure of water vapor in the atmosphere.
   1. mathematically defined as

$$\text{relative humidity} = \frac{\text{partial pressure water in atmosphere}}{\text{equilibrium vapor pressure of water}} \times 100\%$$

   2. when partial pressure of water at a given environmental (or ambient) temperature is equal to the equilibrium pressure, the air is saturated with water vapor and produces 100% relative humidity
   3. at 100% relative humidity, water can't be transferred from body to atmosphere efficiently by evaporation because the condensation rate equals the vaporization rate; body temperature can rise dangerously if environmental temperatures climb under these circumstances

Key Terms conduction, convection, evaporation, relative humidity

6.9 *Attractive Forces and the Structure of Solids*

A. Solids are more ordered than liquids or gases.
   1. some solids have regular atomic arrangements; these are crystalline forms
   2. some solids have a less ordered atomic arrangement; these are amorphous forms

B. Elements in solids may be held together by a variety of kinds of bonds or attractions.
   1. covalently bonded solids
      a. will have very high melting points
      b. called covalent crystals
   2. ionically bonded solids
      a. will have relatively high melting points, but lower than covalent crystals
      b. called ionic crystals
   3. solids joined by weak secondary forces such as dipole-dipole or London forces
      a. will have relatively low melting points
      b. called molecular crystals
   4. melting points of solids provide an indication of the type and strength of attractions between elements in the solid

C. Some solids can vaporize under normal conditions without first melting
   1. this process is called sublimation
   2. the process requires energy equivalent to that needed to both melt and vaporize the substance

Key Terms amorphous, covalent crystal, crystalline solid, ionic crystal, molecular crystal, sublimation

Key Figures and Table

Figure 6.13 A crystalline solid

Figure 6.14 SiO_2 can exist as a crystalline solid or as an amorphous solid (glass)

Table 6.3 Relative strengths of attractive forces in different crystalline solids

Example 6.9 Predicting melting points from molecular formulas

Problem 6.9 Predicting melting points from molecular formulas

ARE YOU ABLE TO. . . AND WORKED TEXT PROBLEMS

Objectives Section 6.1 The Three States of Matter and Transitions Between Them

Are you able to. . .

- define solid, liquid, and gas (Exercises 6.6, 6.7, 6.27, 6.28)
- explain the differences and similarities among solid, liquid, and gas samples of the same substance
- describe the molecular-level view of each state (Exercises 6.1, 6.2, 6.3, 6.4, 6.31, 6.32)
- explain the relationship between the kinetic energy of molecular particles and the state in which they exist (Exercises 6.3, 6.4, 6.5, 6.6)
- explain what a phase transition is
- define melting, freezing, vaporization, condensation, and sublimation
- explain the relationships between any two phase transitions (Exercises 6.9, 6.10, 6.11)
- explain how the kinetic energy of component particles is related to phase transitions (Exercises 6.3, 6.4, 6.30)
- explain the relationship between intermolecular attractive forces and phase transitions or sample state (Exercises 6.31, 6.32)
- define melting point and boiling point (Exercises 6.3, 6.4, 6.11)
- explain why melting and boiling points can be used to identify a pure substance
- explain what happens at the molecular level when a substance melts or boils

Objectives Section 6.2 Attractive Forces Between Molecules

Are you able to. . .

- define secondary force and describe the inherent properties of secondary forces
- identify and describe principal secondary forces
- explain how dipole-dipole interactions, London forces, and hydrogen bonds occur at the molecular level

- explain how molecular polarity influences secondary forces (Exercises 6.17, 6.18, 6.21, 6.22, 6.23, 6.25)
- explain how molecular mass influences secondary forces (Exercises 6.19)
- explain how secondary forces influence physical properties (Exercises 6.15, 6.16, 6.35, 6.36)
- explain the relationship between melting and boiling points and secondary forces (Exercises 6.20, 6.25, 6.26)

Problem 6.1 Which of the following compounds condense to the liquid phase from the gas phase through polar secondary forces?

a) H—H b) CCl_4 (Cl—C—Cl with Cl above and below C) c) H—C≡N d) I—I

Polar secondary forces act between polar molecules. Refer back to Chapter 3 to determine how molecular structure influences polarity. Remember that any two atoms of the same element bonded covalently to each other will produce a linear, nonpolar molecule; H_2 and I_2 are nonpolar diatomic molecules and will condense through nonpolar secondary forces. CCl_4 contains polar bonds between atoms, *but* the tetrahedral shape results in a nonpolar molecule overall and this will condense to a liquid through nonpolar attractions. HCN is definitely polar; there are three different atoms bonded linearly through two different types of covalent bonds.

Objectives Section 6.3 The Hydrogen Bond

Are you able to. . .

- define hydrogen bond
- identify molecules in which hydrogen bonds can occur
- explain the relationship between hydrogen bonding and dipole-dipole attractions
- explain the influence of hydrogen bonding on physical properties
- distinguish between intermolecular and intramolecular bonds and attractions
- describe how hydrogen bonds operate at the molecular level

Problem 6.2 Which of the following molecules can form hydrogen bonds?

a) SiH_4 b) CH_4 c) CH_3NH_2 d) H_2Te e) C_2H_5OH

Remember that in order for hydrogen bonds to form, a molecule must have a hydrogen atom covalently bonded to one of the following: oxygen, nitrogen, or fluorine. Examine the chemical formulas to determine which molecules meet the criterion. $CH_3\mathbf{NH_2}$ and $C_2H_5\mathbf{OH}$ are capable of hydrogen bonding.

Objectives Section 6.4 Secondary Forces and Physical Properties

Are you able to. . .

- explain how secondary forces influence boiling points, melting points, surface tension, and solubility (Exercises 6.43, 6.44)
- explain how the magnitude of energy required to boil or melt a substance can be related to the strength and types of secondary forces present
- analyze molecular structures to predict what types of secondary forces will be present
- predict relative melting and boiling points by analyzing molecular structures
- describe surface tension
- describe the general structure of surfactant molecules
- explain how surfactants influence surface tension
- explain how surfactants are involved in the digestion of fats
- explain the role of surfactants in the lungs
- explain the relationship between secondary forces and solubility (Exercises 6.45, 6.46)
- predict when solutions will and will not form between molecules or ions (Exercises 6.45, 6.46)

Problem 6.3 Arrange the normal boiling points of the following substances from highest to lowest:

a) He b) CH_4 c) I_2 d) HF

First, separate the polar from the nonpolar substances; the polar substances will have relatively higher boiling points. He, CH_4, and I_2 are all nonpolar. HF is polar *and* will form hydrogen bonds because the electronegativity difference between H and F is great enough to allow it. HF, then, will have the highest relative boiling point. The trick now is to rank the nonpolar substances. You need to determine what secondary forces operate in nonpolar substances and what factors influence them. London forces dominate the secondary attractions between nonpolar substances. These are influenced by the number of electrons; that means that heavier atoms or molecules will have larger London forces operating. The ranking from lightest to heaviest is He then CH_4 then I_2.

On the basis of these orderings, the relative boiling points from highest to lowest will be

$$HF > I_2 > CH_4 > He$$

Problem 6.4 Can water and carbon tetrachloride (CCl_4) form a solution? Explain your answer.

No. For a solution to form, both substances must be able to form attractions to each other, which means that they must both be polar or they must both be nonpolar. The secondary forces present in each substance must be of the same type for solutions to form. Water, H_2O, is a polar molecule that is able to form hydrogen bonds. CCl_4 is a nonpolar molecule (refer to Lewis structures, Chapter 3) because of its shape, even though it contains polar bonds; weak London forces are present among molecules. H_2O and CCl_4 will not form a solution because nonpolar molecules cannot offer a more attractive (i.e., stronger) or equally attractive alternative to water molecules than the hydrogen bonds they already share.

Problem 6.5 Explain why the following pairs of substances form or do not form solutions.

(a) CCl_4 and H_2O (b) C_5H_{12} and CCl_4

(a) The explanation for the fact that water and carbon tetrachloride do not form solutions is given in Problem 6.4. (b) C_5H_{12} and CCl_4 are both nonpolar molecules, and weak London forces influence their ability to form solutions. They will be able to mingle at the molecular level because the secondary forces among both substances are the same.

Objectives Section 6.5 The Vaporization of Liquids

Are you able to. . .

- describe the process of vaporization
- explain the relationship between molecular kinetic energy and vaporization
- explain the relationship between evaporation and vaporization
- explain the role of evaporation in the regulation of body temperature (Expand Your Knowledge 6.56)
- describe what is meant by "vapor" and by "vapor pressure" (Exercises 6.33, 6.34, 6.37, 6.38)
- explain how vapor pressure differs in an open and in a closed system
- describe how equilibrium is established between liquid and vapor in a closed system (Exercises 6.39, 6.40)

Objectives Section 6.6 Vapor Pressure and Dynamic Equilibrium

Are you able to..

- explain why equilibrium is a dynamic situation
- explain the relationship between vaporization and condensation
- identify the factors that influence the rate of vaporization and the rate of condensation
- explain why equilibrium develops in a closed system, but not in an open system
- describe the molecular-level activity associated with vaporization and condensation

Objectives Section 6.7 The Influence of Secondary Forces on Vapor Pressure

Are you able to. . .

- explain the relationship between secondary forces and vapor pressure (Exercises 6.35, 6.36, Expand Your Knowledge 6.57)
- predict relative vapor pressure by analyzing secondary forces among molecules
- describe boiling
- explain what happens at the molecular level when boiling occurs

- describe the relationship between atmospheric pressure and boiling point (Exercise 6.42)
- explain how normal boiling points are measured (Exercise 6.41)
- describe the relationship between boiling points and secondary forces
- interpret vapor pressure versus temperature curves

Problem 6.6 Estimate the vapor pressures at 50°C of the three liquids whose pressure-versus-temperature relations appear in Figure 6.12.

To determine vapor pressure, go to the 50° point on the temperature axis. At this point, draw a line straight up until you intersect the curve for ethanol; draw a line straight across to the vapor pressure axis from this point; this is the vapor pressure of ethanol at 50°C; it is about 200 torr. Do the same for hexane and you should get about 390 torr; do the same for chloroform and you should get about 500 torr.

Problem 6.7 Methyl alcohol, CH_3OH, and methyl iodide, CH_3I, are both polar molecules, but methyl iodide's vapor pressure at room temperature is almost four times as great as that of methyl alcohol. What is the molecular basis for the difference?

You should first ask yourself what factors influence vapor pressure: type of secondary forces and temperature both influence vapor pressure. In this case, both samples are at room temperature, so you need only consider the types of secondary forces present. From the molecular structures, you can see that both molecules are polar. What types of polar secondary forces are present for each? CH_3OH has an oxygen atom bonded to a hydrogen atom and can form hydrogen bonds. CH_3I is polar, but will not form hydrogen bonds. The dipole-dipole interactions among CH_3I molecules are weaker than the hydrogen bonds among CH_3OH molecules. Thus, it is easier for CH_3I molecules to escape the liquid and move to the gas phase, and the easier escape results in a higher vapor pressure. It is much harder for CH_3OH molecules to acquire the kinetic energy necessary to overcome hydrogen bonds in the liquid and move into the gas phase.

Problem 6.8 Looking at the molecular structures of ammonia and phosphine shown in the text, predict which has the higher boiling point.

Ammonia molecules can hydrogen bond to one another. Phosphine cannot form hydrogen bonds. Ammonia will have the higher boiling point because more energy will be required to move ammonia molecules from the liquid to the gas phase.

Objectives Section 6.8 Vaporization and the Regulation of Body Temperature

Are you able to. . .

- explain how vaporization regulates temperature in the body
- describe the relationship between evaporation and relative humidity
- explain how relative humidity is determined (Exercises 6.51, 6.52)
- explain why high relative humidity and high temperatures can be deadly

Objectives Section 6.9 Attractive Forces and the Structure of Solids

Are you able to. . .

- describe the molecular structure of solids (Exercises 6.47, 6.48)
- compare and contrast the molecular picture of solids, liquids, and gases
- explain the similarities and differences between a crystalline sample of substance and an amorphous sample of substance (Exercises 6.47, 6.48)
- describe the types of bonding and secondary forces that can be present in solids
- explain the relationship between melting points and the types and strengths of attractions between solid elements (Exercises 6.49, 6.50)
- describe the process of sublimation

Problem 6.9 Arrange the order of melting points for the following solids, from highest to lowest.

a) Br_2 b) I_2 c) F_2 d) Cl_2

All of these molecules are nonpolar because they are all diatomic elements. When molecules are nonpolar, weak London forces are the dominant secondary attractions and the strength of these depends on mass (number of electrons). These molecules can be arranged by mass from highest to lowest, and this order also corresponds to the melting point order.

$$I_2 > Br_2 > Cl_2 > F_2$$

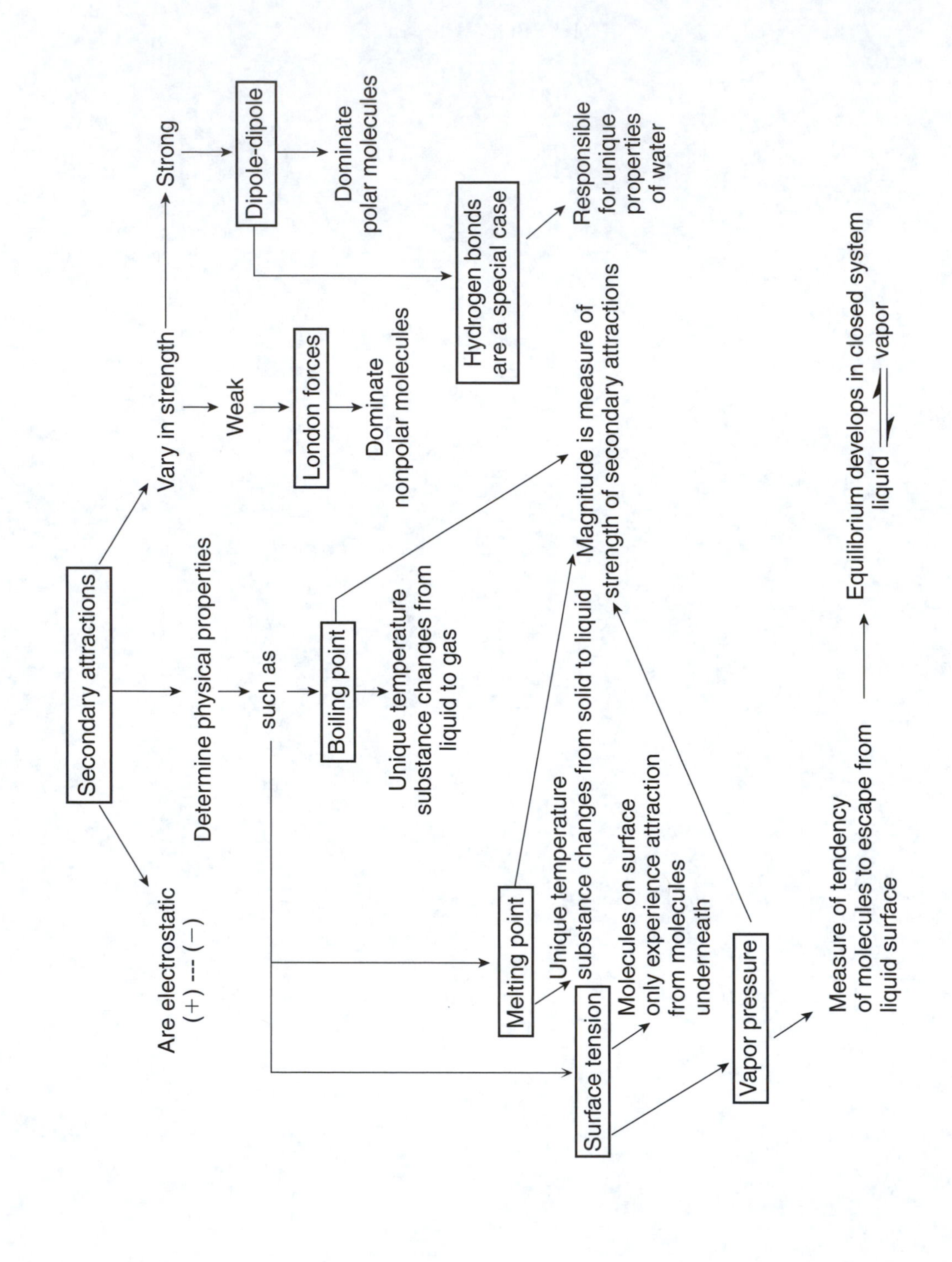
Secondary attractions
Are electrostatic (+) --- (−)
Determine physical properties
such as
Boiling point
Unique temperature substance changes from liquid to gas
Melting point
Unique temperature substance changes from solid to liquid
Surface tension
Molecules on surface only experience attraction from molecules underneath
Vapor pressure
Measure of tendency of molecules to escape from liquid surface
Equilibrium develops in closed system
liquid ⇌ vapor
Magnitude is measure of strength of secondary attractions
Vary in strength
Weak
London forces
Dominate nonpolar molecules
Strong
Dipole-dipole
Dominate polar molecules
Hydrogen bonds are a special case
Responsible for unique properties of water

FILL-INS

Matter can exist in three states: ______. The various states of a sample of matter differ in their ______ and ______ and in the ______ of the atoms or molecules in the sample. Solids have a ______ shape that is ______ under pressure. The atomic-level arrangement can be very ordered, in which case the solid is called a ______, or it can be disordered, in which case the solid is called an ______. Elements in solids can be held together by ______, ______, ______, or ______. The kinetic energy of constituents in a solid sample is ______. Elements are held ______ and have little kinetic energy. Liquids are ______ to compression: their volume changes ______, but they have a ______ that is determined by the container in which they are held. ______ in liquids can be dipole, hydrogen bonds, or weak London forces. Constituent liquid particles are able to ______ and have relatively ______ kinetic energy than those in solids. Gases are ______ to changes in pressure. Gas particles possess enough ______ that they are able to keep far apart from one another. Secondary forces are ______ in determining the properties of gases because the particles are so far apart that they cannot interact.

Matter can be ______ among the three states. These changes are called ______ and can be brought about by either ______ the kinetic energy of sample particles. The kinetic energy of a sample is directly related to the ______. An ______ in temperature results in an increase in kinetic energy for the sample particles and they move around faster and with more energy. A ______ in temperature reduces the kinetic energy of particles; they move more slowly and with less energy.

Phase transitions from solid to liquid are called ______ , and are the result of ______ of solid particles at the atomic level. The transformation of a liquid to a gas is called ______ and it requires that particles possess ______ to free themselves from the surface of a liquid sample. A solid can also be transformed directly into a gas through ______. The energy required to sublime a solid is equal to the ______. Each of these processes can be ______ as heat is removed from a sample. Gas to liquid phase transitions are called ______; liquid to solid changes are called ______.

The amount of heat required to bring about a phase transition is an ______ of a pure substance and can be used to help identify an unknown sample. Pure substances have ______ melting, boiling, and freezing points.

All phase transitions are related to the making or breaking of ______ between molecules. Molecules possessing ______ are able to form electrostatic interactions between ______ areas of different molecules. When a molecule has a ______ covalently bonded to an oxygen, nitrogen, or fluorine atom, the differences in electronegativities between the atoms produce a larger dipole. The interactions between these molecules are relatively ______ than normal dipole attractions and are called ______. They are not bonds in the sense of covalent bonds, but are ______between the relatively positive hydrogen atom of one molecule and the relatively negative oxygen, nitrogen, or fluorine on another molecule. The unique properties of water result from ______. Weak secondary attractions and hydrogen bonding ______ most of the physiological phenomena in our bodies. ______ interact by London forces, which are transitory charge separations that result from electron movement. ______ , the more electrons it has, and the more ______ London forces become.

Secondary forces can be arranged by relative strengths with the ______ being London forces and the being hydrogen bonds. ______ can indicate the type and magnitude of secondary attractions between molecules in a sample. ______ secondary attractions are not easily overcome and produce substances with ______ melting and boiling points and ______ vapor pressure. Substances dominated by London forces will have ______ melting and boiling points and relatively ______ vapor pressures. Surface tension is also determined by the nature of secondary attractions. When there is significant attraction for a ______ of molecules from the others beneath it, a sort of ______ is produced. This can prevent ______ of the surface under the right circumstances. ______ are molecules that contain both ______ areas. They are able to ______ surface tension and play an important part in respiratory function.

Solutions between different types of molecules form when the secondary forces between them are ______. Like dissolves like means that ______ will interact at the molecular level with other ______ molecules, but not with ______ molecules. Likewise, nonpolar molecules interact with other nonpolar molecules, but will not mingle atomically with ______.

Answers

Matter can exist in three states: solid, liquid, or gas. The various states of a sample of matter differ in their shape and volume and in the kinetic energy of the atoms or molecules in the sample. Solids have a fixed shape that is not easily compressed under pressure. The atomic-level arrangement can be very ordered, in which case the solid is called a crystalline solid, or it can be disordered, in which case the solid is called an amorphous solid. Elements in solids can be held together by covalent, ionic, dipole, or London forces. The kinetic energy of constituents in a solid sample is very small. Elements are held rigidly in place and have little kinetic energy. Liquids are resistant to compression: their volume changes very little, but they have a shape that is determined by the container in which they are held. Secondary attractions in liquids can be dipole, hydrogen bonds, or weak London forces. Constituent liquid particles are able to move around each other and have relatively more kinetic energy than those in solids. Gases are very responsive to changes in pressure. Gas particles possess enough kinetic energy that they are able to keep far apart from one another. Secondary forces are not significant in determining the properties of gases because the particles are so far apart that they cannot interact.

Matter can be transformed among the three states. These changes are called phase transitions and can be brought about by either increasing or decreasing the kinetic energy of sample particles. The kinetic energy of a sample is directly related to the temperature. An increase in temperature results in an increase in kinetic energy for the sample particles and they move around faster and with more energy. A decrease in temperature reduces the kinetic energy of particles; they move more slowly and with less energy.

Phase transitions from solid to liquid are called melting and are the result of increasing the kinetic energy of solid particles at the atomic level. The transformation of a liquid to a gas is called vaporization, and it requires that particles possess enough kinetic energy to free themselves from the surface of a liquid sample. A solid can also be transformed directly into a gas through sublimation. The energy required to sublime a solid is equal to the energy required to first melt, then vaporize it. Each of these processes can be reversed as heat is removed from a sample. Gas to liquid phase transitions are called condensations; liquid to solid changes are called freezing.

The amount of heat required to bring about a phase transition is an inherent physical property of a pure substance and can be used to help identify an unknown sample. Pure substances have characteristic melting, boiling, and freezing points.

All phase transitions are related to the making or breaking of secondary attractions between molecules. Molecules possessing dipoles are able to form electrostatic interactions between positive and negative areas of different molecules. When a molecule has a hydrogen atom covalently bonded to an oxygen, nitrogen, or fluorine atom, the differences in electronegativities between the atoms produce a larger dipole. The interactions between these molecules are relatively stronger than normal dipole attractions and are called hydrogen bonds. They are not bonds in the sense of covalent bonds, but are secondary attractions between the relatively positive hydrogen atom of one molecule and the relatively negative oxygen, nitrogen, or fluorine on another molecule. The unique properties of water result from hydrogen bonding. Weak secondary attractions and hydrogen bonding govern most of the physiological phenomena in our bodies. Nonpolar molecules interact by London forces, which are transitory charge separations that result from electron movement. The larger a molecule, the more electrons it has, and the more significant London forces become.

Secondary forces can be arranged by relative strengths with the weakest being London forces and the strongest being hydrogen bonds. Physical properties can indicate the type and magnitude of secondary attractions between molecules in a sample. Strong secondary attractions are not easily overcome and produce substances with higher melting and boiling points and lower vapor pressure. Substances dominated by London forces will have lower melting and boiling points and relatively higher vapor pressures. Surface tension is also determined by the nature of secondary attractions. When there is significant attraction for a surface layer of molecules from the others beneath it, a sort of elastic skin is produced. This can prevent penetration of the surface under the right circumstances. Surfactants are molecules that contain both polar and nonpolar areas. They are able to reduce surface tension and play an important part in respiratory function.

Solutions between different types of molecules form when the secondary forces between them are similar. Like dissolves like means that polar molecules will interact at the molecular level with other polar molecules, but not with nonpolar molecules. Likewise, nonpolar molecules interact with other nonpolar molecules, but will not mingle atomically with polar molecules.

TEST YOURSELF

1. Vegetable oil is a liquid at room temperature and does not mix with water. Margarine, which is mostly vegetable oil, is a solid at room temperature and does not mix with water. What can you say about the nature of secondary forces in vegetable oil and in margarine?
2. Does perfume have a relatively high or a relatively low vapor pressure? Explain your answer.
3. For each situation below, use melting, freezing, vaporization, condensation, or sublimation to describe the phase transition and explain your choice.
   (a) A steel beam is formed from molten iron.
   (b) On a cold day, car windows fog up on the inside.
   (c) The "air" around a gasoline nozzle appears wavy as you pump gas.
   (d) Dry ice (frozen carbon dioxide) looks "smoky" when placed on a lab bench.
   (e) A piece of chocolate softens in your mouth.
4. Describe the molecular activity and ultimate results of each set of circumstances.
   (a) A pot of water is heated at a constant temperature, below the boiling point, without a lid.
   (b) A pot of water is heated at constant temperature, below the boiling point, with a lid.
   (c) A pot of water is left outside in frigid weather.
   (d) A pot of water is heated to its boiling point.
5. What are three methods that can be used to determine whether a sample of clear liquid is water?
6. Design a method to determine the nature of secondary attractions in an unknown sample of white powder. How could you determine its identity?
7. What relative melting points would you expect from a covalent solid and from a molecular solid?
8. Draw the interface between a group of surfactant molecules and water in an open container. Do the same for a group of surfactant molecules in a closed container where there is no air, only water and surfactant molecules.
9. Repeat Problem 8 for a group of surfactant molecules in salad oil. How do you think the roles of surfactants would change in each situation (8 and 9)?
10. Design an experiment to distinguish between two clear liquids: one polar, the other nonpolar.

Answers

1. The fact that neither vegetable oil nor margarine mixes with water indicates that the interactions between the molecules in vegetable oil are nonpolar London attractions. The difference in their states points to weaker London forces in vegetable oil than in margarine. This difference implies that the molecules in margarine have more electrons (are heavier) than the molecules in salad oil. In fact, salad oil has hydrogen added to it to produce margarine. This addition changes both the weight and the shape of the molecules and promotes London forces over larger areas.
2. To be smelled, perfume must be in a gaseous state. Perfume must have a relatively high vapor pressure, as perfume molecules must continuously leave the liquid for the scent to be detected.
3. (a) Freezing. The liquid steel becomes solid steel. (b) Condensation. Warm water vapor from breath becomes liquid on the colder car windows. (c) Vaporization. Liquid gasoline has a high vapor pressure and gasoline molecules easily escape as gases. (d) Sublimation. At atmospheric pressure and room temperature, solid carbon dioxide changes directly to gaseous carbon dioxide without becoming a liquid. (e) Melting. The heat from your body supplies enough energy to break the secondary attractions in the solid.
4. (a) Some molecules will attain sufficient kinetic energy to break loose of the surface. Evaporation will occur from the surface and will continue until all the water is gone. (b) Some molecules will achieve sufficient kinetic energy to break loose of the surface. Eventually, a sufficient number of gas molecules will build up so that they will strike the liquid and reenter. An equilibrium will develop between the liquid and the gas phases. (c) The cold weather will decrease the kinetic energy of the molecules, particularly at the surface, and they will be able to form the hydrogen bonds that correspond to crystallization. Ice will form. (d) Vapor bubbles will form in the liquid as some molecules attain sufficient kinetic energy and the pressure equals atmospheric pressure. Eventually, the vapor bubbles work their way to the liquid surface and escape. Continued boiling will result in all the liquid vaporizing.

5. The boiling and freezing points could be determined because they are unique for water. The density could also be determined. Water has defined densities at specific temperatures.
6. The powder can be stirred into both a polar and a nonpolar liquid. The one in which it dissolves has the same type of secondary forces as are in the powder. Its identity can be determined by measuring its melting point.
7. Covalent crystals consist of elements held in place by covalent bonds. These would exhibit extremely high melting points (refer to Table 6.3). Molecular solids are held together by much weaker secondary attractions ranging from dipole to London forces. These solids would have a range of melting points ranging from very low for London forces to higher for dipole attractions. Relative to covalent solids, however, the melting points of molecular solids are very much lower.
8. In an open container of water, the molecules will align themselves on the surface. The polar part will be in the water and the nonpolar part will be outside the water, in the air. In a closed container, the molecules will aggregate into a sphere. The polar parts will point out into the water and the nonpolar parts will be inside the sphere, away from the water.
9. The situation is the reverse of the above. The nonpolar parts will be in the oil at the surface and the polar parts will be away from the oil. In the closed container the molecules will aggregate with their nonpolar part in the oil and the polar part inside the sphere away from the oil. In Problem 8, the sphere can carry nonpolar substances inside and transport them through an aqueous medium. In Problem 9, the sphere can carry polar substances inside and transport them through a nonpolar medium.
10. Any nonpolar liquid, such as salad oil, will not dissolve in a polar liquid because the nature of the secondary attractions is different. Try putting a small amount of salad oil in each sample. It will dissolve in the nonpolar, but not in the polar liquid.

chapter 7

Solutions

OUTLINE

7.1 General Aspects of Solution Formation

A. Solutions are
   1. mixtures of solvent and solute intermingled at the atomic or molecular level
   2. solvent: the substance in larger amount
   3. solute: the substance in smaller amount
   4. solvent and solute proportions can vary
      a. miscible liquids can mix in all proportions
      b. gas solutes and solvent particles mix in all proportions
   5. solutions can be gas in gas; liquid in liquid; solid in solid; liquid in solid; solid in liquid; gas in liquid; gas in solid; liquid in gas; solid in gas (Table 7.1)

B. Suspensions are
   1. visible particles in solvent: solid in gas; liquid in gas; solid in liquid
   2. colloidal when particles are very small (but not small enough to be solute particles); colloidal particles remain suspended and do not settle out

Key Terms miscible, solute, solution, solvent, suspension
Key Table Table 7.1 Nine types of mixtures

7.2 Molecular Properties and Solution Formation

A. Solution formation depends on secondary forces.
   1. solvent-solute interactions must be similar to solute-solute and solvent-solvent interactions
   2. solute-solvent interaction is called solvation

B. Water as a solvent is
   1. polar
   2. can dissolve polar molecules
      a. like dissolves like
      b. hydration is solvation by water molecules (Fig. 7.1)
   3. does not dissolve nonpolar solutes

Key Terms aqueous solution, hydration, solvation
Key Figure Figure 7.1 Solvation of sodium and chloride ions in water

7.3 Solubility

A. "Solubility" and "soluble" are different (Table 7.2).
   1. solubility is quantitative
      a. refers to amount of a specified solute that can dissolve in a given solvent at a specific temperature
      b. most solutes have a limit of solubility in a given solvent at a specific temperature (Table 7.3)

c. solutions having solute below solubility limit are unsaturated
d. solutions having as much solute as determined by solubility limit are saturated
2. soluble is qualitative; refers to ability of solute to dissolve in given solvent

Key Terms saturated solution, solubility, unsaturated solution
Key Tables
Table 7.2 Soluble and insoluble salts
Table 7.3 Solubility rules for ionic solids in water

7.4 Concentration

A. Concentration measures the amount of solute dissolved in a given amount of solvent (Figure 7.2).

$$\text{Concentration} = \frac{\text{amount of solute}}{\text{specified amount of solution}}$$

B. If you took a sample of a given amount of solution from any part of the solution the amount of solute would be the same as described by the concentration (Fig. 7.3); analogous to density.
C. Concentration expressions can be
1. mass of solute in mass of solution or mass of solute in volume of solution
   a. solute units: grams
   b. solution units: milligrams, grams, kilograms, milliliters, liters
2. number of moles of solute in volume of solution or number of moles of solute in mass of solution
   a. solute units: moles
   b. solution units: milliliters, liters, grams, kilograms
3. number of charges in volume of solution, number of particles in volume of solution, or number of particles in mass of solution
   a. solute units: milliequivalents or equivalents (charge); osmoles (particles)
   b. solution units: milliliters, liters, kilograms

Key Term concentration
Key Figures
Figure 7.2 Visualizing molarity as molecular density
Figure 7.3 Constant concentration of a solution

7.5 Percent Composition

A. Percent composition
1. is a concentration term that refers to the amount of solute in 100 g or in 100 mL of solution
   a. % w/w is the amount of solute (in grams) in 100 g of solution
   b. % v/v is the volume of liquid solute (in milliliters) in 100 mL of solution
   c. % w/v is the amount of solute (in grams) in 100 mL of solution
2. percent composition can be determined by dividing the amount of solute by the amount of solution and multiplying by 100%:

$$\frac{\text{amount solute (in grams or milliliters)}}{\text{amount solution (in grams or milliliters)}} \times 100\% = \text{percent concentration}$$

↑ this term is both solvent and solute amounts

B. Percent composition terms are used
1. to determine the amount of solute in a given amount of solution
2. to determine the amount of solution that would contain a given amount of solute
3. as conversion factors between amount of solute and amount of solution

C. To make a solution of a specified percent concentration:
1. measure out the desired amount of solute in grams or milliliters
2. add solute to an empty volumetric flask or graduated cylinder
3. add water to the desired volume (Fig. 7.4)

4. put stopper in flask and invert several times to mix; if in graduated cylinder, stir to mix
5. if the mass of water is used instead of the volume, then the density of water is used to convert the volume of water to the desired mass of water

Key Terms percent concentration, v/v percent, w/v percent, w/w percent
Key Figure Figure 7.4 Steps in the preparation of a w/v solution

Exercise 7.1 Preparing % w/w solutions
Problem 7.1 Preparation of a % w/w solution
Exercise 7.2 Determining mass of solute in a % w/w solution
Problem 7.2 Determining mass of solute in a given mass of solution from the % w/w
Exercise 7.3 Determining the volume of solution containing a desired mass of solute
Problem 7.3 Determining the volume of solution containing a desired mass of solute
Exercise 7.4 Preparing % v/v solutions
Problem 7.4 Preparing % v/v solutions
Exercise 7.5 Determining mass of solute in a given volume of % w/v solution
Problem 7.5 Determining mass of solute in a given volume of % w/v solution

7.6 Molarity

A. Molarity
1. is a concentration term that refers to the number of moles of solute in 1 L of solution
2. can be determined by dividing the number of moles of solute by the volume of solution in units of liters

$$\frac{\text{Moles of solute}}{\text{Volume of solution in liters}} = \text{molarity}$$

3. is abbreviated M̲ or *M*

B. Molarity is used
1. to determine the number of moles of solute in a given volume of solution
2. to determine the volume of a given solution that would contain a specified number of moles of solute
3. as a conversion factor between number of moles of solute and liters of solution

C. To make a solution of a specified molarity,
1. first determine the weight of solute in grams or kilograms that corresponds to the desired number of moles (use molar mass of solute to convert moles to grams)
2. measure out that amount of solute
3. add solute to empty volumetric flask or graduated cylinder
4. add water to desired volume
5. put stopper on flask and invert several times to mix; if in graduated cylinder, stir to mix

D. Concentration is sometimes expressed as millimoles per milliliter: 1000 mmol = 1 mol.

Key Term molarity

Exercise 7.6 Preparing molar solutions
Problem 7.6 Preparing molar solutions
Exercise 7.7 Determining mols of solute in given volume of solution
Problem 7.7 Determining mols of solute in given volume of solution
Exercise 7.8 Determining volume of a solution that will contain specified amount of solute
Problem 7.8 Determining volume of a solution that will contain specified amount of solute
Exercise 7.9 Calculating the molarity of a solution
Problem 7.9 Calculating the molarity of a solution
Exercise 7.10 Calculating concentrations in millimoles per milliliter
Problem 7.10 Calculating concentrations in millimoles per milliliter

7.7 Dilution

A. Solution preparation sometimes requires taking an amount of a more concentrated solution and diluting it to the desired concentration.
   1. same mass of solute, but in a larger volume of solvent

$$\underset{\text{initial concentration and volume needed to obtain}}{\text{Concentration}_1 \times V_1} = \underset{\text{final concentration and final volume}}{\text{Concentration}_2 \times V_2}$$

   2. volume and masses can be in any units as long as they are the same on both sides of the equation

$$\% \text{ conc}_1 \times \text{vol}_1 = \% \text{ conc}_2 \times \text{vol}_2$$
$$\text{Molarity}_1 \times \text{Vol}_1 = \text{Molarity}_2 \times \text{Vol}_2$$
$$\text{mg } \%_1 \times \text{Vol}_1 = \text{mg } \%_2 \times \text{Vol}_2$$

   3. provides an easy way of preparing solutions with very small amounts of solute

Key Term dilution

Exercise 7.11 Determining the concentration of a diluted solution
Problem 7.11 Determining the concentration of a diluted solution
Exercise 7.12 Preparing a dilute solution
Problem 7.12 Preparing a dilute solution

7.8 Concentration Expressions for Very Dilute Solutions

A Solutions containing a very small amount of solute per volume of solution are very dilute.
B. Concentration expressions for very dilute solutions are used to avoid decimals or scientific notation.
   1. mg % = number of mg of solute per 100 mL of solution
   2. parts per million (ppm) = number of parts of solute per million parts of solution; also expressed as mg of solute per liter of solution
   3. parts per billion (ppb) = number of parts of solute per billion parts of solution; also expressed as micrograms of solute per liter of solution

Key Terms parts per billion, parts per million

Exercise 7.13 Calculating the concentration of solute in a mg % solution
Problem 7.13 Calculating the concentration of solute in a mg % solution

7.9 The Solubility of Solids in Liquids

A. Maximum mass of a solid that can be dissolved in a specified volume of solvent at a given temperature is called the solubility of the solute in the solvent.
B. Solutions containing the maximum amount of solute are saturated.
   1. saturated solutions are equilibrium systems
      solute (solid) $\rightleftharpoons$ solute (dissolved)
   2. solid solute particles are dissolving at the same rate as dissolved solute particles are recrystallizing
C. Exceeding solubility limits in biological fluids can result in the formation of solids that cause problems.

Key Term solubility
Key Figure Figure 7.5 Solubilities of solids increase with increasing temperature

7.10 Insolubility Can Result in a Chemical Reaction

A. Solutions of ions can result in chemical reactions that produce solids (Figs. 7.6, 7.7, Tables 7.2, 7.3).
B. The process in solution can be represented by an ionic equation:
   $A^+(aq) + C^-(aq) + B^+(aq) + D^-(aq) \rightleftharpoons AD(s) + C^-(aq) + B^+(aq)$

C. Ions that do not form precipitate, but stay in solution, are called spectator ions (C^- and B^+, in the example above).

D. The process of precipitate formation can be represented by a net ionic equation: $A^+(aq) + D^-(aq) \rightleftharpoons AD(s)$.

E. Spectator ions are not included in the net ionic equation.

Key Terms complete ionic equation, insoluble, net ionic equation

Key Figures and Table

Figure 7.6 Solution of sodium chloride added to solution of silver nitrate

Figure 7.7 Yellow precipitate of lead iodide forms immediately when colorless aqueous solutions of lead nitrate and potassium iodide are mixed

Table 7.3 Qualitative Solubility Rules for Ionic Solids in Water

Example 7.14 Writing net ionic equations describing the formation of an insoluble precipitate

Problem 7.14 Writing net ionic equations describing the formation of an insoluble precipitate

7.11 Diffusion

A. Diffusion is the random motion of molecules that results in a net movement from an area of higher concentration (more molecules) to an area of lower concentration (fewer molecules).

1. in gases, molecules don't collide with each other often and diffusion occurs quickly; think of a perfume bottle being opened
2. in liquids, solute particles collide frequently with solvent molecules and diffusion occurs more slowly (Fig. 7.8)

B. Rate of diffusion is related to steepness of concentration gradient between areas.

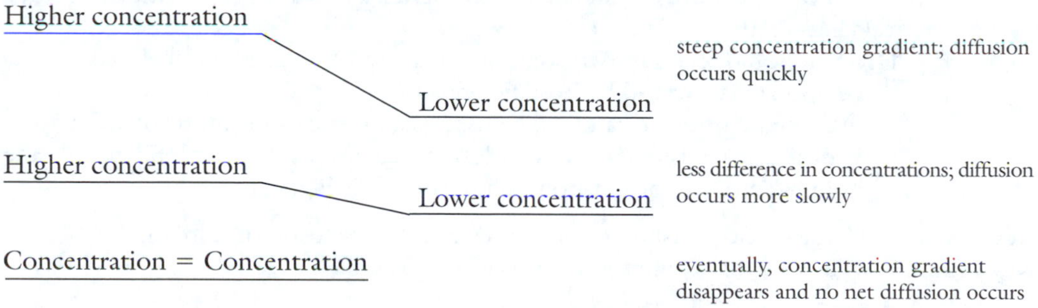

Higher concentration → Lower concentration — steep concentration gradient; diffusion occurs quickly

Higher concentration → Lower concentration — less difference in concentrations; diffusion occurs more slowly

Concentration = Concentration — eventually, concentration gradient disappears and no net diffusion occurs

C. When solutions of different concentrations are separated, net diffusion occurs from the compartment with more solute molecules (higher concentration) to the compartment with fewer molecules (lower concentration) (Fig. 7.9).

1. this net movement is totally due to random motion
2. movement is still occurring in all directions, that is, from area with more solute molecules to area with fewer solute molecules; but *net* diffusion is from higher concentration to lower concentration

Key Terms diffusion, solute

Key Figures and Box

Figure 7.8 Molecular model of a solute diffusing

Figure 7.9 Molecular model of the random motion of solute molecules

Box 7.1 Diffusion and the Cardiovascular System

7.12 Osmosis and Membranes

A. Diffusion can be modified when membranes separate solutions of different concentrations; membranes may have pores of varying sizes or may carry a net charge

1. pores allow smaller solutes to pass through membrane but keep larger solute particles from diffusing through

2. charges on membranes repel solute particles of like charge but allow solute particles of opposite charge to diffuse through
3. membranes can regulate movement of solute particles from one area into another
4. membranes that can regulate are called semipermeable membranes
5. biological membranes are semipermeable and thus very specific with respect to which solute molecules can diffuse through
6. diffusion of water (solvent) is not affected by semipermeable membrane

B. Osmosis is the net movement of water (solvent) through a semipermeable membrane from a area of lower solute concentration to an area of higher solute concentration (Fig. 7.10)
   1. net movement of water causes volume increase in the compartment with the higher solution concentration
   2. weight of water causes an increase in pressure that pushes solvent molecules back across the membrane
   3. liquid volumes stabilize when the influx of solvent molecules by diffusion equals the outflow of solvent molecules by pressure of solution

Key Terms membrane, osmosis, pressure, semipermeable membrane
Key Figure Figure 7.10 U-tube manometer measures osmotic pressure

7.13 Osmotic Pressure

A. Osmotic pressure is generated by a difference in concentration of solute particles between two solutions that are separated by a permeable membrane.
B. Osmotic pressure depends on the number of solute particles, not on their identity.
C. Hydrostatic pressure is generated by a difference in height or volume of liquid; it is mechanical (Fig. 7.10).
   1. when liquid levels in two solutions separated by a semipermeable do not change, then osmotic pressure and hydrostatic pressure are equal
   2. hydrostatic pressure can then be used as a measure of osmotic pressure
   3. hydrostatic pressure "pushes" solutions against a semipermeable membrane and can be used to separate solvent from solute (Box 7.2, Fig. 7.11)

Key Terms hydrostatic pressure, osmotic pressure, reverse osmosis, ultrafiltrate
Key Figure Figure 7.11 Commercial reverse-osmosis unit

7.14 Osmolarity

A. Osmolarity
   1. is a concentration term that refers to the number of solute particles in 1 L of solution
   2. can be determined by dividing the osmoles of solute by the volume of solution in units of liters

$$\frac{\text{Osmoles of solute}}{\text{Volume of solution in liters}} = \text{osmolarity}$$

   3. is abbreviated os*M*

B. Osmolarity is used
   1. to determine the number of osmoles of solute in a given volume of solution
   2. to determine the volume of a given solution that would contain a specified number of osmoles of solute
   3. as a conversion factor between osmoles of solute and liters of solution

C. To determine the number of osmoles produced by 1 mol of solute,
   1. write the dissociation equation for the solute in water.
      a. if solute is a nonelectrolyte, then 1 mol produces 1 osmol

$$C_6H_{12}O_6(s) \rightarrow C_6H_{12}O_6(aq)$$
$$1 \text{ mol glucose} \Rightarrow 1 \text{ osmol glucose}$$

b. if solute is an electrolyte, then 1 mole does not produce 1 osmol

$$NaCl(s) \rightarrow Na^{+}(aq) + Cl^{-}(aq)$$
$$1 \text{ mol NaCl} \Rightarrow 2 \text{ osmol ions}$$

D. Osmotic pressure is determined by multiplying the number of osmoles produced per mole by the number of moles of solute per liter then multiplying that by the gas constant and temperature in kelvins:

$$\Pi = t(n/V)RT$$

↑ t comes from the dissociation expression of 1 mol of solute

Key Terms osmol, osmolarity, osmole

7.15 Osmosis and the Living Cell

A. Solutions in cells have specific concentrations of ions and molecules that determine their osmotic pressures.

B. Three osmotic situations can exist relative to living cells
   1. the solutions inside and outside the cell are of equal concentration and osmotic pressure
      a. this is an isotonic situation
      b. there is no net change in volume of cell
   2. the solution outside the cell is less concentrated than the solution inside the cell
      a. this is a hypotonic situation
      b. there is a net movement of water into the cell, so it swells
      c. the cell may burst, a process called hemolysis in the case of a red blood cell
   3. the solution outside the cell is more concentrated than the solution inside the cell
      a. this is a hypertonic situation
      b. there is a net movement of water out of the cell, so it shrinks
      c. cell may shrivel, a process called crenation in the case of red blood cells

Key Terms crenation, hemolysis, hypertonic, hypotonic, isotonic

Key Figure and Box

Figure 7.12 Osmosis in erythrocytes

Box 7.3 The Osmotic Pressure of Isotonic Solutions

7.16 Macromolecules and Osmotic Pressure in Cells

A. Large molecules that are made of smaller molecular subunits covalently linked together are biological polymers; also called macromolecules (Table 7.4).

B. Macromolecules in biological systems are too large to diffuse through semipermeable membranes; this restriction of diffusion affects osmotic pressure across membrane (Boxes 7.2, 7.4).

C. Macromolecules are colloidal particles.
   1. they are not visible to the naked eye
   2. they do not settle out of solution
   3. a beam of light shining through a colloidal solution can be seen; called the Tyndall effect (Fig. 7.13)

Key Terms colloidal particle, macromolecule, polymer, Tyndall effect

Key Figures, Table, and Boxes

Figure 7.13 Colloidal particles produce the Tyndall effect

Figure 7.14 Typical spherical micelle

Table 7.4 Dimension and masses of biological molecules

Box 7.2 Semipermeability and Urine Formation

Box 7.4 Semipermeability and the Digestive System

Box 7.5 Association Colloids, Micelles, and Protein Structure

ARE YOU ABLE TO. . . AND WORKED TEXT PROBLEMS

Objectives Section 7.1 General Aspects of Solution Formation

Are you able to. . .

- identify the components of a solution (Expand Your Knowledge 7.67)
- explain how a solution forms
- describe different types of mixtures
- explain what is meant by miscible

Objectives Section 7.2 Molecular Properties and Solution Formation

Are you able to. . .

- explain the role of secondary forces in solution formation
- explain what is meant by solvation
- describe an aqueous solution
- explain what is meant by hydration

Objectives Section 7.3 Solubility

Are you able to. . .

- describe the qualitative and quantitative meaning of soluble and solubility
- describe unsaturated and unsaturated solutions
- identify water soluble compounds (Tables 7.2, 7.3)

Objectives Section 7.4 Concentration

Are you able to. . .

- define concentration
- identify the units used to express concentration

Objectives Section 7.5 Percent Composition

Are you able to. . .

- explain what is meant by percent concentration
- distinguish between w/w, w/v, and v/v percent concentration
- calculate the percent concentration of a solution (Exercises 7.45, 7.48, 7.49, 7.57)
- explain how to prepare a solution of specified concentration (Exs. 7.1, 7.4, Probs. 7.1, 7.4, Exercises 7.3, 7.4, 7.8, 7.9, 7.58, Expand Your Knowledge 7.70)
- use concentration terms as conversion factors
- use concentration terms in a variety of calculations (Exs. 7.2, 7.3, 7.5, Probs. 7.2, 7.3, 7.5, Exercises 7.1, 7.2, 7.5, 7.6, 7.7, 7.46, 7.47, 7.56)

Problem 7.1 How would you prepare 200 g of a 5.50% w/w aqueous solution of gelatin?

Think about what you would have to do to prepare an aqueous solution. . .you would first measure some amount of solute, put it into a container and add water (aqueous solution) to the appropriate volume. In this case you know how much solution you want: 200 g; you also know that every 100 g should contain 5.50 g of gelatin:

$$5.50\%\ \text{w/w}\ \textit{means}\ \frac{5.50\ \text{grams of solute}}{100\ \text{grams of solution}}$$

↑ solution weight = weight solute + weight solvent

NOTE: Whenever you see a % sign, you should think "per 100."

A common mistake with this type of problem is for the student to multiply or divide the number in front of the % sign by 100. *Don't do this! Leave the number just as it is!*

$$200\ \cancel{\text{g solution}} \times \frac{5.50\ \text{g gelatin}}{100\ \cancel{\text{g solution}}} = 11.0\ \text{g gelatin}$$

The 200 g of solution must contain 11.0 g of gelatin, so the remainder of the weight is water: 200 g solution −11.0 g gelatin = 189 g water. 11.0 g of gelatin is added to 189 g of water to produce 200 g of a 5.50% w/w gelatin solution. Check yourself.

$$\frac{11.0 \text{ g gelatin}}{200 \text{ g solution}} \times 100 = 5.50\% \text{ gelatin}$$

You can also draw a picture to help you get to the answer:

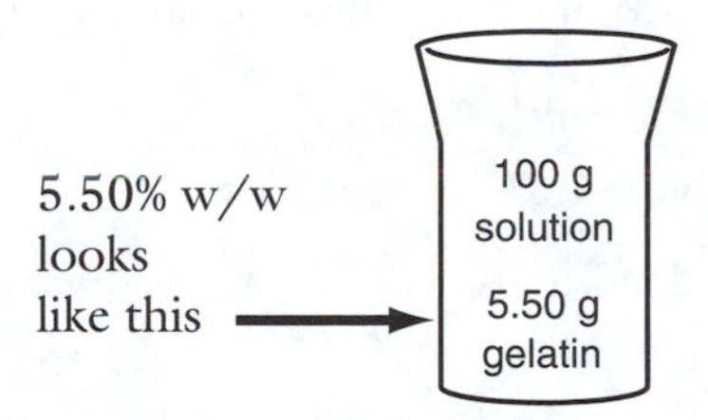

Use this to figure out how many grams of gelatin would be in 200 g of solution.

Problem 7.2 Phenol is a strong germicide. Calculate the number of grams of phenol present in 1.0 g of a 0.20% w/w aqueous solution.

Remember what the % term tells you: 0.20% phenol *means* that every 100 g of the aqueous solution contains 0.20 g of phenol.

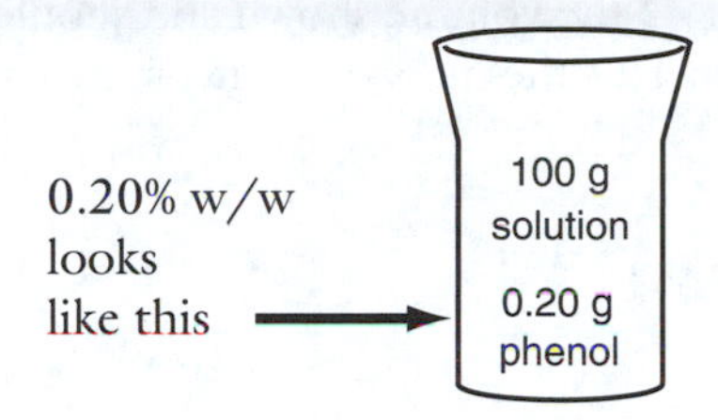

Use this to figure out how many g of phenol would be in 1.0 g of solution.

$$1.0 \cancel{\text{ g solution}} \times \frac{0.20 \text{ g phenol}}{100 \cancel{\text{ g solution}}} = 0.0020 \text{ g phenol}$$

Problem 7.3 How many grams of a 6.30% w/w solution of $CaCl_2$ must be used to obtain 27.6 g of $CaCl_2$?

Draw a picture:

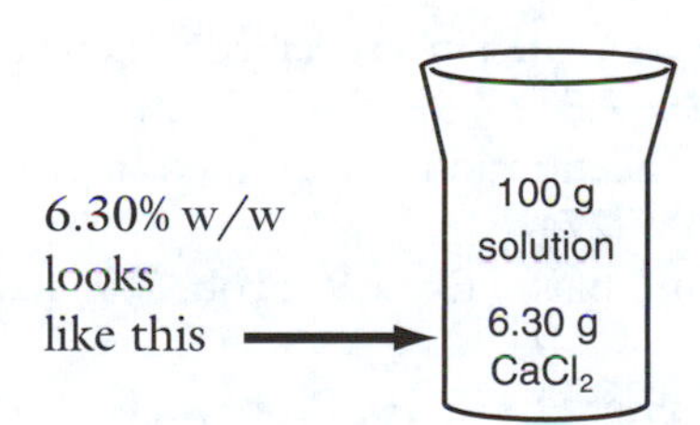

You could start drawing similar 100-g units until you get close to 27.6 g of $CaCl_2$. This gives you a ballpark idea of what the answer should be.

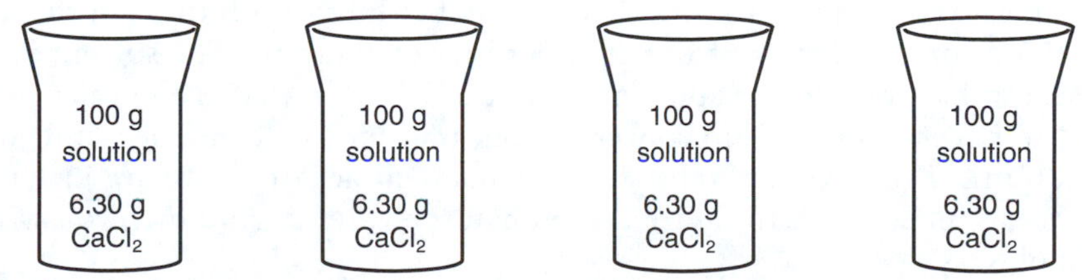

Four of these 100-g units will give you 25.20 g $CaCl_2$; five of the units will give you 31.50 g $CaCl_2$. The answer, then, should be more than 400 g of solution, but less than 500 g of solution.

$$27.6 \cancel{\text{ g } CaCl_2} \times \frac{100 \text{ g solution}}{6.30 \cancel{\text{ g } CaCl_2}} = 438 \text{ g of solution} \quad \text{just as predicted!}$$

Problem 7.4 How would you prepare an aqueous 4.8% v/v solution of acetone?

Remember that 4.8% v/v aqueous means that the solvent is water and that every 100.0 mL of solution contains 4.8 mL of acetone.

$$4.8\%\ \text{v/v}\ \textit{means}\ \frac{4.8\ \text{mL acetone}}{100.0\ \text{mL solution}}$$

↑ solution volume = solute volume + solvent volume

The solution would be 100.0 mL, and 4.8 mL of that would be acetone; the remainder would be water:

$$100.0\ \text{mL solution} - 4.8\ \text{mL acetone} = 95.2\ \text{mL water}$$

95.2 mL of water would be added to 4.8 mL of acetone to give a 4.8% v/v solution.

Problem 7.5 Calculate the number of grams of NaCl contained in 55 mL of a 12% w/v aqueous solution of NaCl.

Draw a picture:

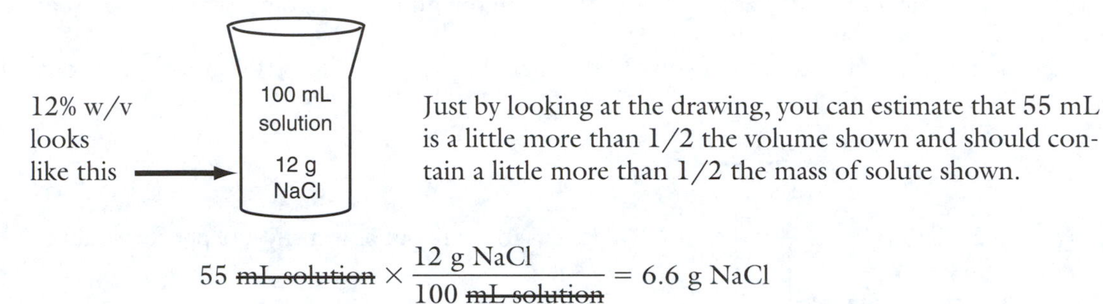

12% w/v
looks
like this
100 mL
solution
12 g
NaCl
Just by looking at the drawing, you can estimate that 55 mL is a little more than 1/2 the volume shown and should contain a little more than 1/2 the mass of solute shown.

$$55\ \cancel{\text{mL solution}} \times \frac{12\ \text{g NaCl}}{100\ \cancel{\text{mL solution}}} = 6.6\ \text{g NaCl}$$

Objectives Section 7.6 Molarity

Are you able to. . .

- define molarity
- prepare a solution of specified molarity (Ex. 7.6, Prob. 7.6, Exercises 7.54, 7.55)
- use molar mass to convert between grams and moles
- use molarity as a conversion factor
- use molarity to calculate the number of moles or mass of solute contained in a specified volume of solution (Ex. 7.7, Prob. 7.7, Exercises 7.15, 7.18, 7.19, 7.24, 7.25, 7.51, 7.52, 7.53, 7.59)
- use molarity to calculate the volume of a solution of specified concentration that contains a given mass of solute (Ex. 7.8, Prob. 7.8, Exercises 7.12, 7.20, 7.21, 7.26, 7.27)
- calculate the molarity of a solution if given a volume and solute mass (Ex. 7.9, Prob. 7.9, Exercises 7.16, 7.17, 7.22, 7.23, 7.50, 7.60)
- calculate the concentration of a solution in millimoles per milliliter (Ex. 7.10, Prob. 7.10)

Problem 7.6 How would you prepare 500 mL of a 0.275 M aqueous solution of KH_2PO_4, and how many grams of KH_2PO_4 would be needed?

Molarity is defined as the number of moles of solute per liter of solution. In this case, the concentration is 0.275 M or 0.275 mol of KH_2PO_4 per liter of solution. The solvent is water. Preparation would require that you measure out an amount of KH_2PO_4 that is equivalent to the number of moles needed to get the appropriate concentration, transfer to a container, and add water until the volume is 500 mL. The problem tells you how many moles of solute are in 1000 mL (1 L). You don't have to do much math to figure that in half the volume, 500 mL, you would have half the number of moles.

$$500\ \text{mL solution} \times \frac{0.275\ \text{mol}\ KH_2PO_4}{1000\ \text{mL solution}} = 0.138\ \text{mol}\ KH_2PO_4$$

This tells you that 0.138 mol KH_2PO_4 in 500 mL is the same concentration as a 0.275 *M* KH_2PO_4 solution. Check this:

$$\frac{0.138 \text{ mol } KH_2PO_4}{0.500 \text{ L}} = \frac{0.275 \text{ mol } KH_2PO_4}{1 \text{ L}}$$

Once you know how many moles are needed, you convert to grams by using the molar mass:

$$0.138 \text{ mol } KH_2PO_4 \times \frac{136.1 \text{ g } KH_2PO_4}{1 \text{ mol } KH_2PO_4} = 18.8 \text{ g } KH_2PO_4$$

18.8 g of KH_2PO_4 are put into a container and water is added until the volume is 500 mL to produce a concentration of 0.275 *M* KH_2PO_4.

Problem 7.7 How many moles of solute are contained in 0.680 L of a 1.35 *M* solution?

The molarity term *means* $\frac{1.35 \text{ mol solute}}{1 \text{ L solution}}$ ← this can be used as a conversion factor to move between moles of solute and volume of solution

$$0.680 \cancel{\text{ L solution}} \times \frac{1.35 \text{ mol solute}}{1 \cancel{\text{ L solution}}} = 0.918 \text{ mol solute}$$

The answer makes sense: Because you have more than one-half liter of the solution, you would expect more than one-half the amount of solute.

Problem 7.8 What volume of a 0.750 *M* solution will contain 1.25 mol of solute?

The molarity term *means* $\frac{0.750 \text{ mol solute}}{1 \text{ L solution}}$ ← this can be used as a conversion factor to move between moles of solute and volume of solution

$$1.25 \cancel{\text{ mol solute}} \times \frac{1 \text{ L solution}}{0.750 \cancel{\text{ mol solute}}} = 1.67 \text{ L solution}$$

The answer makes sense: Because 1 liter contains 0.750 mol, 1.25 mol should be contained in a volume larger than 1 L.

Problem 7.9 What is the molarity of a KCl solution in which 20.6 g of KCl is dissolved in 920 mL of solution?

When calculating molarity, you want to know how many moles of solute are in a specified volume (in liters) of solution. In this case, you are given the mass of the solute in grams, which must be changed to moles. You are also given the volume in milliliters, which must be changed to liters in order for the units to be those of molarity.

you have → $\frac{20.6 \text{ g KCl}}{920 \text{ mL solution}}$ ← change to moles (numerator); ← change to liters (denominator)

$$20.6 \cancel{\text{ g KCl}} \times \frac{1 \text{ mol}}{74.55 \cancel{\text{ g KCl}}} = 0.276 \text{ mol KCl} \qquad 920 \cancel{\text{ mL}} \times \frac{1 \text{ L}}{1000 \cancel{\text{ mL}}} = 0.920 \text{ L}$$

$$\frac{0.276 \text{ mol KCl}}{0.920 \text{ L}} = 0.300\ M$$

Problem 7.10 Express the concentration in millimoles per milliliter of a solution in which 0.259 g of $Ca(OH)_2$ is dissolved in 175 mL of solution.

$$\text{you have} \rightarrow \frac{0.259 \text{ g } Ca(OH)_2 \leftarrow \text{change to moles, then to millimoles}}{175 \text{ mL solution} \leftarrow \text{leave alone}}$$

$$0.259 \text{ g } Ca(OH)_2 \times \frac{1 \text{ mol}}{74.10 \text{ g}} = 0.00350 \text{ mol } Ca(OH)_2$$

$$0.00350 \text{ mol } Ca(OH)_2 \times \frac{1000 \text{ millimol}}{1 \text{ mol}} = 3.50 \text{ millimol } Ca(OH)_2$$

$$\frac{3.50 \text{ mmol } Ca(OH)_2}{175 \text{ mL solution}} = \frac{0.0200 \text{ mmol}}{1 \text{ mL}}$$

Objectives Section 7.7 Dilution

Are you able to. . .

- explain what is meant by dilute solution
- outline the steps for preparing a solution of specified concentration
- explain how to prepare a diluted solution
- use the dilution equation $M_1V_1 = M_2V_2$ to calculate any one variable if given the other three (Exs. 7.11, 7.12, Probs. 7.11, 7.12, Exercises 7.28, 7.29, 7.30, 7.31, 7.32, 7.33, 7.34, 7.35, 7.61, 7.62, 7.63, 7.64, Expand Your Knowledge 7.68)
- recognize the lab situation that a dilution problem represents

Problem 7.11 What is the final volume of a 0.00750 *M* solution prepared by dilution of 75.0 mL of a 0.180 *M* solution?

Think about what is going on: You have 75.0 mL of a solution that contains 0.180 mol of solute (draw a picture to help visualize this). You have to add enough water to change the concentration to 0.0750 mol. The question is, After you have added enough water to dilute the concentration, what will the final volume be? The dilution formula is $M_1V_1 = M_2V_2$.

$$\text{you have} \rightarrow M_1 = 0.180\ M \qquad M_2 = 0.00750\ M$$
$$V_1 = 75.0 \text{ mL} \qquad V_2 = \ ? \leftarrow \text{you want this}$$

Rearrange the dilution equation to solve for V_2:

$$V_2 = \frac{M_1V_1}{M_2} = \frac{(0.180\ \cancel{M})\ (75.0 \text{ mL})}{0.00750\ \cancel{M}} = 1800 \text{ mL} = 1.80 \times 10^3 \text{ mL} = 1.80 \text{ L}$$

This makes sense: The initial molarity is diluted by more than twenty times to get the final molarity, so the volume increase should be a little more than twenty times the original volume.

Problem 7.12 What volume of a 0.730 *M* solution must be obtained to prepare 1.36 L of a 0.270 *M* solution?

Think about what is going on: You want to prepare 1.36 L of a solution that contains 0.270 mol of solute (draw a picture to help visualize this). You have to add enough water to dilute the concentration of a 0.730 *M* solution. The question is, What volume of the more concentrated solution is required to get to the final dilution? The dilution formula is $M_1V_1 = M_2V_2$.

$$\text{you have} \rightarrow M_1 = 0.730\ M \qquad M_2 = 0.270\ M$$
$$V_1 = \ ? \qquad V_2 = 1.36 \text{ L}$$

Rearrange the dilution equation to solve for V_1:

$$V_1 = \frac{M_2 V_2}{M_1} = \frac{(0.270\ \cancel{M})\ (1.36\ \text{L})}{0.730\ \cancel{M}} = 0.503\ \text{L}$$

This makes sense: The initial molarity is more than twice as concentrated as the final dilution, so you would expect to start with less than half the volume.

Objectives Section 7.8 Concentration Expressions for Very Dilute Solutions

Are you able to. . .

- explain what is meant by very dilute solution
- recognize the concentration terms used to describe very dilute solutions
- define mg %
- calculate mg % concentrations (Exercises 7.10, 7.11)
- use mg % as a conversion factor (Ex. 7.13, Prob. 7.13)
- explain what is meant by ppm and ppb
- calculate the ppm or ppb of a solution (Exercise 7.12)
- calculate the concentration of a solution in mass per liter from ppm and ppb data (Exercise 7.13)

Problem 7.13 Assuming that the normal serum calcium concentration is 9.0 mg %, calculate its concentration in grams per milliliter.

$$9.0\ \text{mg}\ \%\ \textit{means}\ \frac{9.0\ \text{mg Ca}}{100\ \text{mL serum}} \quad \begin{matrix}\leftarrow \text{change to grams} \\ \leftarrow \text{leave alone}\end{matrix}$$

$$9.0\ \cancel{\text{mg}}\ \text{Ca} \times \frac{1\ \text{g}}{1000\ \cancel{\text{mg}}} = 0.0090\ \text{g Ca}$$

$$\frac{0.0090\ \text{g Ca}}{100\ \text{mL serum}} = \frac{9.0 \times 10^{-5}\ \text{g Ca}}{\text{mL serum}}$$

Objectives Section 7.9 The Solubility of Solids in Liquids

Are you able to. . .

- describe the dynamics of dissolution that produce a saturated solution
- describe the relationship between dissolution and crystallization of solute in a saturated solution (Exercise 7.36)
- use solubility data to discuss solution phenomena (Exercise 7.37)
- use solubility data to calculate other concentration data (Exercises 7.38, 7.39)
- explain how solubility limits in physiological solutions can be exceeded and what the consequences may be

Objectives Section 7.10 Insolubility Can Result in a Chemical Reaction

Are you able to. . .

- explain how the formation of an insoluble compound occurs in aqueous solution
- determine whether a solute will dissolve in water
- use solubility data to predict whether a reaction will take place between compounds in aqueous solution (Ex. 7.14, Prob. 7.14, Exercises 7.65, 7.66)
- write the ionic equation for the dissolution of a soluble compound (Ex. 7.14, Prob. 7.14, Exercises 7.65, 7.66)
- write the net ionic equation for the formation of an insoluble precipitate (Ex. 7.14, Prob. 7.14, Exercises 7.65, 7.66)

Problem 7.14 Write (a) the complete ionic equation and (b) the net ionic equation for the reaction that takes place when calcium chloride, $CaCl_2$, is added to sodium phosphate, Na_3PO_4. (Refer to Table 7.3.)

To get a picture of what is happening in the solution, you need to determine whether both $CaCl_2$ and Na_3PO_4 are soluble in water. From Table 7.3, you can tell that Na_3PO_4 is soluble, but it doesn't contain any information about $CaCl_2$. Look back in Section 7.10 for information about $CaCl_2$, which is also soluble in water. Once you have determined that both compounds are soluble, you can draw a picture that shows which ions are present with water molecules: Na^+, PO_4^{3-}, Ca^{2+} , and Cl^-. The ionic equations for the dissolution of each of these are

$$CaCl_2\ (s) \rightarrow Ca^{2+}\ (aq) + 2Cl^-\ (aq)$$
$$Na_3PO_4\ (s) \rightarrow 3Na^+\ (aq) + PO_4^{3-}\ (aq)$$

adding these together and balancing to produce the complete ionic equation →

$$3Ca^{2+}(aq) + 6Cl^-(aq) + 6Na^+(aq) + 2PO_4^{3-}(aq) \rightarrow Ca_3(PO_4)_{2(s)} + 6Na^+(aq) + 6Cl^-(aq)$$

From Table 7.3, you see that calcium phosphate, $Ca_3(PO_4)_2$ is an insoluble salt.

The net ionic equation is

$$3Ca^{2+}(aq) + 2PO_4^{3-}(aq) \rightarrow Ca_3(PO_4)_2\ (s)$$

Objectives Section 7.11 Diffusion

Are you able to. . .
- define diffusion
- describe the relationship between diffusion rates and concentration gradients
- describe the random motion of a sample of molecules
- describe diffusive movement between solutions of differing concentrations

Objectives Section 7.12 Osmosis and Membranes

Are you able to. . .
- define membrane and semipermeable
- explain what membrane characteristics affect diffusion
- describe the process of osmosis
- describe the movement of water and solutes across a semipermeable membrane
- describe hydrostatic pressure

Objectives Section 7.13 Osmotic Pressure

Are you able to. . .
- describe osmotic pressure
- explain how osmotic and hydrostatic pressure differ
- predict water movement between compartments of different hydrostatic and osmotic pressures (Expand Your Knowledge 7.69)
- provide biological examples of hydrostatic pressure
- describe reverse osmosis

Objectives Section 7.14 Osmolarity

Are you able to. . .
- define osmolarity
- describe, qualitatively, the relationship between osmotic pressure and osmolarity
- explain the mathematical relationship between osmotic pressure and osmolarity
- determine the number of solute particles produced by a compound
- calculate the number of osmoles produced by one mole of a compound
- calculate the osmolarity of a solution from its molarity or percent concentration (Exercises 7.40, 7.41, Expand Your Knowledge 7.71)

Objectives Section 7.15 Osmosis and the Living Cell

Are you able to. . .

- define isotonic, hypertonic, and hypotonic
- determine the relative tonicity of two solutions
- define hemolysis and crenation
- predict what will happen when two solutions of differing tonicities are separated by a semipermeable membrane (Exercises 7.42, 7.43, 7.44)

Objectives Section 7.16 Macromolecules and Osmotic Pressure in Cells

Are you able to. . .

- define macromolecule and polymer
- describe the activity of colloidal particles
- explain the relationship between colloidal particles and osmotic pressure
- describe the Tyndall effect

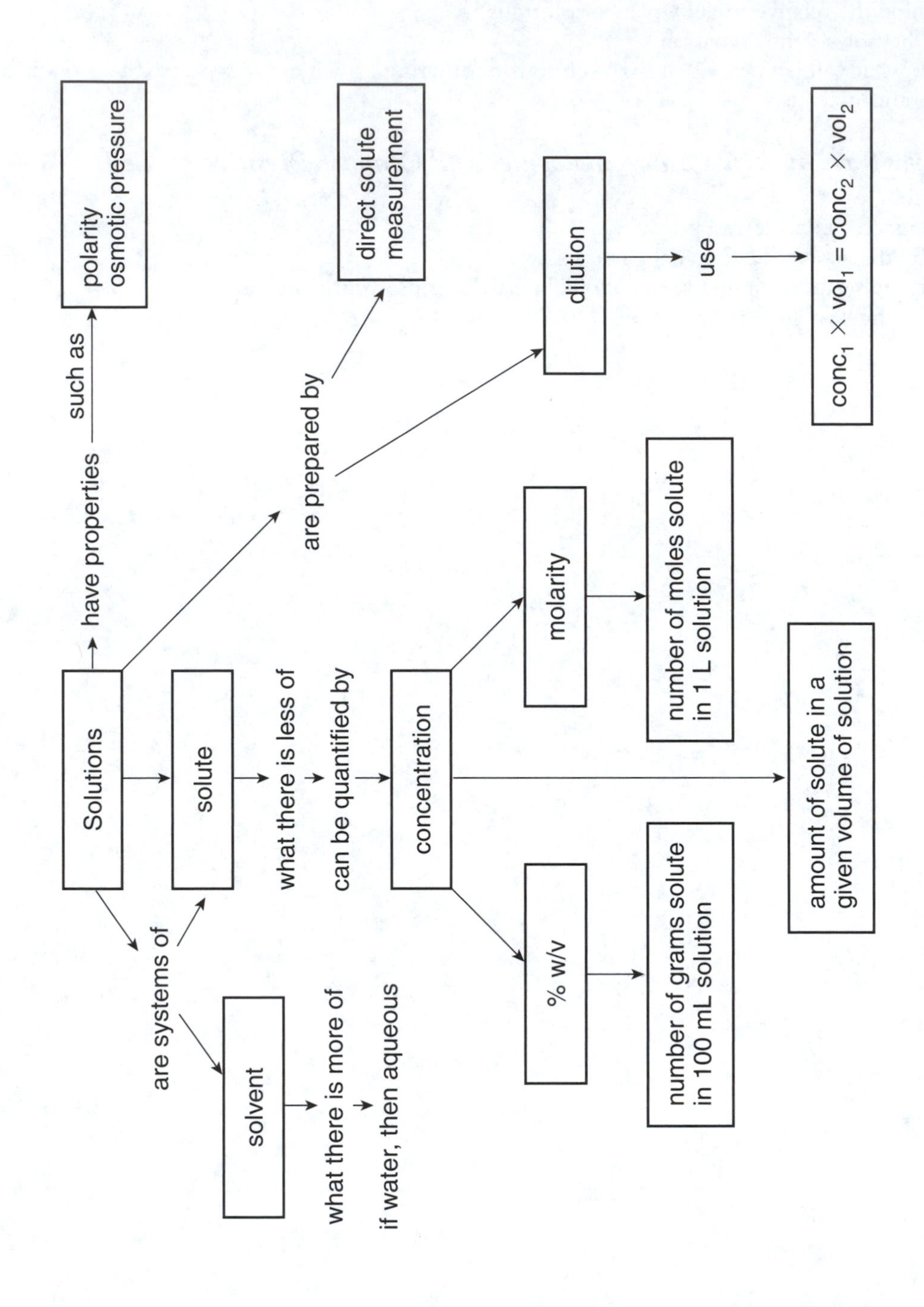
Solutions
have properties
such as
polarity
osmotic pressure
are prepared by
direct solute measurement
dilution
use
$conc_1 \times vol_1 = conc_2 \times vol_2$
are systems of
solute
what there is less of
can be quantified by
concentration
molarity
number of moles solute in 1 L solution
amount of solute in a given volume of solution
% w/v
number of grams solute in 100 mL solution
solvent
what there is more of
if water, then aqueous

FILL-INS

Solutions result when solute and solvent particles are able to intermingle at the ______ level. This intermingling occurs when the nature of the secondary attraction forces between solute particles are ______ to those between solvent particles. ______ can be single ions, polyatomic ions, or small molecules. Polar solutes can form attractions with ______ solvent particles. Nonpolar solutes ______ form attractions with polar solvent molecules and will not form solutions with them. This observation leads to the statement ______. When particles are too large to mingle at the atomic level, but still have the ability to form attractions with solvent molecules, a ______ is formed. Colloidal particles are not visible to the naked eye but cause light passing through a solution to be reflected, a phenomenon called the ______. Colloidal particles ______ settle out of solution, but remain suspended. Proteins in blood are ______ particles. Particles larger than colloidal size are ______ to the eye and can settle out of the mixture. Any medication that needs to be shaken is a ______.

The amount of solute that can be dissolved in a solvent depends on the ______ and ______. Most solutes have a ______ in a given solvent. This is the ______ amount of solute that can be dissolved in that solvent at the specified temperature. When a solution contains ______ than it can hold at a specific temperature, it is unsaturated and more solute can be added and dissolved. When the maximum amount of solute has been added to a solvent, the solution is ______. Any additional solute will ______. The relationship between dissolved and undissolved solid solute is an ______ process when the solution is saturated. Solute particles are ______ at the same rate that dissolved solute particles are ______.

The amount of solute in a volume of solution is quantified by ______ terms. Concentration terms can be used to measure the mass of solute in 100 mL of solution, a ______; the mass of solute in 100 grams of solution, a ______; or the volume of solute in 100 mL of solution, ______. ______ is a concentration term that measures the number of moles of solute in a liter of solution and ______ is a measure of the number of particles of solute in a liter of solution. When there are very small amounts of solute in a solution, the solution is ______ and concentration may be expressed as the number of milligrams of solute per 100 mL of solution or ______; the number of milligrams of solute per liter of solution, also called ______; or as the number of micrograms of solute per liter of solution, also called ______.

Solution preparation always requires that a specified amount of solute be ______ and transferred to a volumetric flask or graduated cylinder. ______ is added until the desired volume is reached. It is easier and more efficient to prepare ______ that can be used as a starting point for more dilute or very dilute solutions. The equation ______ is used to determine the volume of more concentrated solution needed to produce a more dilute solution of given concentration.

Solute particles move by ______ from areas of higher concentration to areas of lower concentration. Diffusion is a result of the ______ motion of particles, and the ______ at which it occurs depends on the magnitude of the concentration differences between areas of a solution. The ______ the concentration difference, the more quickly diffusion occurs. When solutions are separated by semipermeable membranes, as happens in biological systems, the tendency for water to diffuse from more dilute to less dilute solutions produces a ______ against the membrane as the volume increases in the more concentrated solution. The movement of water is called ______. The ______ of the solution is defined as the pressure of solution on one side of a membrane that is enough to push solvent particles back through the membrane and stop a net flow of solvent. Hydrostatic and osmotic pressures are equal when ______ occurs between separated solutions.

Cells can be affected differently by being in solutions that are more, less, or equally as concentrated as the solution in the cell. When a cell is surrounded by a more concentrated solution, water will ______ the cell, causing it to ______. The solution outside the cell is said to be "______." In extreme cases involving red blood cells, the cells may become ______. When a cell is surrounded by a more dilute solution, water ______ the cell, causing it to ______. The solution outside the cell is said to be "______." In extreme cases involving red blood cells, the cells may ______.

Answers

Solutions result when solute and solvent particles are able to intermingle at the atomic level. This intermingling occurs when the nature of the secondary attraction forces between solute particles are very similar to those between solvent particles. Solutes can be single ions, polyatomic ions, or small molecules.

Polar solutes can form attractions with polar solvent particles. Nonpolar solutes cannot form attractions with polar solvent molecules and will not form solutions with them. This observation leads to the statement "like dissolves like." When particles are too large to mingle at the atomic level, but still have the ability to form attractions with solvent molecules, a colloidal solution is formed. Colloidal particles are not visible to the naked eye but cause light passing through a solution to be reflected, a phenomenon called the Tyndall effect. Colloidal particles do not settle out of solution, but remain suspended. Proteins in blood are colloidal particles. Particles larger than colloidal size are visible to the eye and can settle out of the mixture. Any medication that needs to be shaken is a suspension.

The amount of solute that can be dissolved in a solvent depends on the nature of the solute and solvent and the temperature. Most solutes have a limited solubility in a given solvent. This is the maximum amount of solute that can be dissolved in that solvent at the specified temperature. When a solution contains less solute than it can hold at a specific temperature, it is unsaturated and more solute can be added and dissolved. When the maximum amount of solute has been added to a solvent, the solution is saturated. Any additional solute will remain undissolved in the solution. The relationship between dissolved and undissolved solid solute is an equilibrium process when the solution is saturated. Solute particles are dissolving at the same rate that dissolved solute particles are recrystallizing.

The amount of solute in a volume of solution is quantified by concentration terms. Concentration terms can be used to measure the mass of solute in 100 mL of solution, a w/v %; the mass of solute in 100 grams of solution, a w/w %; or the volume of solute in 100 mL of solution, a v/v %. Molarity is a concentration term that measures the number of moles of solute in a liter of solution and osmolarity is a measure of the number of particles of solute in a liter of solution. When there are very small amounts of solute in a solution, the solution is very dilute and concentration may be expressed as the number of milligrams of solute per 100 mL of solution or mg %; the number of milligrams of solute per liter of solution, also called parts per million; or as the number micrograms of solute per liter of solution, also called parts per billion.

Solution preparation always requires that a specified amount of solute be weighed out and transferred to a volumetric flask or graduated cylinder. Water is added until the desired volume is reached. It is easier and more efficient to prepare concentrated solutions that can be used as a starting point for more dilute or very dilute solutions. The equation (concentration 1) × (volume 1) = (concentration 2) × (volume 2) is used to determine the volume of more concentrated solution needed to produce a more dilute solution of given concentration.

Solute particles move by diffusion from areas of higher concentration to areas of lower concentration. Diffusion is a result of the random motion of particles, and the rate at which it occurs depends on the magnitude of the concentration differences between areas of a solution. The greater the concentration difference, the more quickly diffusion occurs. When solutions are separated by semipermeable membranes, as happens in biological systems, the tendency for water to diffuse from more dilute to less dilute solutions produces a hydrostatic pressure against the membrane as the volume increases in the more concentrated solution. The movement of water is called osmosis. The osmotic pressure of the solution is defined as the pressure of solution on one side of a membrane that is enough to push solvent particles back through the membrane and stop a net flow of solvent. Hydrostatic and osmotic pressures are equal when no net movement of solvent occurs between separated solutions.

Cells can be affected differently by being in solutions that are more, less, or equally as concentrated as the solution in the cell. When a cell is surrounded by a more concentrated solution, water will leave the cell, causing it to shrink. The solution outside the cell is said to be "hypertonic." In extreme cases involving red blood cells, the cells may become crenated. When a cell is surrounded by a more dilute solution, water moves into the cell, causing it to swell. The solution outside the cell is said to be "hypotonic." In extreme cases involving red blood cells, the cells may hemolyze.

TEST YOURSELF

1. Consider a container that holds 25 white jelly beans and 100 red jelly beans.
   (a) Which jelly beans are the solvent beans?
   (b) Which jelly beans are the solute beans?
   (c) How many jelly beans are in the solution?
   (d) What is the percent concentration of white jelly beans?

2. If 1 mL of water can hold 1 g of glucose at room temperature,
   (a) what is the solubility of glucose per 100 mL of water?
   (b) what are the units of concentration in this case?
   (c) what is the percent concentration of glucose?
3. In the glucose solution above, at the point of saturation, what is the molarity of the glucose solution?
4. Glucose is a nonelectrolyte: 1 mol of glucose produces 1 osmol in aqueous solution. At the point of saturation, what is the osmolarity of the glucose solution?
5. Water and ethanol (the alcohol in alcoholic beverages) can mix in all proportions. Look up the structures of water and ethanol and explain this observation.
6. Draw the random motion of a molecule that is able to move freely through space.
7. Consider 5% and a 10% glucose solutions.
   (a) Draw the molecular level picture of each solution. What does 100 mL of each solution look like?
   (b) On the basis of your drawing, which solution is more dense?
   (c) Imagine that a semipermeable membrane separates these two solutions. Describe the solute and solvent interactions at the membrane surface on both sides.
   (d) If diffusion of the solvent occurs between the solutions, in which solution will the amount of solvent increase?
   (e) If diffusion occurs between the solutions, in which solution will the concentration of solute decrease? In which will it increase?
   (f) Which solution exerts the greater osmotic pressure, initially?
   (g) How are the osmotic pressures of the solutions related at equilibrium?
8. The red blood cell has a 5% glucose concentration. Relative to this cell, what can you say about the solutions in Question 7?
9. The amount of dissolved minerals in tap water is referred to as the hardness of the water. The concentrations of these minerals are commonly reported in ppm. (a) If a water sample is reported to have a calcium ion concentration of 10 ppm, how many grams of calcium are present in 1 million grams of water? (b) How many milligrams of calcium per liter would this be?

Answers

1. (a) Red jelly beans are solvent beans because there are more of them.
   (b) White jelly beans are solute beans because there are fewer of them.
   (c) 125 jelly beans in the solution
   (d) (25 white jelly beans/125 jelly beans in solution) × 100% = 20% white
2. (a) 100 mL × (1 g glucose/ 1 mL) = 100 g glucose per 100 mL
   (b) the units are grams and mL or w/v
   (c) the percent concentration is 100% w/v
3. $$\frac{100\ \cancel{\text{g glucose}}}{0.100\ \text{L}} \times \frac{1\ \text{mol glucose}}{180.18\ \cancel{\text{g glucose}}} = 5.55\ M$$
4. 5.55 osmol/L
5. Both molecules are polar, and ethanol can form hydrogen bonds with water molecules.
7. (a)

5 grams of solute in 100 mL of solution	10 grams of solute in 100 mL of solution	← this solution is more dense because it has more solute particles per unit volume

   (c) More solvent molecules hit the membrane on the 5% side than on the 10% side; more solute particles hit the membrane on the 10% side.
   (d) The solvent level will increase in the 10% side.
   (e) The solute concentration will decrease in the 10% side and increase in the 5% side.
   (f) The 10% side has the higher osmotic pressure initially.
   (g) The osmotic pressures are equal at equilibrium.
8. The 5% solution is isotonic and the 10% solution is hypertonic to the red blood cell.
9. (a) 10 g of calcium per 1 million grams of water
   (b) 10 mg of calcium per liter of water

chapter 8
Chemical Reactions

OUTLINE

8.1 Reaction Rates

A. A chemical reaction occurs when reactants are changed to products.
1. reactant molecules must collide
2. reactants must be oriented in a favorable spatial position
3. only a small percentage of molecules usually react relative to the number of collisions

B. The concentrations of reactants and products change over time (Fig. 8.1).
1. reactants disappear
2. products appear
3. the speed with which changes take place is called the rate of reaction

C. Reaction rates are affected by several factors.
1. concentration of reactants: larger concentrations mean more collisions between molecules
2. spatial orientation when collisions occur: the chemically active areas of reactant molecules must meet
3. temperature at which reaction takes place: higher temperatures mean more energetic collisions
4. presence of catalysts: catalysts increase reaction rates

D. Reaction rates can be written in terms of reactants or products.
1. rate = number of moles of product appearing per elapsed time
2. rate = number of moles of reactant disappearing per elapsed time
3. rates are negative for the disappearance of a reactant
4. rates are positive for the appearance of a product

Key Terms chemical kinetics, product, reactant, reaction rate
Key Figure Figure 8.1 Chart of change in reactant and product concentrations in a reaction over a period of time

8.2 Reactive Collisions

A. Reactive collisions occur when
1. reactant molecules collide with sufficient energy
   a. the minimum energy needed for a reactive collision is called the activation energy, E_a
   b. this is also the minimum energy needed to form an activated complex
   c. temperature determines energy of collisions (Fig. 8.2); a 10-kelvin temperature increase can double the rate for many reactions
2. reactant molecules collide in a favorable spatial orientation
3. sufficient energy and proper orientation produce an activated complex (Fig. 8.3)
4. activated complex is unstable and quickly becomes product in a reactive collision

B. Activation-energy diagrams show (Fig. 8.4)
   1. energy changes as a reaction proceeds
   2. difference between energy of reactants and activated complex
   3. difference between energy of products and activated complex
   4. overall energy difference between reactants and products

$$\Delta H_{rxn} = E_{forward} - E_{back}$$

   5. energy of products is lower than energy of reactants in an exothermic reaction
      a. the forward reaction in Fig. 8.4
      b. ΔH_{rxn} is negative
      c. heat is lost
   6. energy of products is higher than the energy of reactants in an endothermic reaction
      a. the forward reaction in Fig. 8.5
      b. ΔH_{rxn} is positive
      c. heat is gained

Key Terms activated complex, activation energy, back reaction, endothermic, exothermic, forward reaction, heat of reaction, reactive collision, transaction state

Key Figures and Box

Figure 8.2	The relationship between number of effective collisions, kinetic energy of molecules, and temperature
Figure 8.3	Nonreactive collision due to too little energy; nonreactive collision due to incorrect orientation; reactive collision results when both kinetic energy and orientation are optimal for bond breakage and formation
Figure 8.4	Activation-energy diagram for an exothermic reaction
Figure 8.5	Activation-energy diagram for an endothermic reaction
Box 8.1	Influence of temperature on physiological processes

Example 8.1 Using activation energy data to calculate the heat of a reaction
Problem 8.1 Using activation energy data to calculate the heat of a reaction

8.3 Catalysts

A. Catalysts
   1. increase the rate of a chemical reaction; they lower the activation energy of a reaction by providing a different path by which reactants become products
   2. remain unchanged after the reaction; participate in the reaction, but emerge unchanged
   3. change the speed of the reaction, but not the outcome

Key Terms catalyst, reaction intermediate

8.4 Biochemical Catalysts

A. Biological catalysts
   1. are called enzymes
   2. are usually proteins
   3. have specific three-dimensional structures that interact with specific molecules
B. Enzymes work by
   1. binding a substrate to an active site
      a. the active site is where catalysis takes place
      b. the substrate is the substance converted by the enzyme
      c. the active site and substrate are structurally related; this relationship is called complementarity
      d. secondary forces bind the substrate to the active site
   2. the enzyme-substrate binding produces an activated complex
      a. complex provides a lower energy pathway for product to form
      b. reaction proceeds rapidly

C. Chemical reactions in cells are controlled by changes in enzyme activity and concentration.
D. Enzyme-catalyzed reactions have smaller activation energies than uncatalyzed reactions do (Table 8.1).
   1. reactions with small activation energies are less sensitive to temperature changes (Fig. 8.7)
   2. rates in reactions with large activation energies are more strongly influenced by temperature changes (Fig. 8.8)

Key Terms active site, complementarity, enzyme, substrate
Key Figures and Table
Figure 8.6 Reaction progress for catalyzed and uncatalyzed reactions
Figure 8.7 Effects of activation energy on relative reaction rates as temperature is changed
Table 8.1 Activation energies of some chemical and biological processes

8.5 *Chemical Equilibrium*

A. Chemical equilibrium
   1. is a dynamic state; the rate of the forward reaction is equal to the rate of the reverse reaction

$$A + B \rightleftharpoons C + D$$

   2. the concentrations of products and reactants no longer change with time when a system is at equilibrium
   3. the reactions are reversible; the double arrow indicates reversibility
B. Competition between enzyme-substrate complexes and enzyme-competitor complexes can be written as a reversible reaction.

$$ES + C \rightleftharpoons EC + S$$

   1. competitors are molecules that closely resemble a substrate
   2. competitors compete with substrate molecules for binding sites on enzymes
   3. changes in substrate or competitor concentration can change the type of complex favored
      a. high substrate concentrations favor the formation of ES
      b. high competitor concentrations favor the formation of EC

Key Terms chemical equilibrium, irreversible reaction, reversible reaction
Key Figure Figure 8.8 Equilibrium occurs when the concentrations of products and reactants no longer change over time; it is a dynamic state

8.6 *Equilibrium Constants*

A. At equilibrium the concentrations of reactants and products in a specified equilibrium at a defined temperature remain constant.
B. The relationships between reactant and product equilibrium concentrations can be quantified mathematically using the balanced equation and the equilibrium constant expression

$$aA + bB \rightleftharpoons cC + dD \leftarrow \text{lowercase letters are coefficients in balanced equation}$$

$$\text{equilibrium constant} \rightarrow K_{eq} = \frac{[C]^c\,[D]^d}{[A]^a\,[B]^b}$$

Square brackets, [ ], represent molar concentration of the substance in the brackets.

C. The physical states of all reactants and products must be considered.
   1. gaseous state is designated (*g*)
   2. liquid state is designated (*l*)
   3. solid state is designated (*s*)
   4. aqueous solution is designated (*aq*)
D. The equilibrium constant changes with temperature; therefore, temperature should be specified.
E. Rules for writing equilibrium constants: based on the balanced equation for an equilibrium system.

1. products of a reaction are always in the numerator, with reactants in the denominator
2. [ ] are used to show molar concentrations at equilibrium
3. the coefficient of each component from the balanced equation becomes the exponent to which the molar concentration (at equilibrium) is raised
4. pure liquids and pure solids are not included in the equilibrium-constant expression

F. Equilibrium constant contains both quantitative and qualitative information.
   1. magnitude of the constant describes the relative concentrations of reactants and products at equilibrium
      a. a number larger than 1 means more product than reactant, so equilibrium is favorable to product formation
      b. 1 means that the product and reactant concentrations are equal
      c. a number much smaller than 1 means much more reactant than product, so equilibrium is unfavorable to product formation
   2. equilibrium constants do not contain information about how fast a reaction proceeds

G. Equilibrium constant can be calculated from concentration data.

H. Equilibrium reactant and product concentrations can be calculated by using equilibrium constant (Box 8.2).

Key Terms equilibrium constant, favorable reaction, heterogeneous equilibrium, homogeneous equilibrium, unfavorable reaction

Key Box Box 8.2 Method of calculating concentrations of reaction components by using the equilibrium constant

Example 8.2 Writing equilibrium constant expressions for homogeneous equilibria
Problem 8.2 Writing equilibrium constant expressions for homogeneous equilibria
Example 8.3 Writing equilibrium constant expressions for heterogeneous equilibria
Problem 8.3 Writing equilibrium constant expressions for heterogeneous equilibria
Example 8.4 Estimating relative concentrations of reactants and products at equilibrium
Problem 8.4 Estimating relative concentrations of reactants and products at equilibrium
Example 8.5 Calculating equilibrium constants
Problem 8.5 Calculating equilibrium constants

8.7 Biochemical Reactions Are Connected in Sequences

A. Sequences of reactions are commonly connected in biological systems
   1. cellular metabolism consists of sequences of reactions
   2. these can be shown as

$$A + B \rightleftharpoons C \rightleftharpoons D + E$$

   3. Each reaction has an equilibrium constant:

$$A + B \rightleftharpoons C \quad K_1 = \frac{[C]}{[A][B]}$$

$$C \rightleftharpoons D + E \quad K_2 = \frac{[D][E]}{[C]}$$

   4. these can be linked together to produce the overall reaction:

$$A + B \rightleftharpoons D + E$$

$$\frac{\cancel{[C]}}{[A][B]} \times \frac{[D][E]}{\cancel{[C]}} = \frac{[D][E]}{[A][B]} \leftarrow \text{this is } K_{eq3} \text{ for the overall reaction}$$

$$K_1 \times K_2 = K_3 \leftarrow \text{the } K_{eq}\text{'s of each reaction are multiplied to produce the overall } K_{eq}$$

Example 8.6 Calculating equilibrium constants for a reaction sequence
Problem 8.6 Calculating equilibrium constants for a reaction sequence

8.8 Le Chatelier's Principle

A. Le Chatelier's principle states
   1. If a system in an equilibrium state is disturbed, the system will adjust so as to counteract the disturbance and restore the equilibrium
B. Le Chatelier's principle can be used to make qualitative predictions when
   1. there are changes in concentrations of reactants or products
      a. when a component is added, the system will shift in the direction that consumes some of the additional component

$$A + B \rightleftharpoons C$$

      b. if the concentration of A is increased, the system will shift right to consume some of the excess A
      c. if the concentration of C is increased, the system will shift left to consume some of the excess C
   2. heat is added to or removed from a system
      a. treat heat as a product or a reactant
         i. heat is a reactant in endothermic reactions
         ii. heat is a product in exothermic reactions
      b. for $A \rightleftharpoons B$ + heat, if heat is added to the system, it will shift to the left to absorb the excess heat
      c. for A + heat $\rightleftharpoons$ B, if heat is added to the system, it will shift to the right to absorb excess heat

Key Term Le Chatelier's principle
Key Box Box 8.3 The industrial synthesis of ammonia via the Haber cycle

Example 8.7 Using Le Chatelier's principle to predict changes in a system at equilibrium based on concentration effects
Problem 8.7 Using Le Chatelier's principle to predict changes in a system at equilibrium based on concentration effects
Example 8.8 Using Le Chatelier's principle to predict changes in a system at equilibrium based on temperature effects
Problem 8.8 Using Le Chatelier's principle to predict changes in a system at equilibrium based on temperature effects

ARE YOU ABLE TO... AND WORKED TEXT PROBLEMS

Objectives Section 8.1 Reaction Rates

Are you able to. . .
- describe the relationship between reactants and products in a chemical reaction
- explain what is meant by "reaction rate" of a chemical reaction (Fig. 8.1)
- write the reaction rates for the disappearance of reactant or appearance of product (Exercises 8.1, 8.2)
- explain what negative and positive reaction rates mean
- describe how a chemical reaction occurs
- explain how reactant concentration, spatial orientation, temperature, and the presence of catalysts affect reaction rates (Exercise 8.4)

Objectives Section 8.2 Reactive Collisions

Are you able to. . .
- describe a reactive collision (Fig. 8.3)
- describe the relationship between the kinetic energy of molecules and temperature (Fig. 8.2)
- explain how temperature affects the number of reactive collisions (Exercise 8.3)

- define activation energy
- describe an activated complex
- explain how activation energies and reaction rates are related (Exercise 8.5)
- interpret activation-energy diagrams for both exothermic and endothermic reactions (Figs. 8.4, 8.5, Expand Your Knowledge 8.37)
- calculate the ΔH_{rxn} using activation energies (Ex. 8.1, Prob. 8.1)
- calculate $E_{forward}$ or E_{back} from ΔH_{rxn} (Exercises 8.7, 8.8, 8.35)

Problem 8.1 The activation energies characteristic of the reversible chemical reaction $A + B \rightleftharpoons C$ are $E_{forward} = 74$ kJ and $E_{back} = 68$ kJ. Calculate the ΔH_{rxn} for the forward reaction. Is the forward reaction exothermic or endothermic?

First, write the equation that relates the three variables ΔH_{rxn}, $E_{forward}$, and E_{back}:

$$\Delta H_{rxn} = E_{forward} - E_{back}$$

Substitute the values given in the problem:

$$\Delta H_{rxn} = 74 \text{ kJ} - 68 \text{ kJ} = +6 \text{ kJ}$$

The sign of the heat of reaction is positive, which means that heat has been gained and the reaction is endothermic.

Objectives Section 8.3 Catalysts

Are you able to. . .

- explain the role of a catalyst in a chemical reaction (Expand Your Knowledge 8.36)
- explain the role of a reaction intermediate in a reaction sequence (Expand Your Knowledge 8.36)

Objectives Section 8.4 Biochemical Catalysts

Are you able to. . .

- describe how enzymes speed up biological reactions (Exercise 8.6)
- discuss the specificity of enzymes
- describe the relationship between an enzyme's binding site and its substrate
- describe the relationship between temperature changes, activation energies, and reaction rates (Fig. 8.6)
- explain how activation energy controls the rate of a chemical reaction
- discuss the typical activation energies of enzyme-catalyzed biological processes
- compare the activation energies of enzyme-catalyzed biological processes and uncatalyzed chemical reactions (Fig. 8.7)
- discuss the factors that control chemical reactions in cells

Objectives Section 8.5 Chemical Equilibrium

Are you able to. . .

- explain the dynamics of a system at equilibrium (Fig. 8.7)
- write the chemical equation for an equilibrium system (Exercises 8.11, 8.12)
- differentiate between reversible and irreversible reactions (Exercises 8.9, 8.10)
- describe the equilibrium between an enzyme and its substrate
- describe how enzyme competitors work
- explain how the equilibrium between enzyme-substrate and enzyme-competitor species can be affected by changes in substrate and competitor concentrations

Objectives Section 8.6 Equilibrium Constants

Are you able to. . .

- distinguish between homogeneous and heterogeneous equilibria
- write the equilibrium constant expression for a homogeneous equilibrium reaction (Ex. 8.2, Prob. 8.2, Exercises 8.13, 8.14, 8.23, 8.28, 8.29, 8.32)

- write the equilibrium constant expression for a heterogeneous equilibrium reaction (Ex. 8.3, Prob. 8.3, Exercises 8.13, 8.14, 8.22, 8.24, 8.25, 8.30, 8.40)
- describe the relationship between temperature and equilibrium constants for both exothermic and endothermic reactions
- explain what the magnitude of the equilibrium constant tells you
- explain what the equilibrium constant does not tell you
- estimate relative concentrations of reactants and products at equilibrium (Ex. 8.4, Prob. 8.4, Exercise 8.31)
- calculate the equilibrium constant expression for a reaction from concentration data (Ex. 8.5, Prob. 8.5, Exercises 8.15, 8.16)
- use the equilibrium constant expression to calculate the concentrations of components when a reaction is at equilibrium

Problem 8.2 Write the equilibrium constant for the following balanced equation:

$$H_2(g) + I_2(g) \rightleftharpoons 2\ HI(g)$$

Refer to the rules for writing equilibrium constants:

1. The products of the reaction are in the numerator: product ⇒ 2 HI
2. The reactants of the reaction are in the denominator: reactants ⇒ H_2 and I_2
3. Square brackets are used to show molar concentrations at equilibrium
4. The coefficient of each product and reactant in the balanced equation becomes the power to which the molar concentration is raised:

$$\frac{[HI]^2}{[H_2][I_2]}$$

5. Pure solids and liquids do not appear in the equilibrium expression. This rule does not apply in this example because all components are gases.

The equilibrium-constant expression for the reaction is

$$K_{eq} = \frac{[HI]^2}{[H_2][I_2]}$$

Problem 8.3 Write the equilibrium constant for the following reaction:

$$2\ C(s) + O_2(g) \rightleftharpoons 2\ CO(g)$$

Refer to the rules for writing equilibrium constants:

1. The products of the reaction are in the numerator: product ⇒ 2 CO
2. The reactants of the reaction are in the denominator: reactants ⇒ 2 C and O_2
3. Pure solids and liquids do not appear in the equilibrium expression: C is a solid, so only O_2 and CO, the gases, will appear in the equilibrium-constant expression
4. Square brackets are used to show molar concentrations at equilibrium
5. The coefficient of each product and reactant in the balanced equation becomes the power to which the molar concentration is raised:

$$\frac{[CO]^2}{[O_2]}$$

The equilibrium-constant expression for the reaction is

$$K_{eq} = \frac{[CO]^2}{[O_2]}$$

Problem 8.4 Estimate the concentration of product with respect to reactant at equilibrium for the hypothetical reaction A ⇌ B, for which $K_{eq} = [B] / [A] = 1.0$.

The expression for K_{eq} is

$$K_{eq} = \frac{[\text{product}]^x}{[\text{reactant}]^y}$$

For K_{eq} to equal 1, the numerator and denominator must be equal; therefore, the concentrations of reactant and products at equilibrium must be equal:

At equilibrium, [A] = [B]

Problem 8.5 In the reaction 2 HI(g) ⇌ H_2(g) + I_2(g), the equilibrium concentrations at 25°C are $[H_2] = [I_2] = 0.0011\ M$ and $[HI] = 0.0033\ M$. Calculate the equilibrium constant.

The equation for the equilibrium constant is

$$K_{eq} = \frac{[0.0011][0.0011]}{[\ 0.0033]^2}$$

The numerical value of K_{eq} is 0.11.

Objectives Section 8.7 Biochemical Reactions Are Connected in Sequences

Are you able to. . .

- write the equilibria constants for a sequence of reactions
- calculate the equilibrium constant for a reaction sequence (Ex. 8.6, Prob. 8.6)

Problem 8.6 Consider a set of three hypothetical reactions connected by common intermediates:

$$\begin{aligned} A &\rightleftharpoons B \quad K_{eq} = 1 \times 10^{-3} \\ B &\rightleftharpoons C \quad K_{eq} = 1 \times 10^{2} \\ C &\rightleftharpoons D \quad K_{eq} = 1 \times 10^{3} \end{aligned}$$

Calculate the overall equilibrium constant for the following reaction: A ⇌ D.

The reaction sequence produces a final K_{eq} equal to [D]/[A]. If the sequence is written in terms of products and reactants to produce the overall reaction A ⇌ D, the following is obtained:

$$\frac{\cancel{[B]}}{[A]} \times \frac{\cancel{[C]}}{\cancel{[B]}} \times \frac{[D]}{\cancel{[C]}} = \frac{[D]}{[A]}$$

$$\uparrow \qquad \uparrow \qquad \uparrow \qquad \uparrow$$

$$K_1 \times K_2 \times K_3 = K_4$$

Substitution of the values given for each K results in:

$$(1 \times 10^{-3}) \times (1 \times 10^{2}) \times (1 \times 10^{3}) = 1 \times 10^{2}$$

K_{eq} for A ⇌ D is large: 1×10^2.

Objectives Section 8.8 Le Chatelier's Principle

Are you able to. . .

- state Le Chatelier's principle (Expand Your Knowlege 8.39)
- apply Le Chatelier's principle to an equilibrium system
- predict changes in an equilibrium system when concentrations of reactants or products are changed (Ex. 8.7, Prob. 8.7, Exercises 8.17, 8.19, 8.20, 8.26, 8.33, Expand Your Knowlege 8.38)
- write an equilibrium reaction that includes heat as a product or reactant
- predict changes in an equilibrium system when heat is added to or removed from the system (Ex. 8.8, Prob. 8.8, Exercises 8.18, 8.19, 8.20, 8.26, 8.34)

Problem 8.7 Suppose some H_2(g) is added to the following reaction: 2 HI(g) ⇌ H_2(g) + I_2(g). What would the effect on the equilibrium mixture be?

The stress introduced into the system is an increase in mass on the right-hand side of the equation. The system will respond by trying to consume some of the mass, which results in a shift to the left and the production of more HI.

Problem 8.8 Consider the following equilibrium system:

$$CaCO_3(s) + 158\ kJ \rightleftharpoons CaO(s) + CO_2(g)$$

In which direction will the equilibrium shift if heat is (a) added to the system or (b) removed from the system?

Treat the addition and removal of heat just like it is a product or a reactant. Adding heat to a system will produce a shift in the direction that consumes some of the heat. Removing heat from a system will produce a shift in the direction that produces heat. (a) Adding heat in this system produces a shift to the right; more CaO and CO_2 are produced. (b) Removing heat from the system produces a shift to the left; more $CaCO_3$ is produced.

MAKING CONNECTIONS

Draw a concept map or write a paragraph describing the relation among the components of each of the following groups:

(a) reaction rate, product, reactant
(b) reaction rate, concentration, spatial orientation, temperature, catalyst
(c) negative reaction rate, positive reaction rate, product, reactant
(d) kinetic energy, activated complex, effective collisions
(e) endothermic reaction, exothermic reaction, products, reactants
(f) enzyme, catalyst, reaction rate
(g) equilibrium, product, reactant, reaction rate
(h) homogeneous equilibrium, heterogeneous equilibrium, equilibrium constant
(i) equilibrium constant, product concentration, reactant concentration
(j) Le Chatelier's principle, equilibrium system, concentration change

FILL-INS

A chemical reaction has occurred when a ____ is transformed into a _____. In the course of a chemical reaction no mass is lost, but _____ are rearranged. The rate at of a chemical reaction is described in terms of the _____ of reactant or by the _______ of product per ______. A reaction rate is negative for the _____ and positive for the _____. Chemical reaction rates depend on the _____of reactants, the _____of reactants when they collide, the _____ at which the reaction takes place, and the presence of _____.

In order for a chemical reaction to be effective, the reactants must _____ with sufficient energy and in the correct _____ so that an _____ forms. The minimum energy required for a reactive collision is called the _____, abbreviated _____. The activated complex is not _____ and returns to reactant or turns into product very rapidly. The _____ associated with a chemical reaction can be depicted in an activation energy diagram. The energy difference between the reactants and products is called the _____, abbreviated _____. When the energy of reactants is greater than the energy of products, heat must have been _____and ΔH is _____; the reaction is called _____. When the energy of reactants is less than the energy of products, heat must have been _____ and ΔH is _____; the reaction is called _____.

Biological reactions are accelerated due to the presence of_____. These biological catalysts assist in the conversion of reactant to product and are very specific for _____. The structure of an enzyme's active site is spatially _____ to the substrate in the way that a key is to a lock. The _____ formed due to attractive forces between enzyme and substrate is an _____ that provides a lower _____ pathway for the reaction.

The reversible reaction between an enzyme and its substrate is called an _____ reaction. Equilibrium systems consist of a _____ and a _____reaction that occur _____ and at the same _____. The equilibrium constant for a reversible reaction provides _____ information about the _____ of products and reactants when the system reaches equilibrium. When an equilibrium constant is much larger than 1, the concentration of products is _____ than reactants and the reaction is called _____. When an equilibrium constant is much smaller than 1, the concentration of reactants is _____ than products and the reaction is called _____. When a sequence of reactions is connected, the _____ of one reaction become the _____ of the next. In this case the _____ can be combined into one overall constant for the process.

Disturbances to a system at equilibrium cause it to shift in _____ways. The application of Le Chatelier's principle allows for the qualitative prediction of _____. A system at equilibrium will shift so as to _____. When increases are made in reactant concentration, the system will shift _____ the reactant and in doing so _____. Both reactant and product concentrations will be _____ from the original state when equilibrium is reestablished. Heat can be treated as a _____ and added to either side of the equilibrium equation. In exothermic reactions, heat is on the _____. In endothermic reactions heat is on _____.

Answers

A chemical reaction has occurred when a reactant is transformed into a product. In the course of a chemical reaction no mass is lost, but atoms are rearranged. The rate at of a chemical reaction is described in terms of the disappearance of reactant or by the appearance of product per unit time. A reaction rate is negative for the disappearance of reactant and positive for the appearance of product. Chemical reaction rates depend on the concentration of reactants, the spatial orientation of reactants when they collide, the temperature at which the reaction takes place, and the presence of catalysts.

In order for a chemical reaction to be effective, the reactants must collide with sufficient energy and in the correct orientation so that an activated complex forms. The minimum energy required for a reactive collision is called the activation energy, abbreviated Ea. The activated complex is not stable and returns to reactant or turns into product very rapidly. The energy changes associated with a chemical reaction can be depicted in an activation energy diagram. The energy difference between the reactants and products is called the heat of reaction, abbreviated ΔH. When the energy of reactants is greater than the energy of products, heat must have been lost and ΔH is negative; the reaction is called exothermic. When the energy of reactants is less than the energy of products, heat must have been gained and ΔH is positive; the reaction is called endothermic.

Biological reactions are accelerated due to the presence of enzymes. These biological catalysts assist in the conversion of reactant to product and are very specific for substrate molecules. The structure of an enzyme's active site is spatially complementary to the substrate in the way that a key is to a lock. The enzyme-substrate complex formed due to attractive forces between enzyme and substrate is an activated complex that provides a lower activation energy pathway for the reaction.

The reversible reaction between an enzyme and its substrate is called an equilibrium reaction. Equilibrium systems consist of a forward and a backward reaction that occur simultaneously and at the same rate. The equilibrium constant for a reversible reaction provides quantitative information about the relative amounts of products and reactants when the system reaches equilibrium. When an equilibrium constant is much larger than 1, the concentration of products is greater than reactants and the reaction is called favorable. When an equilibrium constant is much smaller than 1, the concentration of reactants is greater than products and the reaction is called unfavorable. When a sequence of reactions is connected, the products of one reaction become the reactants of the next. In this case the individual equilibrium constants can be combined into one overall constant for the process.

Disturbances to a system at equilibrium cause it to shift in predictable ways. The application of Le Chatelier's principle allows for the qualitative prediction of how a system will respond to a stress. A system at equilibrium will shift so as to counteract the disturbance and restore equilibrium. When increases are made in reactant concentration, the system will shift to reduce the reactant and in doing so will produce more product. Both reactant and product concentrations will be increased from the original state when equilibrium is reestablished. Heat can be treated as a reactant or product and added to either side of the equilibrium equation. In exothermic reactions, heat is on the product side. In endothermic reactions heat is on the reactant side.

TEST YOURSELF

1. Explain what each symbol stands for in the equations below:
   (a) $O_2(g) + 2H_2\,(g) \rightarrow 2H_2O$
   (b) $H_2CO_3\,(aq) \rightleftharpoons HCO_3^-\,(aq) + H^+\,(aq)$
   (c) $NaOH\,(aq) + HCl\,(aq) \rightarrow NaCl\,(aq) + H_2O$
2. Translate the equations above into sentences.

3. Draw a molecular level view of what happens to a chemical reaction when:
   (a) the concentration of reactants is increased
   (b) the temperature of the reaction is increased
4. Draw an activation energy diagram for the following reaction: reactants A and B combine to form the product AB. The reaction is exothermic. The reaction can also occur in the presence of a catalyst. Label all parts of the diagram including the transition state, heat of reaction, activation energy, and change of activation energy in the presence of a catalyst.
5. The equilibrium reaction between H_2 and I_2 is written as: $H_2(g) + I_2(g) \Rightarrow 2\ HI(g)$
   Examine the following equilibrium expressions for the reaction and determine which one is correct. Explain why the others are not correct.
   (a) $[2HI]/[H_2][I_2]$
   (b) $[HI]^2/[H_2]\ [I_2]$
   (c) $[H_2]\ [I_2]/[HI]^2$
   (d) $[HI]^2/[2H][2I]$
6. A reaction sequence is comprised of the following:

$$X \rightleftharpoons Y \rightleftharpoons Z \rightleftharpoons A$$

(rxn 1) (rxn 2) (rxn 3)

   (a) write the equilibrium expression for each individual reaction
   (b) determine the equilibrium constant for the overall reaction
   (c) If the K_{eq} values for the rxns are as follows: rxn 1, $K_{eq} = 1 \times 10^{-5}$; rxn 2, $K_{eq} = 1 \times 10^{-2}$; rxn 3, $K_{eq} = 1 \times 10^{-4}$; what is the numerical value of the equilibrium constant for the overall reaction?
7. Use Le Chatelier's Principle to explain the following:
   (a) adding reactant to a system at equilibrium results in an increase in product concentration
   (b) adding heat to an endothermic reaction at equilibrium causes an increase in product concentration
   (c) adding heat to an exothermic reaction causes the reduction of product concentration
   (d) removing product from a system at equilibrium causes a reduction in reactant concentration

Answers

1. (a) O_2 represents 2 atoms of oxygen bonded together, which makes it one molecule of oxygen. $2H_2$ represents two molecules of hydrogen; each molecule consists of two atoms of hydrogen bonded together. $2H_2O$ represents two molecules of water; each molecule contains one oxygen atom bonded to two hydrogen atoms. $\rightarrow$ means a chemical reaction occurs in only one direction: reactants to products. (g) means that these molecules are in the gaseous state.
   (b) H_2CO_3 represents a molecule of carbonic acid, which consists of two hydrogen atoms, one carbon atom, and three oxygen atoms bonded to one another. HCO_3^- represents one bicarbonate anion; the carbonic acid reactant has lost a proton to form the bicarbonate anion. H^+ represents a proton; a hydrogen cation. (aq) means that the reactants and products are dissolved in water or hydrated. $\rightleftharpoons$ means that the reaction is reversible: reactants form products and products reform reactants.
   (c) NaOH represents one formula unit of sodium hydroxide that contains Na^+ and OH^- in a one to one ratio. HCl represents one molecule of hydrochloric acid. NaCl represents one formula unit of sodium chloride that contains Na^+ and Cl^- in a one to one ratio. H_2O represents one molecule of water that contains one oxygen and two hydrogen atoms bonded to one another. $\rightarrow$ means the chemical reaction occurs in one direction. (aq) means that these species are dissolved in water or hydrated.
2. (a) One molecule of oxygen gas reacts with two molecules of hydrogen gas to produce two molecules of water.
   (b) One molecule of carbonic acid dissolved in water dissociates into proton and a bicarbonate ion; the proton and a bicarbonate ion can recombine to produce one molecule of carbonic acid.

(c) One formula unit of sodium hydroxide dissolved in water reacts with one molecule of hydrochloric acid dissolved in water to produce one formula unit of sodium chloride dissolved in water and one molecule of water.

3. (a) The molecular level view of increased reactant concentration shows more reactants and more collisions between reactants.
   (b) The molecular level view when temperature is increased shows more frequent and more energetic collisions between reactants because they are moving faster.

4.

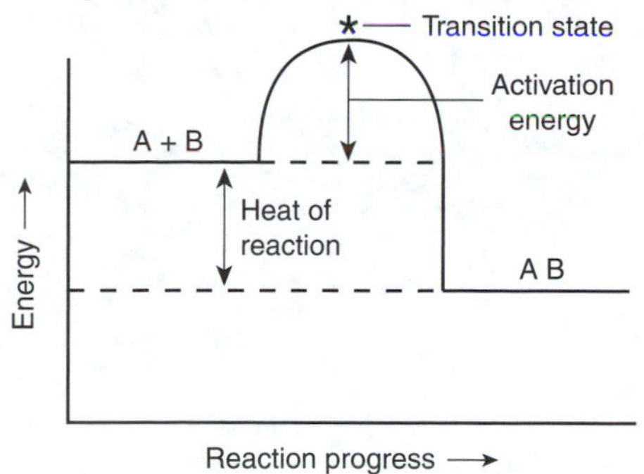

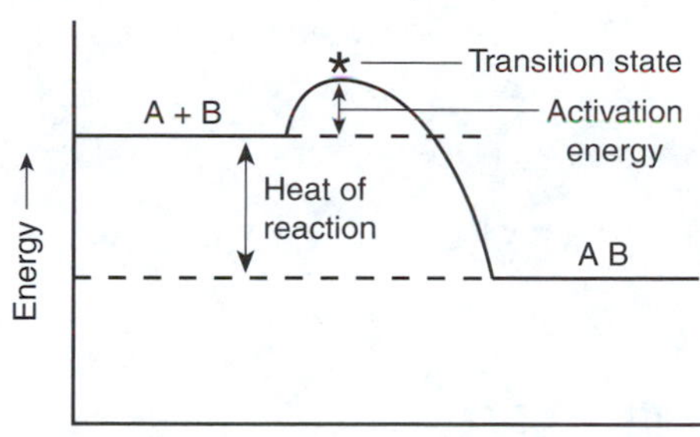

5. (b) is correct. The form in which an equilibrium expression is written puts the reactant concentrations into the numerator and the product concentrations into the denominator. The coefficients of the balanced equation become the exponents to which that species' concentration is raised. The subscripts are not changed. (a) does not raise the product concentration of HI to its coefficient. (c) puts products in the denominator and reactants in the numerator instead of the other way. (d) changes the identities of the reactants, I_2 and H_2, by putting subscripts as coefficients.

6. (a) $K_{eq1} = [Y]/[X]$ $\quad K_{eq2} = [Z]/[Y]$ $\quad K_{eq3} = [A]/[Z]$
   (b) $K_{eq4} = \frac{[Y]}{[X]}\frac{[Z]}{[Y]}\frac{[A]}{[Z]} = \frac{[A]}{[X]}$
   (c) $(10^{-5})(10^{2})(10^{4}) = 10^{1} = 10$

7. (a) In order to maintain the equilibrium constant at a given temperature, when more reactant is added the system shifts to reduce some of the reactant and produce some more product to maintain the ratio of product to reactant that defines the equilibrium ratio.
   (b) The shift in the system will be in the direction that relieves the stress. In this case, a shift to products is due to the fact that heat is a reactant in an endothermic reaction; it is absorbed as product is formed. The production of more product will cause the absorption of more heat.
   (c) Adding heat to an exothermic system is like increasing the product, heat. The system shifts to absorb extra heat, which is toward the reactant side, thus decreasing the product concentration.
   (d) Removing product decreases the numerator concentration in the equilibrium expression. In order to maintain a constant ratio at a defined temperature, the system shifts to replace some product, thus decreasing the reactant concentration.

chapter 9

Acids, Bases, and Buffers

OUTLINE

Introduction: Properties of Acids and Bases

A. Acids are substances that
   1. produce hydrogen ions when dissolved in water
   2. turn blue litmus paper red
   3. neutralize bases to produce water and salts

B. Bases are substances that
   1. produce hydroxide ions when dissolved in water
   2. turn red litmus paper blue
   3. neutralize acids to produce water and salts

Key Terms acid, base, litmus, neutralize, salt

Key Figure and Table

Figure 9.1 Red and blue litmus paper treated with acid, base, and pure water

Table 9.1 Properties of acids and bases

9.1 Water Reacts with Water

A. One water molecule reacts with another water molecule.
   1. water molecule that accepts proton forms hydronium ion, H_3O^+; sometimes given proton symbol, H^+
   2. water molecule that donates proton is left as hydroxide ion, OH^-
   3. only 1 in 1 billion water molecules is in this form

B. This equilibrium is represented as

$$H_2O(l) + H_2O(l) \rightleftharpoons H_3O^+(aq) + OH^-(aq)$$

C. The ion-product represents the equilibrium constant, K_w.

$$K_w = [H_3O^+][OH^-]$$

   1. in pure water at 25°C, $K_w = 1.0 \times 10^{-14}$
   2. K_w is a constant as long as temperature remains constant

D. In pure water, both hydronium ion and hydroxide ion concentrations are 10^{-7} moles per liter.

$$[H_3O^+] = [OH^-] = 1.0 \times 10^{-7}\ M$$

   1. in neutral solutions, the concentration of hydronium ions equals the concentration of hydroxide ions

$$[H_3O^+] = [OH^-]$$

2. in acidic solutions, the concentration of hydronium ions is greater than the concentration of hydroxide ions

$$[H_3O^+] > [OH^-]$$

3. in basic solutions the concentration of hydroxide ions is greater than the concentration of hydronium ions

$$[OH^-] > [H_3O^+]$$

Key Terms hydronium ion, hydroxide ion, ion-product of water

9.2 Strong Acids and Strong Bases

A. Strong acids dissociate completely into ions (Table 9.2).
   1. one of the ions is always a proton, H^+
   2. in water, all strong acid molecules are dissociated
   3. all the protons donated by the acid are associated with water molecules as H_3O^+
B. Strong bases dissociate completely into ions.
   1. all strong bases in Table 9.2 produce OH^-
   2. in water, all strong bases in Table 9.2 are dissociated
C. Acid or base strength is a measure of the extent to which the acid or base dissociates; strength has nothing to do with the concentration of the substance.
   1. solutions with a high concentration of acid or base are called concentrated
   2. solutions with a low concentration of acid or base are called dilute
D. Hydronium and hydroxide ion concentrations can be calculated from strong acid or strong base solution concentrations
   1. (molarity of acid solution) × (number of H^+ per mole of acid) = concentration of hydronium ion in solution
   2. (molarity of base solution) × (number of OH^- per mole of base) = concentration of hydroxide ion in solution
E. Hydronium and hydroxide ion concentrations can be calculated by using K_w.
   1. $K_w = [H_3O^+][OH^-]$
   2. $[H_3O^+] = K_w/[OH^-]$
   3. $[OH^-] = K_w/[H_3O^+]$

Key Terms dissociation, strong acid, strong base

Key Figure and Table

Figure 9.2 Reactions of a strong acid and a weak acid of the same concentrations

Table 9.2 Names and formulas of all the strong acids and bases

Example 9.1 Calculation of hydronium ion concentration from the molarity of a strong acid solution

Problem 9.1 Calculation of hydronium ion concentration from the molarity of a strong acid solution

Example 9.2 Calculation of hydroxide ion concentration from the molarity of a strong base solution

Problem 9.2 Calculation of hydroxide ion concentration from the molarity of a strong base solution

Example 9.3 Calculation of hydroxide ion concentration from the molarity of a strong base solution of a divalent cation

Problem 9.3 Calculation of hydroxide ion concentration from the molarity of a strong base solution of a divalent cation

Example 9.4 Calculation of the hydronium ion concentration of a strong base solution using the ion-product of water

Problem 9.4 Calculation of the hydronium ion concentration of a strong base solution using the ion-product of water

Example 9.5 Calculation of the hydroxide ion concentration of a strong acid solution using the ion-product of water

Problem 9.5 Calculation of the hydroxide ion concentration of a strong acid solution using the ion-product of water

9.3 *A Measure of Acidity: pH*

A. Acidity of a solution is measured by the concentration of hydronium ions in moles per liter; range is generally from 1.0 *M* to 1.0×10^{-14} *M* (Table 9.3).

B. pH scale expresses hydronium ion concentration without the use of scientific notation.

1. $pH = -\log [H_3O^+]$
   says that the pH of a solution is equal to the negative logarithm of the hydronium ion concentration (in moles per liter)
2. pH scale ranges from 0 to 14

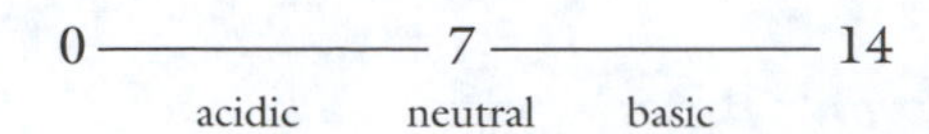

3. lower pH values correspond to higher hydronium ion concentrations (i.e., lower hydroxide ion concentrations), and higher pH values correspond to lower hydronium ion concentrations (i.e., higher hydroxide ion concentrations) (Figure 9.4)
4. logarithmic relation means that changing pH by 1 unit changes the hydronium ion concentration by a factor of ten

C. Mathematical definition can be used to calculate

1. pH of a solution if the hydronium concentration is known

$$pH = -\log [H_3O^+]$$

2. hydronium ion concentration of a solution if the pH is known

$$[H_3O^+] = 10^{-pH}$$

D. pH and pOH are related by

$$pH + pOH = 14$$

Key Terms antilogarithm, pH, p*K*, pOH

Key Figures and Table

Figure 9.3 A pH meter measures and displays hydronium ion concentration as pH; lemon juice and orange juice are weak acids

Figure 9.4 pH values and their corresponding hydronium ion concentrations

Figure 9.5 pH ranges of some common aqueous solutions

Table 9.3 Hydronium ion concentrations with corresponding pH values

Example 9.6 Calculation of the pH of a strong base solution using hydroxide ion concentration
Problem 9.6 Calculation of the pH of a strong base solution using hydroxide ion concentration
Example 9.7 Calculation of the pH of a strong acid solution using hydronium ion concentration
Problem 9.7 Calculation of the pH of a strong acid solution using hydronium ion concentration
Example 9.8 Calculation of the pH of a strong base solution using pOH
Problem 9.8 Calculation of the pH of a strong base solution using pOH
Example 9.9 Calculation of the pOH of a strong acid solution using pH
Problem 9.9 Calculation of the pOH of a strong acid solution using pH
Example 9.10 Using the pH to calculate of the concentration of hydronium ion in a strong acid solution
Problem 9.10 Using the pH to calculate of the concentration of hydronium ion in a strong acid solution

9.4 *Weak Acids and Weak Bases*

A. Weak acids do not dissociate completely into ions (Table 9.4).

1. in a water solution, there are a few hydronium ions, many undissociated weak acid molecules, lots of water molecules
2. most weak organic acids contain carboxyl groups, —COOH, which donate H^+, leaving carboxylate anions, $—COO^-$

B. Weak bases do not react completely with water to form hydroxide ions (Table 9.4).
   1. in a water solution, there are a few hydroxide ions, many unreacted weak base molecules, lots of water molecules
   2. most weak organic bases are derivatives of the weak base ammonia, NH_3

C. Most acids and bases in biological systems are weak acids and bases.

D. Reaction of weak acids and bases with water
   1. weak acids donate protons to water molecules, creating hydronium ions

$$RCOOH(aq) + H_2O(l) \rightleftharpoons RCOO^-(aq) + H_3O^+(aq)$$

Equilibrium lies far to the left

   2. most weak bases accept protons from water molecules, leaving hydroxide ions

$$NH_3(aq) + H_2O(l) \rightleftharpoons NH_4^+(aq) + OH^-(aq)$$

Equilibrium lies far to the left

E. Quantitative aspects of acid-base equilibria
   1. dissociation constants K_a (acid) and K_b (base) provide information about relative amounts of dissociated and undissociated acid or base
   2. as K_a or K_b get larger, the number of dissociated ions increases, which means the acid or base gets stronger (Table 9.4)

$$K_a = \frac{[H_3O^+][A^-]}{[HA]}$$ ←as this gets bigger, more ions have formed and K_a increases

   3. as pK_a or pK_b get larger, the acid or base get weaker
   4. K_a and K_b are determined from the equilibrium expressions for the dissociation reactions
   5. K_a's are used to calculate the pHs of weak acid and base solutions (Box 9.1)

Key Terms K_a, K_b, weak acid, weak base
Key Table Table 9.4 Values of K_a, pK_a, K_b, pK_b for some weak acids and weak bases

9.5 *Brønsted-Lowry Theory of Acids and Bases*

A. Brønsted-Lowry theory of acids and bases
   1. acids donate protons
   2. bases accept protons
   3. water is both a Brønsted-Lowry acid and a Brønsted-Lowry base

B. Conjugate acids and bases
   1. proton transfer requires a conjugate acid-base pair
   2. after the acid donates a proton, the acid becomes a base because it can take a proton back
   3. after the base accepts a proton, the base becomes an acid because it can give the proton up

$$HA + H_2O \rightleftharpoons H_3O^+ + A^-$$

↑ (HA) ... ↑ (A^-)
acid ← conjugate pair → base

C. Relation between acid and base dissociation constants
   1. $K_a \times K_b = K_w$
   2. if K_b is known, K_a can be calculated
   3. if K_a is known, K_b can be calculated

Key Terms Brønsted-Lowry acid, Brønsted-Lowry base, conjugate acid, conjugate base, K_w
Key Table Table 9.5 Values of K_a, pK_a, K_b, pK_b for selected conjugate acid-base pairs

Example 9.11 Deriving K_b for a conjugate acid-base pair by using K_a and K_w
Problem 9.11 Deriving K_b for a conjugate acid-base pair by using K_a and K_w
Example 9.12 Deriving pK_b for a conjugate acid-base pair by using pK_a and pK_w
Problem 9.12 Deriving pK_b for a conjugate acid-base pair by using pK_a and pK_w

9.6 *Dissociation of Polyprotic Acids*

A. Acids may donate more than one proton.
 1. if two protons can be donated, acid is diprotic
 2. if three protons can be donated, acid is triprotic
 3. dissociation reactions can be written for the ionization of each proton
 4. dissociation constants are usually different for each proton; for example, first proton may be donated more readily and have a larger K_a than the second proton

B. Dissociation of phosphoric acid, H_3PO_4, a triprotic acid
 1. dissociations can be written sequentially
  a. $H_3PO_4 \rightleftharpoons H_2PO_4^- + H_3O^+$
   $\uparrow$ $\nwarrow$
   acid ← conjugate pair → base
  b. $H_2PO_4^- \rightleftharpoons HPO_4^{2-} + H_3O^+$
   $\uparrow$ $\nwarrow$
   acid ← conjugate pair → base
  c. $HPO_4^{2-} \rightleftharpoons PO_4^{3-} + H_3O^+$
   $\uparrow$ $\nwarrow$
   acid ← conjugate pair → base
 2. $H_2PO_4^-$ and HPO_4^{2-} are amphoteric, acting as both acids and bases; water is amphoteric and thus can donate or accept a proton
 3. each dissociation has its own K_a
 4. K_a's become smaller with each dissociation; most H_3O^+ in solution comes from dissociation of the first proton

C. Dissociation of carbonic acid, H_2CO_3, a diprotic acid
 1. dissociations can be written sequentially
  a. $H_2CO_3 \rightleftharpoons HCO_3^- + H_3O^+$
   $\uparrow$ $\nwarrow$
   acid ← conjugate pair → base
  b. $HCO_3^- \rightleftharpoons CO_3^{2-} + H_3O^+$
   $\uparrow$ $\nwarrow$
   acid ← conjugate pair → base

Key Terms diprotic acid, polyprotic acid, triprotic acid

9.7 *Salts and Hydrolysis*

A. Conjugate bases of weak acids are also the anions of salts.

EXAMPLE: Acetate, carbonate, phosphate

B. Conjugate acids of weak bases are cations of salts.

EXAMPLE: ammonium ion

C. These salts react with water through hydrolysis.
 1. if OH^- is produced, ion is basic
 2. if H_3O^+ is produced, ion is acidic

D. Rules for determining the acidity or basicity of a salt solution.
 1. the anions/cations of strong acids and bases form neutral aqueous solutions
 2. the anions of weak acids form basic solutions
 3. the cations of weak bases form acidic solutions
 4. if salts of both weak acid and weak base are present, pH depends on which dissociation constant is larger
  a. if K_a is greater than K_b, the solution is acidic
  b. if K_b is greater than K_a, the solution is basic
  c. if K_a is about the same as K_b, the solution is neutral

Key Terms hydrolysis, hydrolyze, litmus, salt

Example 9.13 Prediction of the acidity, basicity, or neutrality of a salt solution
Problem 9.13 Prediction of the acidity, basicity, or neutrality of a salt solution

9.8 *Buffers and Buffered Solutions*

A. Buffered solutions contain buffer systems that resist pH changes.
B. A buffer system is a conjugate acid-base pair.
   1. the conjugate acid neutralizes added base and forms a weaker base

$$HA + \underset{\text{stronger base}}{OH^-} \rightarrow H_2O + \underset{\text{weaker base}}{A^-}$$

   2. the conjugate base neutralizes added acid and forms a weaker acid

$$A^- + \underset{\text{stronger acid}}{H^+} \rightarrow \underset{\text{weaker acid}}{HA}$$

   3. buffer action is maximized when the concentrations of both the conjugate acid and the conjugate base are about equal
   4. buffering capacity is increased as concentrations of conjugate acid and conjugate base are increased
C. Quantitative aspects of buffer systems
   1. pH of a buffered system is calculated by using the Henderson-Hasselbalch equation (Box 9.4)

$$pH = \underset{\substack{\uparrow \\ \text{determined by the identity of the conjugate acid in the buffer system}}}{pK_a} + \log\frac{[\text{proton acceptor}] \leftarrow \text{conjugate base}}{[\text{proton donor}] \leftarrow \text{conjugate acid}}$$

   2. when the concentration of conjugate base is equal to the concentration of conjugate acid, the ratio is 1. The log of 1 is zero, so pH is equal to the pK_a
   3. conjugate base/conjugate acid concentration ratios much larger or smaller than 1 produce much greater changes in pH (Table 9.6)
   4. ratio of conjugate acid to conjugate base concentrations determines the extent to which the system can resist pH changes
D. Buffer systems are chosen to work at specific pH's as dictated by clinical requirements.

Key Terms buffered solution, buffer system, Henderson-Hasselbalch equation
Key Table Table 9.6 Effect on pH of conjugate acid/conjugate base ratios

Example 9.14 Using the Henderson-Hasselbalch equation to calculate the pH of a buffered solution from the concentrations of a weak acid and its salt
Problem 9.14 Using the Henderson-Hasselbalch equation to calculate the pH of a buffered solution from the concentrations of a weak acid and its salt

9.9 *Buffer System of the Blood*

A. Carbon dioxide maintains the acid-base balance of the blood.
   1. carbon dioxide is produced as a waste product by metabolic processes and must be removed
   2. carbon dioxide is transported in the blood as H_2CO_3, carbonic acid
   3. carbon dioxide is exhaled as gas from the lungs
   4. carbon dioxide is moved down its concentration gradient

$$CO_2\text{ (lungs)} < CO_2\text{ (blood)} < CO_2\text{ (tissues)}$$

   5. carbon dioxide removal depends on the ventilation rate

B. Acidosis and alkalosis
   1. blood pH must be maintained at 7.4

$$7.4 = 6.3 + \log \frac{[HCO_3^-] \leftarrow 10}{[CO_2 + H_2CO_3] \leftarrow 1}$$ this approximate ratio must be maintained to produce a pH of ~7.4

      a. HCO_3^- concentration is under control of kidneys; the metabolic part
      b. CO_2/H_2CO_3 concentrations are under control of lungs; the respiratory part
   2. when the concentration of HCO_3^- increases, the ratio increases above 10 and the pH increases
      a. produces metabolic alkalosis
      b. can be caused by continuous vomiting or excessive antacid intake
   3. when the concentration of CO_2 plus H_2CO_3 decreases, the ratio increases above 10 and the pH increases
      a. produces respiratory alkalosis
      b. can be caused by increased ventilation rate, hyperventilation
   4. when the concentration of CO_2 plus H_2CO_3 increases, the ratio decreases below 10 and the pH decreases
      a. produces respiratory acidosis
      b. can be caused by lung disease or low ventilation rate, hypoventilation
   5. when the concentration of HCO_3^- decreases, the ratio decreases below 10 and the pH decreases
      a. produces metabolic acidosis
      b. can be caused by uncontrolled diabetes

Key Terms acidosis, alkalosis, metabolic acidosis, metabolic alkalosis, respiratory acidosis, respiratory alkalosis, ventilation rate

9.10 Titration

A. Titration is an analytical technique that uses the neutralization reaction between an acid and a base to determine the concentration of acid or base in a solution (Figure 9.8).
   1. volume of solution (either acid or base)of unknown concentration is measured very accurately; indicator is added; indicator molecule color depends on pH (Figures 9.9 and 9.10)
   2. solution of known concentration (either acid or base) is added in small volume until all acid-base ions have neutralized each other to produce water (equivalence point) (Figure 9.10)
   3. complete neutralization is signaled by color change of indicator (end point)
   4. concentration of unknown solution can be calculated when the amount of acid or base used has been determined

EXAMPLE: to calculate the concentration of an acid solution, base of known concentration is added

(volume OH^- added) × (molar concentration of OH^- added) = number of moles of OH^- used by acid

EXAMPLE: to calculate the concentration of a base solution, acid of known concentration is added

(volume H^+ added) × (molar concentration of H^+ added) = number of moles of H^+ used by base

B. Neutralization reactions may produce a salt, gas, or both along with water

Key Terms end point, equivalent point, indicator, neutralization reaction, standard solution, titration

Key Box 9.5 Use of indicator's in determining pH

Example 9.15 Using data from a neutralization reaction between a strong acid and a strong base to calculate the acid concentration of a strong acid solution

Problem 9.15 Calculation of volume of strong base needed to neutralize a given volume of strong acid of known concentration

9.11 Normality

A. Normality (N) is a measure of the concentration of charged units (equivalents) per volume

$$\text{Normality} = \text{equivalents/liter}$$

$$1 \text{ equivalent} = 1 \text{ mol of charge}$$

neutralization reactions take place when 1 equivalent of acid reacts with 1 equivalent of base

B. Equivalents per mole of substance can be determined by writing the dissociation or ionization reaction of 1 mol of the substance

$$HCl \rightarrow H^+ + Cl^-$$

1 mol HCl produces 1 mol H^+ and 1 mol Cl^- ⇒ 1 mol HCl produces 1 equivalent of acid and 1 equivalent of Cl^-

$$H_2CO_3 \rightleftharpoons 2H^+ + CO_3^{2-}$$

1 mol of H_2CO_3 produces 2 equivalents of acid and two equivalents of CO_3^{2-}

$$Ca \rightarrow Ca^{2+} + 2\,e^-$$

1 mol of Ca produces two equivalents of Ca^{2+}

C. Equivalent mass relates the number of equivalents produced per mole of substance and the weight of 1 mol of substance.
1 mol Ca weighs ~40.00 g
40.00 g produces 2 equivalents
1 equivalent Ca^{2+} weighs ~20.00 g

D. Equivalent mass of an acid is the formula weight of the acid divided by the number of dissociable protons per mole of the acid.

$$\frac{H_2CO_3 \Rightarrow 62.0 \text{ g per mole}}{2 \text{ eq} \quad \Rightarrow 2 \text{ protons per mole}} \qquad \text{Equivalent mass} = 31.0 \text{ g}$$

E. Equivalent mass can be used to express the amount, in grams, of a given acid required to act with an amount, in grams, of a given base.
1 mol HCl produces 1 equivalent acid ⇒ 1 mol HCl weighs 36.5 g
1 mol NaOH produces 1 equivalent base ⇒ 1 mol NaOH weighs 40.0 g
therefore, 40.0 g of NaOH will neutralize 36.5 g of HCl

Key Terms equivalent, equivalent mass, normality

Example 9.16 Preparation of an acidic solution of a given normality when the acid is triprotic
Problem 9.16 Preparation of an acidic solution of a given normality when the acid is diprotic
Example 9.17 Calculation of the molar concentration of ions using normality data
Problem 9.17 Calculation of the molar concentration of ions using normality data

ARE YOU ABLE TO... AND WORKED TEXT PROBLEMS

Objectives Introduction Properties of Acids and Base

Are you able to...

- define acid
- define base
- list the properties of acids (Table 9.1)
- list the properties of base (Table 9.1)
- explain why litmus is used to determine acidity
- explain what neutralization means

Objectives Section 9.1 Water Reacts with Water

Are you able to...

- explain how water molecules react with each other to form ions
- write the equilibrium expression for the ionization of water
- identify the hydronium ion
- identify the hydroxide ion
- define the term K_w
- provide the value of K_w in pure water at 25°C (Exercise 9.1)
- provide the values for hydronium and hydroxide ion concentrations in pure water at 25°C
- use K_w to calculate hydroxide and hydronium ion concentrations (Exercise 9.2)
- explain the relation between hydronium and hydroxide ion concentrations and the acidity, basicity, or neutrality of a solution

Objectives Section 9.2 Strong Acids and Strong Bases

Are you able to...

- define strong acid (Exercise 9.6)
- write the dissociation reaction for a strong acid
- identify common strong acids (Table 9.2, Exercise 9.6)
- define strong base (Exercise 9.5)
- write the dissociation reaction for a strong base
- identify common strong bases (Table 9.2, Exercise 9.5)
- distinguish between the strength of an acid or base and its concentration
- calculate the hydroxide ion concentration of a strong base solution (Exs. 9.2, 9.3, Probs. 9.2, 9.3, Exercises 9.9, 9.10, 9.11)
- calculate the hydronium ion concentration of a strong acid solution (Ex. 9.1, Prob. 9.1, Exercises 9.7, 9.8)
- calculate the hydronium or hydroxide ion concentration of a solution using K_w (Exs. 9.4, 9.5, Probs. 9.4, 9.5, Exercises 9.12, 9.13, 9.14)

Problem 9.1 What is the concentration of hydronium ion in a 0.050 *M* aqueous solution of HCl? What is the total ionic concentration?

Table 9.2 shows that HCl is a strong acid. All hydronium ions in a strong acid solution result from protons donated by the acid. Strong acids dissociate completely into ions. Write the dissociation reaction for 1 mol of HCl to determine the ratio of HCl to H^+ in solution:

$$HCl \rightarrow H^+ + Cl^- \qquad \text{1 mol HCl produces 1 mol } H^+.$$

$0.050\ M\ HCl \rightarrow 0.050\ M\ H^+ + 0.050\ M\ Cl^-$ The concentration of hydronium ions is 0.050 *M*. The total concentration of ions is $0.050\ M + 0.050\ M = 0.100\ M$.

Problem 9.2 What is the concentration of hydroxide ion in a 0.050 *M* aqueous solution of NaOH? What is the total ionic concentration?

Table 9.2 shows that NaOH is a strong base. All hydroxide ions in a strong base solution come from hydroxide donated by the base. Strong bases dissociate completely into ions. Write the dissociation reaction for 1 mol of NaOH to determine the ratio of NaOH to OH^- in solution:

$$NaOH \rightarrow Na^+ + OH^- \qquad \text{1 mol NaOH produces 1 mol } OH^-.$$

$0.050\ M\ NaOH \rightarrow 0.050\ M\ Na^+ + 0.050\ M\ OH^-$ The concentration of hydroxide ions is 0.050 *M*. The total concentration of ions is $0.050\ M + 0.050\ M = 0.100\ M$.

Problem 9.3 What is the concentration of hydroxide ion in a 0.030 *M* aqueous solution of $Sr(OH)_2$? What is the total ionic concentration?

Table 9.2 shows that strontium hydroxide is a strong base. All hydroxide ions in a strong base solution come from hydroxide donated by the base. Strong bases dissociate completely into ions. Write the dissociation reaction for 1 mol of $Sr(OH)_2$ to determine the ratio of $Sr(OH)_2$ to OH^- in solution:

$$Sr(OH)_2 \rightarrow Sr^{2+} + 2\ OH^- \qquad \text{1 mol } Sr(OH)_2 \text{ produces 2 mol } OH^-.$$

$0.030\ M\ Sr(OH)_2 \rightarrow 0.030\ M\ Sr^{2+} + 0.060\ M\ OH^-$ The concentration of hydroxide ions is $0.060\ M$. The total concentration of ions is $0.030\ M + 0.060\ M = 0.090\ M$.

Problem 9.4 Calculate the hydronium ion concentration in an aqueous $0.00500\ M$ KOH solution.

Table 9.2 shows that KOH is a strong base. All hydroxide ions in a strong base solution come from hydroxide donated by the base. Strong bases dissociate completely into ions. Write the dissociation reaction for 1 mol of KOH to determine the ratio of KOH to OH^- in solution:

$KOH \rightarrow K^+ + OH^-$ 1 mol KOH produces 1 mol OH^-.

$0.00500\ M\ KOH \rightarrow 0.00500\ M\ K^+ + 0.00500\ M\ OH^-$ The concentration of hydroxide ions is $0.00500\ M$.

The relation between hydroxide and hydronium ions in an aqueous solution is through the ion product of water: $K_w = [H_3O^+][OH^-]$,

K_w is a constant: $1.00 \times 10^{-14}\ M$ and $[OH^-]$ was already determined: $0.00500\ M$

Substitute the values in the equation for K_w to determine the $[H_3O^+]$:

$$1.00 \times 10^{-14}\ M / 0.00500\ M = 2.00 \times 10^{-12}\ M\ H_3O^+$$

$[H_3O^+] = 2.00 \times 12^{-12}\ M$ This is a small value, which makes sense for a strong base solution.

Problem 9.5 Calculate the hydroxide ion concentration in an aqueous $0.00500\ M\ HNO_3$ solution.

Table 9.2 shows that HNO_3 is a strong acid. All hydronium ions in a strong acid solution come from protons donated by the acid. Strong acids dissociate completely into ions. Write the dissociation reaction for 1 mol of HNO_3 to determine the ratio of HNO_3 to H^+ in solution:

$HNO_3 \rightarrow H^+ + NO_3^-$ 1 mol HNO_3 produces 1 mol H^+.

$0.00500\ M\ HNO_3 \rightarrow 0.00500\ M\ H^+ + 0.00500\ M\ NO_3^-$ The concentration of hydronium ions is $0.00500\ M$.

The relation between hydroxide and hydronium ions in an aqueous solution is through the ion-product of water: $K_w = [H_3O^+][OH^-]$

K_w is a constant: $1.00 \times 10^{-14}\ M$ and $[H_3O^+]$ was determined as: $0.00500\ M$

Substitute the values in the equation for K_w to determine the $[OH^-]$:

$$1.00 \times 10^{-14}\ M / 0.00500\ M = 2.00 \times 10^{-12}\ M\ OH^-$$

$[OH^-] = 2.00 \times 12^{-12}\ M$ This is a small value, which makes sense for a strong acid solution.

Objectives Section 9.3 A Measure of Acidity: pH

Are you able to...

- explain the relations between the acidity of a solution and the hydronium ion concentration (Table 9.3)
- explain the pH scale
- identify acidic, basic, and neutral pH values
- describe the mathematical relation between the pH of a solution and the hydronium ion concentration
- calculate the pH of a solution by using the hydronium ion concentration (Ex. 9.7, Prob. 9.7, Exercises 9.17, 9.18, 9.20, 9.21, 9.61)
- calculate the pH of a solution by using the hydroxide ion concentration (Ex. 9.6, Prob. 9.6, Exercises 9.19, 9.22, 9.64)
- calculate the hydronium ion concentration of a solution from the pH (Ex. 9.10, Prob. 9.10, Exercise 9.70)
- explain the relation between pH and pOH
- calculate pH and pOH of solutions (Exs. 9.8, 9.9, Probs. 9.8, 9.9, Exercises 9.19, 9.20, 9.62, 9.63, 9.71)

Problem 9.6 Calculate the hydronium ion concentration and the pH of an aqueous $0.010\ M$ KOH solution.

KOH is a strong base, so the hydroxide ion concentration will equal the KOH concentration: $0.010\ M$ OH^- (see Problem 9.2). The hydronium ion concentration is calculated by using the K_w of water, $1.00 \times$

10^{-14} *M*, and the $[OH^-]$ (see Problem 9.4). The pH can be calculated from the hydronium ion concentration: pH = $-\log [H_3O^+]$

$$K_w/[OH^-] = [H_3O^+] \qquad 1.00 \times 10^{-14}\ M / 0.010\ M = 1.0 \times 10^{-12}\ M$$

$[H_3O^+] = 1.0 \times 10^{-12}\ M$ pH $= -\log[1.0 \times 10^{-12}\ M] = 12.00$ A pH of 12 makes sense for a strong base solution.

Problem 9.7 Calculate the pH of an aqueous 0.0037 *M* HCl solution.

HCl is a strong acid so the hydronium ion concentration equals the HCl concentration, 0.0037 *M* (see Problem 9.1). The pH can be calculated from the hydronium ion concentration. The relation between hydronium ion and pH is: pH $= -\log[H_3O^+]$

$$pH = -\log[0.0037\ M] = 2.43$$ This pH makes sense for a strong acid solution.

Problem 9.8 What is the pH of a 0.00010 *M* solution of TlOH?

TlOH is a strong base, so the hydroxide ion concentration equals the TlOH concentration, 0.00010 *M* (see Problem 9.2) . The pOH can be calculated from the hydroxide ion concentration: pOH $= -\log [OH^-]$, and the pH can be calculated from the relation between pH and pOH: pH + pOH = 14.00

$$pOH = -\log[0.00010\ M] = 4.00$$

$$14.00 - 4.00 = pH \qquad pH = 10.00$$ A basic pH makes sense for a strong base solution.

Problem 9.9 What is the pOH of an aqueous 0.0020 *M* HNO_3 solution?

HNO_3 is a strong acid, so the hydronium ion concentration equals the HNO_3 concentration, 0.0020 *M* (see Problem 9.1). The pH can be calculated from the hydronium ion concentration: pH $= -\log [H_3O^+]$, and the pOH can be calculated from the relation between pH and pOH: pH + pOH = 14.00

$$pH = -\log[0.0020\ M] = 2.70$$ An acidic pH makes sense for a strong acid solution.

$$14.00 - 2.70 = pOH \qquad pOH = 11.30$$

Problem 9.10 The pH of an aqueous HI solution was found to be 2.46. Calculate the molar concentration of the acid.

The relation between pH and H_3O^+ can be written as $[H_3O^+] = 10^{-pH}$.

$$[H_3O^+] = 10^{-2.46}$$

Following the steps outlined in Example 9.10, $[H_3O^+] = 0.0035\ M$.

Objectives Section 9.4 Weak Acids and Weak Bases

Are you able to...

- define weak acid (Exercise 9.23)
- write the dissociation reaction for a weak acid
- describe the particles that are present in a weak acid solution
- identify common weak inorganic and organic acids (Exercise 9.23)
- define weak base (Exercise 9.24)
- write the dissociation reaction for a weak base
- describe the particles that are present in a weak base solution
- identify common weak bases (Exercise 9.24)
- write the equilibrium reaction for a weak acid and water (Exercises 9.31, 9.32, 9.33, 9.34)
- write the equilibrium equation for a weak base and water (Exercises 9.31, 9.32, 9.33, 9.34)
- define K_a and K_b
- explain what information is contained in the magnitude of K_a and K_b
- recognize relative strengths of acids and bases from their K_a and K_b values (Table 9.4, Exercises 9.25, 9.26)

Objectives Section 9.5 Brønsted-Lowry Theory of Acids and Bases

Are you able to…

- explain the Brønsted-Lowry theory of acids and bases
- recognize Brønsted-Lowry acids and bases
- identify conjugate acid-base pairs (Exercises 9.27, 9.28)
- explain the relation between K_a, K_b, and K_w (Exs. 9.11, 9.12, Probs. 9.11, 9.12)
- use the relation between K_a, K_b, and K_w to calculate K_a or K_b
- explain the relation between pK_a and K_a and pK_b and K_b (Exercises 9.31, 9.32, 9.33, 9.34)
- recognize the relation between pK_a or pK_b and the relative strengths of acids and bases (Table 9.5)
- calculate pK_a and pK_b values (Exercises 9.31, 9.32, 9.33, 9.34)

Problem 9.11 Show how the K_b value for cyanate is derived from the K_a value of cyanic acid.

Table 9.5 shows that the K_a for cyanic acid is 2.19×10^{-4}. The relation between K_a and K_b is:

$$K_a \times K_b = K_w \qquad K_w \text{ is a constant: } 1.00 \times 10^{-14}\ M$$

$$K_b = K_w / K_a = 1.00 \times 10^{-14}\ M / 2.19 \times 10^{-4}\ M = 4.57 \times 10^{-11}$$

Table 9.5 shows the value of K_b for cyanate to be 4.57×10^{-11}.

Problem 9.12 Show how the pK_b value for cyanate is derived from the pK_a value of cyanic acid.

Table 9.5 shows the pK_a for cyanic acid is 3.66. The relation between pK_a and pK_b is:

$$pK_a + pK_b = 14.00$$

$$pK_b = 14.00 - pK_a = 14.00 - 3.66 = 10.34$$

Table 9.5 shows the value of pK_b for cyanate to be 10.34.

Objectives Section 9.6 Dissociation of Polyprotic Acids

Are you able to…

- define diprotic, triprotic, and polyprotic acid (Exercise 9.72)
- recognize polyprotic acids
- write the dissociation reactions for each proton of a polyprotic acid (Exercises 9.35, 9.36)
- write the equilibria expressions for proton dissociations of polyprotic acids (Exercises 9.37, 9.38)
- recognize the relation between K_a for a proton and the degree to which it is dissociated (Exercises 9.39, 9.40)
- use phosphoric acid as a model for discussing the dissociation of a triprotic acid
- use carbonic acid as a model for discussing the dissociation of a diprotic acid
- identify amphoteric substances (Exercises 9.3, 9.4)

Objectives Section 9.7 Salts and Hydrolysis

Are you able to…

- identify conjugate bases that are also anions of salts
- identify conjugate acids that are also cations of salts
- write the hydrolysis reactions of salts and water
- write the equilibria reactions for anions and cations with water (Exercises 9.29, 9.30)
- determine the acidity, basicity, or neutrality of a salt solution (Ex. 9.13, Prob. 9.13, Exercises 9.41, 9.42, 9.43, 9.44, 9.65, 9.73)
- explain the relationship between K_a and K_b and the pH of a salt solution

Problem 9.13 Predict whether the following salts will produce an acidic, basic, or neutral aqueous solution: $MgCl_2$, KNO_3, $SnCl_2$, NH_4NO_3.

The first step is to tabulate the information so that you can see the identity and nature of each anion and cation. Identify the group number or nature of the cation and the acid (Table 9.2) from which the anion is derived. Refer to the rules for determining acidity, basicity, or neutrality of ions in the text.

Salt	Cation	Nature	From rules	Anion	Nature	From rules
$MgCl_2$	Mg^{2+}	Group II/from strong base, $Mg(OH)_2$	neutral	Cl^-	from strong acid, HCl	neutral
KNO_3	K^+	Group I/from strong base, KOH	neutral	NO_3^-	from strong acid, HNO_3	neutral
$SnCl_2$	Sn^{2+}	Group IV/from weak base, $Sn(OH)_2$	acidic	Cl^-	from strong acid, HCl	neutral
NH_4NO_3	NH_4^+	from weak base, NH_3	acidic	NO_3^-	from strong acid, HNO_3	neutral

Results: $MgCl_2$ and KNO_3 will be neutral solutions. $SnCl_2$ and NH_4NO_3 will be acidic solutions.

Objectives Section 9.8 Buffers and Buffered Solutions

Are you able to...

- describe the components of a buffer system (Exercise 9.75)
- explain how a buffer system works (Exercise 9.45)
- identify buffer systems (Exercise 9.74)
- explain the relation between buffer capacity and the concentrations of buffer components (Exercise 9.46)
- calculate the pH of a buffered system by using the Henderson-Hasselbalch equation (Box 9.8, Ex. 9.14, Prob. 9.14, Exercises 9.47, 9.48, 9.49, 9.50, 9.66)
- explain how the ratio of anion : acid concentration affects the pH of a buffered system (Table 9.6, Exercise 9.67)
- explain the relation between the pH and pK_a of a buffered solution

Problem 9.14 Calculate the pH of an aqueous solution consisting of 0.0060 *M* acetic acid and 0.0080 *M* sodium acetate.

Recognize that the solution consists of a weak acid and its salt. The pH of weak acid solutions is determined by the relation: pH = pK_a + log [anion]/[acid]. The pK_a of the weak acid is used (from Table 9.4 or Table 9.5), and the RATIO of the anion to the acid is what determines the overall pH.

pK_a acetic acid = 4.76 [anion] = [acetate] = 0.0080 *M* [acid] = [acetic acid] = 0.0060 *M*

$$\begin{aligned} pH &= 4.76 + \log [0.0080\ M]/[0.0060\ M] \\ &= 4.76 + \log 1.33 \\ &= 4.76 + 0.125 = 4.89 \end{aligned}$$

Objectives Section 9.9 Buffer System of the Blood

Are you able to...

- explain the role of carbon dioxide in maintaining the acid-base balance of the blood
- describe the relation between carbon dioxide in the lungs, carbon dioxide in the blood, and carbon dioxide in the tissues
- explain the relation between carbon dioxide and carbonic acid
- define acidosis and alkalosis
- recognize blood pH values that are acidotic and alkalotic
- explain how metabolic and respiratory changes affect the pH of blood
- write the equilibria reactions for carbon dioxide in the lungs and blood and for carbon dioxide and carbonic acid
- show how metabolic and respiratory changes affect the preceding equilibria

Objectives Section 9.10 Titration

Are you able to...

- explain titration as an analytical technique
- explain how titrations can be used to determine the concentration of an acidic or basic solution

- list the criteria for a successful analytical procedure
- explain what a neutralization reaction is
- write neutralization reactions for acids and bases
- explain what an indicator is equivalence and the role that it plays in acid-base titrations (Box 9.4)
- distinguish between equivalence point and end point
- calculate the concentration of an acid or base solution from titration data (Ex. 9.15, Prob. 9.15, Exercises 9.51, 9.52, 9.53, 9.54, 9.68)

Problem 9.15 What volume of a 0.0200 *M* KOH standard solution is required to neutralize 35.0 mL of a 0.0150 *M* solution of HNO_3?

Neutralization reactions take place in a 1 to 1 ratio: 1 equivalent of base (OH^-) neutralizes 1 equivalent of acid (H^+). You need to first determine how many equivalents of acid are in 35.0 mL of acid solution. The molar amount of acid is equal to the equivalent amount because 1 mol H^+ = 1 equivalent. The number of moles of acid can be calculated from the volume of acid and the concentration of the acid; remember to keep the volume units consistent:

$$\text{volume of acid} \times \text{molarity of acid} = \text{number of moles of acid}$$

$$(35.0 \text{ mL acid}) \times (0.0150 \text{ mol H}^+/1000 \text{ mL}) \quad = \quad 0.000525 \text{ mol H}^+ = 0.000525 \text{ equivalent H}^+$$

0.000525 equivalent of acid require 0.000525 equivalent of base. The volume of base needed can be calculated using the concentration data and the number of equivalents of base needed. 1 mol OH^- equals 1 equivalent base. The concentration of the standard solution is 0.0200 *M* KOH, which means that it is also 0.0200 equivalent base/liter (0.0200 mol KOH ⇒ 0.0200 mol OH^- ⇒ 0.0200 equivalent base; refer to Problem 9.2).

$$(0.000525 \text{ equivalent base needed}) \times 1000 \text{ mL}/0.0200 \text{ equivalent base} = 26.3 \text{ mL KOH needed}$$

Objectives Section 9.11 Normality

Are you able to...

- define normality, equivalent, equivalent weight
- determine the number of equivalents per 1 mol of substance
- determine the equivalent weight of a substance
- calculate the number of equivalents and weight of an acid or base that will completely neutralize a given amount of base or acid
- calculate the normality of a solution (Exercises 9.57, 9.58, 9.60, 9.69)
- prepare a solution of specified normality (Ex. 9.16, Prob. 9.16, Exercises 9.55, 9.56)
- use normality data to calculate the concentrations of ions in a solution (Ex. 9.17, Prob. 9.17, Exercise 9.59)

Problem 9.16 How would you prepare a 1.00 **N** solution of the diprotic acid $(CH_2)_2(COOH)_2$, succinic acid?

A 1.00 **N** solution of succinic acid means that there is 1.00 equivalent of acid per 1.00 liter of solution. The equivalent weight of a polyprotic acid can be calculated by dividing the molar mass of the acid by the number of dissociable protons. Succinic acid has 2 —COOH groups, therefore, it is diprotic.

$$\text{formula weight succinic acid}/2 = \text{equivalent weight of succinic acid}$$

118.10 g/2 = 59.05 g   59.1 grams of succinic acid dissolved in enough water to make 1 liter of solution will produce a 1.00 **N** solution.

Problem 9.17 The concentrations of potassium and magnesium ions in a sample of blood plasma are given as K^+ = 5.0 meq/L and Mg^{2+} = 3.0 meq/L. Calculate the molar concentrations of these ions.

Molar concentration means that you want to know the number of moles of these ions per liter of solution. The problem tells you how many milliequivalents of each ion are in 1 L of solution. The problem requires that you change milliequivalents into moles. The relation between milliequivalents and moles can be determined on the basis of 1 mol of each ion. The charge on each ion tells you the number of equivalents produced by 1 mol of the element. The equivalent unit can be rewritten as milliequivalents: 1 eq = 1000 meq

1 mol K^+ produces 1000 meq (1 eq)   1 mol Mg^{2+} produces 2000 meq (2 eq)

$$(5.0 \text{ meq } K^+) \times 1 \text{ mol } K^+/1000 \text{ meq} = 0.0050 \text{ mol } K^+ \text{ in 1 L of solution}$$

$$(3.0 \text{ meq } Mg^{2+}) \times 1 \text{ mol } Mg^{2+}/ 2000 \text{ meq} = 0.0015 \text{ mol } Mg^{2+} \text{ in 1 L of solution}$$

MAKING CONNECTIONS

Construct a concept map or write a paragraph describing the relation among the components of each of the following groups:

(a) acid, base, proton, pH
(b) acid, base, aqueous solution
(c) acid, base, neutralization
(d) hydronium ion, hydroxide ion, K_w
(e) pH scale, hydronium ion, hydroxide ion
(f) K_w, pH, hydronium ion, hydroxide ion
(g) hydronium ion concentration, pH scale, acidity, basicity
(h) hydronium ion concentration, pH unit, acidity
(i) strong acid, strong base, dissociation, pH
(j) weak acid, weak base, dissociation, pH
(k) dissociation constant, acid strength, acid concentration
(l) molarity, normality, equivalents, polyprotic acid, monoprotic acid
(m) equivalents, acid, base
(n) acid concentration, base concentration, molarity, equivalents
(o) conjugate acid, conjugate base, buffer system
(p) acid dissociation constant, polyprotic acid
(q) buffer system, pH , buffer capacity
(r) amphoteric molecule, acid, base, water
(s) acid, base, salt, hydrolysis, neutralization
(t) buffer system, Henderson-Hasselbalch equation, pH
(u) buffer system, acid/anion ratio, pH
(v) acidosis, alkalosis, blood pH, carbonic acid/bicarbonate ion
(w) titration, neutralization reaction, concentration of acid, indicator
(x) equivalence point, end point, indicator
(y) equivalence point, end point, hydroxide ion, hydronium ion

FILL-INS

Acids and bases are categorized on the basis of their properties. Acids turn blue litmus paper _____ and bases turn red litmus paper _____. Acids and bases _____ each other in chemical reactions. In pure water, 1 of every _____ water molecules is ionized to form a _____, which is characteristic of acids, and a _____, which is characteristic of bases. At 25°C the concentration of these ions in pure water is _____ each and their product, called the _____, is _____. The ion-product of water is a _____ and is given the symbol _____. The relative amounts of hydronium and hydroxide ions determine the _____ of a solution. When hydronium ion concentration _____ hydroxide ion concentration, the solution is neutral. When hydronium ion concentration _____ hydroxide ion concentration, the solution is acidic. When hydronium ion concentration _____ hydroxide ion concentration, the solution is basic.

The degree to which an acid or base produces these ions in solution is a measure of the _____ of the acid or base. Strong acids and bases _____ in water and produce _____. Weak acids and bases _____ in water and produce _____. Weak acids give up _____ to water, leaving behind _____ that can accept a proton and are _____. These pairs are called _____. The conjugate base is what's left after the acid has _____. These pairs satisfy the definition of _____ acids and bases. The _____ of weak acids and bases can be determined by looking at the value of the acid or base dissociation constant, K_a or K_b. The _____, the more ions are produced in a solution.

_____ is a measure of the concentration of hydronium ions in solution and is taken as a measure of the solution's acidity. The pH scale ranges from _____ to _____ with _____ being the most acidic and _____ being the most basic. A change of one pH unit represents a _____ change in hydronium ion concentration. The pH of solutions of strong acids or bases can be determined by using the relation _____, whereas the pH of solutions of weak acids or bases is determined by using the relation _____. The pH and pOH of solutions are related by: _____.

Polyprotic acids have _____. Each proton dissociation reaction has its own _____ and an equilibrium reaction can be written for each dissociation. Some anions of polyprotic acids are _____, capable of both donating and accepting a proton. _____ is also amphoteric.

Acids and bases produce _____ when they react. The extent to which the salt reacts with water to form acidic, basic, or neutral solutions can be determined by _____. In cases in which the cation produces an acidic solution and the anion produces a basic solution, the magnitude of _____ determines which will dominate.

_____ are weak-acid/conjugate base or weak base/conjugate acid pairs that work to resist large changes in pH. The acid component neutralizes added base by _____, creating a water molecule and leaving a _____ behind. The base component works by _____, creating a _____. The _____ anion to acid in the buffer pair determines the extent to which the system can prevent large pH changes. The maximum buffer capacity is achieved when the concentrations are _____ so the ratio is _____. The farther from the ideal ratio, the _____ the buffer. In the blood, _____ and _____ work as a buffer pair. Carbonic acid concentration is determined by the amount of _____. The _____ control bicarbonate ion concentration. When the ratio of bicarbonate to carbonic acid is _____, the blood pH is maintained at the required pH of _____. Respiratory and metabolic conditions that _____ produce pH values either below 7.4, a condition called _____, or pH values above 7.4, a condition called _____. Either of these conditions can be harmful.

The concentrations of acids or bases in solutions can be determined by a technique called, which relies on the fact that _____ reacts with _____ in a neutralization reaction. When all acid and base ions are neutralized, the _____ has been reached. Any additional acid or base produces a pH change that can be detected by using _____ whose _____ depends on the pH of the solution. The point at which the color change occurs is called the _____ of the titration. The _____ of a solution gives the concentration of equivalents of acid or base per liter of solution.

Answers

Acids and bases are categorized on the basis of their properties. Acids turn blue litmus paper red and bases turn red litmus paper blue. Acids and bases neutralize each other in chemical reactions. In pure water, 1 of every billion water molecules is ionized to form a hydronium ion, which is characteristic of acids, and a hydroxide ion, which is characteristic of bases. At 25° the concentration of these ions in pure water is 1×10^{-7} *M* each and their product, called the ion-product of water, is 1×10^{-14}. The ion-product of water is a constant and is given the symbol K_w. The relative amounts of hydronium and hydroxide ions determine the acidity, basicity, or neutrality of a solution. When hydronium ion concentration equals hydroxide ion concentration, the solution is neutral. When hydronium ion concentration is greater than hydroxide ion concentration, the solution is acidic. When hydronium ion concentration is less than hydroxide ion concentration, the solution is basic.

The degree to which an acid or base produces these ions in solution is a measure of the strength of the acid or base. Strong acids and bases ionize completely in water and produce many ions. Weak acids and bases ionize incompletely in water and produce few ions. Weak acids give up protons to water, leaving behind anions that can accept a proton and are weak bases. These pairs are called conjugate acid-base pairs. The conjugate base is what's left after the acid has donated a proton. These pairs satisfy the definition of Brønsted-Lowry acids and bases. The extent of ionization of weak acids and bases can be determined by looking at the value of the acid or base dissociation constant, K_a or K_b. The larger the K_a or K_b, the more ions are produced in a solution.

pH is a measure of the concentration of hydronium ions in solution and is taken as a measure of the solution's acidity. The pH scale ranges from 0 to 14, with 0 being the most acidic and 14 being the most basic. A change of one pH unit represents a ten-fold change in hydronium ion concentration. The pH of solutions of strong acids or bases can be determined by using the relation $pH = -\log [H_3O^+]$, whereas the pH of solutions of weak acids or bases is determined by using the relation $pH = pK_a + \log [anion]/[acid]$. The pH and pOH of solutions are related by: $pH + pOH = 14$.

Polyprotic acids have more than one dissociable proton. Each proton dissociation reaction has its own K_a and an equilibrium reaction can be written for each dissociation. Some anions of polyprotic acids are amphoteric, capable of both donating and accepting a proton. Water is also amphoteric.

Acids and bases produce salts when they react. The extent to which the salt reacts with water to form acidic, basic, or neutral solutions can be determined by looking at the acid and base from which the anion and cation are derived. In cases in which the cation produces an acidic solution and the anion produces a basic solution, the magnitude of K_a and K_b determines which will dominate.

Buffer systems are weak acid/conjugate base or weak base/conjugate acid pairs that work to resist large changes in pH. The acid component neutralizes added base by donating a proton to the base, creating a water molecule and leaving a weaker base behind. The base component works by accepting added protons, creating a weaker acid. The ratio of anion to acid in the buffer pair determines the extent to which the system can prevent large pH changes. The maximum buffer capacity is achieved when the concentrations are about equal so the ratio is close to 1. The farther from the ideal ratio, the less effective the buffer. In the blood, carbonic acid and bicarbonate ions work as a buffer pair. Carbonic acid concentration is determined by the amount of carbon dioxide in the lungs. The kidneys control bicarbonate ion concentration. When the ratio of bicarbonate to carbonic acid is 20 : 1, the blood pH is maintained at the required pH of 7.4. Respiratory and metabolic conditions that cause the ratio to deviate from this ratio produce pH values either below 7.4, a condition called acidosis, or pH values above 7.4, a condition called alkalosis. Either of these conditions can be harmful.

The concentrations of acids or bases in solutions can be determined by a technique called titration, which relies on the fact that 1 equivalent of acid reacts with 1 equivalent of base in a neutralization reaction. When all acid and base ions are neutralized, the equivalence point has been reached. Any additional acid or base produces a pH change that can be detected by using an indicator whose color depends on the pH of the solution. The point at which the color change occurs is called the end point of the titration. The normality of a solution gives the concentration of equivalents of acid or base per liter of solution.

TEST YOURSELF

1. You are given three clear solutions and are told to identify which is acid, which is base, and which is water. Explain how you would do this.
2. $Mg(OH)_2$, in milk of magnesia, neutralizes excess stomach acid. $NaHCO_3$, sodium bicarbonate, is a traditional treatment for acid stomach. How could you determine which of these bases is a more effective neutralizer of excess stomach acid?
3. Explain how acids with K_as of 10, 1, 0.1, and 0.001 differ from one another.
4. How are the dissociation reactions for a strong acid or base and a weak acid or base different?
5. Write the dissociation reaction for acetic acid. Write the equilibrium constant for the dissociation.
6. Write the dissociation reaction for water, and identify the conjugate acid-base pairs.
7. Answer the following questions below for the generic acid H_3A:
   (a) How many dissociable protons are in the acid?
   (b) Write the dissociation reaction for each proton.
   (c) How many equivalents will 1 mol of the acid produce?
   (d) How many moles of NaOH would be required to neutralize 1 mol of the acid?
   (e) How could you use K_a values for each proton to determine which ionizes to the greatest degree?
8. HCl and NaOH neutralize each other when they undergo a reaction.
   (a) What type of acid is HCl? What type of base is NaOH? What are the products of the reaction?
   (b) What is the mole ratio in which they react? What is the weight ratio in which they react?
   (c) What type of solution will form from salts derived from HCl and NaOH?
   (d) A 1 **N** solution of HCl would require an equal volume of what normality NaOH solution for complete neutralization to occur?

9. Write the Henderson-Hasselbalch equation for carbonic acid/bicarbonate ion and use it to answer the following questions:
   (a) The blood pH of a person with metabolic acidosis is 7.1. What happened in the equation to cause this decrease?
   (b) The blood pH of a person with respiratory acidosis is 7.2. What happened in the equation to cause this decrease?
   (c) The blood of a person with metabolic alkalosis is 7.8. What happened in the equation to cause this increase?
   (d) The blood of a person with respiratory alkalosis is 7.6. What happened in the equation to cause this increase?

Answers

1. There are several ways in which you could distinguish among the three clear solutions. You could put a drop of each on both red and blue litmus paper. The basic solution will cause the red litmus to turn blue but will have no effect on the blue litmus. The acid solution will cause the blue litmus to turn red but will have no effect on the red litmus. Water will cause no color change in either paper. You could also test each with indicators designed to change color in specific pH regions. Water is neutral with a pH of 7. The acidic solution has a pH below 7 and the basic solution has a pH above 7.
2. You could titrate equal masses of $Mg(OH)_2$ and $NaHCO_3$ with HCl solution to determine which substance neutralized the greater volume of HCl solution.
3. The K_a value is derived from the equilibrium constant equation for $HA \rightleftharpoons H^+ + A^-$. $K_a = [H^+][A^-]/[HA]$. A K_a above 1 means that the numerator is greater than the denominator and there are more ions than undissociated acid molecules in the solution. When the numerator and denominator are equal, the ratio is 1 and there are equal numbers of ions and undissociated acid molecules. The greater the denominator becomes, the more undissociated acid molecules there are. As the K_a gets smaller, the denominator gets larger and the numerator gets smaller. The acids differ in their strengths with $K_a = 10$ being the strongest and $K_a = 0.001$ being the weakest of the group.
4. Dissociation reactions for strong acids and bases are written in one direction: $HCl \rightarrow H^+ + Cl^-$ and $NaOH \rightarrow Na^+ + OH^-$. Dissociation reactions for weak acids and bases are equilibrium reactions that occur in both directions simultaneously: $NH_3 + H_2O \rightleftharpoons NH_4^+ + OH^-$ and $H_2CO_3 \rightleftharpoons H^+ + HCO_3^-$.
5. From Table 9.4, the formula for acetic acid is $HC_2H_3O_2$. Because acetic acid is an organic acid, another way to write the formula is CH_3COOH, which clearly shows one acid group with one dissociable proton (—COOH).
   The dissociation reaction is $CH_3COOH \rightleftharpoons CH_3COO^- + H^+$
   The equilibrium constant can be written as $K_a = [CH_3COO^-][H^+]/[CH_3COOH]$
6. $H_2O + H_2O \rightleftharpoons H_3O^+ + OH^-$
   H_2O is a conjugate acid that donates a proton to form the conjugate base OH^-
   H_3O^+ is a conjugate acid that can donate a proton to form the conjugate base H_2O
   Water is amphoteric and can act both as an acid and as a base.
7. (a) 3; from the formula, the acidic (dissociable) protons are written first.
   (b) $H_3A \rightleftharpoons H_2A^- + H^+$
   $H_2A^- \rightleftharpoons HA^{2-} + H^+$
   $HA^{2-} \rightleftharpoons A^{3-} + H^+$
   (c) 1 mol of the acid produces 3 equivalents
   (d) Because 1 mol of base neutralizes 1 equivalent of acid and H_3A produces 3 eq of acid, 3 eq of base would be needed. 1 mol NaOH produces 1 eq of base, therefore, 3 mol of NaOH would be needed to completely neutralize 1 mol H_3A.
   (e) If K_a values for each proton are very close to each other, then you can say that all protons are ionized almost simultaneously. If the K_a values differ greatly, the largest indicates the proton that ionizes to the greatest extent and the smallest the proton that ionizes least.

8. (a) HCl is a strong acid and NaOH is a strong base. Strong acids and bases react to form salts and water.

$$HCl + NaOH \rightarrow NaCl + H_2O$$

   (b) They react in a 1 to 1 mole ratio; the weight of 1 mol of HCl is ~36 grams and the weight of 1 mol of NaOH is ~40 grams.
   (c) The rules governing the formation of such solutions say that cations and anions formed from strong acids and bases will form neutral aqueous solutions.
   (d) A 1 **N** solution of HCl would have 1 equivalent per liter of acid and would require an equal volume of a 1 **N** solution of NaOH to react completely.
9. $pH = pK_a + \log [anion]/[acid]$ for carbonic acid, the pK_a is 6.35 (Table 9.4); the anion is HCO_3^-, bicarbonate ion

   The normal pH of blood is 7.4, which results from a bicarbonate : carbonic acid ratio of 20 : 1; the plasma concentration of HCO_3^- is about 0.026 *M* and that of H_2CO_3 is about 0.0013 *M*

$$pH = 6.1 + \log 20/1 = 6.1 + 1.3 = 7.4$$

When the ratio moves away from 20 :1, the pH changes. Changes in concentration of either HCO_3^- or H_2CO_3 can cause pH fluctuations:

Change	Affect on ratio	Affect on pH	Physiological condition
HCO_3^- increases	increases	increases	metabolic alkalosis
HCO_3^- decreases	decreases	decreases	metabolic acidosis
H_2CO_3 increases	decreases	decreases	respiratory acidosis
H_2CO_3 decreases	increases	increases	respiratory alkalosis

In this equation, HCO_3^- concentration is under kidney control and H_2CO_3 concentration is under lung control

$$HCO_3^- = 10 \quad \leftarrow \text{metabolic component (kidney)}$$

$$H_2CO_3 = 1 \quad \leftarrow \text{respiratory component (lungs)}$$

   (a) In metabolic acidosis, a pH of 7.1 results when the HCO_3^- decreases. This decrease can be due to kidney disease or diabetes.
   (b) In respiratory acidosis, a 7.2 results when H_2CO_3 increases. This increase can be due emphysema, pneumonia, or other lung problems.
   (c) In metabolic alkalosis, a pH of 7.8 results when HCO_3^- increases. This increase can be due to prolonged vomiting or excessive intake of antacids.
   (d) In respiratory alkalosis, a pH of 7.6 results when H_2CO_3 decreases due to prolonged hyperventilation.

chapter 10

Chemical and Biological Effects of Radiation

OUTLINE

10.1 Electromagnetic Radiation Revisited

A. Travels from source through space in form of wave
  1. Has characteristic frequency (Fig. 10.1)
  2. Speed of electromagnetic radiation, regardless of frequency, is constant (3×10^8 m/s)
  3. Speed of radiation = wavelength of radiation × frequency of radiation ($c = \lambda + \nu$)

B. Electromagnetic radiation has energy
  1. Einstein proposed that light consists of particles called photons.
  2. Energy of photon is proportional to frequency: $E = h\nu$ where h is Planck's constant.

10.2 Radioactivity

A. Radioactivity is an inherent property of some elements.

B. Radioactivity is a result of instability in a nucleus.
  1. caused by repulsive forces between protons
  2. all elements with atomic numbers greater than 83 are radioactive
  3. some isotopes of lighter elements are radioactive

C. Nuclear processes require specific notation.
  1. isotopes are elements having the same number of protons (atomic number) but differ in the number of neutrons; their mass numbers differ
     a. nuclide is a generic term for isotopes
     b. radionuclide is used for radioactive isotopes
  2. isotopes are designated by using isotopic notation
     a. mass number as superscript and atomic (proton) number as subscript

$$^{\text{mass number}}_{\text{proton number}}X \quad \text{for example, } ^{14}_{6}C$$

     b. element name followed by hyphen (proton number is implied by element)

Carbon−14

Key Terms isotope, nucleon, nuclide, radioactivity, radionuclide

10.3 Radioactive Emissions

A. Radioactive emission is the emission of energy or particles from a radioactive nucleus.
  1. often results in changes in mass number, atomic number, or both
  2. represented by nuclear equations that account for nucleons present before and after nuclear processes

B. Three major types of radioactive emissions are (Table 10.1)
   1. alpha-particle, α
      a. helium nucleus with no electrons
      b. designated ${}^{4}_{2}He$
      c. emission from nucleus reduces the atomic number by 2 and the mass number by 4
   2. beta-particle, β
      a. energetic electron
      b. designated ${}^{0}_{-1}e$
      c. emission from nucleus increases the atomic number by 1 but has no effect on the mass number
      d. produced in nucleus when neutron is converted into a proton
   3. positron
      a. has all properties of electrons but has positive instead of negative charge
      b. designated ${}^{0}_{1}e$
      c. emission from the nucleus decreases the atomic number by 1 but has no effect on the mass number
      d. produced in nucleus when proton is converted into neutron
      e. positron contact with electron produces 2 gamma rays and annihilation of both positron and electron
   4. gamma-ray, γ
      a. high-energy photon
      b. designated ${}^{0}_{0}\gamma$
      c. emission from nucleus does not change mass number or atomic number

C. Transmutation is a nuclear process.
   1. nuclide is transformed into a new element
   2. can be spontaneous or artificially induced; bombarding other elements with high-energy nuclear particles creates all elements above atomic number 92
   3. represented by nuclear emission equations

$$\text{radioactive isotope} \rightarrow \text{emitted particle} + \text{emission product}$$

   4. sum of the subscripts (number of protons) on right side equals sum of subscripts on left side
   5. sum of superscripts on right side equals sum of superscripts on left side

Key Terms alpha-particle, beta-particle, gamma ray, nuclear equation, photon, positron, transmutation

Example 10.1 Writing a nuclear equation for α emission
Problem 10.1 Writing a nuclear equation for α emission
Example 10.2 Writing a nuclear equation for β emission
Problem 10.2 Writing a nuclear equation for β emission
Example 10.3 Writing a nuclear equation for positron emission
Problem 10.3 Writing a nuclear equation for positron emission

10.4 Radioactive Decay

A. Radioactive decay is the emission of nuclear radiation by a radioactive isotope.
   1. results in the formation of daughter nuclei
   2. if daughter nuclei are stable, decay stops
   3. if daughter nuclei are unstable, decay continues
      a. series from unstable isotope to stable isotope is called nuclear decay series or nuclear disintegration series
      b. three decay series are found in nature
         i. uranium-238 → lead-206

        ii. uranium-235 → lead-207
        iii. thorium-232 → lead-208
    4. decay stops when stable nuclei are reached
B. Rate of decay is an inherent characteristic of a radioactive isotope
    1. defined as the loss of a constant amount of the original material per unit time
    2. measured in terms of half-life, $t_{\frac{1}{2}}$ (Table 10.2), which is the period of time that it takes for 50% of original material to decay (Figure 10.2)
    3. amount of material left after specific number of half-lives can be calculated by (Box 10.1)

$$N/N_0 = (\tfrac{1}{2})^n$$

← n = number of half-lives
N = material left after n half-lives have passed
N_0 = material before decay begins

Key Terms daughter nuclei, disintegration series, half-life, nuclear decay series, radioactive decay

Example 10.4 Determining the relation between half-life and percent decay
Problem 10.4 Determining the relation between half-life and percent decay
Example 10.5 Determining the amount of radioisotope remaining after a given number of half-lives
Problem 10.5 Determining the amount of radioisotope remaining after a given number of half-lives
Example 10.6 Determining the amount of radioisotope remaining after a given number of half-lives
Problem 10.6 Determining the amount of radioisotope remaining after a given number of half-lives

10.5 Effects of Radiation

A. Study of effects of radiation on matter is called radiochemistry or radiation chemistry.
B. Effects of radiation on biological matter are due to energy of particles and length of exposure (Table 10.3).
    1. penetrating ability is a measure of how energetic a particle is
        a. α-particle penetrates tissue to a depth of a 1 mm
            i. large mass and charge cause many intense atomic collisions
            ii. knocks electrons from valence shells and produces ions
            iii. leaves dense trail of ions in its path
        b. β-particle penetrates human tissue to a depth of 5–10 mm
            ii. small mass and high charge cause fewer and less-intense collisions
            ii. leaves thin trail of ions in its path
        c. γ-rays pass through human tissue
        d. penetrating-ability scale is from most energetic to least energetic: $\gamma > \beta > \alpha$
C. Chemical effects of radiation
    1. free radicals and ions are produced along penetration path; free radicals have unpaired electrons and are highly reactive
    2. primary radiation event can lead to secondary chemical process
        a. radiation generates chemical species that initiates a series of chemical reactions
        b. these reactions can trigger more harmful reactions; biomolecules can be inactivated or altered
D. Radiation sickness is a result of exposure to radiation (Table 10.5).
    1. symptoms can range from mild to lethal; nausea and drop in white blood cell count are typical
    2. tissues that undergo rapid division are affected early
        a. DNA and protein damage affects cell reproduction
        b. bone marrow, intestinal lining, children's tissues are most vulnerable
    3. other tissues may suffer long-term consequences
        a. somatic cells may suffer cancer
        b. mutations may show up in later generations owing to chromosomal damage in ovaries and testes
E. Radiation exposure can be minimized by
    1. using radiation-absorbing barrier

2. remaining far away from radiation; radiation intensity falls off very rapidly as distance from source is increased

$$I \propto 1/d^2$$

3. limiting time of exposure

Key Box 10.2 The ozone layer and radiation from space

Key Terms electron-volt, free radical, kilojoule, penetrating power, primary radiation event, radiation chemistry, radiation sickness, radiochemistry, secondary chemical process

Example 10.7 Calculating the effect of increasing distance on radiation intensity
Problem 10.7 Calculating the effect of increasing distance on radiation intensity

10.6 Detection of Radioactivity

A. Path of ions left behind as nuclear emissions penetrate matter can be detected.
  1. photographic film is the most common method
    a. film badges of varying thickness identify type of radiation
    b. film becomes progressively darker as exposure increases
  2. Geiger counter uses electronics to detect ions created when radiation passes through matter (Figure 10.3)
    a. photon or particle creates ion when it hits collection tube; the buildup of these ions produces a current
    b. current is measured or translated into a click
  3. scintillation counters contain substances that emit light when struck by an energetic photon or particle (Figure 10.4); more flashes per unit time mean greater amount of radiation

Key Terms Geiger counter, scintillation counter

10.7 Measuring Radioactivity

A. The units of measurement used to report radioactivity or radiation exposure depend on the information required (Table 10.4).
  1. curie (Ci) is the most common unit and measures the number of atoms that decay per second, called the disintegrations per second (dps)
    a. 1 Ci = 3.7×10^{10} dps
    b. millicuries (mCi), microcuries (μCi), and picocuries (pCi) also are common units
  2. SI unit of radioactivity is the becquerel (Bq); 1 Bq = 1 dps
  3. roentgen (R) is a unit that measures the exposure to X-rays or γ-rays by quantifying the number of ions produced by radiation passing through air; 1 R = 2.1×10^9 units of charge/cm^3 of dry air
  4. rad measures the amount of energy absorbed in matter as a result of exposure to any form of radiation
  5. roentgen equivalent for man (rem) is used to measure biological damage
    a. rads $\times$ RBE = rems
    b. RBE is relative biological effectiveness and depends on
      i. type of tissue irradiated
      ii. dose rate
      iii. total dose
      iv. approximately 1 for X-rays, γ-rays, and β-particles and 10 for α-particles, protons, and neutrons
  6. sievert (Sv) measures the effect of absorbed radiation on different kinds of tissue under different circumstances; 1 Sv = 100 rem

B. Background radiation from natural sources bathes life on earth (Table 10.5).
  1. sources may be cosmic rays or naturally radioactive elements in rock and soil
    a. about 50% from radon (Box 10.4)
    b. typical exposure is about 350 millirems per year

2. sources may be manmade, such as X-rays, radiotherapy, radioisotopes; exposure is about 65 millirem per year

Key Tables and Box

Table 10.4	Units of radiation, their symbols, and definitions
Table 10.5	Effects of short-term exposure to radiation
Box 10.4	Radon: a major health hazard

Key Terms background radiation, becquerel, curie, rad, RBE, rem, roentgen, sievert

10.8 Applications

A. Radioisotopes can be used as diagnostic agents or as therapeutic agents; considerations include
   1. emission characteristics
      a. diagnostic radioisotopes must have significant penetrating ability (Table 10.2); γ-emitters are best
      b. therapeutic radioisotopes must be α- or β-emitters (Table 10.6)
   2. intensity of radiation; shorter half-life means more intense radiation
   3. rate of excretion
      a. if radioisotope is excreted too slowly, unwanted damage may result
      b. if radioisotope is excreted too quickly, the desired effect may not be achieved
   4. decay products should not be radioactive
   5. benefits must outweigh risks

B. Radionuclide can be introduced into the body in several ways
   1. orally: solution can be used when radioisotope naturally concentrates in specific cells
   2. injection: solution can be used when radioisotope naturally concentrates in specific cells
   3. insertion: metal or plastic package can be inserted at or near target site

C. Radioisotopes are used to label elements for studies such as the metabolism of glucose.
   1. radioactive and nonradioactive forms of the same element have the same chemical properties and will behave the same in the body
   2. radioactive forms of an element label a compound
   3. radiolabeled compounds can be detected; autoradiography uses photographic film to find radiolabeled compounds in undisrupted cells and tissues (Figure 10.5)

D. Radiation passing through body can be used to generate images.
   1. basic idea is that radiation directed into body is absorbed differently by different types of tissue; the emitted radiation can be measured, and computers turn the information into an internal picture
   2. CAT scans (computer assisted tomography)
      a. X-rays are absorbed by body and emission is detected
      b. circular array of detectors around body provide data that are transformed into a picture of internal structures
   3. MRI (magnetic resonance imaging)
      a. strong electromagnetic field surrounds body and interacts with protons (hydrogen nuclei) to different degrees, depending on the molecular environment
      b. detectors around body provide data that are transformed into a picture of soft internal tissue structures
   4. PET (positron emission tomography; Figure 10.6)
      a. positron-emitter placed in body produces γ-rays
      b. arrays of detectors around body provide data that are transformed into picture of internal structures
      c. especially useful for brain scans

Key Figures

Figure 10.6	Electrons deflected by nuclei of target atoms produce x-rays
Figure 10.7	Intensity of filtered x-radiation as function of increasing voltage

Key Terms autoradiography, Bremsstrahlung, CAT scan, K-shell X-rays, MRI, PET scan, radioisotope, radiolabel

10.9 Nuclear Reactions

A. Nuclear processes in which nuclei are split or fused are called fission and fusion, respectively.
   1. fission
      a. uranium bombarded with neutrons splits into smaller nuclei
      b. process produces a great deal of energy owing to exothermic chain reaction that grows
      c. basis for nuclear reactors; cadmium rods absorb neutrons to control reaction (Figure 10.7)
   2. fusion
      a. nuclei of two light elements fuse to form heavier element through a high-energy collision
      b. process produces a great deal of energy
      c. basis for energy generation in stars

B. Energy changes associated with nuclear reactions are millions of times as great as those associated with chemical reactions.
   1. there are changes in mass in nuclear reactions that do not occur in chemical reactions
   2. changes in mass are associated with energy changes through Einstein's equation
      a. $E = mc^2$
         m = mass c = velocity of light, 3×10^8 m/s ←very large factor
      b. $E = \Delta m \times c^2$
         The difference in mass, Δm, between product and reactant nuclear masses is proportional to the energy difference

Key Figures
Figure 10.10 Control mechanism in a nuclear reactor
Figure 10.11 Heat transfer in a nuclear reactor

Key Terms nuclear fission, nuclear fusion, nuclear reactor

10.9 Nuclear Energy and the Biosphere

A. Damage can occur through the use of nuclear devices
   1. nuclear generators can accidentally release radioisotopes such as iodine-131, which concentrates in the thyroid gland
   2. nuclear weapons release
      a. strontium-90, which mimics calcium in the body and concentrates in bone
      b. cesium-137, which mimics potassium in the body and concentrates in cells

ARE YOU ABLE TO . . . AND WORKED TEXT PROBLEMS

Objectives Section 10.2 Radioactivity

Are you able to. . .
- define radioactivity (Exercise 10.50)
- describe the nuclear conditions that favor the production of radioactivity
- identify radioactive elements
- use isotopic designation to identify radionuclides

Objectives Section 10.3 Radioactive Emissions

Are you able to. . .
- describe radioactive emission
- identify the major types of radioactive emissions
- identify the characteristics of alpha-particles, beta-particles, positrons, and gamma-rays (Table 10.1, Exercises 10.1, 10.2, 10.3, 10.4)
- write the symbols for alpha-particle, beta-particle, positron, and gamma-ray
- explain how radioactive emissions affect the nuclear composition of an atom
- explain transmutation (Chem. Conn. 10.51)
- write nuclear equations for the emission of alpha-particles, beta-particles, positrons, and gamma-rays (Exercises 10.5, 10.6, 10.7, 10.8)

Problem 10.1 Radium-223 is a radioactive α-emitter. Write the nuclear equation for this emission event and identify the product.

An α-particle is a helium nucleus with the symbol He. Radium-223 (Ra) has an atomic number of 88.

Superscript accounting: 223 – 4 = 219, the mass number of the emission product

Subscript accounting: 88 – 2 = 86, the atomic number of the emission product, which identifies it as radon, Rn. The nuclear equation for the emission is:

$$^{223}_{88}\text{Ra} \rightarrow \, ^{4}_{2}\text{He} + \, ^{219}_{86}\text{Rn}$$

Problem 10.2 Radium-230 is a radioactive β-emitter. Write the nuclear equation for this emission event and identify the product.

A β-particle is an electron with the symbol $_{-1}^{0}\text{e}$. It has no mass, but it increases the positive charge in the nucleus by 1 when it is emitted. Radium-230 (Ra) has an atomic number of 88.

Superscript accounting: 230 – 0 = 230, the mass number of the emission product

Subscript accounting: 88 – (–1) = 89, the atomic number of the emission product, which identifies it as actinium, Ac. The nuclear equation for the emission is:

$$^{230}_{88}\text{Ra} \rightarrow \, ^{0}_{-1}\text{e} + \, ^{230}_{89}\text{Ac}$$

Problem 10.3 Sodium-21 is a radioactive positron emitter. Write the nuclear equation for this emission event and identify the product.

A positron is like an electron except that it has a positive instead of negative charge with the symbol $^{0}_{1}\text{e}$ It has no mass, but it decreases the positive charge in the nucleus by 1 when it is emitted.

Sodium-21 (Na) has an atomic number of 11.

Superscript accounting: 21 – 0 = 21, the mass number of the emission product

Subscript accounting: 11 – (+1) = 10, the atomic number of the emission product, which identifies it as neon, Ne. The nuclear equation for the emission is:

$$^{21}_{11}\text{Na} \rightarrow \, ^{0}_{1}\text{e} + \, ^{21}_{10}\text{Ne}$$

Objectives Section 10.4 Radioactive Decay

Are you able to. . .

- explain what is meant by radioactive decay
- describe the factors of radioactive decay
- describe a decay series (Exercises 10.11, 10.12)
- define the term half-life
- explain the relation between half-life and the amount of radioactive material in a sample
- calculate the amount of radioactive material remaining after a defined number of half-lives (Exercises 10.14, 10.43, 10.44, 10.47)
- calculate the age of something by using half-life data (Box 10.1, Exercise 10.13)

Problem 10.4 Estimate how much of a radioisotope will be left after six half-lives.

End of half-life	Amount left (%)
0	100.00
1	50.00 (half the original amount)
2	25.00 (half of 50%)
3	12.50 (half of 25%)
4	6.250 (half of 12.50%)
5	3.125 (half of 6.25%)
6	1.563 (half of 3.125%)

Another way to do it is:

$$\left(\frac{1}{2}\right) \times \left(\frac{1}{2}\right) \times \left(\frac{1}{2}\right) \times \left(\frac{1}{2}\right) \times \left(\frac{1}{2}\right) \times \left(\frac{1}{2}\right) = \left(\frac{1}{64}\right) = 0.0156 \text{ (multiply by 100 for percent = 1.56\%)}$$

A third way to do it is

$$\left(\frac{1}{2}\right)^6 = 0.01562 \text{ (multiply by 100 for percent = 1.56\%)}$$

Problem 10.5 Calculate the percentage of radioisotope remaining after 5.0 half-lives

End of half-life	Amount left (%)
0	100.00
1	50.00 (half the original amount)
2	25.00 (half of 50%)
3	12.50 (half of 25%)
4	6.250 (half of 12.50%)
5	3.125 (half of 6.25%)

Another way to do it is:

$$\left(\frac{1}{2}\right) \times \left(\frac{1}{2}\right) \times \left(\frac{1}{2}\right) \times \left(\frac{1}{2}\right) \times \left(\frac{1}{2}\right) = \left(\frac{1}{32}\right) = 0.03125 \text{ (multiply by 100 for percent = 3.125\%)}$$

A third way to do it is:

$$\left(\frac{1}{2}\right)^5 = 0.03125 \text{ (multiply by 100 for percent = 3.125\%)}$$

Problem 10.6 Calculate the percentage of radioisotope remaining after 3.8 half-lives
A simple way to do this is:

$$\left(\frac{1}{2}\right)^{3.8} = 0.07179 \text{ (multiply by 100 for percent = 7.179\%)}$$

Does the answer make sense? After 4 half-lives, there is 6.25% of sample left and, after 3 half-lives, there is 12.50% of sample left. 3.8 is closer to 4 and falls between 3 and 4. 7.18% is closer to 6.25 and falls between 12.50 and 6.25%. The answer makes sense.

Objectives Section 10.5 Effects of Radiation

Are you able to. . .

- define the term radiochemistry
- describe the factors that contribute to biological radiation damage
- explain how the energy of a particle and its penetrating ability are related (Table 10.3, Expand Your Knowledge 10.52)
- describe the penetrating ability of the major nuclear emission products (Exercise 10.15)
- describe how each of the major emission products interacts with matter (Exercises 10.16, 10.17, 10.18, 10.19, 10.20)
- describe the chemical effects of radiation (Exercise 10.21)
- explain the role played by free radicals in biological damage (Exercises 10.22, 10.50)
- relate radiation sickness to exposure to radiation (Table 10.5, Exercise 10.23)
- describe symptoms associated with various levels of radiation sickness
- describe some of the biological damage done by radiation (Expand Your Knowledge 10.53)
- explain how radiation exposure can be minimized (Exercise 10.26)

- explain how radiation intensity and distance from source are related
- calculate the effect of changing distance on radiation intensity (Exercises 10.25, 10.48)

Problem 10.7 If the distance between a source and a target is 6 m, how far should the source be moved from the target to decrease the radiation intensity to $\frac{1}{4}$ its current value?

Using the format in Example 10.7 to find $d_{2\ new}$, we have

$$\frac{I_{new}}{I_{old}} = \frac{d^2_{old}}{d^2_{new}} \qquad \frac{\frac{1}{4}}{1} = \frac{(6\ \text{m})^2}{d^2_{new}} \qquad \text{so } d^2 = 144\ \text{m}^2 \text{ and } d_{new} = 12\ \text{m}$$

Target should be moved 12 m from source to decrease the radiation intensity by $\frac{1}{4}$; a doubling of the distance causes the intensity to decrease by a factor of 4.

Objectives Section 10.6 Detection of Radioactivity

Are you able to. . .

- explain why radiation can be detected
- explain how radiation can be detected
- describe how a film badge is used to detect radiation
- describe how a Geiger counter is used to detect radiation (Exercise 10.27)
- describe how a scintillation counter is used to detect radiation (Exercise 10.28)

Objectives Section 10.7 Measuring Radioactivity (Table 10.4)

Are you able to. . .

- identify the curie and becquerel as units used to measure the number of atomic disintegrations per second
- identify the roentgen as the unit used to measure the number of ions produced by radiation passing through air
- identify the rad, rem, and sievert as the units used to measure biological damage (Exercises 10.29, 10.30, 10.31,10.32)
- describe the relation between the various units of measurement (Exercises 10.45, 10.46)
- explain the factors that affect relative biological effectiveness
- explain what is meant by background radiation (Exercise 10.24)
- identify sources of background radiation

Objectives Section 10.8 Applications

Are you able to. . .

- describe the characteristics of radioisotopes used as diagnostic agents (Table 10.2, Exercises 10.33, 10.34, 10.39, 10.40, Expand Your Knowledge 10.54)
- describe the characteristics of radioisotopes used as therapeutic agents (Tables 10.2, 10.6, Exercise 10.35, 10.36, Expand Your Knowledge 10.54)
- explain how radionuclides can be introduced into the body
- explain how radiolabeling works
- define autoradiography
- explain the basis for generating internal images with the use of energy
- explain how CAT scans , MRIs, and PET scans are used diagnostically (Exercises 10.37, 10.38, 10.41, 10.42)

Objectives Section 10.9 Nuclear Reactions

Are you able to. . .

- describe the processes of fusion and fission (Expand Your Knowledge 10.55)
- describe how energy is generated in nuclear reactions
- compare the energy changes associated with chemical reactions with nuclear reactions

Objectives Section 10.10 Nuclear Energy and the Biosphere

Are you able to. . .

- describe the effect of nuclear energy production and weapons on the biosphere
- identify some of the damaging isotopes associated with nuclear reactions and explain their effects on biological systems

MAKING CONNECTIONS

Construct concept maps or write paragraphs describing the relation among the components of each of the following groups:

(a) radioactivity, atomic number, nuclear stability
(b) nuclear process, isotopic notation, product, reactant
(c) isotopic notation, mass number, atomic number
(d) nuclear emission, α-particle, nucleon
(e) nuclear emission, β-particle, nucleon
(f) nuclear emission, positron, nucleon
(g) nuclear emission, γ-ray, nucleon
(h) nuclear emission, atomic number, mass number
(i) α-particle, β-particle, positron, γ-ray, energy
(j) radioactive decay, half-life, daughter nuclei, decay series
(k) rate of decay, half-life, percent decay
(l) biological matter, radiation effects, radiation measurement
(m) radiation effects, particle energy, exposure time, penetrating ability
(n) film badge, lab coat, lead shield, radiation effects, γ-rays
(o) radiation exposure, free radical formation, biological effects
(p) radiation sickness, rem, symptoms
(q) radiation intensity, distance, biological effects
(r) radioisotopes, half-life, penetrating ability, diagnostic agent, therapeutic agent
(s) PET, CAT scan, MRI, radiation
(t) α-particle, inhalation, external exposure, biological damage
(u) radioisotope, chemical property, biological activity
(v) fusion, fission, energy, nuclear reaction

FILL-INS

Radioactivity is a result of _____ in a nucleus due to the repulsive forces between _____. Although mediated by the presence of _____ in smaller atoms, all elements with atomic numbers greater than _____ are radioactive. Lighter elements also have radioactive _____. Radioisotopes are designated by one of two notations: the elemental symbol with the _____ as a superscript and the _____ as the subscript, both to the left of the element, or the element name followed by _____. When a particle or energy is emitted from a radioactive nucleus, the term _____ describes the event. The most common emissions are: _____. Each of these emissions can be represented by a _____. The emission of an α-particle results in a product whose atomic number is _____ than the original element and whose mass number is _____. Beta emission produces an element whose atomic number is _____ than the original element, but there is no change in the _____. When a _____ is emitted, the product element's atomic number is one less than the original element, but there is no change in the mass number. _____ changes neither atomic nor mass number. The nuclear process by which one element is changed into another is called _____. Radioactive decay occurs when a radioisotope emits radiation that results in the formation of _____. If these nuclei are _____, the decay stops. If the daughters also are radioactive, _____. The rate at which radioactive isotopes decay is an inherent property resulting in the loss of a _____ of the material over a unit of time. The half-life of a radioisotope is the period of time that it takes for _____ of the original material to decay. The _____ of a radioisotope can be used to estimate how much of a radioactive sample will remain after a given amount of time, or to determine the age of a sample containing radioactive material. The effects of radiation on biological material depend on the _____ and _____. Highly energetic radiation penetrates _____, whereas less-energetic radiation _____. As radiation penetrates the tissue, it interacts with molecules, causing the formation of _____. Free radicals are _____ species with very reactive _____. These molecular fragments _____ with whatever they meet.

The _____ can lead to secondary chemical processes that trigger more harmful reactions resulting in the inactivation or alteration of _____. _____ structures change, and inhibited bioactivity of these molecules is an example of the consequences of radiation damage. Overexposure to radiation causes _____. The symptoms of radiation sickness can range from _____. __________ are typical symptoms of radiation sickness. Tissues that _____ are early victims of radiation damage, whereas the effects on _____ may not be seen for generations. Radiation exposure can be minimized by using barriers such as _____ and _____. Increasing distance from a radiation source also minimizes exposure because the _____. For every doubling of distance form a radiation source, the intensity falls off by a factor of _____. _____ the time of exposure will help. The _____ left behind as nuclear emissions penetrate matter can be detected by various methods. _____ darken as exposure to radiation increases and can also be used to distinguish _____. Geiger counters detect _____ created when radiation passes through matter and register the radiation as _____. Scintillation counters _____ when struck by radiation.

Different units of radiation are used, depending upon _____ from the measurements. The _____, abbreviated _____, is the most common unit and measures the _____, called the _____. The SI unit of radioactivity is the _____, abbreviated _____, which equals _____. The roentgen, abbreviated R, is a unit that measures _____. The rad measures _____ and is combined with a factor called _____ to produce the _____, abbreviated _____, which is used to measure biological damage. The RBE depends on the _____. The sievert, abbreviated Sv, measures the _____. Background radiation from natural sources bathes life on earth owing to _____ or to _____. A typical exposure due to background radiation is about _____ per year. Other sources such as _____ cause exposure of about 65 millirems per year. _____ can be used as diagnostic agents or as therapeutic agents. Diagnostic radioisotopes must have _____ penetrating ability, whereas therapeutic radioisotopes must be _____ or _____. The _____ of radiation is important; it must _____ to be detected if it is a diagnostic agent. As a therapeutic agent, the intensity is directed to _____ period of time with all attempts made to minimize damage to healthy tissue. If the radioisotope is excreted _____, unwanted damage may occur, but, if it is excreted _____, the desired effect may not be achieved, so the rate of excretion is important. Additionally, decay products should not be _____. Radionuclides can be introduced into the body _____. A solution can be used when a radioisotope _____, but _____ if natural concentration doesn't occur. Radioactive and nonradioactive forms of the same element have the same _____ and will behave the same in the body. This allows radioactive forms of an element to be used _____ a compound. Radiolabeled compounds can be detected in undisrupted cells and tissues by using _____. Radiation passing through body can be used to _____. The basic idea is that radiation directed into body is _____. The emitted radiation can be _____, and computers turn the information into a _____. _____ detect emissions after X-rays are absorbed. _____ images are created when protons in the body interact with a strong electromagnetic field. _____ relies on the emission of γ-rays from a positron emitter that is placed in the body.

_____ occurs when uranium bombarded with neutrons splits into smaller nuclei in a process that produces a lot of _____ owing to an exothermic chain reaction that grows. _____ is an energy-generating nuclear process in which nuclei of _____ fuse to form _____ through a high-energy collision. Energy changes associated with nuclear reactions are _____ as those associated with chemical reactions. There are changes in _____ in nuclear reactions that do not occur in chemical reactions. These changes in mass are associated with energy changes through Einstein's equation _____. Damage to the environment can occur through the use of nuclear devices such as _____. They can accidentally release radioisotopes such as _____, which concentrates in the thyroid gland. Nuclear weapons release strontium-90, which mimics _____ in the body and concentrates in _____, and cesium-137, which mimics _____ in the body and concentrates in cells.

Answers

Radioactivity is a result of instability in a nucleus due to the repulsive forces between protons. Although mediated by the presence of neutrons in smaller atoms, all elements with atomic numbers greater than 83 are radioactive. Lighter elements also have radioactive isotopes. Radioisotopes are designated by one of two notations: the elemental symbol with the mass number as a superscript and the atomic number as the subscript, both to the left of the element, or the element name followed by a hyphen and the mass number. When a particle or energy is emitted from a radioactive nucleus, the term radioactive emission describes the event. The most common emissions are: α-particle, β-particle, positron, and γ-ray. Each of these emissions can be represented by a nuclear equation. The emission of an α-particle results in a product whose atomic number is two units smaller than the original element and whose mass number is smaller by four. Beta emission produces an element whose atomic number is 1 unit larger than the original element, but there is no change in the mass number. When a positron is emitted, the product element's atomic number is one less than the original element, but there is no change in the mass number. Gamma-ray emission changes neither atomic nor mass number. The nuclear process by which one element is changed into another is called transmutation. Radioactive decay occurs when a radioisotope emits radiation that results in the formation of daughter nuclei. If these nuclei are stable, the decay stops. If the daughters also are radioactive, decay continues until stable nuclei result. The rate at which radioactive isotopes decay is an inherent property resulting in the loss of a constant amount of the material over a unit of time. The half-life of a radioisotope is the period of time that it takes for 50% of the original material to decay. The half-life of a radioisotope can be used to estimate how much of a radioactive sample will remain after a given amount of time, or to determine the age of a sample containing radioactive material. The effects of radiation on biological material depend on the energy of the radiation and the length of exposure. Highly energetic radiation penetrates through body tissue and bone, whereas less energetic radiation penetrates to a lesser degree or not at all. As radiation penetrates the tissue, it interacts with molecules, causing the formation of free radicals and ions. Free radicals are unstable species with very reactive unpaired electrons. These molecular fragments interact indiscriminately with whatever they meet.

The primary radiation event can lead to secondary chemical processes that trigger more harmful reactions resulting in the inactivation or alteration of biomolecules. Protein and DNA structures change, and inhibited bioactivity of these molecules is an example of the consequences of radiation damage. Overexposure to radiation causes radiation sickness. The symptoms of radiation sickness can range from negligible to lethal. Nausea, vomiting, and a drop in white blood cell count are typical symptoms of radiation sickness. Tissues that undergo rapid division are early victims of radiation damage, whereas the effects on reproductive cells may not be seen for generations. Radiation exposure can be minimized by using barriers such as lab coats for α-particles and lead shields for γ-rays. Increasing distance from a radiation source also minimizes exposure because the intensity of radiation falls off as the distance from the source increases. For every doubling of distance form a radiation source, the intensity falls off by a factor of four. Limiting the time of exposure will help. The path of ions left behind as nuclear emissions penetrate matter can be detected by various methods. Film badges darken as exposure to radiation increases and can also be used to distinguish penetration levels. Geiger counters detect ions created when radiation passes through matter and register the radiation as a series of clicks. Scintillation counters emit light when struck by radiation.

Different units of radiation are used, depending on the information sought from the measurements. The curie, abbreviated Ci, is the most common unit and measures the number of atoms that decay per second, called the disintegrations per second (dps). The SI unit of radioactivity is the becquerel, abbreviated Bq, which equals 1 dps. The roentgen, abbreviated R, is a unit that measures the exposure to X-rays or γ-rays by quantifying the number of ions produced by radiation passing through a cubic centimeter of dry air. The rad measures amount of energy absorbed in matter as a result of exposure to any form of radiation and is combined with a factor called relative biological effectiveness to produce the roentgen equivalent for man, abbreviated rem, which is used to measure biological damage. The RBE depends on the type of tissue irradiated, the dose rate, and the total dose of the radiation. The sievert, abbreviated Sv, measures the effect of absorbed radiation on different kinds of tissue under different circumstances. Background radiation from natural sources bathes life on earth owing to the presence of cosmic rays or to naturally radioactive elements in rock and soil. A typical exposure due to background radiation is about 350 millirems per year. Other sources such as X-rays, radiotherapy, and radioisotopes cause exposure of about 65 millirems per year. Radioisotopes can be used as diagnostic agents or as therapeutic agents. Diagnostic radioisotopes must have significant penetrating ability, whereas therapeutic radioisotopes must be α- or β-emitters. The intensity of radiation is important; it must be intense enough to be detected if it is a diagnostic agent. As a therapeutic agent, the intensity is directed to specific diseased tissue for a designated period of time with all attempts made to minimize damage to healthy tissue. If the radioisotope is excreted too slowly, unwanted damage may occur, but, if it is excreted too quickly, the desired effect may not be achieved, so the rate of excretion is important. Additionally, decay products should not be radioactive. Radionuclides can be introduced into the body orally, by injection, or as a package. A solution can be used when a radioisotope naturally concentrates in specific cells, but a package must be inserted at or near the target site if natural concentration doesn't occur. Radioactive and nonradioactive forms of the same element have the same chemical properties and will behave the same in the body. This allows radioactive forms of an element to be used to label a compound. Radiolabeled compounds can be detected in undisrupted cells and tissues by using autoradiography. Radiation passing through body can be used to generate images. The basic idea is that radiation directed into body is absorbed differently by different types of tissue. The emitted radiation can be measured, and computers turn the information into an internal picture. CAT scans (computer-assisted tomography) detect emissions after X-rays are absorbed. MRI (magnetic resonance imaging) images are created when protons in the body interact with a strong electromagnetic field. PET (positron emission tomography) relies on the emission of γ-rays from a positron emitter that is placed in the body.

Fission occurs when uranium bombarded with neutrons splits into smaller nuclei in a process that produces a lot of energy owing to an exothermic chain reaction that grows. Fusion is an energy-generating nuclear process in which nuclei of two light elements fuse to form heavier elements through a high-energy collision. Energy changes associated with nuclear reactions are millions of times as great as those associated with chemical reactions. There are changes in mass in nuclear reactions that do not occur in chemical reactions. These changes in mass are associated with energy changes through Einstein's equation $E = mc^2$. Damage to the environment can occur through the use of nuclear devices such as nuclear generators. They can accidentally release radioisotopes such as iodine-131, which concentrates in the thyroid gland. Nuclear weapons release strontium-90, which mimics calcium in the body and concentrates in bone, and cesium-137, which mimics potassium in the body and concentrates in cells.

TEST YOURSELF

1. Explain why the following designations do not identify isotopes.
   (a) C-6 (b) cobalt (c) Ra (d) Sr^{2+}
2. Connect one or more of the following radioactive emissions with each of the statements below: α-particle, β-particle, positron, γ-ray
   (a) has the ability to penetrate through bone and tissue
   (b) its emission results from the conversion of a neutron into a proton
   (c) can be used as a therapeutic agent to kill diseased tissue
   (d) its emission results from the conversion of a proton into a neutron
   (e) can be stopped by thin cloth or paper
3. Explain the difference between transmutation and nuclear reactions.
4. Write generic equations for the emission of an α-particle, a β-particle, and a positron.
5. As a radiation technician, what precautions should you take to both monitor and minimize your exposure to radiation? What parts of your body are particularly vulnerable to radiation damage?

6. Give conversion factors for the following units of measure:
   (a) curies and becquerels
   (b) sieverts and rads
   (c) energy in kj and rem (use Table 10.4)
7. Using what you know about the periodic table, isotopes, and chemical behavior explain the following statements:
   (a) strontium-90 mimics calcium
   (b) cesium-137 mimics potassium
   (c) iron-59 can be used to assess bone marrow function
8. What characteristics of each of the following radionuclides make them suitable for the application?
   (a) barium-131 is used to detect bone tumors (b) cobalt-60 is used to fight cancer
9. On the basis of an average X-ray delivering 15 mrem per frame, how many X-rays would you have to have in 1 year to achieve a dose that would temporarily decrease your white blood cell count? How many would you need to achieve a possibly lethal dose?
10. Why are α- or β-particles more dangerous if they are inside the body than if they are outside the body?

Answers

1. (a) carbon-6: a six, the atomic number, should not follow C. It should be followed by the mass number of the isotope. The element name is used instead of the symbol. (b and c) do not identify the isotope, only the element. (d) Sr^{2+} is an ionic designation, not an isotopic designation.
2. (a) γ-ray
   (b) β-particle
   (c) α- or β-particle
   (d) positron
   (e) α-particle
3. Transmutation is the term used to refer to the spontaneous or induced changes in the nucleus of an element as seen in decay processes. The term nuclear reaction is used to describe a process such as fusion or fission in which one or more reactant atoms take part in the production of different atoms.
4. $^{\text{mass number}}_{\text{atomic number}}X \rightarrow {}^{4}_{2}He + {}^{\text{mass number}\ -4}_{\text{atomic number}\ -2}Y$ α-particle

   $^{\text{mass number}}_{\text{atomic number}}X \rightarrow {}^{0}_{-1}\beta + {}^{\text{mass number}}_{\text{atomic number}\ +1}Y$ β-particle

   $^{\text{mass number}}_{\text{atomic number}}X \rightarrow {}^{0}_{1}e + {}^{\text{mass number}}_{\text{atomic number}\ -1}Y$ positron

5. Radiation or film badges are worn to monitor amount and type of radiation exposure. Lab coats protect against α-particles, but high-energy X-rays require standing behind a special barrier designed to protect the body. The reproductive organs and bone marrow are especially vulnerable.
6. (a) 1 Ci = 3.7×10^{10} dps and 1 Bq = 1 dps

   $$\frac{1\ \text{Bq}}{1\ \cancel{\text{dps}}} \times \frac{3.7 \times 10^{10}\ \cancel{\text{dps}}}{1\ \text{Ci}} = \frac{3.7 \times 10^{10}\ \text{Bq}}{1\ \text{Ci}}$$

   (b) 1 Sv = 100 rem sieverts and 1 rem = 1 rad × RBE

   $$\frac{100\ \cancel{\text{rem}}}{1\ \text{Sv}} \times \frac{1\ \text{rad} \times \text{RBE}}{1\ \cancel{\text{rem}}} = \frac{100\ \text{rad} \times \text{RBE}}{1\ \text{Sv}}$$

(c) 1 rad = 0.01 J energy absorbed/kg matter   1 rem = 1 rad × RBE

$$\frac{100 \text{ rad}}{0.01 \text{ J}} \times \frac{1 \text{ rem}}{1 \text{ rad} \times \text{RBE}} = \frac{1 \text{ rem}}{0.01 \text{ J} \times \text{RBE}}$$

converting J into kJ: $0.01 \text{ J} \times \frac{1 \text{ kJ}}{1000 \text{ J}} = 1 \times 10^{-5} \text{ kJ}$

7. (a) Sr and Ca are both Group II elements. Their valence configurations are similar, which can "trick" the body into incorporating Sr where calcium is used (that is, bone).
   (b) Cs and K are both Group I elements, and their valence configurations are similar: +1 ions. The body accepts Cs where K is used because it "looks" electronically similar. The body does not distinguish between radioactive and nonradioactive isotopes of an element. The chemical properties of both isotopes are the same.
   (c) Iron is used to build red blood cells in bone marrow, and the body will incorporate radioactive iron there as well.

8. (a) Barium-131 is used to detect bone tumors because it is a Group II element and has electronic properties similar to those of calcium; so it will be absorbed and used in the areas, such as bone, where calcium is absorbed and used. It is a good diagnostic isotope because it emits γ-rays that can be detected by an external monitor.
   (b) Cobalt-60 is used to fight cancer because it is a β-and γ-emitter. Inside the body, emissions kill diseased (as well as healthy) cells.

9. A dose that would temporarily decrease your white blood cell count is 25,000 mrem (25 rem), which corresponds to approximately 1700 X-rays. A lethal dose is 500,000 mrem (500 rem), which corresponds to 33,000 X-rays.
10. Alpha- or beta-particles are slowed down inside the body, which means that they are confined longer and emit radiation over a longer period of time and thus can damage surrounding cells and tissue. Because they do not penetrate skin well, from the outside they are not very harmful. However, if they are inhaled and find entry, then a lot of damage can be done in a localized area where they are confined.

chapter 11

Saturated Hydrocarbons

OUTLINE

Introduction Saturated Hydrocarbons

A. Inorganic chemistry is the study of compounds that do not contain carbon with the exceptions of
   1. carbon monoxide and carbon dioxide
   2. carbonates (compounds containing the CO_3^{2-} ion) and bicarbonates (compounds containing the HCO_3^- ion)
   3. cyanides (compounds containing CN^- ion)
B. Organic chemistry is the study of carbon-containing compounds.
   1. some contain only carbon and hydrogen (Figure 11.1)
   2. some contain carbon, hydrogen, and one or more other elements, such as oxygen, nitrogen, sulfur, phosphorus, fluorine, chlorine, bromine, and iodine
C. Organic compounds account for more than 10 million chemical compounds.
   1. biomolecules such as carbohydrates, lipids, proteins, nucleic acids, vitamins, and hormones
   2. energy-producing fuels such as gasoline, petroleum, kerosene, and heating and cooking gases
   3. medicines such as aspirin, penicillin, and anesthetics
   4. synthetic plastics, textiles, and rubbers
   5. active ingredients in soaps, cosmetics, hygiene products, detergents, and polishes
   6. prosthetics such as artificial hips, heart valves, and dentures

Key Terms inorganic chemistry, inorganic compound, organic chemistry, organic compound

11.1 Molecular and Structural Formulas

A. The molecular formula for a compound indicates the identity and number of atoms of elements that constitute the compound.
B. Molecular formulas provide no information about the way in which atoms are connected to each other.
C. Structural formulas show how atoms are connected to each other.
D. The combining power of an atom tells how many bonds it can form with other atoms (Table 11.1).
   1. this is the number of bonds that will allow the element to achieve a valence octet (or duet for H)
   2. for nonmetals (except S and P) in organic compounds, the group number subtracted from eight gives the combining power
   3. some atoms can form multiple bonds
E. For compounds that contain carbon and hydrogen only, the Lewis structure is the same as the structural formula.

F. The molecular formula implies that atoms placed at the right of a carbon atom are connected to that carbon.

Key Terms combining power, molecular formula, structural formula
Key Table Table 11.1 Combining powers of elements present in organic compounds

Example 11.1 Identification of correct molecular formulas
Problem 11.1 Which of the following molecular formulas are correct and which are incorrect?

11.2 Families of Organic Compounds

A. Organic compounds are arranged in families (Table 11.2).
  1. simplifies the task of organizing millions of compounds
  2. members of each family have similar chemical characteristics
  3. members of each family may have common physical characteristics
  4. similarities result from the presence of specific functional groups

B. Functional groups are specific atoms, bonding arrangements, or groups of atoms bonded in a certain way.
  1. functional groups determine the physical and chemical characteristics of the molecules of which they are a part
  2. organic compounds may contain more than one functional group

C. The major functional groups are (Table 11.2)
  1. alkanes: contain only C–C and C–H single bonds
  2. alkenes: contain at least one C–C double bond
  3. alkynes: contain at least one C–C triple bond
  4. alcohols: contain the hydroxyl group, –OH, bonded to a C atom. *Note:* don't confuse –OH, the functional group, with OH^-, the hydroxide ion
  5. ethers: contain an oxygen atom bonded directly to two carbon atoms
  6. amines: contain an N atom, which can be bonded to one, two, or three carbon atoms
  7. aldehydes: contain a carbonyl group, –C=O, the carbon in which is also bonded to another carbon atom and to a hydrogen atom
  8. ketones: contain a carbonyl group, –C=O, the carbon in which is also bonded to two other carbon atoms
  9. carboxylic acids: contain a carbonyl group, the carbon in which is also bonded to an –OH group
  10. esters: contain a carbonyl group, the carbon in which is also bonded to an oxygen atom that is bonded to another carbon atom
  11. amides: contain a carbonyl group, the carbon in which is also bonded to a nitrogen atom

D. Hydrocarbons contain only hydrogen and carbon.
  1. there may be single, double, or triple bonds between the carbon atoms
  2. if there are only single bonds between carbon atoms, the hydrocarbon is saturated; each carbon atom is bonded to the maximum number of other atoms, four
  3. if there are double or triple bonds between carbon atoms, the hydrocarbon is unsaturated; some carbon atoms are bonded to fewer than four other atoms
  4. unsaturated hydrocarbons are chemically more reactive than saturated hydrocarbons

E. The ratios of each type of atom are distinctive for each family.

Key Terms alcohol, aldehyde, alkane, alkene, alkyne, amide, amine, amino group, carbonyl group, carboxylic acid, ester, ether, family, functional group, hydrocarbon, hydroxyl group, ketone, organic family, saturated hydrocarbon, unsaturated hydrocarbon

Key Figure and Table
Figure 11.1 Hydrocarbon families
Table 11.2 Families of organic compounds

Example 11.2 Identifying organic families by using Table 11.2
Problem 11.2 Identifying organic families by using Table 11.2.

11.3 Alkanes

A. Alkanes are hydrocarbons in which the carbon atoms are singly bonded to other carbon atoms
B. The general molecular formula for alkanes is: $C_nH_{(2n+2)}$ (Table 11.3).
   n = whole-number integer greater than 0

$$\begin{array}{cc} C & : H \\ \uparrow n: & \uparrow(2n+2) \end{array}$$

C. Alkane structure can be represented in a variety of ways.
   1. ball-and-stick models show bonds between all atoms in a molecule and give a sense of the three-dimensional structure
   2. space-filling models approximate the relative sizes of atoms in a molecule and the way in which they are connected
D. Carbon-to-carbon bonds in alkanes
   1. each carbon forms four bonds: tetravalent
   2. the geometry of the bonds around each carbon is tetrahedral
      a. each bond angle is 109.5 degrees
      b. tetrahedral geometry is a result of hybridization of *s* and *p* orbitals
         i. one 2*s* orbital and three 2*p* orbitals "mix"
         ii. valence electrons are redistributed: one per each sp^3 orbital
   3. each bond has similar properties; single bonds formed using sp^3 orbitals are called sigma bonds
   4. there is free rotation around single bonds
      a. produces a variety of conformations due to different atomic orientations
      b. bonds are always rotating
      c. preferred conformations in larger molecules affect physiological functions

Key Terms alkane, conformation, free rotation, orbital hybridization theory, sp^3 orbital, tetrahedral bond angle, tetrahedral geometry

Key Box, Figures, and Table

Box 11.1	Natural gas and petroleum
Figure 11.2	Ball-and-stick and space-filling models of alkanes
Figure 11.3	Tetrahedral sp^3-hybridized carbon
Figure 11.4	Electronic configuration of sp^3-hybridized orbitals
Figure 11.5	Formation of CH_4 from hydrogen *s* orbitals and carbon sp^3 orbitals
Figure 11.6	Ball-and-stick models of butane conformations
Table 11.3	Formulas and properties of normal alkanes

11.4 Types of Structural Formulas

A. Organic molecules can be drawn in a variety of ways.
   1. expanded structural formula shows every atom and every bond in a molecule
   2. condensed structural formula does not show bonds but implies bonds by the way in which the structure is written:

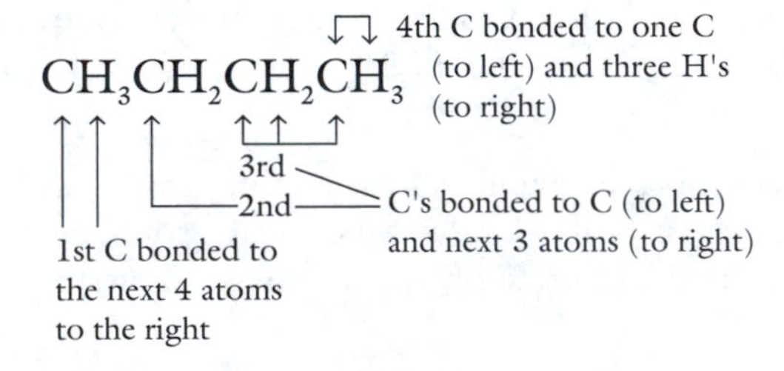

   3. skeletal structures show only carbon atoms and assume all other bonds to be to hydrogen
   4. line structures do not show any atoms; carbon atoms are assumed to be at the beginning and end of lines and at any intersection of lines

5. carbon atoms can be linked in straight chains or branched chains

Key Terms condensed structural formula, expanded structural formula, line structural formula, skeleton formula

Example 11.3 Drawing different structural formula representations
Problem 11.3 Drawing different structural formula representations

11.5 Constitutional Isomers of Alkanes

A. All isomers of a compound have the same chemical formula.
B. Isomers of a compound may differ in the way in which atoms are connected to each other.
   1. structural isomers or constitutional isomers
   2. physical and chemical properties differ
   3. the greater the number of carbon atoms in a formula, the greater the number of possible isomers
   4. there must be at least four carbon atoms in a formula for constitutional isomers to exist
   5. carbon atoms may connect in a straight chain or in branches; lengths of the longest carbon chain may differ

Key Terms branched alkane, connectivity, constitutional isomer, isomer, longest chain, straight chain alkane, structural isomer, unbranched alkane

Key Figures
Figure 11.7 Ball-and-stick model of isobutane
Figure 11.8 Ball-and-stick models of pentane isomers

Example 11.4 Drawing constitutional isomers
Problem 11.4 Drawing constitutional isomers

11.6 Naming Alkanes

A. Carbon atoms in a molecule may be primary, 1°, secondary, 2°, tertiary, 3°, or quaternary, 4°.
   1. a primary carbon is attached to one other carbon atom
   2. a secondary carbon is attached to two other carbon atoms
   3. a tertiary carbon is attached to three other carbon atoms
   4. a quaternary carbon is attached to four other carbon atoms
B. Alkanes are named by using IUPAC nomenclature rules (Figure 11.9).
   1. suffix -ane tells you that there are only single bonds between carbon atoms in the hydrocarbon
   2. the number of carbon atoms in the longest continuous chain of the molecule determines the prefix before the suffix -ane

meth-	1	hex-	6
eth-	2	hept-	7
prop-	3	oct-	8
but-	4	non-	9
pent-	5	dec-	10

C. The longest continuous carbon chain is the skeleton from which other alkyl groups are hung.

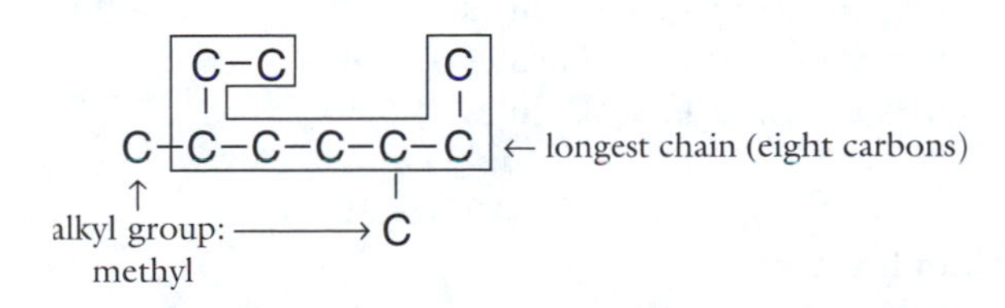

D. Alkyl groups are identifiable clusters of atoms that are attached to a parent chain of carbon atoms. The number of the carbon to which an alkyl group is attached and its name identify alkyl groups. Alkyl groups are **not molecules;** there is an implied unattached bond at one carbon (Table 11.4).

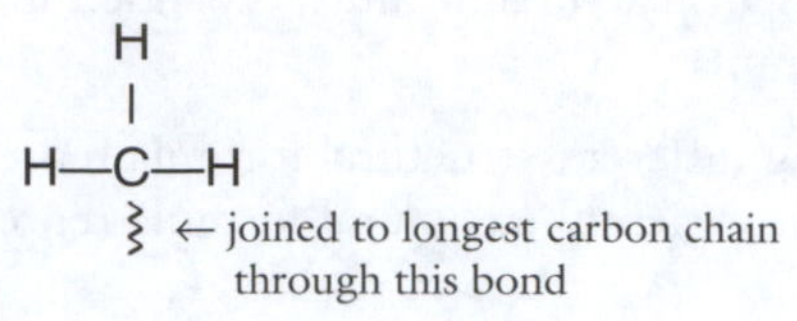

IUPAC Rules for Naming Alkanes

1. Identify the longest continuous carbon chain; atoms do not have to be linked in a straight line in a drawing. The longest chain is the one through which you can draw a continuous line without lifting the pencil.
2. Count the number of carbons in the chain and add the -ane ending to the appropriate prefix. The result is the base name, and it tells you the number of carbon atoms in the longest chain. It does not tell you anything about the groups attached to the longest chain.
3. Identify the substituents on the chain. Each substituent has a name and an address: the name is the alkyl name, and the address is the number of the carbon in the chain to which it is attached. The carbon atoms in the chain are numbered from the direction that gives any substituents the lowest possible numbers.
4. If there is more than one of the same type of alkyl group, use prefixes before the name.

   di- 2 tri- 3 tetra- 4 penta- 5 hexa- 6
5. Number the carbon atoms in the chain and assign a position to each alkyl group. Place the number to which the substituent group is attached in front of the alkyl name and separate it with a hyphen.
6. Alphabetize the alkyl groups and arrange them in front of the base name. When alphabetizing, ignore prefixes except iso-. Use hyphens to separate words from numbers, and commas to separate numbers.

3,6—	di	methyl	oct	ane
numbers↑ of carbon	↑prefix indicating two of the same group	↑identity of alkyl	↑prefix indicating number of carbons in longest chain	↖suffix indicating only single bonds between carbon atoms

Key Terms alkyl group, -ane suffix, butyl, ethyl, isobutyl, isopropyl, IUPAC nomenclature system, methyl, n- prefix, neo- prefix, normal alkane, parent chain, primary carbon, propyl, quaternary carbon, *s*-butyl, sec-butyl, secondary carbon, simple alkyl group, substituents, tertiary carbon, *t*-butyl, tert-butyl

Key Figure and Table
Figure 11.9 IUPAC nomenclature example
Table 11.4 Alkyl groups

Example 11.5 Identifying 1°, 2°, 3°, and 4° carbons
Problem 11.5 Identifying 1°, 2°, 3°, and 4° carbons
Example 11.6 Naming alkanes by using IUPAC rules
Problem 11.6 Naming alkanes by using IUPAC rules
Example 11.7 Naming alkanes by using IUPAC rules
Problem 11.7 Naming alkanes by using IUPAC rules

11.7 Cycloalkanes

A. Carbon atoms can form rings.
B. Alkanes that contain rings are called cycloalkanes.
C. Line structures can be used to represent rings (Table 11.5).

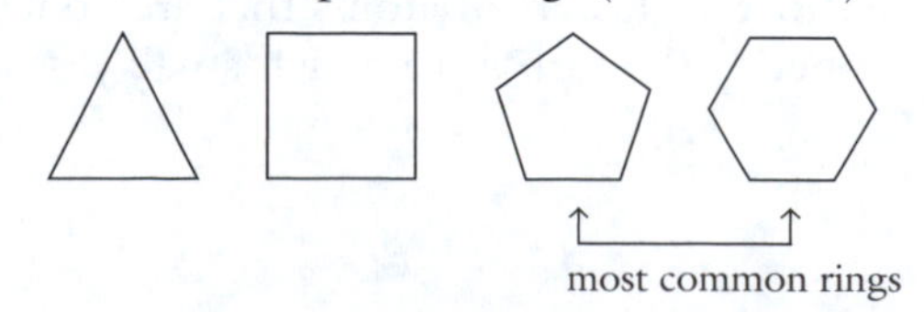

D. Cycloalkanes are named on the basis of the number of carbons in the ring structure.

IUPAC Rules for Naming Cycloalkanes

1. The number of carbons in the ring determines the base name, just as in straight-chain alkanes. The prefix cyclo- is attached to the base name to indicate the presence of a ring in the molecule.
2. Substituents are named as indicated for straight-chain alkanes.
3. Carbon atoms are numbered starting with a carbon containing a substituent and proceeding around the ring in the direction that provides the lowest possible numbers for substituents.
4. Substituent numbers and names are placed in front of the base name, as indicated for straight-chain alkanes.

Key Terms acyclic, cycloalkane, open-chain molecule
Key Table and Box
Table 11.5 Cycloalkanes
Box 11.2 Stability and shape of cycloalkanes

Example 11.8 Naming cycloalkanes by using IUPAC rules
Problem 11.8 Naming cycloalkanes by using IUPAC rules
Example 11.9 Drawing constitutional isomers of cycloalkanes
Problem 11.9 Drawing constitutional isomers cycloalkanes

11.8 Cis-Trans Stereoisomerism in Cycloalkanes

A. Due to restricted rotation; substituents on cycloalkanes are "stuck" in an orientation on the ring and cannot be moved except through a chemical reaction.
   1. produces geometric isomers or stereoisomers that differ in configuration
   2. sometimes called diastereomers
   3. substituents may sit above or below the plane of the ring relative to one another
   4. cis-trans isomerism occurs when two different substituents are attached to two different ring carbons

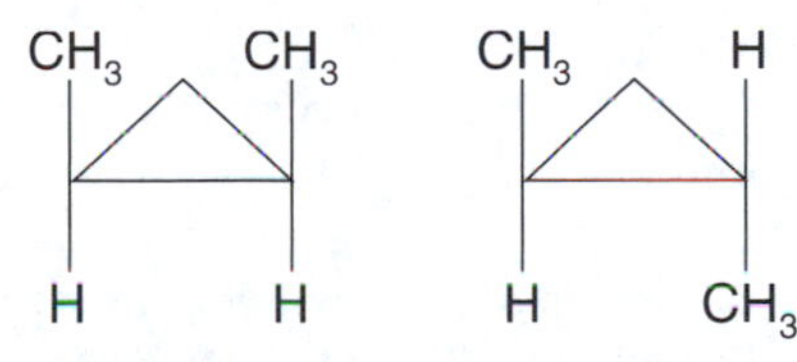

   5. a *cis*- isomer has two substituents on different carbons positioned on the same side of the ring
   6. a *trans*- isomer has two substituents on different carbons positioned on opposite sides of the ring
   7. *cis*- and *trans*-isomers have different chemical and physical properties

B. The prefix *cis*- or *trans*- is used in front of the IUPAC name of the cycloalkane.

Key Terms *cis*-isomer, configuration, diastereomers, geometric isomer, restricted rotation, stereocenter, stereoisomer, *trans*-isomer

Example 11.10 Identifying and drawing cis-trans isomers
Problem 11.10 Identifying and drawing cis-trans isomers

11.9 Physical Properties of Alkanes and Cycloalkanes

A. Physical properties are determined by the nature and magnitude of secondary forces in molecules (Table 11.3).
   1. melting point
      a. temperature at which substance goes from the solid to the liquid state
      b. is unique for each substance
      c. magnitude is indicative of strength of attractions between molecules
         i. higher melting point means that it takes more energy to separate molecules, which implies that stronger attractive forces exist between them

ii. lower melting point implies that the molecules are easily separated with a smaller amount of energy input

a. generally, alkanes and cycloalkanes have the lowest melting points of organic families

b. melting points increase within a family as the number of carbon atoms increases

2. boiling point
   a. temperature at which substance goes from the liquid to the gas state
   b. is unique for each substance
   c. magnitude is indicative of strength of attractions between molecules
      i. higher boiling point means that it takes more energy to separate molecules, which implies that stronger attractive forces exist between them
      ii. lower boiling point implies that the molecules are separated with a relativley small amount of energy input
   a. generally, alkanes and cycloalkanes have the lowest boiling points of organic families
   b. boiling points increase within a family as the number of carbon atoms increases
3. density
   a. stronger attractions between molecules imply tighter packing and result in higher densities
   b. cyclic molecules can pack more tightly than molecules with straight chains; molecules with straight chains can pack more tightly than branched-chain molecules

B. Secondary forces are the result of interactions between molecules.
   1. hydrogen bonds can form when –O–H and –N–H groups are present in the molecule
   2. dipole attractions can form when polar bonds and molecular geometry produce polar molecules
   3. London forces are always present owing to transitory fluctuations of electron field around molecules; these forces are dominant in non-polar molecules
   4. Strength order is: hydrogen bonding stronger than dipole attractions stronger than London forces
   5. The order of secondary-force strength can be reliably predicted only for a small number of compounds with very similar structures

C. The strength and type of secondary forces between molecules depends on
   1. family to which molecule belongs
      a. determines types of atoms and presence of polar bonds
      b. hydrocarbons are nonpolar
   2. molecular mass
      a. in the same family, as carbon atoms are added, secondary forces increase
      b. for straight-chain alkanes, as carbon chain length increases, strength of London forces increases
   3. molecular shape
      a. within the same family, if molecular mass is similar, more tightly packed molecules have higher boiling points: cyclic > straight-chain > branched chain (Figure 11.10)
      b. shape affects surface area over which electronic interactions can occur

Key Terms boiling point, density, melting point, solubility
Key Figure Figure 11.10 Straight-chain, branched, and cyclic molecules differ in their packing ability
Key Box 11.3 Health hazards and medical uses of alkanes

Example 11.11 Predicting relative boiling points of various compounds
Problem 11.11 Predicting relative boiling points of various compounds
Example 11.12 Predicting relative solubilities of various compounds
Problem 11.12 Predicting relative solubilities of various compounds

11.10 Chemical Properties of Alkanes and Cycloalkanes

A. Saturated hydrocarbons are relatively unreactive because of stability of C–H bonds

B. Organic reactions are written in the following notation:
   1. reactants on left side of arrow
   2. conditions (heat, light) and reagents (halogen) are written above the arrow

3. product OF INTEREST is written on right; organic notation often omits products other than organic products
4. organic reactions are often left unbalanced

$$\text{Reactant} \xrightarrow[\text{reagent}]{\text{conditions}} \text{product of interest}$$

C. Saturated hydrocarbons react with oxygen in combustion reactions.
   1. reaction breaks all C–C and C–H bonds to produce water and carbon dioxide
   2. reaction is exothermic and produces heat
   3. each carbon atom is oxidized from the lowest oxidation state (bonded to hydrogen or carbon) to the maximum oxidized state; all four bonds are with oxygen
   4. complete combustion oxidizes all carbons completely (two oxygen atoms per carbon atom)

$$\text{Hydrocarbon} + O_2 \rightarrow CO_2 + H_2O$$

   5. incomplete combustion produces some carbon monoxide, CO, instead of carbon dioxide, CO_2

$$\text{Hydrocarbon} + O_2 \rightarrow CO + CO_2 + H_2O$$

D. Saturated hydrocarbons can undergo replacement of a hydrogen atom by a halogen atom; called halogenation.
   1. reaction replaces a hydrogen with F, Cl, or Br atom
   2. when a halide replaces a hydrogen, the hydrocarbon becomes an alkyl halide
   3. replacement of one hydrogen is monohalogenation, of two is dihalogenation, of many is polyhalogenation
   4. halogenation reactions produce a variety of halogenated products
   5. halogenation reactions require heat or ultraviolet light:

$$\text{Hydrocarbon} \xrightarrow[\text{halogen}]{\text{heat or UV}} \text{Alkyl halide}$$

E. Alkyl halides are named by using IUPAC rules for alkanes.
   1. halide substituent is given number and identity: fluoro-, chloro-, bromo-, and iodo-
   2. halide name and number are inserted alphabetically before base name

Key Terms alkyl halide, combustion, complete combustion, halogenation, incomplete combustion, monohalogenation, oxidation reaction, oxidation state, polyhalogenation

Key Boxes 11.4 The greenhouse effect and global warming
11.5 Applications of alkyl halides and some problems they create

Example 11.13 Writing equations for the combustion of hydrocarbons
Problem 11.13 Writing equations for the combustion of hydrocarbons
Example 11.14 Writing equations for the halogenation of hydrocarbons
Problem 11.14 Writing equations for the halogenation of hydrocarbons
Example 11.15 Writing equations for the halogenation of hydrocarbons
Problem 11.15 Writing equations for the halogenation of hydrocarbons
Example 11.16 Naming alkyl halides
Problem 11.16 Naming alkyl halides

ARE YOU ABLE TO... AND WORKED TEXT PROBLEMS

Objectives Introduction Saturated Hydrocarbons

Are you able to...

- explain the difference between organic and inorganic chemistry (Exercise 11.39)
- identify the carbon-containing compounds that are not organic
- list the elements most commonly found in organic molecules
- describe some uses of organic compounds

Objectives Section 11.1 Molecular and Structural Formulas

Are you able to...

- describe the information contained in a molecular formula
- explain how molecular formulas differ from structural formulas
- describe the ways in which the elements commonly found in organic compounds form bonds (Exercise 11.1)
- identify the number of bonds that these elements form (Table 11.1, Exercise 11.2)
- determine the correctness of a molecular formula (Ex. 11.1, Prob. 11.1, Exercises 11.3, 11.4, 11.5, 11.6, 11.40)

Problem 11.1 Which of the following molecular formulas are correct and which are incorrect?

(a) CH_5N

This formula implies that five H's and one N are connected to a central carbon atom. Because carbon can form only four bonds, this formula is incorrect. However, if C is connected to N and the remaining H's are distributed, three to C and two to N, then C has four bonds, N has three bonds, and each H has only one. This formula could be rewritten as CH_3NH_2 to show more clearly how the atoms are bonded to each other.

(b) CH_5O

Carbon can form four bonds, O usually forms two bonds, and each H can form one bond. No arrangement will satisfy these requirements; therefore this formula is incorrect.

(c) C_2H_5Cl

Carbon can form four bonds; chlorine and hydrogen atoms form one bond each. If the molecule is arranged such that the carbons are bonded to each other, three hydrogen atoms are attached to one carbon, and two hydrogen atoms and the chlorine atom are attached to the other carbon, all bonding requirements are satisfied. This formula can be rewritten as CH_3CH_2Cl.

Objectives Section 11.2 Families of Organic Compounds

Are you able to...

- define functional group
- explain the significance of functional groups
- identify major functional groups (Table 10.2, Ex. 11.2, Prob. 11.2, Exercises 11.7, 11.8)
- categorize organic molecules according to the functional groups present (Ex. 11.2, Prob. 11.2, Exercises 11.9, 11.10)
- identify hydrocarbons
- distinguish between saturated and unsaturated hydrocarbons
- identify organic families by examining the atomic ratios of the molecular formula

Problem 11.2 Identify the family of the following compounds by referring to Table 11.2.

When considering the family to which an organic compound belongs, look for groupings of atoms that are characteristic and isolate these groups. The C=O functional group is found in several organic families. In this case, follow the bonds from the carbonyl carbon and identify the atoms to which it is attached.

bonded to carbon on both sides

$$CH_3—\underset{\underset{O}{||}}{C}—CH_3$$

carbonyl carbon

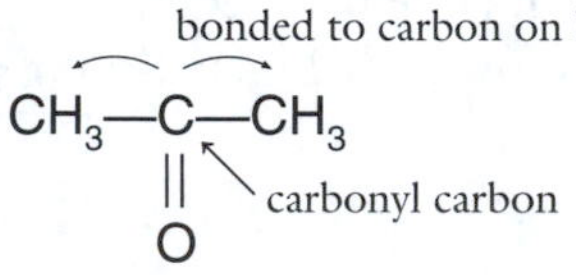

(a) Carbonyl in a ketone

Follow the bonds from the carbonyl carbon and identify the atoms to which it is attached:

$$CH_3CH_2—\overset{\overset{O}{||}}{C}—CH_2CH_3$$

↑carbonyl carbon

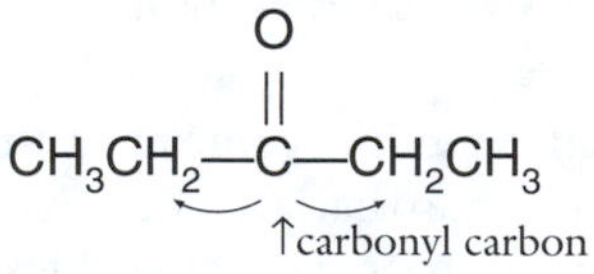

(b) Ester

Follow the bonds from the carbonyl carbon and identify the atoms to which it is attached:

$$CH_3CH_2—\underset{\underset{O}{||}}{C}—O—CH_2CH_2CH_3$$

carbonyl carbon

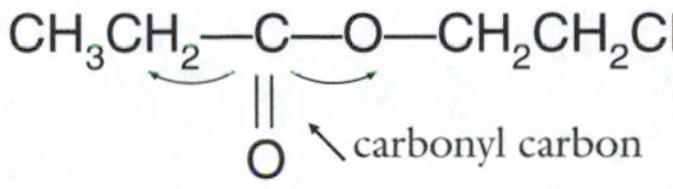

(c) Amide

Follow the bonds from the carbonyl carbon and identify the atoms to which it is attached:

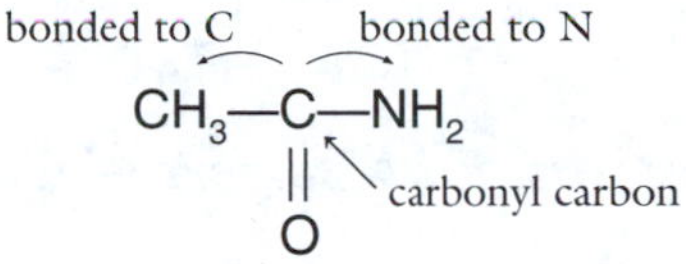

bonded to C   bonded to N

$$CH_3—\underset{\underset{O}{||}}{C}—NH_2$$

carbonyl carbon

(d) Amine

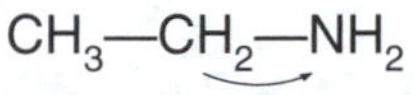

$CH_3—CH_2—NH_2$ can be looked at as NH_3 having one H replaced by hydrocarbon chain

Objectives Section 11.3 Alkanes

Are you able to. . .

- identify alkanes
- recognize alkanes by the carbon: hydrogen ratio of a molecular formula
- provide the correct number of hydrogens for a given number of carbons in an alkane
- explain the advantages and limitations of the various models of alkane structure
- describe the geometry of tetravalent carbons (Exercise 11.12)
- explain how *s* and *p* orbitals hybridize in tetravalent carbon (Exercise 11.11)
- identify sigma bonds
- describe rotational activity around bonds and conformations in alkanes

Objectives Section 11.4 Types of Structural Formulas

Are you able to. . .

- draw organic molecules by using expanded structural formulas (Ex. 11.3, Prob. 11.3, Exercises 11.13, 11.41)
- draw organic molecules by using condensed structural formulas (Ex. 11.3, Prob. 11.3, Exercises 11.16, 11.42)
- draw organic molecules by using skeletal structural formulas (Ex. 11.3, Prob. 11.3, Exercise 11.14)
- draw organic molecules by using line structural formulas (Ex. 11.3, Prob. 11.3, Exercise 11.15)
- draw various conformations of an alkane
- write the molecular formula of a molecule from a structural representation (Exercise 11.47)

Problem 11.3 Draw different structural formulas representing hexane, C_6H_{14}. Include at least two different conformations, one expanded structural formula, one condensed structural formula, one skeleton structural formula, and one line structural formula.

Conformation 1

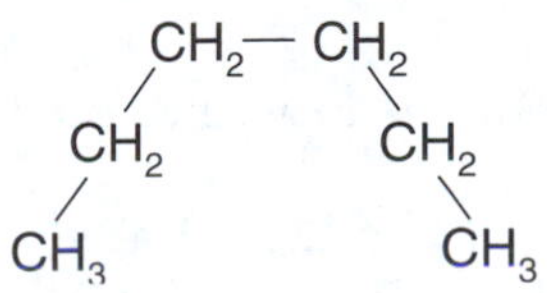

CH_3—CH_2—CH_2—CH_2—CH_2—CH_3 (drawn as:)

```
      CH2— CH2
     /        \
  CH2          CH2
  /              \
CH3               CH3
```

Conformation 2

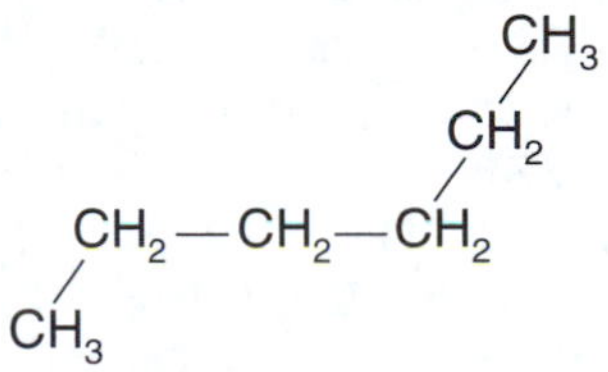

```
                      CH3
                      /
                    CH2
                    /
  CH2—CH2—CH2
  /
CH3
```

Condensed formula

$CH_3CH_2CH_2CH_2CH_2CH_3$

or

$CH_3(CH_2)_4CH_3$

Skeletal structure

C—C—C—C—C—C

Line structure

Expanded structural formula

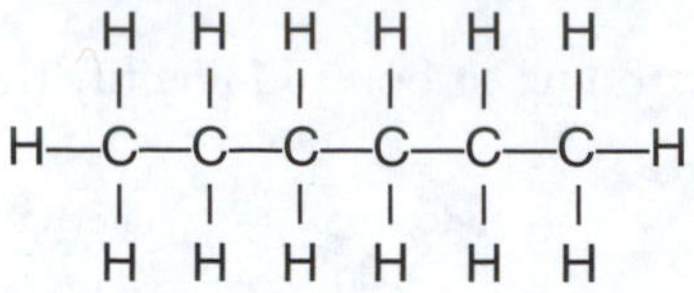

```
   H  H  H  H  H  H
   |  |  |  |  |  |
H—C—C—C—C—C—C—H
   |  |  |  |  |  |
   H  H  H  H  H  H
```

Objectives Section 11.5 Constitutional Isomers of Alkanes

Are you able to. . .

- explain what is meant by the term structural or constitutional isomer
- describe how structural isomers are the same and how they differ from one another (Exercises 11.17, 11.18, 11.43)
- show how connectivity differs in constitutional isomers (Ex. 11.4, Prob. 11.4, Exercises 11.19, 11.20)
- identify the longest continuous carbon chain in a molecule
- distinguish between branched and unbranched hydrocarbons

Problem 11.4 Draw the different constitutional isomers of C_6H_{14}.

(a) $CH_3CH_2CH_2CH_2CH_2CH_3$

(b) 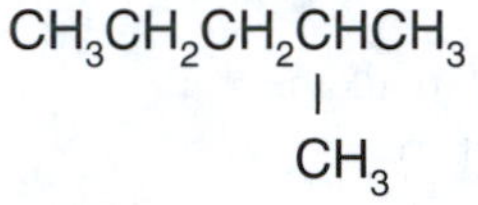
```
CH3CH2CH2CHCH3
            |
            CH3
```

(c) 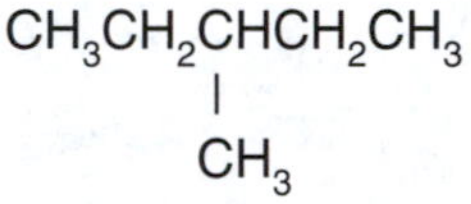
```
CH3CH2CHCH2CH3
        |
        CH3
```

(d) 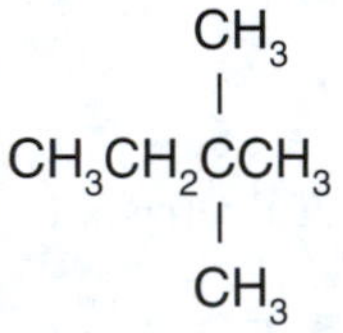
```
        CH3
        |
CH3CH2CCH3
        |
        CH3
```

(e) 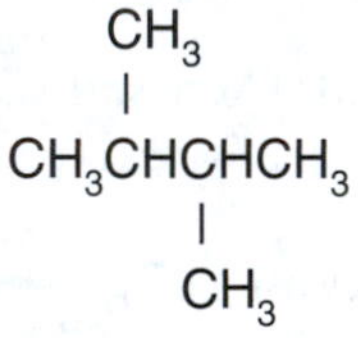
```
   CH3
   |
CH3CHCHCH3
      |
      CH3
```

Objectives Section 11.6 Naming Alkanes

Are you able to...

- assign a prefix to the longest carbon chain in an alkane
- identify the number of carbons in the longest chain from an alkane name
- identify common alkyl groups (Table 11.4)
- recognize primary, 1°, secondary, 2°, tertiary, 3°, or quaternary, 4°, carbons in an alkane (Ex. 11.5, Prob. 11.5, Exercises 11.24, 11.27)
- name an alkane by using IUPAC rules (Ex. 11.6, Prob. 11.6, Exercises 11.21, 11.22, 11.44)
- draw the structure of an alkane from its IUPAC name (Ex. 11.7, Prob. 11.7, Exercise 11.23)

Problem 11.5 Indicate the 1°, 2°, 3°, and 4° carbons in the following compound.

```
                  4°↘   ↙1°
                     C(CH3)3
 1°↓    2°↓    3°↓   |
CH3—CH2—CH—CH—CH2—CH2—CH2—CH3
                |     ↑   ↑___↑___↑   ↑
               CH3    3°      2°      1°
           1°↗
```

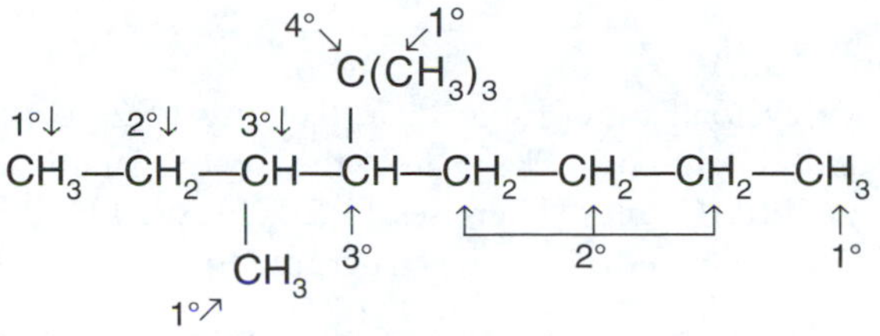

Problem 11.6 Give the IUPAC name for each of the following compounds.
(a) 5-Ethyl-2,3,3,8-tetramethylnonane

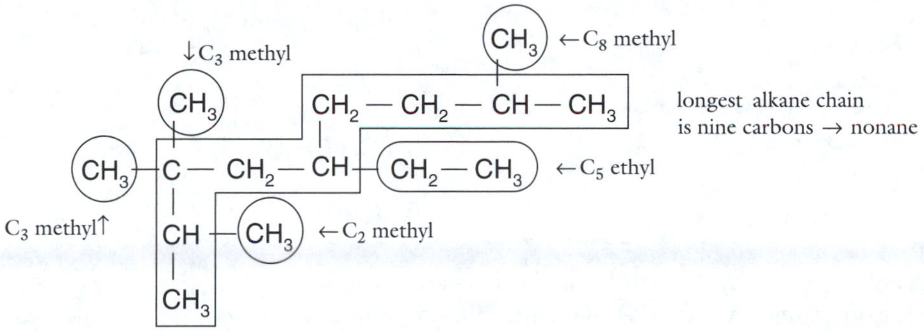

```
                                     (CH3) ←C8 methyl
         ↓C3 methyl                    |
        (CH3)         CH2 — CH2 — CH — CH3      longest alkane chain
          |            |                        is nine carbons → nonane
(CH3) — C — CH2 — CH — (CH2 — CH3)  ←C5 ethyl
C3 methyl↑ |
         CH — (CH3)  ←C2 methyl
          |
         CH3
```

(b) 4-Isopropyl-3-methylheptane

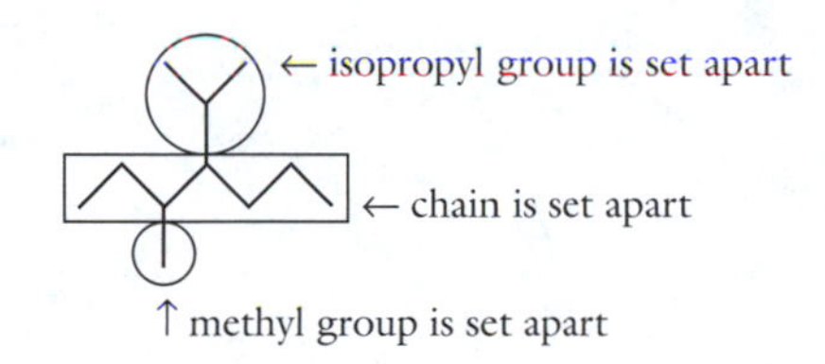

Problem 11.7 Draw the structural formula of each of the following compounds.
(a) 4-ethyl-3-methylhexane

Start with the base to generate the longest carbon chain: hexane tells you that there are six carbons in the longest chain; -ane tells you that there are only single bonds between carbon atoms. The substituents are a methyl group at the number 3 carbon and an ethyl group at the number 4 carbon.

```
              CH3
               |
CH3—CH2—CH—CH—CH2—CH3
                   |
                 CH2CH3
```

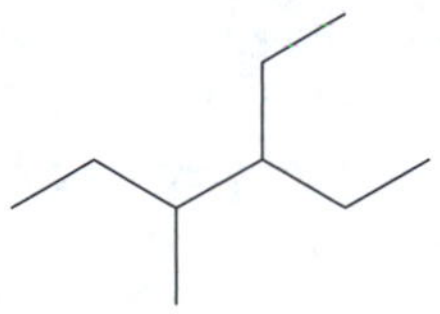

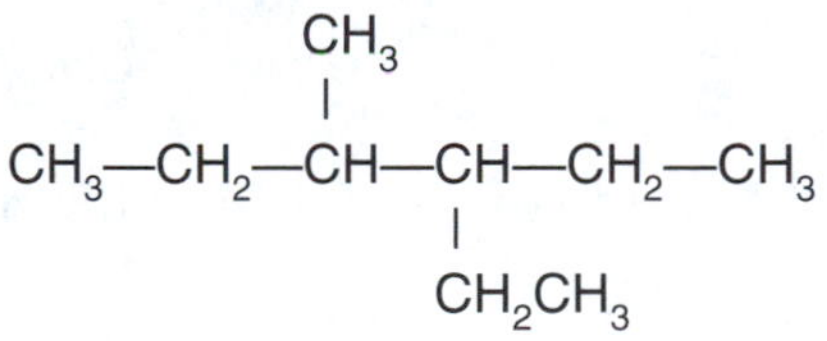

(b) 4-*t*-butyl-3-methyloctane

The base chain is octane, eight carbons with only single bonds between them. There is a methyl group at the number 3 carbon and a *t*-butyl group at the number 4 carbon.

```
        CH3
         |
CH3—CH2—CH—CH—CH2—CH2—CH2—CH3
             |
           C(CH3)3
```

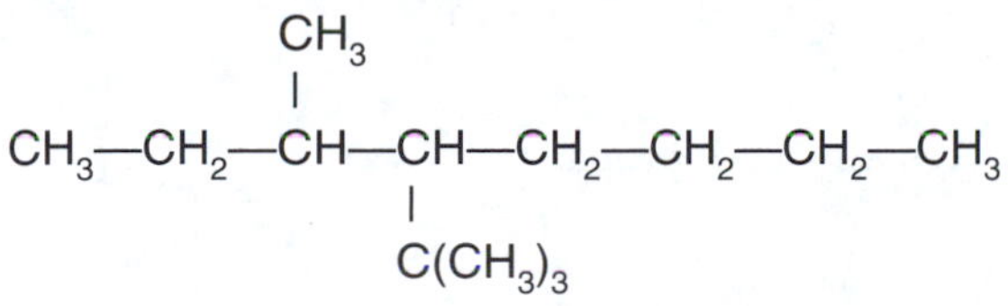

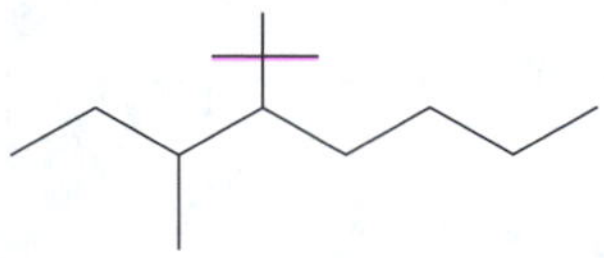

Objectives Section 11.7 Cycloalkanes

Are you able to. . .

- identify cycloalkanes (Table 11.5)
- draw structural formulas for cycloalkanes (Table 11.5, Exercise 11.25)
- distinguish between cyclo- and straight-chain alkanes from their molecular formulas (Exercises 11.45, 11.46)
- name cycloalkanes by using IUPAC rules (Exercises 11.26, 11.29, 11.30)
- use the IUPAC name of a cycloalkane to draw its structure (Exercises 11.28, 11.29, 11.30)

Problem 11.8 Give the IUPAC name to each of the following compounds.

(a) 1-*t*-Butyl-3-ethyl-5-isopropylcyclohexane

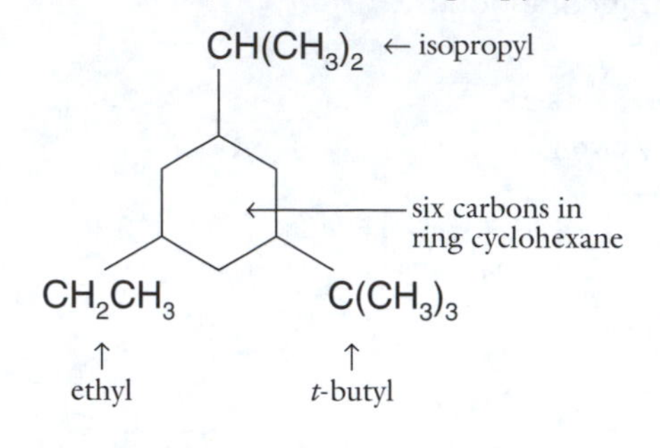

(b) 1-*s*-Butyl-1,3-dimethylcyclobutane

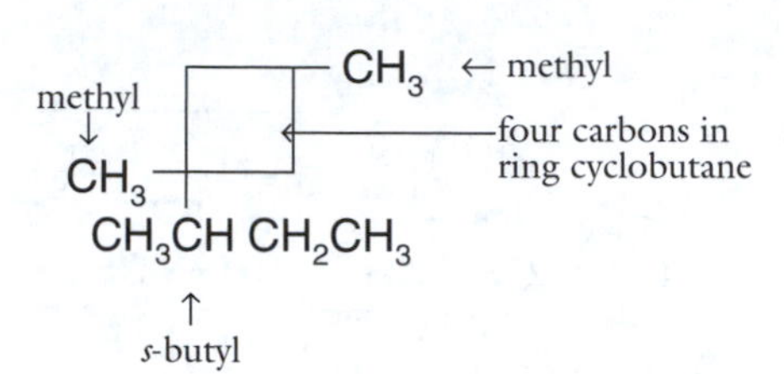

Problem 11.9 How many constitutional isomers are possible for C_6H_{12}? Draw one structural formula for each isomer and give the IUPAC name of each isomer.

(a) Cyclohexane

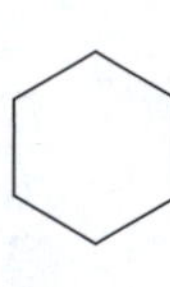

(b) Methylcyclopentane

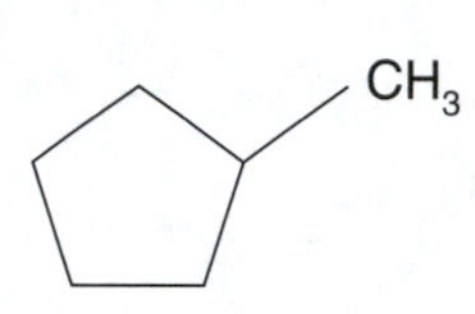

(c) 1,2-Dimethylcyclobutane

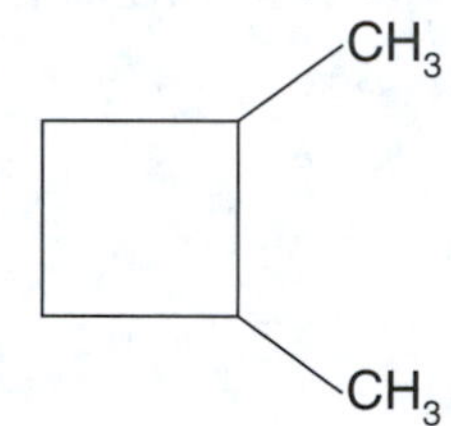

(d) 1,3-Dimethylcyclobutane

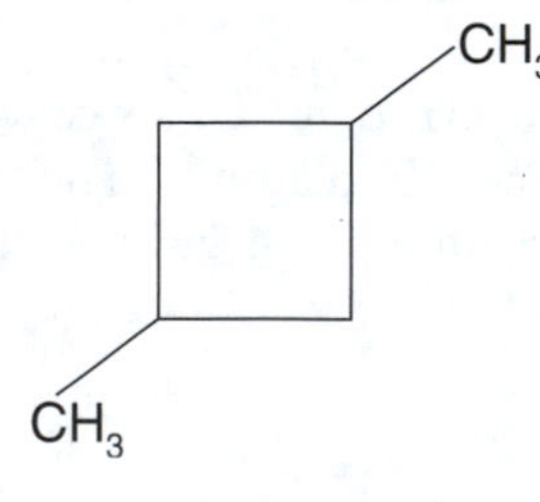

(e) 1,1-Dimethylcyclobutane

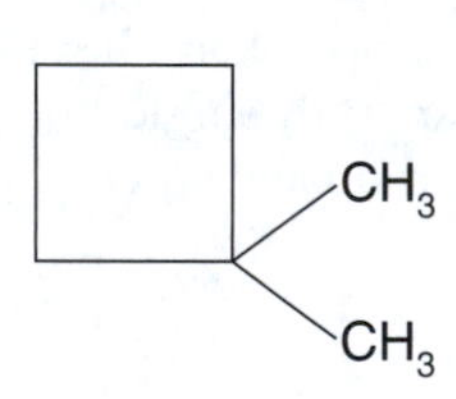

(f) 1,1,2-Trimethylcyclopropane

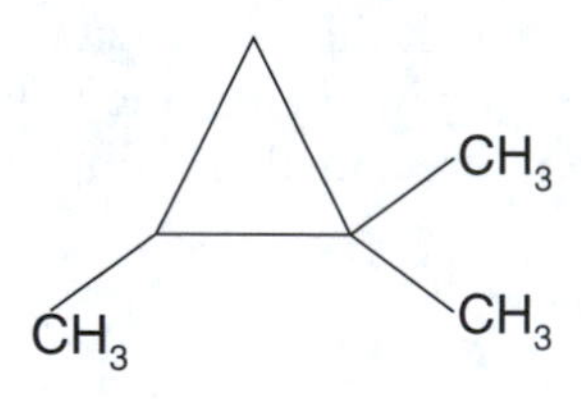

(g) 1,2,3-Trimethylcyclopropane

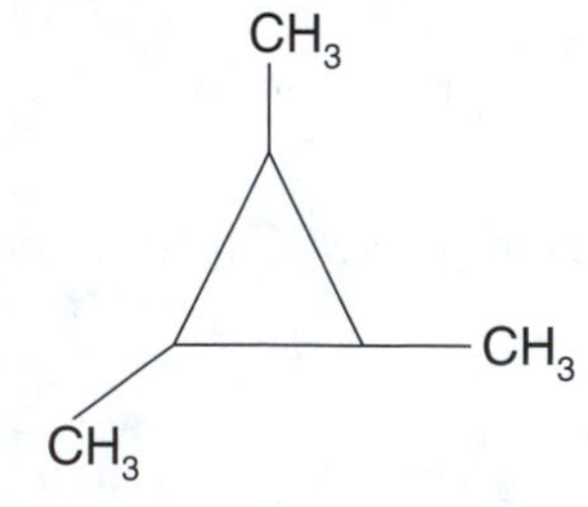

(h) 1-Ethyl-2-methylcyclopropane

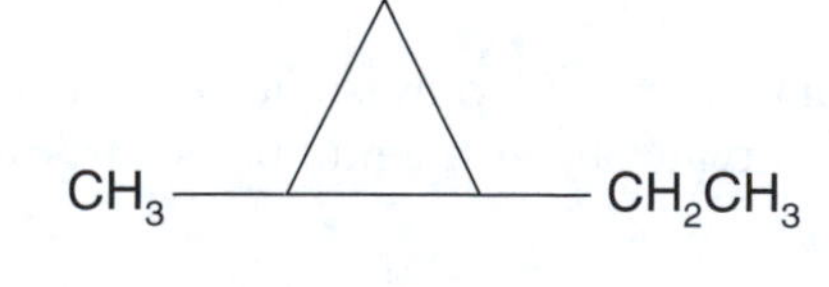

(i) Propylcyclopropane

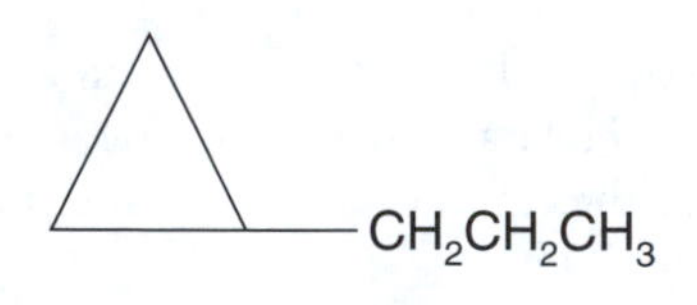

(j) Isopropylcyclopropane

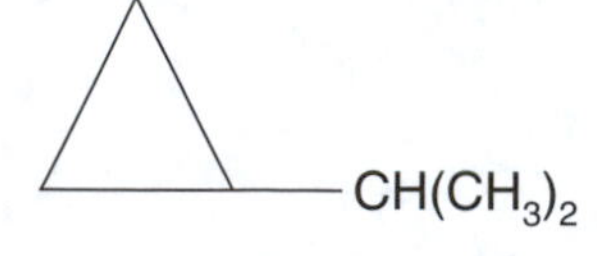

(k) 1-Ethyl-1-methylcyclopropane

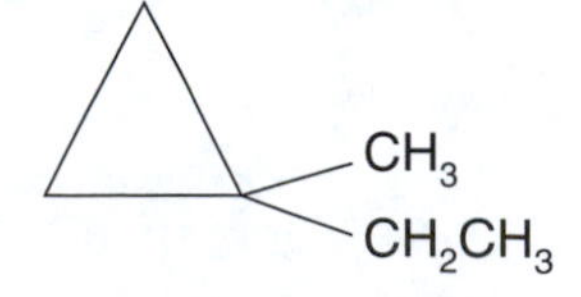

Objectives Section 11.8 Cis-Trans Stereoisomerism in Cycloalkanes

Are you able to...

- explain how cis-trans isomerism arises in cycloalkanes
- distinguish between *cis-* and *trans-*isomers (Exercise 11.32)
- recognize molecules in which cis-trans isomerism is possible (Exercise 11.31)
- draw cis-trans isomers for a given molecule
- explain how cis-trans isomers differ from constitutional isomers (Exercise 11.32)
- explain the difference between configuration and conformation

Problem 11.10 Which of the following compounds exist as *cis-* and *trans-*isomers? If *cis-* and *trans-*isomers are possible, show one structural drawing for each isomer.

(a) 1,3-Dimethylcyclopentane; yes

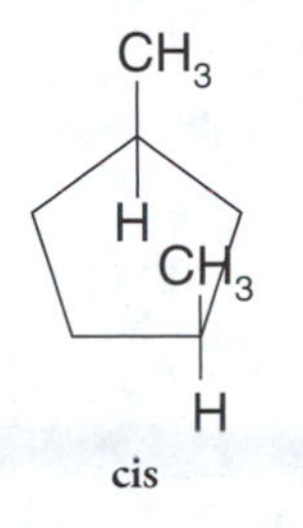

cis

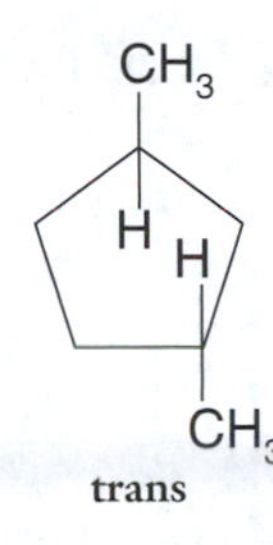

trans

(b) 1-Ethyl-1,2,2-trimethylcyclopropane; no

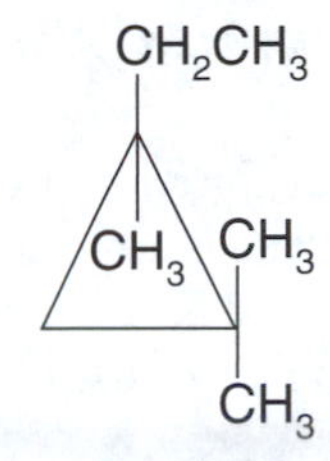

(c) 1,4-dimethylcyclohexane; yes

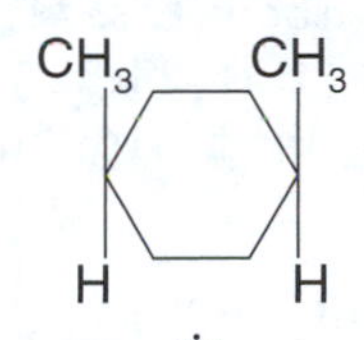

cis

CH_3 H

H CH_3

trans

Objectives Section 11.9 Physical Properties of Alkanes and Cycloalkanes

Are you able to. . .

- define melting point, boiling point, and density
- explain the differences between hydrogen bonds, dipole attractions, and London forces
- identify secondary forces in operation in molecules by analysis of structures
- explain how secondary forces in molecules are related to melting point, boiling point, and density (Ex. 11.11, Prob. 11.11, Exercises 11.33, 11.34)
- explain trends in melting and boiling points within a family of organic compounds (Exercise 11.33)
- predict relative melting and boiling points within a family (Exercises 11.33, 11.34)
- identify the factors that affect the strength and type of secondary forces between molecules (Exs. 11.11, 11.12, Probs. 11.11, 11.12, Exercise 11.34)
- relate geometry and shape to melting and boiling points (Exercise 11.33)

Problem 11.11 Which compound has the higher boiling point in each of the following pairs. Explain each answer in regard to the difference in secondary forces in the two compounds.

(a) cyclopentane and 2-methylbutane: both have the same molecular formula, C_5H_{12}, and therefore have the same mass, so mass is not a factor. One compound has a ring structure, and the other is a branched alkane, which points to stacking as the determining factor in boiling points. The cyclic structure is constrained to its "platelike" geometry, whereas the branched alkane rotates freely. The cyclic structure is able to pack together in a way that puts more surface area in closer proximity to other molecular surfaces than the branched alkane, and has the higher boiling point.

(b) nonane and 3-methylnonane: nonane is one carbon smaller than 3-methylnonane, and therefore, has a slightly smaller molecular mass. Nonane has a straight chain, whereas 3-methylnonane is branched. The molecular mass of 3-methylnonane favors it having the higher boiling point; however, the branched nature of the molecule implies fewer surface attractions than the straight-chain of nonane has.

It is hard to predict this one and you would have to look at reference values or do a boiling point experiment to determine which of the two compounds may have the highest boiling point. 3-Methylnonane has one more carbon than nonane because it has one extra carbon atom.

(c) Cyclobutane and cycloheptane: the greater molecular mass of cycloheptane makes it the candidate for the higher boiling point.

Problem 11.12 Does NaCl dissolve in hexane? Why?

No. NaCl is an ionic compound whose structure is based on electrostatic attractions between oppositely charged ions. Hexane is a nonpolar compound with London forces between molecules. NaCl will not dissolve in hexane, because there are no attractive forces between molecules of hexane and ions such as Na^+ and Cl^-.

Objectives Section 11.10 Chemical Properties of Alkanes and Cycloalkanes

Are you able to. . .

- explain the relative stability of saturated hydrocarbons
- write chemical reactions for organic compounds
- describe a combustion reaction
- write a balanced reaction for the combustion of a hydrocarbon (Ex. 11.13, Prob. 11.13, Exercises 11.37, 11.49)
- write a reaction for the halogenation of a saturated hydrocarbon (Exs. 11.14, 11.15, Probs. 11.14, 11.15, Exercises 11.35, 11.36, 11.38, 11.48, 11.49, 11.50, 11.51, 11.52)
- describe the various products formed from the halogenation of a saturated hydrocarbon (Exs. 11.14, 11.15, Probs. 11.14, 11.15, Exercises 11.35, 11.36, 11.38, 11.48, 11.49, 11.50, 11.51, 11.52)
- use IUPAC rules to name the alkyl halides formed from halogenation reactions (Ex. 11.16, Prob. 11.16)

Problem 11.13 Write balanced equations for the combustion of (a) cyclopentane and (b) 2,2,3-trimethylhexane.

(a) Draw the structure for cyclopentane to determine the number of carbons that will be oxidized to carbon dioxide in a combustion reaction and the number of hydrogen atoms that will become part of water molecules.

$$2\ C_5H_{10} + 15\ O_2 \rightarrow 10\ CO_2 + 10\ H_2O$$

(b) Draw the structure of 2,2,3-trimethylhexane to determine the chemical formula of the molecule.

$$C_9H_{20} + 14\ O_2 \rightarrow 9\ CO_2 + 10\ H_2O$$

Problem 11.14 Write the equation for the chlorination of ethane.

Ethane is C_2H_6. One of the hydrogen atoms will be replaced by a chlorine to produce C_2H_5Cl.

```
     H  H                        H  H
     |  |         Cl2            |  |
  H—C—C—H      ————————→      H—C—C—Cl
     |  |       UV/heat          |  |
     H  H                        H  H
```

$$C_2H_6 \xrightarrow[\text{UV/heat}]{Cl_2} C_2H_5Cl$$

Problem 11.15 Write the equation for the monobromination of 2-methylpropane.

C_4H_{10} is the molecular formula for 2-methylpropane. One hydrogen atom will be replaced by a bromine to produce C_4H_9Br. There are two places in which substitution can occur: at a primary carbon or at a tertiary carbon.

$$CH_3CH(CH_3)CH_3 \xrightarrow[\text{UV/heat}]{Br_2} CH_3C(CH_3)(Br)CH_3 \quad \text{and} \quad CH_3CH(CH_3)CH_2Br$$

bromination at a 3° carbon (first product); bromination at a 1° carbon (second product)

Problem 11.16 Give the IUPAC and common names for the following compounds.

(a) 1-Bromo-2-methylpropane or *s*-butylbromide

(b) *cis*-1-Bromo-2-methylcyclopentane; no common name

MAKING CONNECTIONS

Construct concept maps or write a paragraph describing the relation among the components of each of the following groups:

(a) saturated hydrocarbon, unsaturated hydrocarbon, alkane, carbon, hydrogen
(b) molecular formula, molecular structure, atomic connectivity
(c) group number, valence, combining power
(d) combining power, octet, multiple bond
(e) functional group, atom, property
(f) alkane, alkene, alkyne, hydrocarbon
(g) alcohol, hydroxyl, hydroxide ion
(h) carbonyl, aldehyde, ketone, carboxylic acid
(i) halogen, hydrocarbon, functional group
(j) alkane, structure, geometry, carbon orbitals
(k) alkane, free rotation, sigma bond, conformation
(l) conformation, constitutional isomer, molecular formula
(m) molecular formula, condensed structural formula, expanded structural formula, line structure
(n) straight-chain alkane, branched alkane, isomer
(o) parent chain, substituent, alkyl group, alkane
(p) primary carbon, secondary carbon, tertiary carbon, or quaternary carbon
(q) cycloalkane, isomer, molecular formula
(r) melting point, boiling point, density, secondary forces
(s) London forces, dipole attractions, hydrogen bonds, physical properties
(t) organic family, boiling point, melting point, carbon chain
(u) polar bond, nonpolar bond, secondary force
(v) organic reaction, product, reactant
(w) combustion, hydrocarbon, oxidation
(x) halogenation, alkyl halide, alkane

FILL-INS

Organic chemistry is the study of _____ compounds. Some inorganic compounds such as _____ also contain carbon but are not classified as organic. Organic compounds are molecular and can be identified by a _____. The molecular formula tells exactly _____ and _____ of each kind there are in a molecule, but it does not convey any information about _____. The shape of organic molecules is important because isomers have different _____. Structural formulas convey information about the _____ in a molecule. There are a few variations, such as _____ structures. Each shows clearly _____ in a molecule. The way in which atoms connect depends on how many _____ they are able to form. In organic molecules, carbon and hydrogen are the dominant atoms, with carbon being _____ and forming _____ and hydrogen always forming _____. Other common elements such as nitrogen, oxygen, sulfur, and phosphorus will form the number of bonds required to _____. The _____ of these elements (with the exception of S and P) subtracted from eight will tell you the _____ of each element. Some elements, such as carbon, nitrogen, and oxygen, are able to form _____ bonds.

Organic molecules are classified according to the _____, which are called _____ and which give the _____ to the families of organic compounds. Saturated hydrocarbons contain only single bonds between atoms and are called _____. An alkane can be recognized from its molecular formula, which contains carbon and hydrogen in a ratio of _____. In alkanes, all carbons are _____ hybridized and bond through _____ to _____ other atoms. This produces _____ geometry around each carbon atom. Atoms _____ around single bonds, which gives rise to a variety of conformations of the molecule. Because alkanes are built up by the addition of carbon atoms, it becomes clear that if a molecule contains more than three carbon atoms, there will be more than _____. These molecules will have _____ that can be distinguished from each other by naming them according to IUPAC rules. Alkanes can be found in ring forms that are called_____. These configurations give rise to yet another form of isomer, _____. *cis*-Isomers differ from *trans*-isomers by _____ the plane of a ring. They can also be distinguished from each other by using _____ rules. Alkanes differ in their _____ properties. Melting points and boiling points for the first ten straight-chain alkanes follow trends that are explained by examining _____ between molecules. Alkanes are nonpolar, and the extent of weak _____ determines the relative strength of intermolecular attractions. As molecular mass increases for straight-chain alkanes there is _____ over which London forces operate and it takes _____ energy to separate molecules. Alkanes are nonpolar and will not dissolve in _____. Saturated hydrocarbons are relatively _____ chemically. They do burn in the presence of oxygen to produce _____. Incomplete combustion produces _____, which can be deadly. _____ is a reaction in which a hydrogen on the hydrocarbon is replaced by a halogen. Organic reactions are usually written to show only the _____. They are generally not balanced, and any special _____ are identified above or below the reaction arrow.

Answers

Organic chemistry is the study of carbon-containing compounds. Some inorganic compounds such as carbonates and cyanides also contain carbon but are not classified as organic. Organic compounds are molecular and can be identified by a molecular formula. The molecular formula tells exactly what type(s) of atom(s) and how many of each kind there are in a molecule, but it does not convey any information about the way in which the atoms are connected. The shape of organic molecules is important because isomers have different physical and chemical properties. Structural formulas convey information about the way in which atoms are connected in a molecule. There are a few variations, such as expanded, condensed, skeletal, and line structures. Each shows clearly how atoms bond to one another in a molecule. The way in which atoms connect depends on how many covalent bonds they are able to form. In organic molecules, carbon and hydrogen are the dominant atoms, with carbon being tetravalent and forming four bonds and hydrogen always forming one bond. Other common elements such as nitrogen, oxygen, sulfur, and phosphorus will form the number of bonds required to fill their valence shells. The group number of these elements (with the exception of S and P) subtracted from eight will tell you the combining power of each element. Some elements, such as carbon, nitrogen, and oxygen, are able to form multiple bonds.

Organic molecules are classified according to the presence of certain atoms or groups of atoms, which are called functional groups and which give the chemical and physical characteristics to the families of organic compounds. Saturated hydrocarbons contain only single bonds between atoms and are called alkanes. An alkane can be recognized from its molecular formula, which contains carbon and hydrogen in a ratio of n C to $2n + 2$ H atoms. In alkanes, all carbon atoms are sp^3 hybridized and bond through sigma bonds to four other atoms. This produces tetrahedral geometry around each carbon atom. Atoms rotate freely around single bonds, which gives rise to a variety of conformations of the molecule. Because alkanes are built up by the addition of carbon atoms, it becomes clear that if a molecule contains more than three carbon atoms there will be more than one way for the atoms to connect. These molecules will have constitutional or structural isomers that can be distinguished from each other by naming them according to IUPAC rules. Alkanes can be found in ring forms that are called cycloalkanes. These configurations give rise to yet another form of isomer, cis-trans isomers. *cis*-Isomers differ from *trans*-isomers by the placement of designated groups of atoms relative to one another above and below the plane of a ring. They can also be distinguished from each other by using IUPAC rules. Alkanes differ in their physical and chemical properties. Melting points and boiling points for the first ten straight-chain alkanes follow trends that are explained by examining the nature and magnitude of the secondary forces between molecules. Alkanes are nonpolar, and the extent of weak London forces determines the relative strength of intermolecular attractions. As molecular mass increases for straight-chain alkanes, there is greater surface area over which London forces operate and it takes increasingly more energy to separate molecules. Alkanes are nonpolar and will not dissolve in polar or ionic compounds. Saturated hydrocarbons are relatively inert chemically. They do burn in the presence of oxygen to produce carbon dioxide and water. Incomplete combustion produces carbon monoxide, which can be deadly. Halogenation is a reaction in which a hydrogen atom on the hydrocarbon is replaced by a halogen atom. Organic reactions are usually written to show only the reactants and products of interest. They are generally not balanced, and any special reaction conditions, reagents, or catalysts are identified above or below the reaction arrow.

TEST YOURSELF

1. Examine the names associated with the following structures below and identify the mistakes. Use IUPAC rules to correctly name each molecule.
   (a) 3-Propylpentane (b) 1-Ethyl-2-methylcyclobutane (c) 3-Chloro-4-methylpentane

   (a) $CH_3CH_2CHCH_2CH_2CH_3$ with CH_2CH_3 on the third carbon

   (b) cyclopentane ring bearing CH_2CH_3 and CH_3 on adjacent carbons

   (c) $CH_3CHCHCH_2CH_3$ with CH_3 on the second carbon and Cl on the third carbon

2. Consider the following molecules: dibromocyclopentane and dibromopentane.
   (a) Write the molecular formula for each molecule. How do they compare?
   (b) What types of isomers are possible for each molecule?
   (c) Draw and name the various isomers of each molecule.
3. Explain the implications for structure of free rotation versus restricted rotation around single bonds.
4. Draw the expanded structure of the following molecules.

   (a) $CH_3CH_2CH_2CH_2CH_3$ (b) $(CH_3)_3CCH_2CH_3$
5. Explain why the formula $CH_3CH_2CHClCH_3$ cannot represent C–H–H–H–C–H–H–C–H–Cl–C–H–H–H.
6. How could you use physical properties to distinguish between cyclopentane and pentane?
7. Isobutane has a melting point of −159°C and a boiling point of −12°C, whereas normal butane has a melting point of −138°C and a boiling point of 0°C.
   (a) Write the molecular formula for each molecule.
   (b) Draw the structure of each molecule.
   (c) Use the structures to explain the differences in melting and boiling points.

Answers

1. (a) 3-ethylhexane (b) 1-ethyl-2-methylcyclopentane (c) 3-chloro-2-methylpentane
2. (a) The molecular formula for dibromocyclopentane is $C_5H_8Br_2$. The molecular formula for dibromopentane is $C_5H_{10}Br_2$. They differ by two hydrogen atoms. Cis-trans isomers (stereoisomers) exist for dibromocyclopentane. Constitutional or structural isomers exist for dibromopentane.

Isomers of dibromocyclopentane

1,1-dibromocyclopentane *cis*-1,2-dibromocyclopentane *trans*-1,2-dibromocyclopentane

cis-1,3-dibromocyclopentane *trans*-1,3-dibromocyclopentane

Isomers of dibromopentane

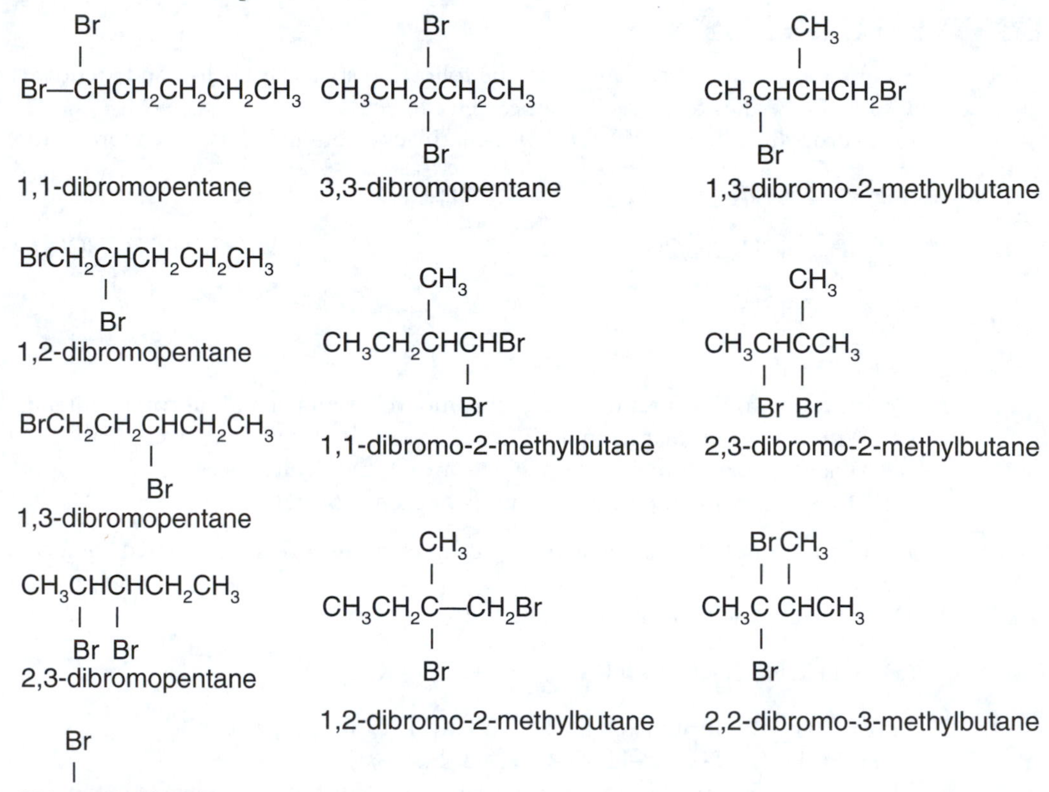

Br
|
$CH_3CCH_2CH_2CH_3$
|
Br

2,3-dibromopentane

$$\begin{array}{c} CH_3 \\ | \\ CH_3C\text{—}CHBr_2 \\ | \\ CH_3 \end{array}$$ 1,1-dibromo-2,2-dimethylpropane

$$\begin{array}{c} CH_3 \\ | \\ BrCH_2CCH_2Br \\ | \\ CH_3 \end{array}$$ 1,3-dibromo-2,2-dimethylpropane

3. Free rotation around single bonds, as seen in alkanes, allows the molecules to adopt a variety of conformations that vary in their linear character. Atoms can be in different spatial orientations as the molecule rotates. Free rotation also means that the molecule is constantly in motion around C–C bonds. Restricted rotation, as seen in cycloalkanes, forces the molecule into positions that cannot be changed except by breaking a bond (chemical reaction). The spatial orientations of atoms and substituents on rings are fixed. There is limited flexibility in a ring structure.

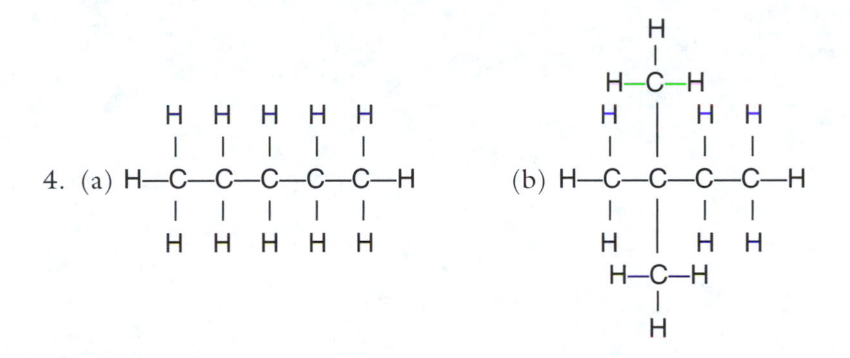

5. The structure shows carbon as monovalent and divalent, forming one and two bonds, and hydrogen and Cl as divalent, forming two bonds. Carbon is tetravalent, always forming four bonds and both hydrogen and chlorine are always monovalent, forming only one bond.
6. Cyclopentane, C_5H_{10}, and pentane, C_5H_{12}, are very close in molecular mass. The major structural difference is the ring versus the straight chain. The packing of rings should be more dense than that of the straight chain. You could predict that the density of cyclopentane will be higher than that of pentane, and the melting and boiling points should be relatively higher than those of pentane, owing to the increased surface area over which London forces operate.

	Cyclopentane	**Pentane**
MW (g/mol)	70.14	72.15
MP (°C)	– 93.88	– 129.72
BP (°C)	49.26	36.07
Density (g/mL)	0.7457	0.6262

7. (a) The molecular formula for both isobutane and normal butane is C_4H_{10}.
   (b) Isobutane — Normal butane

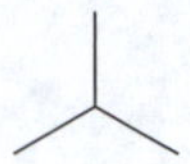

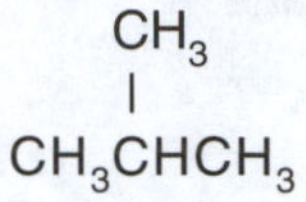

$$\begin{array}{c} CH_3 \\ | \\ CH_3CHCH_3 \end{array}$$

$$CH_3CH_2CH_2CH_3$$

   (c) The branched nature of isobutane's structure limits the surface area over which there can be attractions between molecules, which means that it takes less energy to separate molecules. Note that both compounds have very low melting and boiling points and both are gases at room temperature.

chapter 12

Unsaturated Hydrocarbons

OUTLINE

Introduction

A. Unsaturated hydrocarbons contain C–C multiple bonds.
   1. alkenes contain C–C double bonds
   2. alkynes contain C–C triple bonds
   3. aromatics contain benzene rings (six-carbon ring with three double bonds)
   4. alkanes, alkenes, and alkynes are aliphatic compounds (do not contain benzene rings)
   5. multiple bonds increase chemical reactivity

Key Terms aliphatic, alkene, alkyne, aromatic

12.1 Alkenes

A. Alkenes are unsaturated hydrocarbons containing at least one C–C double bond.
B. The ratio of carbon to hydrogen is n C : $2n$ H (Table 12.1).
   1. two fewer hydrogens than alkane with the same number of C atoms: C_nH_{2n+2}
   2. cycloalkanes have the same C:H ratio as that of alkenes with one C–C double bond
C. Alkenes are nonpolar and not soluble in water.
D. Alkene boiling points (Table 12.1) are similar to those of alkanes.

Key Terms alkene, unsaturated hydrocarbon
Key Table Table 12.1 Formulas and Properties of 1-Alkenes

Example 12.1 Using molecular formulas to determine the hydrocarbon family to which a molecule belongs
Problem 12.1 Using molecular formulas to determine the hydrocarbon family to which a molecule belongs

12.2 Bonding in Alkenes

A. Atomic orbitals used by carbon atoms to form C—C double bonds are sp^2-hybridized (Fig. 12.1)
   1. form from mixing of one s and two p orbitals
   2. one bond is a sigma bond along the internuclear axis (Fig. 12.2)
   3. one bond is a pi bond above and below the internuclear axis (Fig. 12.2)
   4. rotation is restricted about the bond
   5. produces bond angles of 120 degrees at doubly bonded carbons (Fig. 12.4)
   6. results in trigonal planar geometry (Figs. 12.4, 12.5)
B. Carbon-carbon double bonds are more chemically reactive than are carbon-carbon single bonds.

Key Terms pi bond, sigma bond, trigonal, trigonal planar

Key Figures

Figure 12.1	Formation of sp^2 orbitals
Figure 12.2	Double-bond formation in ethene
Figure 12.3	Pi-bond formation in ethene
Figure 12.4	Trigonal bond angles in alkene double bond
Figure 12.5	Ball-and-stick and space-filling models of ethene

12.3 Constitutional Isomers of Alkenes

A. Constitutional isomers of alkenes result from (Fig. 12.6)
  1. varied connectivity of carbon atoms
  2. placement of double bond

Key Term constitutional isomer
Key Figure Figure 12.6 Structural isomers of butane and butene

Example 12.2 Drawing constitutional isomers of alkenes
Problem 12.2 Drawing constitutional isomers of alkenes

12.4 Naming Alkenes

A. Double bond must be contained in the base chain, which may not be the longest continuous chain in the molecule.

IUPAC Rules for Naming Alkenes

1. Identify the longest continuous carbon chain containing the double bond.
2. The prefix identifying the number of carbons in the chain is attached to the suffix -ene.
3. The number of the first carbon containing the double bond is put in front of the alkene name.
   a. Number carbons to give double bond the lowest possible number.
   b. If there are multiple double bonds, identify the number of the first carbon for each bond and insert prefix di, tri, tetra, penta, or hexa before -ene suffix.
4. Substituents are named and numbered as for alkanes.
5. Cyclic alkenes are named as cycloalkenes; number 1 carbon is the first one of a double bond that produces the lowest number for any substituents on the ring.

Key Terms alkene, cycloalkene

Example 12.3 Naming alkenes by using IUPAC rules
Problem 12.3 Naming alkenes by using IUPAC rules

12.5 Cis-Trans Stereoisomerism in Alkenes

A. Cis-trans isomers have different physical and chemical properties (Boxes 12.2, 12.3).
  1. Cis-trans isomers have different shapes (Figure 12.7).
  2. Cis-trans isomers have different physiological properties.

Key Terms cis isomer, configuration, connectivity, restricted rotation, trans isomer, trigonal stereocenter
Key Figure Figure 12.7 Ball-and-stick and space-filling models of *cis*- and *trans*-2-butene
Key Boxes Box 12.2 Vision and cis-trans isomerism
Box 12.3 Cis-trans isomers and pheromones

Example 12.4 Identifying cis-trans isomerism in molecules
Problem 12.4 Identifying cis-trans isomerism in molecules

12.6 Addition Reactions of Alkenes (Box 12.4)

A. Pi bonds in alkenes are reactive.
  1. can be broken to add one substituent to each of the originally double-bonded carbon atoms
  2. produces saturated carbons
  3. addition of hydrogen, F_2, Cl_2, Br_2, HCl, HBr, HI, H_2O takes place

B. Addition of symmetric reagents
   1. symmetric reagents are H_2 (H–H), F_2 (F–F), Cl_2 (Cl–Cl), and Br_2 (Br–Br) (Table 12.2)
      a. H_2 adds with catalyst, Ni or Pt, present at ambient temperatures; Ni or Pt is shown above reaction arrow
      b. hydrogenation is a reduction reaction
      c. halogenation occurs at ambient temperatures
      d. bromination can be used as a test for the presence of double bonds in a molecule (Fig. 12.8)
   2. one atom adds to C_1 of double bond other atom adds to C_2 of double bond

$$R''-\underset{\displaystyle H}{\underset{|}{C_1}}=\underset{\displaystyle H}{\underset{|}{C_2}}-R' \xrightarrow{X_2} R''-\overset{\displaystyle X}{\overset{|}{\underset{\displaystyle H}{\underset{|}{C}}}}-\overset{\displaystyle X}{\overset{|}{\underset{\displaystyle H}{\underset{|}{C}}}}-R'$$

C. Addition of asymmetric reagents
   1. asymmetric reagents are HCl (H–Cl), HBr (H–Br), HI (H–I), H_2O (H–OH) (Table 12.2)
      a. hydrohalogenation adds H to one C of double bond and halogen to the other C of double bond
      b. hydration adds H to one C of double bond and –OH to the other C of double bond
         i. hydration requires acid catalyst—for example, H_2SO_4
         ii. H^+ is shown above reaction arrow

D. Addition reactions may be selective or nonselective.
   1. when more than one product is possible, selective reactions form a major and a minor product
   2. when more than one product is possible, nonselective reactions form about equal amounts of all products
   3. asymmetric reagents added to asymmetric alkenes are selective
      a. addition of asymmetric reagents to asymmetric alkenes follows Markovnikov's rule (Box 12.5)
      b. H adds to carbon of double bond that already has the most hydrogen atoms bonded to it

Key Terms addition reaction, asymmetric alkene, asymmetric reagent, catalytic hydrogenation, halogenation, hydration, hydrohalogenation, major product, Markovnikov's rule, minor product, symmetric alkene, symmetric reagent

Key Figure, Table, and Boxes

Figure 12.8 1-Hexene gives a positive test with Br_2
Table 12.2 Addition reactions of alkenes
Box 12.4 Mechanism of alkane additin reations
Box 12.5 Carbocation stability and the Markovnikov rule

Example 12.5 Writing reactions for the addition of a symmetric reagent to an alkene
Problem 12.5 Writing reactions for the addition of a symmetric reagent to an alkene
Example 12.6 Identifying alkenes by bromine addition
Problem 12.6 Identifying alkenes by bromine addition
Example 12.7 Writing reactions for the addition of an asymmetric reagent to an alkene
Problem 12.7 Writing reactions for the addition of an asymmetric reagent to an alkene

12.7 Addition Polymerization

A. Polymers are long chains of repeating units called monomers (Box 12.6, Table 12.3).
   1. polymers may be synthetic—for example, nylon (Table 12.3)
   2. polymers may be natural—for example, carbohydrates

B. Addition polymerization is similar to addition reaction.
   1. pi bonds break
   2. monomer units add to each side of double bond through sigma bonds
   3. requires a polymerization catalyst
   4. will not take place if other addition reagents are present

Key Terms addition polymerization, monomer, polymer, polymerization, repeat unit
Key Table Table 12.3 Structures and Uses of Addition Polymers

Example 12.8 Writing equations for the polymerization of an alkene
Problem 12.8 Writing equations for the polymerization of an alkene

12.8 The Oxidation of Alkenes

A. Alkenes undergo oxidation.
   1. combustion oxidizes all hydrocarbons to carbon dioxide and water

$$H_2C{=}CH_2 + 3\,O_2 \longrightarrow + 2\,CO_2 + 2\,H_2O$$

   2. selective oxidation places oxygen on carbon in the double bond

$$>C{=}C< \xrightarrow{[O]} >C{=}O$$

      a. permanganate, MnO_4^-, dichromate, $Cr_2O_7^{2-}$, oxygen, O_2, and ozone, O_3, are oxidizing agents
      b. alcohols, aldehydes, ketones, and carboxylic acids are products of oxidation reactions
      c. selective oxidation with the use of permanganate or dichromate can be used as a test for alkenes
         i. purple permanganate solution forms brown precipitate, MnO_2
         ii. orange dichromate solution changes to green solution, Cr^{3+} ion

Key Terms combustion, dichromate, oxidation, ozone, permanganate

Example 12.9 Writing reactions for the combustion of an alkene
Problem 12.9 Writing reactions for the combustion of an alkene

12.9 Alkynes

A. Alkynes are hydrocarbons with at least one triple bond.
   1. ratio of carbon to hydrogen is n C : $(2n - 2)$ H in compounds with one triple bond
   2. alkyne carbons are *sp* hybridized
      a. one sigma bond along the internuclear axis
      b. one pi bond above and below and one pi bond on either side of the internuclear axis
      c. bond angles are 180 degrees between triple-bonded carbons
      d. produces linear geometry around triple-bonded carbons
B. Alkyne triple bonds are more reactive than single or double bonds.
   1. both pi bonds can undergo addition reactions
   2. react with the same reagents as do alkenes: halogens, water, hydrogen halides
   3. use twice as much reagent as do alkenes

IUPAC Rules for Naming Alkynes

1. Identify the longest continuous carbon chain containing the triple bond.
2. The prefix identifying the number of carbons in the chain is attached to the suffix -yne.
3. The number of the first carbon containing the triple bond is put in front of the alkyne name.
   a. Number carbons to give triple bond the lowest possible number.
   b. If there are multiple triple bonds, identify the number of the first carbon for each bond and insert the prefix di, tri, tetra, penta, or hexa, before -yne suffix.
4. Substituents are named and numbered as for alkanes.

Key Term alkyne

Example 12.10 Writing reactions for the addition of an asymmetric reagent to an alkyne
Problem 12.10 Writing reactions for the addition of an asymmetric reagent to an alkyne

12.10 Aromatic Compounds

A. Aromatic compounds contain benzene rings (Fig. 12.10); formula is C_6H_6.

B. Benzene rings
   1. six-membered carbon rings
   2. three alternating double bonds in ring (Box 12.7)

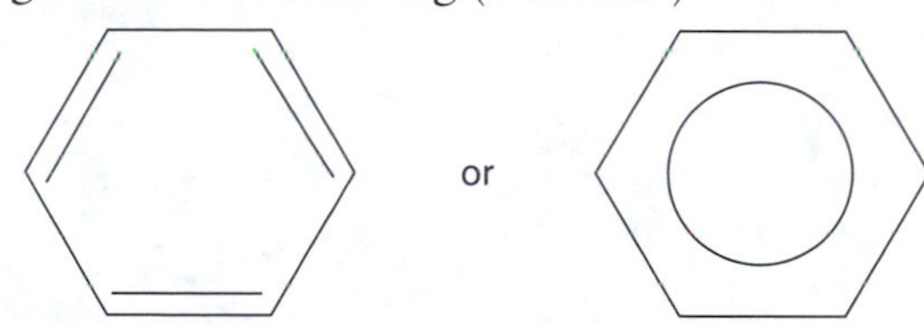

   3. all carbons lie in the same plane
   4. structure is chemically very stable
      a. halogens, hydrogen, hydrogen halides, water, and acids do not add to double bonds
      b. combustion reactions produce less energy than those of aliphatic hydrocarbons
      c. hydrogens on the ring can undergo substitution reactions; no double bonds are broken

Key Terms aliphatic, aromatic, aromaticity, benzene
Key Figure Figure 12.10 Space-filling model of benzene and diagram showing trigonal bond angles

12.11 Isomers and Names of Aromatic Compounds

A. Benzene rings with one hydrogen replaced by a substituent are named as substituted benzenes.
   1. common names are often used
   2. substituents are named followed by the suffix -benzene

B. Two or three substituents are named and designated relative to each other.
   1. ortho substituents are on adjacent carbons
   2. meta substituents are separated by one carbon
   3. para substituents are separated by two carbons

C. In a molecule having both aliphatic and benzene parts, the benzene part becomes a substituent of the aliphatic chain if the aliphatic part dominates.
   1. benzene as a substituent is called phenyl
   2. phenyl group is also written C_6H_5–
   3. phenyl or substituted phenyl substituents are called aryl groups
   4. benzyl groups are $C_6H_5CH_2$–

D. Ring positions are numbered in the same way as cycloalkanes if ortho, meta, or para is not used.

Example 12.11 Naming aromatic compounds by using IUPAC rules
Problem 12.11 Naming aromatic compounds by using IUPAC rules

12.12 Reactions of Aromatic Compounds

A. Benzene undergoes substitution reactions.
   1. halogenation takes place in the presence of a metal halide, $FeCl_3$ or $AlCl_3$; H on ring is replaced by halogen
   2. nitration takes place in the presence of an acid, H_2SO_4; an $-NO_2$ group replaces H on ring
   3. sulfonation takes place in the presence of an acid, H_2SO_4; H on ring is replaced by $-SO_3H$
   4. alkylation takes place in the presence of a metal halide, $FeCl_3$ or $AlCl_3$; an alkyl group replaces H on ring

B. Oxidation of aliphatic parts of aromatic compounds can take place in the presence of hot acidic $KMnO_4$ or $K_2Cr_2O_7$.

C. Combustion of aromatic compounds produces carbon dioxide and water.

Example 12.12 Writing addition reactions for aromatics
Problem 12.12 Writing addition reactions for aromatics

Example 12.13 Writing reactions for the addition of halogens to various hydrocarbons
Problem 12.13 Writing reactions for the addition of halogens to various hydrocarbons

ARE YOU ABLE TO... AND WORKED TEXT PROBLEMS

Objectives Introduction

Are you able to...

- identify alkanes
- identify alkenes
- identify alkynes
- identify aromatics
- explain the differences among alkanes, alkenes, alkynes, and aromatic compounds
- define aliphatic

Objectives Section 12.1 Alkenes

Are you able to...

- identify the structural features of alkenes
- identify alkenes by molecular formula (Table 12.1, Ex. 12.1, Prob. 12.1, Exercises 12.1, 12.2, 12.3, 12.4)
- describe the physical properties of alkenes (Table 12.1)

Problem 12.1 An unknown compound has the molecular formula C_7H_{16}. Is the compound an alkane, a cycloalkane, or an alkene?

The ratio of carbon to hydrogen is 7 C : 16 H. $n = 7$ and 16 is $2(n) + 2$: $2(7) + 2 = 16$, making the formula C_nH_{2n+2}, which corresponds to an alkane.

Objectives Section 12.2 Bonding in Alkenes

Are you able to...

- describe the bonding in alkenes (Exercises 12.7, 12.56)
- explain the geometry that results from *sp*² hybridization
- identify the bond angles in alkenes (Exercise 12.8)
- explain how double bonds restrict rotation in alkenes

Objectives Section 12.3 Constitutional Isomers of Alkenes

Are you able to...

- describe the variety of isomers possible in alkenes (Exercises 12.13, 12.14, 12.43, 12.45, 12.46)
- draw structural formulas for alkene isomers (Ex. 12.2, Prob. 12.2, Exercises 12.9, 12.10, 12.11, 12.12, 12.42)
- provide the molecular formula from structural drawings (Exercises 12.5, 12.6, 12.13, 12.14, 12.53)

Problem 12.2 Draw structural formulas for all constitutional isomers of alkenes with the molecular formula C_6H_{12}. Show only one structural formula for each isomer.

$CH_3CH_2CH_2CH_2CH=CH_2$

$CH_3CH_2CH_2CH=CHCH_3$

$CH_3CH_2CH=CHCH_2CH_3$

CH_3 (on C)
$CH_3CH_2CH_2C=CH_2$

CH_3 (on CH)
$CH_3CH_2CHCH=CH_2$

CH_3 (on CH)
$CH_3CHCH_2CH=CH_2$

CH_3 (on C)
$CH_3CH_2CH=CCH_3$

CH_3 (on C)
$CH_3CH_2C=CHCH_3$

CH_3 (on CH)
$CH_3CHCH=CHCH_3$

CH_3 CH_3 (on CH and C)
$CH_3CHC=CH_2$

CH_3 CH_3 (on each C)
$CH_3C=CCH_3$

CH_2CH_3 (on C)
$CH_3CH_2C=CH_2$

CH_3, CH_3 (above and below the quaternary C)
$CH_2=CHCCH_3$

Objectives Section 12.4 Naming Alkenes

Are you able to...

- name alkenes using IUPAC rules (Ex. 12.3, Prob. 12.3, Exercises 12.16, 12.44)
- draw an alkene structure from its name (Exercise 12.15)

Problem 12.3 What is the IUPAC name for each of the following compounds?

(a) 4-Methyl-1-hexene

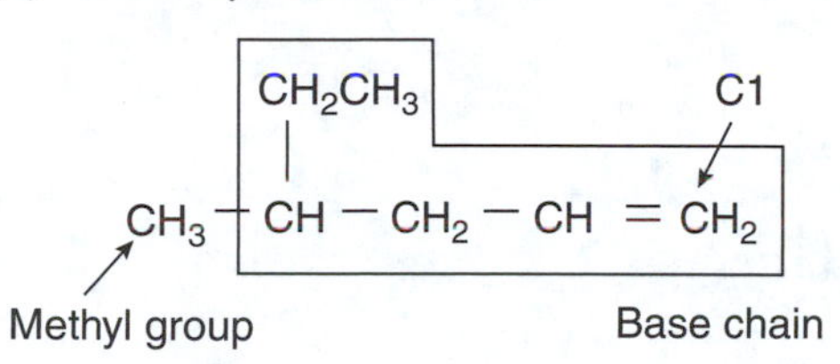

(b) 1-Chloro-4-methyl-2-pentene

Base chain, C2, C1, Chloro

$CH_3 - CH - CH = CH - CH_2 - Cl$

CH_3 ← Methyl

(c) 3-*t*-butyl-2-Isopropylcyclopentene

CH_3, C — CH_3 ← *t*-Butyl group, CH_3

C1

CH ← Isopropyl group

H_3C CH_3

(d) 2-Methyl-1,4-pentadiene

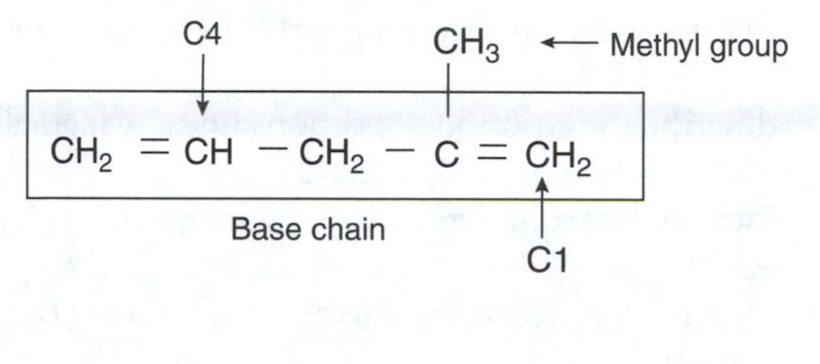

Objectives Section 12.5 Cis-Trans Stereoisomerism in Alkenes

Are you able to...

- explain why cis-trans isomerism is possible in alkenes
- identify cis and trans isomers of alkenes (Ex. 12.4, Prob. 12.4, Exercises 12.17, 12.19, 12.20, 12.43, 12.46)
- draw structures for cis and trans isomers of alkenes (Exercise 12.18)

Problem 12.4 Which of the following compounds exist as a pair of cis and trans isomers?

(a) 2-Methyl-2-pentene No

$(CH_3)(CH_3)C = C(H)(CH_2CH_3)$ — Same (CH_3 and CH_3 on the same carbon)

(b) 3-Methyl-2-pentene Yes

cis: $(H)(CH_3)C = C(CH_3)(CH_2CH_3)$

trans: $(CH_3)(H)C = C(CH_3)(CH_2CH_3)$

cis trans

(c) 4-Methyl-2-pentene Yes

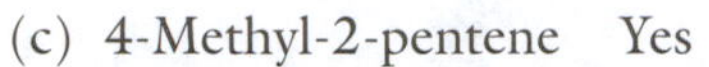

```
CH3        H                                   CH3
   \      /                                   /
    C = C                   CH3             CH
   /      \                    \           /  \
H          CH - CH3              C = C       CH3
           |                    /      \
           CH3                 H        H

   trans                           cis
```

Objectives Section 12.6 Addition Reactions of Alkenes

Are you able to...

- explain the relative reactivity of alkenes
- describe how alkenes undergo addition reactions
- identify symmetric reagents that add to alkenes (Exs. 12.5, 12.7, Probs. 12.5, 12.7, Exercises 12.21, 12.22)
- identify asymmetric reagents that add to alkenes (Exs. 12.5, 12.7, Probs. 12.5, 12.7, Exercises 12.21, 12.22)
- describe the reaction conditions under which alkenes undergo reactions (Exs. 12.5, 12.7, Probs. 12.5, 12.7, Exercises 12.21, 12.22)
- identify hydrohalogenation, hydrogenation, hydration, and halogenation reactions (Exs. 12.5, 12.7, Probs. 12.5, 12.7, Exercises 12.21, 12.22)
- identify symmetric and asymmetric alkenes (Exs. 12.5, 12.7, Probs. 12.5, 12.7, Exercises 12.21, 12.22, 12.50)
- predict the major and minor products of addition reactions (Exs. 12.5, 12.7, Probs. 12.5, 12.7, Exercises 12.21, 12.22, 12.49, 12.50, 12.51, 12.57)
- recognize when Markovnikov's rule applies (Exs. 12.5, 12.7, Probs. 12.5, 12.7, Exercises 12.21, 12.22, Expand Your Knowledge 12.61)
- use the results of addition reactions to determine the identity of an unknown (Ex. 12.6, Prob. 12.6, Exercises 12.23, 12.24, 12.48, 12.49, 12.54, 12.55, 12.63)

Problem 12.5 Complete each of the following reactions by writing the structure of the organic product. If no reaction takes place, write NR.

(a) Reaction shows a symmetric alkene ($CH_3CH{=}CHCH_3$) plus a symmetric reagent (H_2) in the presence of a catalyst (Ni). Hydrogen adds to alkenes in the presence of a catalyst, so the product is $CH_3CH_2CH_2CH_3$, butane.

```
H        H                      H   H
 \      /           H2          |   |
  C = C           ------>   H - C - C - H
 /      \           Ni          |   |
CH3      CH3                   CH3 CH3
```

(b) Reaction shows an asymmetric alkene ($CH_3CH{=}CH_2$) and a symmetric reagent (Br_2) with no special reaction conditions. Bromine adds to alkenes under ambient conditions. The product is $CH_3CHBrCH_2Br$, 1,2-dibromopropane.

```
H        H                      Br  H
 \      /           Br2         |   |
  C = C           ------>   H - C - C - H
 /      \                       |   |
CH3      H                     CH3  Br
```

Problem 12.6 An unknown compound is pentane, cyclopentane, or 2-pentene. The red color of bromine is decolorized when bromine is added to the unknown. Identify the unknown.

The red color of bromine disappears when it adds across the double bond of an alkene. The only compound of those listed that contains double bonds is 2-pentene. The unknown must be 2-pentene.

Problem 12.7 Complete each of the following reactions by writing the structure of the organic product. If no reaction takes place, write NR.

(a) Asymmetric alkene ($CH_2{=}CHCH_2CH_3$) and symmetric reagent (Cl_2) with no special reaction conditions. Chlorine will add to an alkene under ambient conditions. The product is $CH_2ClCHClCH_2CH_3$.

$$(H)_2C{=}C(H)(CH_2CH_3) \xrightarrow{Cl_2} H{-}C(Cl)(H){-}C(Cl)(CH_2CH_3){-}H$$

(b) Asymmetric alkene ($CH_2{=}CHCH_3$) and an asymmetric reagent (H–OH). Hydration requires an acid catalyst and one is specified, so the reaction will take place. Addition pattern follows Markovnikov's rule.

$$(H)_2C{=}C(H)(CH_3) \xrightarrow[H^+]{H_2O} \underset{\text{major}}{H{-}C(H)(H){-}C(OH)(CH_3){-}H} + \underset{\text{minor}}{H{-}C(OH)(H){-}C(H)(CH_3){-}H}$$

(c) Asymmetric alkene ($CH_2{=}CHCH_2CH_3$) and an asymmetric reagent (HCl) with no special reaction conditions. Hydrohalogenation will take place under ambient conditions. There are two products: a major product, $CH_3CHClCH_2CH_3$; and a minor product, $CH_2ClCH_2CH_2CH_3$. Additon pattern follows Markovnikov's rule.

Objectives Section 12.7 Addition Polymerization

Are you able to...

- define the terms monomer and polymer
- describe how addition polymerization takes place
- describe the reaction conditions under which addition polymerization takes place
- write the product of addition polymerization (Ex. 12.8, Prob. 12.8, Exercises 12.25, 12.26, 12.52)

Problem 12.8 Write the equation for the polymerization of propene to form polypropylene.

Propene is $CH_3CH{=}CH_2$. The repeat unit, or monomer, of polypropylene is $+\!\!(CH_2{-}CH(CH_3))\!\!+$.

$$(H)_2C{=}C(H)(CH_3) \xrightarrow{\text{catalyst}} (CH_2{-}CH(CH_3))_n$$

Objectives Section 12.8 Oxidation of Alkenes

Are you able to...

- describe the oxidation reactions that alkenes undergo
- write combustion reactions for alkenes (Ex. 12.9, Prob. 12.9, Exercises 12.27, 12.28)
- write selective oxidation reactions for alkenes
- describe how selective oxidation can be used to identify alkenes (Exercises 12.29, 12.30, 12.58, 12.59, 12.63)

Problem 12.9 Give the balanced equation for the combustion of 2-octene.

The molecular formula for 2-octene is C_8H_{16}. All carbons will be oxidized to carbon dioxide and all hydrogens will be incorporated into water molecules. The products are CO_2 and H_2O.

$$C_8H_{16} + 12\ O_2 \rightarrow 8\ CO_2 + 8\ H_2O$$

Objectives Section 12.9 Alkynes

Are you able to...

- identify alkynes by their molecular formulas
- explain the bonding in alkynes
- describe the geometry of alkynes
- explain the reactivity of alkynes
- describe how alkynes undergo addition reactions
- predict products of alkyne addition reactions (Ex. 12.10, Prob.12.10, Exercise 12.32)
- name alkynes by using IUPAC rules
- draw alkyne structures from the IUPAC names (Exercise 12.31)

Problem 12.10 Show the sequential reaction of 2-butyne with excess HCl. Show only the major product for each addition.

The alkyne is asymmetric and the reagent is asymmetric. Addition to each pi bond is in accord with Markovnikov's rule. $CH_3C{\equiv}CH$ has two pi bonds. The first addition produces an alkene, with H bonding to the triple-bonded carbon that already has a hydrogen: $CH_3CCl{=}CH_2$. The next addition forms an alkane. The hydrogen adds to the double-bonded carbon that has more hydrogens: $CH_3CCl_2CH_3$.

$$CH_3-C\equiv C-H \xrightarrow{HCl} (Cl)(CH_3)C=CH_2$$

$$(Cl)(CH_3)C=C(H)(H) \xrightarrow{HCl} Cl-C(Cl)(CH_3)-C(H)(H)-H$$

Objectives Section 12.10 Aromatic Compounds

Are you able to...

- define aromatic and aromaticity
- recognize aromatic compounds by their molecular formulas
- recognize aromatic compounds by their structures (Exercises 12.33, 12.34)
- describe the structure of benzene
- explain the stability of benzene rings

Objectives Section 12.11 Isomers and Names of Aromatic Compounds

Are you able to...

- name substituted benzene compounds by using IUPAC rules (Exercise 12.44)
- recognize some common names for benzene compounds
- recognize aromatic compounds and isomers (Exercises 12.35, 12.36, 12.41, 12.42, 12.43, 12.45, 12.60)
- name organic compounds that contain benzene substituents (Exercises 12.37, 12.38)
- draw structural formulas for substituted benzenes from their names (Ex. 12.11, Prob. 12.11)
- derive the molecular formula of an aromatic compound from its structure (Exercise 12.53)

Problem 12.11 Draw structural formulas for the following compounds.

(a) *p*-Ethylbenzoic acid or 4-ethylbenzoic acid

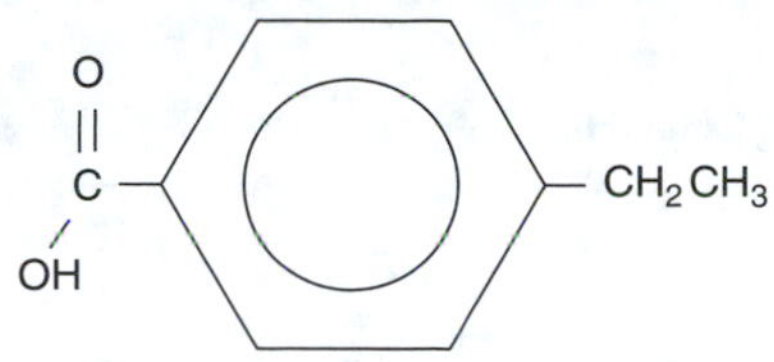

(b) 2,4-Dibromotoluene

Br

$-CH_3$

Br

(c) 4-Phenyl-1-butene

$CH_2=CH$

CH_2CH_2

(d) *m*-Isopropylphenol or 3-isopropylphenol

OH

CH_3

C

H

CH_3

(e) 1-Chloro-3-ethyl-4-propylbenzene

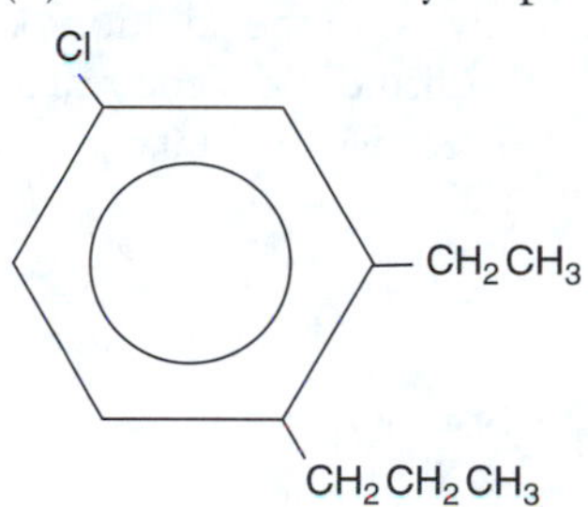

(f) Benzyl chloride

CH_2-Cl

Objectives Section 12.12 Reactions of Aromatic Compounds

- describe the conditions under which benzene undergoes substitution reactions
- describe the halogenation of benzene
- describe the nitration of benzene
- describe the sulfonation of benzene
- describe the alkylation of benzene
- describe the oxidation of benzene compounds
- describe the combustion of benzene
- write substitution reactions for benzene compounds
- predict the products of benzene substitution (Ex. 12.12, Prob. 12.12, Exercises 12.39, 12.40, 12.57)
- write oxidation reactions for benzene compounds
- predict the products of benzene oxidation
- write combustion reactions for benzene compounds
- predict the products of the combustion of benzene compounds

Problem 12.12 Write equations to show the product(s) formed in each of the following reactions. If no reaction takes place, write NR. If more than one product is formed, indicate the major and minor products.

(a) Benzene + CH_3CH_2Cl

NR. Benzene substitution of an alkylchloride requires the presence of a metal halide catalyst.

(b) Benzene + CH_3CH_2Cl + $AlCl_3$

The metal halide catalyst $AlCl_3$ catalyzes the substitution of a hydrogen on the benzene ring by the alkyl group, CH_3CH_2–, with concurrent production of HCl:

$$C_6H_6 \xrightarrow[AlCl_3]{CH_3CH_2Cl} C_6H_5CH_2CH_3 + HCl$$

(c) Isopropylbenzene + hot acidic $K_2Cr_2O_7$

The aliphatic isopropyl chain is oxidized to carbon dioxide and water. The C atom attached to the benzene ring is oxidized to a carboxylic acid.

$$C_6H_5CH(CH_3)_2 \xrightarrow[H^+,\ heat]{K_2Cr_2O_7} C_6H_5COOH + 2\,CO_2 + 3\,H_2O$$

Problem 12.13 Bromine is added to a mixture of cyclohexane, cyclohexene, and benzene in the presence of ultraviolet light but in the absence of $FeCl_3$. Which compounds undergo reaction? Write equations to show the product(s).

Halogen substitution in alkanes requires UV light, so the cycloalkanes will react to produce a bromine-substituted cycloalkane. Alkenes react easily with halogens even without UV or catalysts, so the addition of bromine takes place across the double bond, producing a bromine-substituted cycloalkane. Benzene reacts with halogens only in the presence of a metal halide catalyst, which is absent. No reaction will take place with bromine.

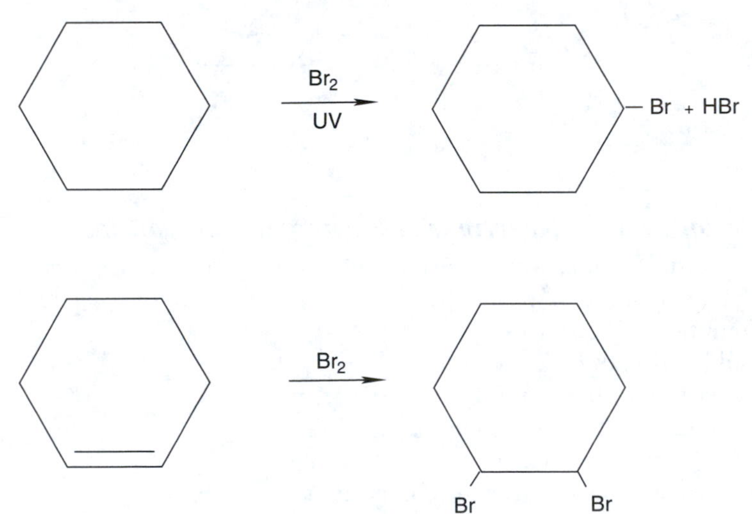

MAKING CONNECTIONS

Construct concept maps or write paragraphs describing the relation among the members of each of the following groups:

(a) alkane, alkenes, alkyne, aromatic
(b) molecular formula, single bond, double bond, triple bond, aromatic
(c) saturated hydrocarbon, unsaturated hydrocarbon, molecular formula
(d) constitutional isomer, alkane, alkene, alkyne, aromatic

(e) rotation, single bond, double bond, triple bond, ring structure
(f) isomer, alkane, alkene, alkyne, aromatic, ring structure
(g) addition reaction, alkane, alkene, alkyne, aromatic
(h) substitution reaction, alkane, alkene, alkyne, aromatic
(i) reaction condition, reactants, products, reaction equation
(j) chemical reactivity, single bond, double bond, triple bond
(k) molecular geometry, carbon orbital hybridization, bonds
(l) molecular shape, conformation isomer, constitutional isomer, stereoisomer
(m) IUPAC rules, alkane, alkene, alkyne, aromatic
(n) -ane, -ene, -yne-, -benzene
(o) ortho-, meta-, para-, benzene
(p) base chain, substituent, isomer
(q) molecular formula, isomer, structure

FILL-INS

Unsaturated hydrocarbons can be divided into families, depending on the type of _____ in the molecule. Molecules in which carbon doubly bonds to carbon are called _____. Alkenes can be identified by the fact that, for every carbon atom in the molecule, there are _____ hydrogen atoms. This ratio can also be applied to _____. The presence of double bonds makes alkenes more _____ than alkanes. The weak _____ bond can be broken and one _____ can be added to each of the formerly unsaturated carbons, making them _____. Alkenes are also _____ and do not dissolve in _____. The _____ geometry of alkene carbons results from the sp^2 hybridization of carbon orbitals, producing bond angles of _____. Rotation about double bonds is _____, but many isomers may be possible owing to _____ as well as _____. Cis-trans isomers are possible when _____. Cis-trans isomers have different shapes and differ in their _____. The pi bond of alkenes is weak enough that it can be broken under _____ conditions in the presence of _____. _____ takes place in the presence of a Ni or Pd catalyst. _____ requires an acid catalyst. When _____ reagents are added to _____ alkenes, Markovnikov's rule states that _____. Alkene _____ can add to form _____ in polymerization reactions. Oxidation of alkenes occurs as _____ or as _____ in the presence of oxidizing agents. Complete combustion produces _____. Selective oxidations place _____on carbons in the double bond. Selective oxidations change the _____ from that of an alkene to an _____. Alkynes are hydrocarbons with at least one _____ between carbon atoms. The ratio of carbon to hydrogen is: _____. The hybridization of carbon orbitals in alkynes is *sp*, which produces _____ geometery. The triple bond is composed of a strong _____ and _____ , which are chemically very similar to the pi bonds in _____. Alkynes react with the _____ as do alkenes and undergo _____ reactions that use _____ the amount of reagent as do alkenes. Aromatic compounds contain _____. Benzene rings are very _____ and undergo substitution reactions for _____ but, unlike alkenes, do not participate in _____.

Answers

Unsaturated hydrocarbons can be divided into families, depending on the type of carbon–carbon bonds in the molecule. Molecules in which carbon doubly bonds to carbon are called alkenes. Alkenes can be identified by the fact that, for every carbon atom in the molecule, there are twice as many hydrogen atoms. This ratio can also be applied to cyclic alkanes. The presence of double bonds makes alkenes more reactive than alkanes. The weak pi bond can be broken and one atom or group of atoms can be added to each of the formerly unsaturated carbons, making them saturated. Alkenes are also nonpolar and do not dissolve in water. The trigonal planar geometry of alkene carbons results from the sp^2 hybridization of carbon orbitals, produce bond angles of 120 degrees. Rotation about double bonds is restricted, but many isomers may be possible owing to constitutional variation as well as placement of the double bond. Cis-trans isomers are possible when there are two different groups attached to each carbon in the double bond. Cis-trans isomers have different shapes and differ in their physical and chemical properties. The pi bond of alkenes is weak enough that it can be broken under ambient conditions in the presence of halogens and hydrogen halides. Hydrogenation takes place in the presence of a Ni or Pd catalyst. Hydration requires an acid catalyst. When asymmetric reagents are added to asymmetric alkenes, Markovnikov's rule states that the carbon with the most hydrogens will get the hydrogen and the other carbon will get the remaining fragment. Alkene monomers can add to form polymers in polymerization reactions. Oxidation of alkenes occurs as combustion or as selective oxidation in the presence of oxidizing agents. Complete combustion produces carbon dioxide and water. Selective oxidations place oxygen on carbons in the double bond. Selective oxidations change the functional group from that of an alkene to an alcohol, aldehyde, ketone, or carboxylic acid. Alkynes are hydrocarbons with at least one triple bond between carbon atoms. The ratio of carbon to hydrogen is: for every n carbons, there are $2n - 2$ hydrogens. The hybridization of carbon orbitals in alkynes is *sp*, which produces linear geometry. The triple bond is composed of a strong sigma bond and two weak pi bonds, which are chemically very similar to the pi bonds in alkenes. Alkynes react with the same reagents as do alkenes and undergo addition reactions that use twice the amount of reagent as do alkenes. Aromatic compounds contain benzene rings. Benzene rings are very stable and undergo substitution reactions for hydrogens on the ring but, unlike alkanes, do not participate in addition reactions.

TEST YOURSELF

1. From the following formulas, develop the carbon : hydrogen ratios and categorize the compounds as alkanes, alkenes, or alkynes.
   (a) $CH_3CH_2CH_2CH_2CH_3$
   (b) $CH_3CHCHCH_2CH_2CH_3$
   (c) $CH_3CH_2CCCH_3$
   (d) $(CH_3)_3CCH_2CH_2CHCHCH_3$
2. Draw the geometric shapes associated with alkanes, alkenes, and alkynes. How do the three-dimensional shapes differ?
3. Build each of the molecules in Exercise 1 by using a model kit, toothpicks and round marshmallows or candies, or straws and Styrofoam balls. How does the rotational ability vary as multiple bonds are added to a hydrocarbon?
4. Draw the structures of each of the following molecules:
   (a) *cis*-1,2-dimethylcyclopentane
   (b) 2-ethyl-1-pentene
   (c) *trans*-2-butene
   (d) *o*-dibromobenzene
   (e) 3-methyl-1-butyne
5. Write the generic reaction for each of the following reactions; ignore reaction conditions; identify major and minor products.
   (a) addition of an asymmetric reagent to a symmetric alkene
   (b) addition of an asymmetric reagent to an asymmetric alkene
   (c) addition of a symmetric reagent to a symmetric alkene
   (d) addition of a symmetric reagent to an asymmetric alkene
   (e) a symmetric alkene polymerization
   (f) an asymmetric alkene polymerization
   (g) addition of an asymmetric reagent to a symmetric alkyne
   (h) addition of an asymmetric reagent to an asymmetric alkyne
6. Benzene and cyclohexene are both six-membered ring compounds. Draw the structures of each and write the molecular formulas. Benzene's melting point is 5.5°C, its boiling point is 80.0°C, and its density is 0.8787 g/mL. Cyclohexene's melting point is −103.50°C, its boiling point is 82.89°C, and its density is 0.8102 g/mL. What do these values tell you about the nature of the secondary forces between benzene molecules and between cyclohexene molecules? How do the chemical properties of the two compare?

Answers

1. (a) 5 C and 12 H $\Rightarrow n$ C: $2n + 2$ H, alkane
   (b) 6 C and 12 H $\Rightarrow n$ C : $2n$, H alkene
   (c) 5 C and 8 H $\Rightarrow n$ C : $2n - 2$ H, alkyne
   (d) 9 C and 18 H $\Rightarrow n$ C : $2n$ H, alkene

2. Alkane — Tetrahedral; Alkene ($C=C$) — Trigonal planar; Alkyne ($-C\equiv C-$) — Linear

   The bond angles in a tetrahedral configuration are 109.5°; the bond angles in a trigonal planar configuration are 120°; the bond angles in a linear configuration are 180°.
3. As double bonds are added to a molecule, the rotational ability is decreased. Alkanes are able to rotate freely about all single bonds, which gives rise to many conformational isomers. The positions of atoms relative to each other can be changed with no breakage of chemical bonds as the molecule twists and turns.
4. (a) *cis*-1,2-dimethylcyclopentane

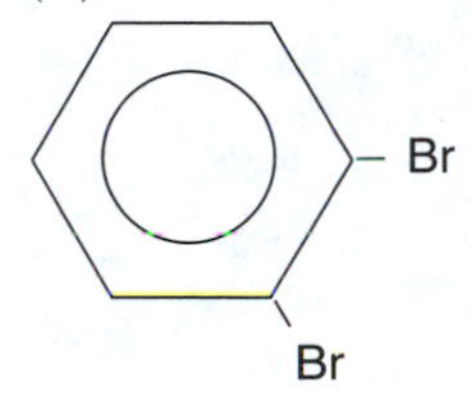

   (b) 2-Ethyl-1-pentene

   $(H)_2C=C(CH_2CH_3)(CH_2CH_2CH_3)$

   (c) *trans*-2-Butene

   $H(CH_3)C=C(CH_3)H$

   (d) *o*-Dibromobenzene

   Br, Br

   (e) 3-Methyl-1-butyne

   $H-C\equiv C-CH(CH_3)-CH_3$

5. (a) Addition of an asymmetric reagent to a symmetric alkene

   $$R'''R''C=CR'''R'' \xrightarrow{XY} R'''-CXR''-CYR''-R'''$$

   (b) Addition of an asymmetric reagent to an asymmetric alkene

   $$HR''C=CR'''R' \xrightarrow{HX} H-CHR''-CXR'-R'''$$

   (c) Addition of a symmetric reagent to a symmetric alkene

   $$HR'C=CHR' \xrightarrow{X_2} H-CXR'-CXR'-H$$

   (d) Addition of a symmetric reagent to an asymmetric alkene

   $$R^3R^4C=CR^1R^2 \xrightarrow{X_2} R^4-CXR^3-CXR^2-R^1$$

(e) A symmetric alkene polymerization

$$\mathrm{(H)(R')C{=}C(H)(R')} \xrightarrow{\text{catalyst}} \left(\begin{array}{cc} \mathrm{H} & \mathrm{H} \\ | & | \\ \mathrm{C} - & \mathrm{C} \\ | & | \\ \mathrm{R'} & \mathrm{R'} \end{array}\right)_n$$

(f) An asymmetric alkene polymerization

$$\mathrm{(R^4)(R^3)C{=}C(R^1)(R^2)} \xrightarrow{\text{catalyst}} \left(\begin{array}{cc} \mathrm{R^4} & \mathrm{R^1} \\ | & | \\ \mathrm{C} - & \mathrm{C} \\ | & | \\ \mathrm{R^3} & \mathrm{R^2} \end{array}\right)_n$$

(g) Addition of an asymmetric reagent to a symmetric alkyne

$$\mathrm{R'-C{\equiv}C-R'} \xrightarrow{\mathrm{XY}} \mathrm{(X)(R')C{=}C(Y)(R')}$$

(h) Addition of an asymmetric reagent to an asymmetric alkyne

$$\mathrm{H-C{\equiv}C-R'} \xrightarrow{\mathrm{HX}} \mathrm{(H)(H)C{=}C(X)(R')}$$

6. Benzene, C_6H_6 Cyclohexene, C_6H_{10}

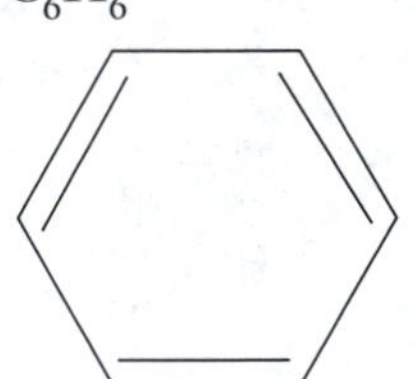

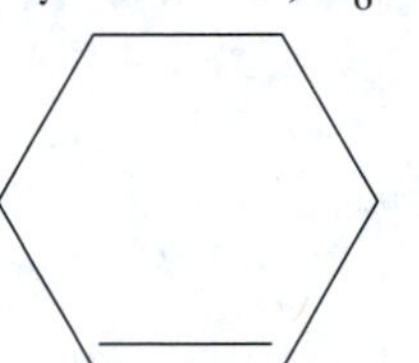

Benzene is a liquid at room temperature and cyclohexene is a gas at room temperature, which implies that the secondary forces between benzene molecules are stronger than are those between cyclohexene molecules. Cyclohexene readily undergoes addition reactions. Benzene does not undergo addition reactions. Rather, in the presence of appropriate catalysts, benzene undergoes substitution reactions.

chapter 13

Alcohols, Phenols, Ethers, and Their Sulfur Analogues

OUTLINE

Introduction

A. When oxygen is added to hydrocarbons, new functional groups are formed. If the oxygen atom is connected by single bonds only, alcohols, phenols, and ethers result.
   1. alcohols contain an –OH group attached to the hydrocarbon chain
   2. phenols contain an –OH group attached to a benzene ring
   3. ethers contain an –O– attached on either side to carbon atoms; consider an O atom to be inserted into the hydrocarbon chain

Key Terms alcohol, ether, phenol

13.1 Structural Relations of Alcohols, Phenols, and Ethers

A. Oxygen atoms in alcohols, ethers, and phenols are sp^3 hybridized (Fig. 13.1).
   1. O has two non-bonding pairs of valence electrons called lone pairs
   2. only two sp^3-hybridized electrons are available for bonding; O is divalent
   3. bond angle around oxygen atoms is close to tetrahedral

B. Alcohols, phenols, and ethers can be thought of as organic derivatives of water (Table 13.1).

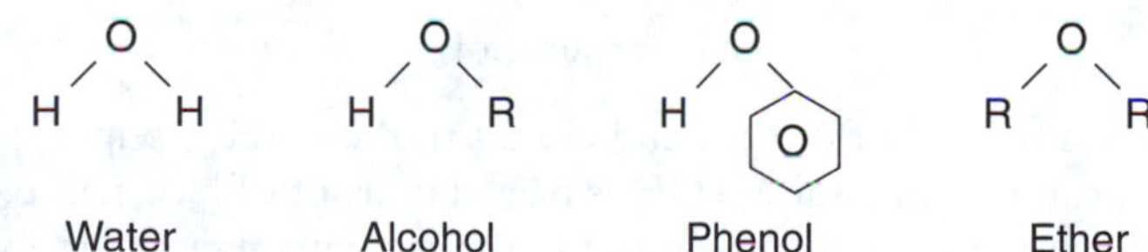

C. When one hydrogen atom on a hydrocarbon chain is replaced with an –OH group, an alcohol is formed; –OH is a hydroxyl group. DO NOT CONFUSE WITH OH^-, HYDROXIDE ION.

D. When an –OH group replaces a hydrogen on a benzene ring, a phenol is formed.

E. Insertion of O into a hydrocarbon chain produces an ether; –O– is bonded to two different carbons; C may be part of aromatic or aliphatic fragment.

Key Terms alcohol, divalent, ether, hydroxyl group, lone pair electrons, nonbonding pair, phenol, tetravalent

Example 13.1 Distinguishing between alcohols, phenols, and ethers
Problem 13.1 Distinguishing between alcohols, phenols, and ethers

13.2 Constitutional Isomerism in Alcohols

A. Two structural features produce isomers in alcohols.
   1. constitutional isomers from carbon skeleton variations
   2. varied placement of –OH group on carbon chain
B. Alcohols must have three or more carbon atoms for isomers to exist.
C. Molecular formula of alcohols tells if molecule has single bonds, double bonds, or ring.
   1. C-to-H ratio is maintained as for alkane if the alcohol has no double bond or ring: C_nH_{2n+2}
   2. C-to-H is maintained as for alkene or cycloalkane if the alcohol has double bond or ring: C_nH_{2n}
   3. Molecular formula of alcohol is also ether molecular formula

$$CH_3CH_2OH \text{ or } CH_3\text{–}O\text{–}CH_3 \quad \text{both are } C_2H_6O$$

Key Term constitutional isomer

Example 13.2 Drawing constitutional isomers of alcohols
Problem 13.2 Drawing constitutional isomers of alcohols
Example 13.3 Determining the general molecular formula for alcohols from structural drawings
Problem 13.3 Determining the general molecular formula for alcohols from structural drawings

13.3 Classifying and Naming Alcohols

A. Alcohols can be classified as primary, secondary, or tertiary, depending on the type of carbon to which –OH is bonded.
   1. primary alcohols: –OH bonded to a carbon that is bonded to one other carbon atom
   2. secondary alcohol: –OH bonded to a carbon that is bonded to two other carbon atoms
   3. tertiary alcohol: –OH bonded to a carbon that is bonded to three other carbon atoms
B. Alcohols are named according to IUPAC rules.

IUPAC Rules for Naming Alcohols

1. The longest continuous carbon chain must contain the –OH group.
2. The chain is named in the same manner as for alkanes, but the -ane ending is replaced with an -ol ending.

$$\text{alkane} \Rightarrow \text{alkanol}$$

3. The chain is numbered from the end closest to the –OH group.
4. Number of carbon containing –OH is placed in front of base name.
5. Substituents are identified and located in the same manner as for alkanes.
6. If a ring structure has an –OH attached, it is a cycloalcohol.
   a. cycloalkane ⇒ cycloalkanol
   b. ring is numbered starting at the C attached to –OH and proceeding in direction that gives lowest numbers for other substituents.
7. If more than one –OH group is on chain or ring, each –OH position is specified by placing C number as prefix before the base name and a prefix is inserted before the -ol ending indicating the number of –OH groups:

$$2\ \text{–OH} \Rightarrow \text{C number, C number, -alkanediol}$$
$$3\ \text{–OH} \Rightarrow \text{C number, C number, C number, -alkanetriol}$$

Key Terms alkanediol, alkanetriol, cycloalcohol, primary alcohol, secondary alcohol, tertiary alcohol

Example 13.4 Classification of 1°, 2°, and 3° alcohols
Problem 13.4 Classification of 1°, 2°, and 3° alcohols
Example 13.5 Naming alcohols by using IUPAC rules
Problem 13.5 Naming alcohols by using IUPAC rules

13.4 Physical Properties of Alcohols

A. Physical properties of alcohols arise from hydrogen bonding (Table 13.3).
   1. –OH is polar
      a. electronegativity difference between O and H is great enough to produce a polar bond
      b. electrons in bond spend more time near O; oxygen has partial negative charge
      c. hydrogen has partial positive charge
   2. alcohol molecules can form hydrogen bonds with each other (Fig. 13.2); produces higher melting and boiling points than for alkanes of similar molecular weight
   3. alcohol molecules can form hydrogen bonds with water (Fig. 13.3); allows small-chain alcohols to dissolve in water

B. As size of hydrocarbon chain increases, polar area becomes less influential and nonpolar secondary forces dominate (Table 13.3).
   1. alcohols with fewer than three carbons are water soluble
   2. alcohols with four or five carbons are slightly soluble
   3. alcohols with more than five carbons are not water soluble unless they contain more than one –OH group; one –OH group can make as many as three carbons soluble

Key Terms electronegativity, hydrogen bond

Example 13.6 Explaining physical properties of alcohols
Problem 13.6 Predicting and explaining properties of alcohols relative to other molecules

13.5 Acidity and Basicity of Alcohols

A. Alcohols are very weakly amphoteric.
   1. very weak acidity is evidenced by ability to react with active metals to form H_2; alcohols will not react with strong bases
   2. weak basicity evidenced by ability to accept protons from strong acids; fewer than 1 of 1000 alcohol molecules are protonated in presence of strong acid
   3. alcohols will not change the color of either red or blue litmus paper

Key Terms amphoteric, weak acid, weak base

13.6 Dehydration of Alcohols to Alkenes

A. Alcohols can be converted into alkenes by abstraction of a water molecule from the alcohol molecule (Box 13.4).
   1. called dehydration and is an intramolecular reaction

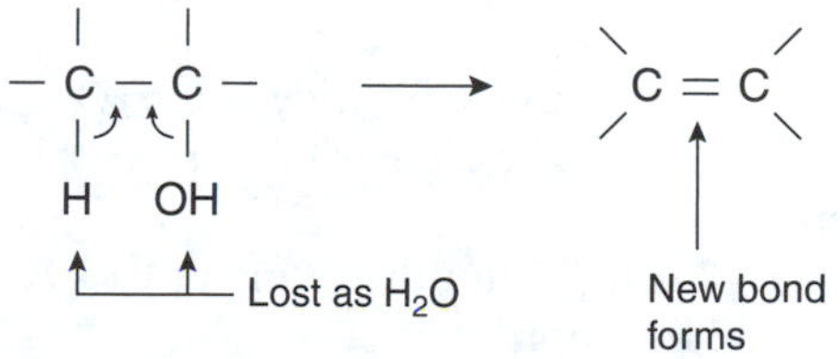

$$-\underset{\text{H}}{\overset{|}{\underset{|}{\text{C}}}}-\underset{\text{OH}}{\overset{|}{\underset{|}{\text{C}}}}- \longrightarrow \;>\text{C}=\text{C}<$$

Lost as H_2O

New bond forms

   2. alcohol must be heated in the presence of a strong acid catalyst

$$\text{Alcohol} \rightarrow \text{alkene} + H_2O$$

   3. dehydration occurs between –OH on one C and –H on an adjacent C

4. dehydration of alcohol is reverse of hydration of alkene
   a. alcohol ⇌ alkene

$$-\underset{\text{H}}{\overset{|}{\underset{|}{\text{C}}}}-\underset{\text{OH}}{\overset{|}{\underset{|}{\text{C}}}}- \underset{+H_2O}{\overset{-H_2O}{\rightleftharpoons}} \;>\text{C}=\text{C}<$$

   b. can take advantage of equilibrium to produce product of choice
5. dehydration is selective (Box 13.4)
   a. more highly substituted double bonds are favored owing to increased stability
   b. Most stable   Least stable

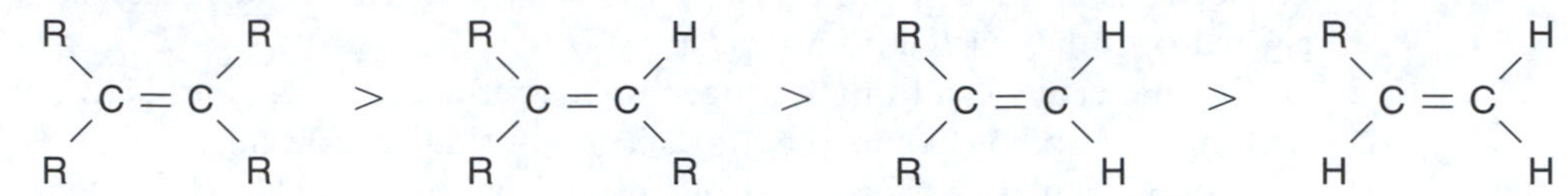

Key Terms dehydration, hydration, intramolecular reaction
Key Box 13.4 Mechanism of Dehydration of Alcohols

Example 13.7 Prediction and writing of dehydration reactions of alcohols
Problem 13.7 Prediction and writing of dehydration reactions of alcohols
Example 13.8 Prediction and writing of dehydration reactions of alcohols
Problem 13.8 Prediction and writing of dehydration reactions of alcohols

13.7 *Oxidation of Alcohols*

A. Alcohols can undergo combustion or selective oxidation, depending on reaction conditions.
B. Combustion reactions produce carbon dioxide and water.
C. Selective oxidation affects only the carbon to which the –OH is attached.
   1. transforms an alcohol into an aldehyde, ketone, or carboxylic acid
   2. primary alcohols are oxidized in two steps
      a. first step: the carbon to which –OH is attached is oxidized to a carbonyl, –C=O

$$\text{R}-\underset{\text{H}}{\overset{\text{H}}{\text{C}}}-\text{OH} \xrightarrow{[\text{O}]} \text{R}-\overset{\text{H}}{\text{C}}=\text{O}$$

      b. first step creates an aldehyde
      c. second step: the H attached to the carbonyl carbon is oxidized to an –OH, creates a carboxyl group

$$\text{R}-\overset{\text{H}}{\text{C}}=\text{O} \xrightarrow{[\text{O}]} \text{R}-\overset{\text{OH}}{\text{C}}=\text{O}$$

      d. second step creates a carboxylic acid, 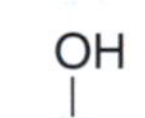
      e. reaction conditions can be controlled to stop oxidation at the first step with mild oxidizing agent
      f. permanganate, MnO_4^-, and dichromate, $Cr_2O_7^{2-}$, are favored oxidizing agents for carboxylic acid formation
   3. secondary alcohols are oxidized in one step
      a. the C to which –OH is attached is oxidized to a carbonyl, –C=O

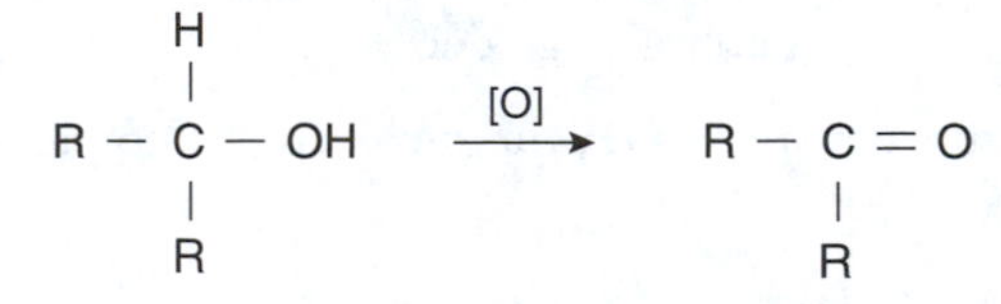

b. produces a ketone
c. groups on carbonyl carbon can be aliphatic or aromatic
4. Tertiary alcohols do not undergo oxidation, because the carbon to which –OH is attached is not bonded to a hydrogen atom that can participate in the reaction

$$R_3C-OH \xrightarrow{[O]} \text{no reaction}$$

D. Selective oxidation with dichromate or permanganate can be used to identify primary and secondary alcohols.
1. positive permanganate tests show change from purple solution to brown precipitate, MnO_2
2. positive dichromate tests show change from orange solution to green solution, signaling presence of Cr^{3+}
3. can be used to distinguish 1° and 2° alcohols from 3° alcohols because 3° do not react
4. 1° and 2° alcohols can be distinguished from ethers because ethers do not react
5. tests will not distinguish 1° and 2° alcohols from alkenes, alkynes, and phenols

E. Dichromate oxidation is the basis for Breathalyzer tests used by police officers to determine alcohol levels of drivers.

Key Terms aldehyde, carbonyl, carboxylic acid, combustion, ketone, selective oxidation

Example 13.9 Prediction and writing of oxidation reactions of alcohols
Problem 13.9 Prediction and writing of oxidation reactions of alcohols
Example 13.10 Identification of an unknown alcohol by using oxidation reactions
Problem 13.10 Identification of an unknown alcohol by using oxidation reactions

13.8 Phenols

A. When a benzene ring has an –OH group attached to one of the carbons in the ring, it is a phenol.
B. When an –OH is attached to a carbon-carbon double bond, it is called an enol.
C. Phenols are named as derivatives of the parent compound.

IUPAC Rules for Naming Phenols

1. The C to which –OH is attached is carbon number one.
2. Substituents are named and numbered so that the lowest possible numbers are given.
3. Numbers and substituent names are placed before the base name -phenol.

D. –OH bonds in phenols are more polar than alcohols.
1. electron-withdrawing effect of benzene ring increases polarization
2. O develops larger partial negative charge
3. H develops larger partial positive charge
4. produces stronger hydrogen bonding
   a. greater hydrogen bonding results in higher melting and boiling points than those of alcohols
   b. greater hydrogen bonding results in greater solubility in water compared with alcohols

E. Phenols are weak acids.
1. phenols are more acidic than alcohols
2. hydronium ion concentrations are about 100 times as great as corresponding alcohols at same concentration
3. phenols turn blue litmus red, unlike alcohols

4. greater acidity allows phenols to react with strong bases such as NaOH
5. electron-withdrawing effect of benzene ring stabilizes $C_6H_5–O^-$, phenoxide ion; charge is dispersed over phenyl ring

F. Phenols do not undergo dehydration.
G. Phenols do undergo oxidation, which is the basis for antioxidant activity.

Key Terms charge dispersal, enol, phenol, phenoxide ion, phenyl ring

Example 13.11 Naming phenols by using IUPAC rules
Problem 13.11 Naming phenols by using IUPAC rules
Example 13.12 Representing hydrogen bonding in phenols
Problem 13.12 Representing hydrogen bonding in phenols

13.9 Ethers

A. Oxygen is attached by single bonds to two different carbons; C may be part of aliphatic or aromatic group.
B. Ethers can be named by identifying the groups on each side of the O as alkyl groups and placing these groups in front of the word ether; used only for simple ethers.
C. More complex ethers are named by using IUPAC rules where ether is named as part of another organic family.
   1. the most complex group determines the base name
   2. simpler groups and O to which they are attached are called alkoxyl or aryloxy groups
D. Ethers are isomers of alcohols containing the same number of carbons.
E. Polar –O–C– bonds in ethers result in stronger secondary forces between molecules than is seen in alkanes
   1. low-molecular mass ethers have higher boiling points than those of corresponding alkanes
   2. effect falls off as molecular mass of ether increases
   3. ethers can form hydrogen bonds between the O of ether and the H of a water molecule; results in moderate water solubility
F. Ethers undergo combustion and halogenation reactions.
G. Ethers do not undergo reactions with acids, bases, or oxidizing agents; makes them useful as solvents in which to carry out other chemical reactions.

Key Terms alkoxyl, aryloxy, ether

Example 13.13 Naming ethers by using IUPAC rules
Problem 13.13 Naming ethers by using IUPAC rules
Example 13.14 Drawing constitutional isomers of ethers
Problem 13.14 Drawing constitutional isomers of ethers
Example 13.15 Representation of hydrogen bonding in ethers
Problem 13.15 Predicting properties of ethers relative to other compounds

13.10 Formation of Ethers by Dehydration of Alcohols

A. Ethers can be formed from alcohols by an intermolecular dehydration.
   1. two alcohol molecules react to produce an ether
      a. –OH from one alcohol molecule and H from a different alcohol molecule form water

$$R-\text{(OH)}+\text{(H)}-OR' \xrightarrow{-H_2O} R-OR' + H_2O$$

      b. acid catalyst is required
      c. temperature of the reaction determines whether an ether or an alkene is major product
         i. at 180°C the alkene is the major product
         ii. at 140°C the ether is the major product
         iii. only 1° alcohols form ethers under these conditions
         iv. 2° and 3° alcohols form alkenes

Key Terms dehydration, intermolecular reaction

Example 13.16 Writing dehydration reactions of alcohols
Problem 13.16 Writing dehydration reactions of alcohols

13.11 Thiols

A. Thiols are sulfur analogues of alcohols.

1. –SH is called a mercapto or sulfhydryl group
2. thiols are also called mercaptans

B. Thiol properties are different from those of alcohols.

1. thiols cannot form hydrogen bonds; the difference in electronegativities between S and H is not enough to create a polar bond
2. thiols have lower boiling points than those of alcohols because of weaker intermolecular secondary forces
3. thiols are stronger acids than alcohols but weaker acids than phenols
4. thiols have disagreeable odors

C. Thiols undergo a variety of chemical reactions.

1. thiols are easily oxidized to form disulfides

$$\mathrm{R-S-H} \xrightarrow{[\mathrm{O}]} \mathrm{R-S-S-R}$$

2. disulfides are easily reduced back to thiols

$$\mathrm{R-S-S-R} \xrightarrow{\mathrm{H_2}} \mathrm{R-S-H} + \mathrm{R-S-H}$$

3. thiols react with heavy metal ions to form salts that are insoluble in water

$$2\,\mathrm{R-S-H} + \mathrm{M^{2+}} \longrightarrow \mathrm{R-S-M-S-R} + 2\mathrm{H^+}$$

D. Thiols are named in the same manner as alcohols, except that a -thiol ending replaces the -ol ending.

Key Terms disulfide, heavy metal, mercaptan, mercapto group, sulfhydryl group, thiol

Example 13.17 Writing reaction equations for thiols
Problem 13.17 Writing reaction equations for thiols

ARE YOU ABLE TO... AND WORKED TEXT PROBLEMS

Objectives Introduction

Are you able to...

- identify alcohols (Exercises 13.1, 13.2, 13.35, 13.36, 13.37)
- identify phenols (Exercises 13.1, 13.2, 13.35, 13.36, 13.37)
- identify ethers (Exercises 13.1, 13.2, 13.35, 13.36, 13.37)

Objectives Section 13.1 Structural Relations of Alcohols, Phenols, and Ethers

Are you able to...

- discuss the divalent nature of oxygen
- discuss the hybridization of orbital around the oxygen atom in alcohols, ethers, and phenols (Fig. 13.1)
- explain the relation between water and alcohols, phenols, and ethers (Table 13.1)
- identify the functional group in alcohols (Ex. 13.1, Prob. 13.1, Exercises 13.1, 13.2, 13.35, 13.36, 13.37)

- identify the functional group in phenols (Ex. 13.1, Prob. 13.1, Exercises 13.1, 13.2)
- identify the functional group in ethers (Ex. 13.1, Prob. 13.1, Exercises 13.1, 13.2)

Problem 13.1 Which of the following structures is an alcohol, a phenol, an ether, or something else?
(1) alcohol; –OH group is bonded to a saturated carbon atom
(2) ether; O is bonded to two different carbons
(3) phenol; –OH is bonded to a carbon that is part of a benzene ring
(4) aldehyde; C=O is a carbonyl bonded to a C and an H

Objective Section 13.2 Constitutional Isomerism in Alcohols

Are you able to…
- identify the constitutional isomers possible in alcohols (Exercise 13.40)
- use the molecular formula of an alcohol to determine if it has single bonds, double bonds, or a ring (Exercises 13.6, 13.39)
- draw the constitutional isomers of a given alcohol from its molecular formula (Ex. 13.2, Prob. 13.2, Exercises 13.3, 13.4, 13.38)
- determine the general formula of an alcohol from structural drawings (Ex.13.3, Prob. 13.3, Exercise 13.5)

Problem 13.2 How many alcohol structures are possible for the molecular formula $C_5H_{12}O$? Draw one structural formula for each isomer.

Eight alcohol isomers are possible for the molecular formula $C_5H_{12}O$:

$CH_3CH_2CH_2CH_2CH_2OH$

OH
|
$CH_3CH_2CH_2CHCH_3$

OH
|
$CH_3CH_2CHCH_2CH_3$

CH_3
|
$CH_3CH_2CHCH_2OH$

OH
|
$CH_3CH_2CCH_3$
|
CH_3

OH
|
$CH_3CHCHCH_3$
|
CH_3

CH_3
|
CH_3CCH_2OH
|
CH_3

$HOCH_2CH_2CHCH_3$
|
CH_3

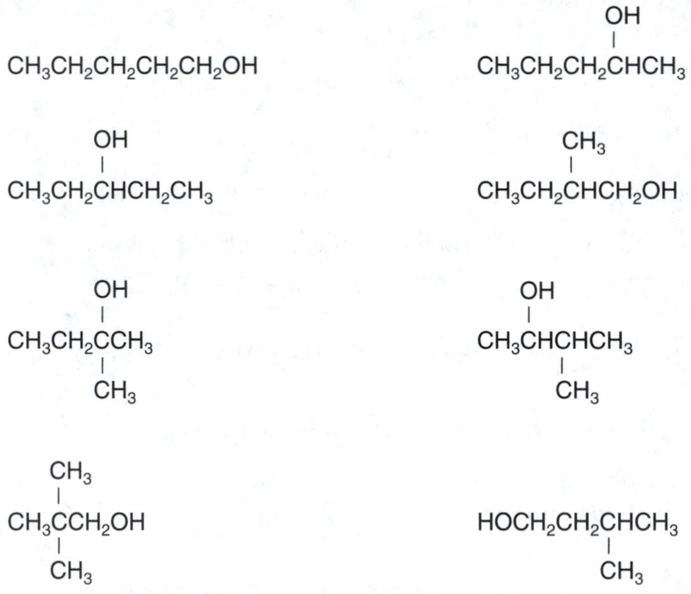

Problem 13.3 Show that alcohols 1 and 2 follow the general formulas $C_nH_{2n+2}O$ and $C_nH_{2n}O$, respectively.
(1) $C_5H_{12}O \Rightarrow n = 5, 2n+2 = 12$ (2) $C_5H_{10}O \Rightarrow n = 5, 2n = 10$

Objective Section 13.3 Classifying and Naming Alcohols

Are you able to …
- classify alcohols as primary, secondary, or tertiary (Ex. 13.4, Prob. 13.4, Exercises 13.7, 13.8)
- name alcohols by using IUPAC rules (Table 13.2, Ex. 13.5, Prob. 13.5, Exercise 13.7)
- draw the structure of an alcohol from its IUPAC name (Exercises 13.8, 13.41)
- identify polyfunctional alcohols

Problem 13.4 Classify the following alcohols as 1°, 2°, or 3° alcohols.

(a) –OH attached to C attached to three other C's, 3°

(b) –OH attached to C attached to one other C, 1°

(c) –OH attached to C attached to two other C's, 2°

(d) –OH attached to C attached to two other C's, 2°

Problem 13.5 Name the following compounds by using IUPAC rules.

(a) 2,5-Dimethyl-3-hexanol

Alcohol

OH

Base chain

$CH_3 - CH - CH - CH_2 - CH - CH_3$

C1

CH_3

C_3

CH_3

Methyl

Methyl

(b) *cis*-2-*t*-Butyl-cyclohexanol

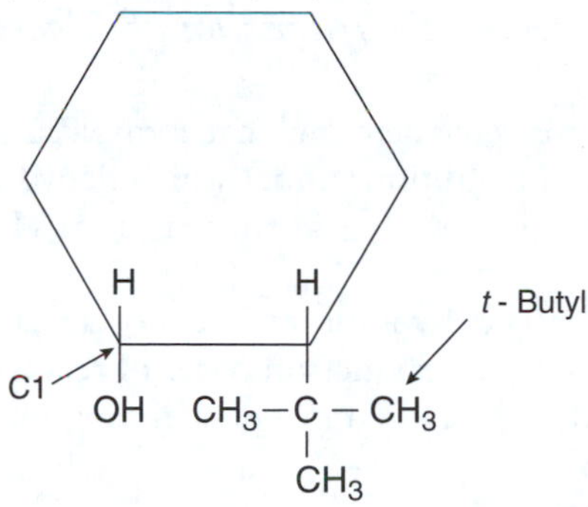

(c) 3-Methyl-1,2-butanediol

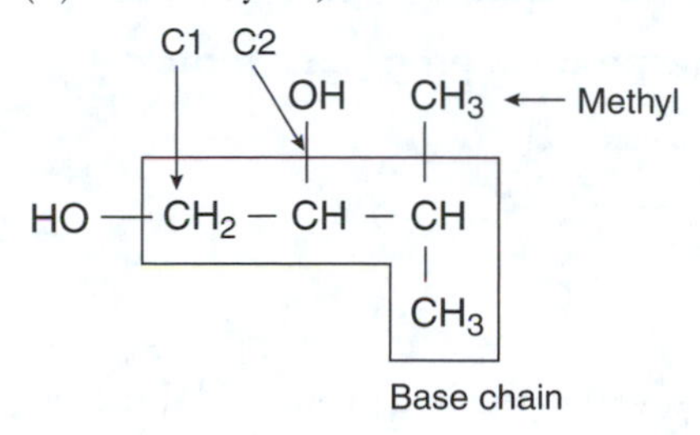

(d) 4-Isopropyl-1-heptanol

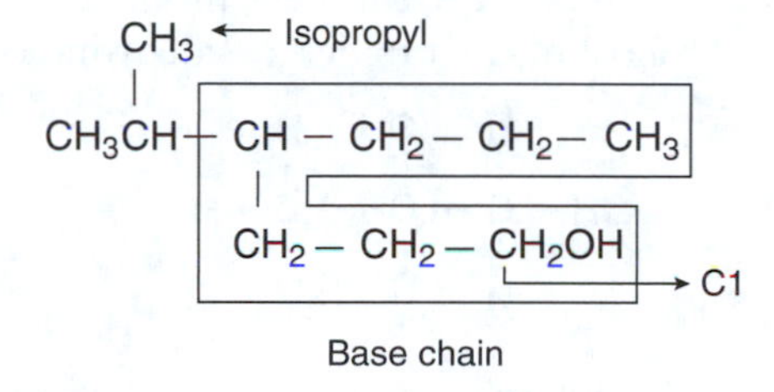

Objectives Section 13.4 Physical Properties of Alcohols

Are you able to...

- describe the structural features of alcohols (Exercises 13.35, 13.36, 13.37, 13.55, 13.56)
- explain how hydrogen bonding takes place between alcohol molecules (Fig. 13.2, Exercises 13.10, 13.43)
- explain the physical properties of alcohols in regard to secondary forces (Ex. 13.6, Prob. 13.6, Exercises 13.9, 13.42, 13.44, Expand Your Knowledge 13.59)
- explain the relative differences in physical properties between alcohols and hydrocarbons (Table 13.3, Ex. 13.6, Prob. 13.6)
- explain the solubility trends of alcohols in water (Fig. 13.2, Exercise 13.42)

Problem 13.6

(a) 1,2-Ethanediol has the higher boiling point because the presence of two –OH groups and a shorter carbon chain allows for greater hydrogen bonding between molecules than is seen in propanol, which has only one –OH group per molecule and a longer carbon chain.

(b) Butanol has a higher boiling point than *t*-butanol because the shape of *t*-butanol approximates a sphere, whereas the straight chain approximates a cylinder. The cylindrical surfaces attract each other over a larger area, so both hydrogen bonding and weak secondary forces operate to a greater extent.

(c) Butane is a nonpolar hydrocarbon and is subject to the same type of secondary attractive forces as is hexane molecules, so it is more soluble in hexane. Propanol's secondary forces include hydrogen bonds, which are not compatible with weaker London forces.

Objective Section 13.5 Acidity and Basicity of Alcohols

Are you able to...

- describe the amphoteric nature of alcohols
- explain why alcohols are weak acids (Exercises 13.11, 13.14)
- explain why alcohols are weak bases (Exercises 13.12, 13.13)
- write acid-base reactions for alcohols (Exercises 13.13, 13.14)

Objective Section 13.6 Dehydration of Alcohols to Alkenes

Are you able to...

- explain how alcohols undergo dehydration to form alkenes (Box 13.4, Expand Your Knowledge 13.60, 13.61)
- describe the reaction conditions under which dehydration takes place (Exercises 13.15, 13.16)
- predict the products of dehydration reactions (Exs. 13.7, 13.8, Probs. 13.7, 13.8, Exercises 13.15, 13.16)
- write equations for the dehydration of alcohols (Exs. 13.7, 13.8, Probs. 13.7, 13.8, Exercise 13.16)
- rate possible dehydration products in order of relative stability
- explain why a product is favored in a dehydration reaction (Exs. 13.7, 13.8, Probs. 13.7, 13.8, Exercises 13.15, 13.16)

Problem 13.7 Which of the following alcohols undergoes dehydration in the presence of a strong acid catalyst? If dehydration takes place, show the product.

(a) 2-Propanol is a 2° alcohol, so it undergoes dehydration. Only one product, propene, is formed.

$$H_3C-CH(OH)-CH_3 \xrightarrow{H^+} H_2C=CH(CH_3) + H_2O$$

(b) 1-Butanol is a 1° alcohol, so it undergoes dehydration. There is only one possible product, 1-butene.

$$H_3C-CH_2-CH_2-CH_2OH \xrightarrow{H^+} H_3C-CH_2-CH=CH_2 + H_2O$$

(c) *t*-Butyl alcohol is a tertiary alcohol and each adjacent carbon has a hydrogen that can react with the –OH group. There is only one possible product, 2-methyl-1-propene.

$$(CH_3)_3C-OH \xrightarrow{H^+} (CH_3)_2C=CH_2 + H_2O$$

Problem 13.8 Which of the following alcohols undergoes dehydration in the presence of a strong acid catalyst? If dehydration takes place, show the product.

(a) 3-Pentanol is a 2° alcohol and undergoes dehydration. There is only one product, 2-pentene.

$$H_3C-CH_2-CH(OH)-CH_2-CH_3 \xrightarrow[-\,H_2O]{H^+} H_3C-CH=CH-CH_2-CH_3$$

trans-2-Pentene predominates over *cis*-2-pentene

(b) 1-Methylcyclopentanol is a tertiary alcohol that undergoes dehydration to produce either 2-cyclopentyl-1-ethene or 1-methylcyclopentene. 1-methylcyclopentene is the major product because it is the most highly substituted alkene.

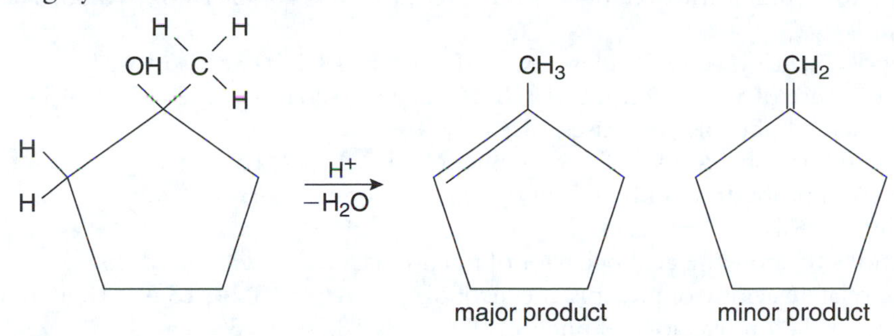

Objective Section 13.7 Oxidation of Alcohols

Are you able to...

- explain how oxidation takes place in alcohols
- identify alcohols that undergo selective oxidation (Exercises 13.17, 13.18)
- identify the conditions under which selective oxidation of alcohols takes place (Ex. 13.9, Prob. 13.9)
- predict the products of selective oxidation of alcohols (Ex. 13.9, Prob. 13.9, Exercises 13.46, 13.47)
- write reactions for the selective oxidation of alcohols (Ex. 13.9, Prob. 13.9)
- predict the products of the combustion of an alcohol (Exercise 13.48)
- write reactions for the combustion of an alcohol (Exercise 13.48)
- explain how oxidation reactions can be used to identify alcohols
- identify unknown alcohols on the basis of the results of oxidation reactions (Ex. 13.10, Prob. 13.10, Exercises 13.19, 13.20, 13.49, 13.50, 13.51, 13.52, 13.53, 13.54, Expand Your Knowledge 13.57)

Problem 13.9 Which of the following alcohols undergo oxidation with dichromate or permanganate? Show the product if oxidation takes place.

(a) 2-Propanol is a 2° alcohol and undergoes selective oxidation to form a ketone.

(b) Phenylmethanol is a 1° alcohol and will undergo selective oxidation first to an aldehyde and then to a carboxylic acid.

(c) *t*-Butyl alcohol is a 3° alcohol and does not undergo oxidation.

Problem 13.10 An unknown compound is either 2-butanol or 1-butanol. When a few drops of dichromate solution (orange) are added to it, the mixture turns green. What is the identity of the unknown? Explain your answer.

2-Butanol is a 2° alcohol and 1-butanol is a 1° alcohol. Both undergo oxidation with dichromate, so this test cannot be used to distinguish between the two.

Objective Section 13.8 Phenols

Are you able to...

- identify the structural features of phenols (Exercises 13.35, 13.36, 13.37, 13.55, 13.56)
- identify enols
- name phenols by using IUPAC rules (Ex. 13.11, Prob. 13.11, Exercise 13.21)
- draw the structure of a phenol from its IUPAC name (Exercises 13.22, 13.38, 13.41)
- identify isomers of phenols (Exercises 13.38, 13.40)
- describe the nature of secondary forces in phenols (Ex. 13.12, Prob. 13.12)
- explain physical properties of phenols in regard to secondary forces
- explain why phenols are weak acids
- write reactions to show the acid behavior of phenols (Exercises 13.23, 13.25)
- discuss the relative acidity of phenols and alcohols (Exercises 13.24, 13.45, 13.49)
- describe the oxidation reactions of phenols (Exercises 13.26, 13.53, 13.54)
- predict the products of the combustion of phenols (Exercise 13.25)
- write reactions for the combustion of phenols (Exercise 13.25)

Problem 13.11 Name the following compounds by using IUPAC rules.

(a) 4-Butyl-3-*t*-butylphenol

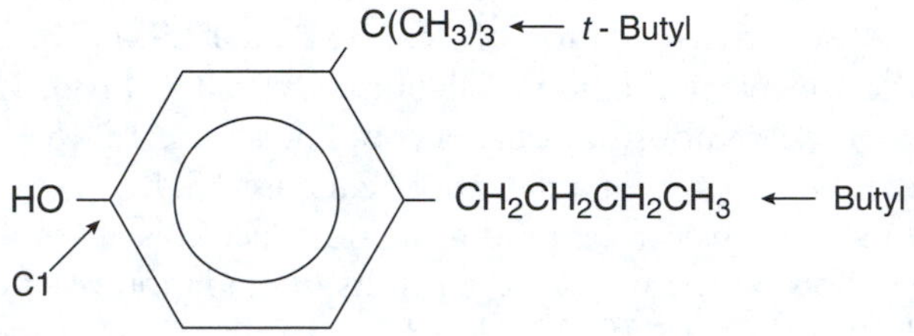

$C(CH_3)_3$ ← *t*-Butyl

HO — C1 — $CH_2CH_2CH_2CH_3$ ← Butyl

(b) 3-Bromo-2-methylphenol

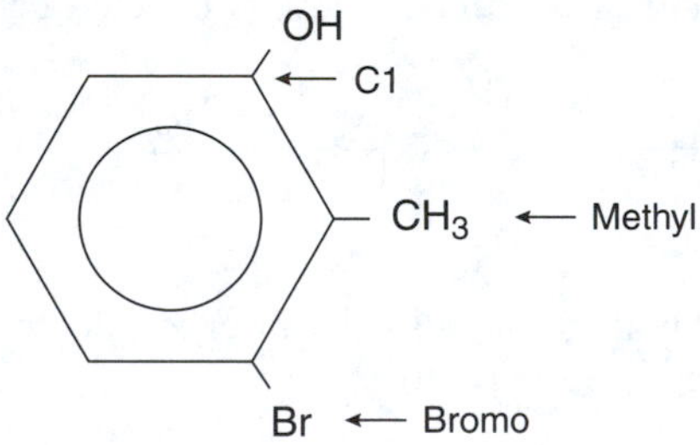

OH ← C1

CH_3 ← Methyl

Br ← Bromo

Problem 13.12 Which of the following representations correctly describe the hydrogen bonding that is responsible for the moderate solubility of phenol in water?

1. Incorrect; shows attraction between two partially positive areas
2. Correct; shows attraction between a partial negative area and a partial positive area
3. Correct; shows attraction between a partial negative area and a partial positive area
4. Incorrect; shows attraction between two partially negative areas

Objectives Section 13.9 Ethers

Are you able to...

- identify the structural features of ethers (Exercises 13.35, 13.36, 13.37, 13.55, 13.56)
- name ethers by using common names and IUPAC rules (Ex. 13.13, Prob. 13.13)
- draw the structure of an ether from its IUPAC name (Exercises 13.27, 13.41)
- identify alkoxy and aryloxy groups
- describe the nature of secondary forces in ethers (Ex. 13.15)

- explain trends in physical properties on the basis of secondary forces (Prob. 13.15, Exercises 13.29, 13.30)
- identify the reactions in which ethers participate (Expand Your Knowledge 13.62, 13.63)
- write an equation for the combustion of an ether (Exercise 13.48)
- predict the products of the combustion of an ether (Exercise 13.48)
- identify isomers of ethers (Exercise 13.40)
- draw constitutional isomers of ethers (Ex. 13.14, Prob. 13.14, Exercise 13.28, 13.38, 13.39)

Problem 13.13 Name the following ethers.
(a) Butyl *t*-butyl ether

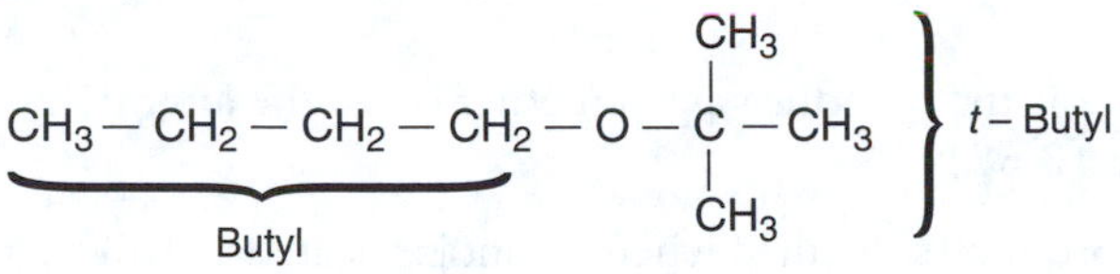

$CH_3-CH_2-CH_2-CH_2-O-C(CH_3)_2-CH_3$

Butyl; *t*-Butyl

(b) 3-Butoxy-3-methyl-1-butene (butyl group on one side and 3-methyl-1-butene on other side)

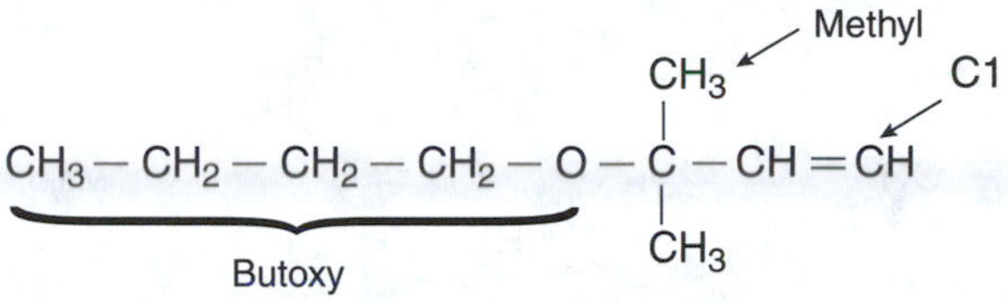

$CH_3-CH_2-CH_2-CH_2-O-C(CH_3)(CH_3)-CH=CH$

Butoxy; Methyl; C1

(c) *s*-Butyl isopropyl ether

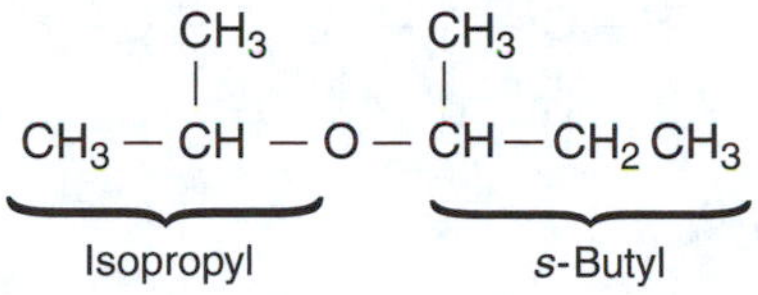

$CH_3-CH(CH_3)-O-CH(CH_3)-CH_2CH_3$

Isopropyl; *s*-Butyl

Problem 13.14 Draw structural formulas of the alcohols that are isomeric with the ethers in Example 13.14. $C_4H_{10}O$ is the molecular formula.

1-Butanol: $CH_3CH_2CH_2CH_2OH$

2-Butanol or *s*-butanol: $CH_3CH_2CH(OH)CH_3$

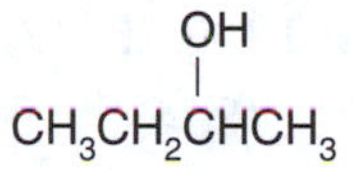

Isobutyl alcohol: $HO-CH_2-CH(CH_3)-CH_3$

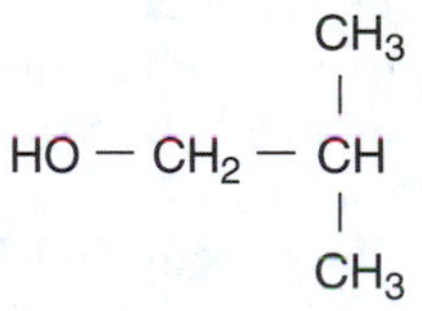

t-Butyl alcohol: $CH_3-C(CH_3)_2-OH$

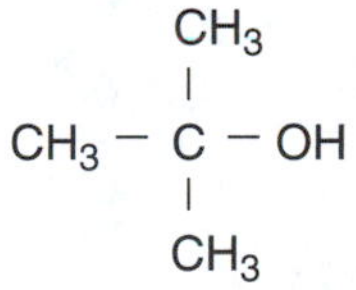

Problem 13.15 Indicate which compound in each of the following pairs has the higher value of the specified property. If two compounds in the pair have nearly the same value, indicate that fact. Explain your answers.
(a) *p*-Methylphenol has higher boiling point. Alcohols can hydrogen bond to each other, whereas ethers have weaker dipole attractions between them.
(b) Both have polar areas but will dissolve in heptane because the ratio of nonpolar to polar areas is large and about the same for the two compounds. However, dipropyl ether is more soluble than 1-hexanol in heptane because these two compounds attract by weak London forces. The secondary forces in 1-hexanol are strong hydrogen-bonding attractions; there are only very weak attractions between 1-hexanol and heptane.

(c) Both are only moderately soluble in water because the effects of the nonpolar areas about equal the effects of the polar areas. 1-Butanol may have slightly greater solubility in water because it can form stronger hydrogen bonds with water than diethylether.

Objectives Section 13.10 Formation of Ethers by Dehydration of Alcohols

Are you able to...

- explain how ethers are formed from the dehydration reaction of alcohols
- describe the conditions under which ether is formed (Ex. 13.16, Prob. 13.16)
- write dehydration reactions for alcohols (Ex. 13.16, Prob. 13.16, Exercise 13.31)
- predict the major and minor products of dehydration of alcohols (Ex. 13.16, Prob. 13.16, Exercise 13.31)
- identify the reactant alcohols that produce a given ether or alkene after dehydration (Exercise 13.32, Expand Your Knowledge 13.58)

Problem 13.16 Give the product(s) formed when ethanol is heated at 180°C in the presence of an acid catalyst. Indicate the major and minor products if more than one product is formed.

Ethanol is a primary alcohol. At 180°C, the major product is the alkene ethene. The ether diethylether is a minor product.

```
    H   H                       H         H
    |   |                        \       /
 H— C — C — OH     180°C          C = C              Major
    |   |         ------>        /       \
    H   H                       H         H

                             H   H       H   H
                             |   |       |   |
                          H— C — C — O— C — C —H    Minor
                             |   |       |   |
                             H   H       H   H
```

Objectives Section 13.11 Thiols

Are you able to...

- identify the structural features of thiols (Exercise 13.35)
- explain the relation between thiols and alcohols
- describe the physical properties of thiols
- identify the chemical reactions in which thiols participate
- write equations for reactions of thiols (Ex. 13.17, Prob. 13.17, Exercise 13.34)
- predict the products formed from thiol oxidation, reduction, and reaction with heavy metals (Ex. 13.17, Prob. 13.17, Exercise 13.34)
- name thiols by using IUPAC rules
- draw the structure of a thiol from its IUPAC name (Exercise 13.33)

Problem 13.17 Write equations for each of the following reactions.

(a) neutralization of 1-propanethiol with NaOH produces water and the sodium salt of the thiol: $CH_3CH_2CH_2S^-Na^+ + H_2O$.

```
    H   H   H                             H   H   H
    |   |   |                             |   |   |
 H— C — C — C — SH + NaOH   ------>   H— C — C — C — S⁻ Na⁺ + H₂O
    |   |   |                             |   |   |
    H   H   H                             H   H   H
```

(b) The reduction of diethyl disulfide produces two molecules of ethanethiol.

```
    H   H           H   H                  H   H               H   H
    |   |           |   |                  |   |               |   |
 H— C — C — S — S — C — C —H   (H)    H— C — C — SH  +  HS — C — C —H
    |   |           |   |     ----->       |   |               |   |
    H   H           H   H                  H   H               H   H
```

(c) The reaction of 1-propanethiol with Hg^{2+} produces the salt $CH_3CH_2CHS–Hg–SCHCH_2CH_3$

```
    H   H   H                         H   H   H                       H   H   H
    |   |   |                         |   |   |                       |   |   |
H - C - C - C - S - H + Hg2+ ----> H - C - C - C - S - Hg - S - C - C - C - H + 2H+
    |   |   |                         |   |   |                       |   |   |
    H   H   H                         H   H   H                       H   H   H
```

MAKING CONNECTIONS

Draw concept maps or write a paragraph describing the relation among the components of each of the following groups:

(a) alkane, alcohol, ether
(b) phenol, alcohol, ether
(c) primary alcohol, secondary alcohol, tertiary alcohol
(d) *sp*³ carbon, *sp*³ oxygen, rotation
(e) *sp*³ carbon, *sp*³ oxygen, molecular geometry
(f) water, alcohol, ether, phenol
(g) hydrogen bond, alcohol, solubility
(h) hydrogen bond, ether, phenol, solubility
(i) alcohol, ether, phenol, boiling point
(j) alcohol, ether, phenol, secondary forces
(k) oxidation, primary alcohol, secondary alcohol, tertiary alcohol
(l) dehydration, alcohol, alkene, ether
(m) isomer, alcohol, ether
(n) acidity, alcohol, phenol
(o) oxidation, permanganate, dichromate
(p) combustion, alcohol, ether, phenol
(q) thiol, water, alcohol
(r) thiol, oxidation, reduction, disulfide
(s) thiol, salts, heavy metals

FILL-INS

The introduction of oxygen into a hydrocarbon ring produces new _____. Oxygen may be _____ or _____ bonded to carbon. When oxygen is singly bonded to either one or two carbons, three functional groups are possible: _____. Replacing the oxygen with a sulfur produces the _____ family. Alcohols, characterized by _____ group, can be classified, according to the type of carbon to which the _____ is bonded. Saturated carbons bonded to one other carbon are _____ and alcohols with the –OH group bonded to them are called _____. In _____ the –OH groups are attached to secondary carbons, and in _____ the –OH groups are attached to tertiary carbons. There are no quaternary alcohols, because that would imply that carbon was bonded to _____ atoms, which is not possible because carbon _____. Alcohols, phenols, and ethers are structurally similar to water because _____. The hydrogen on water has been replaced with _____ group, given the symbol _____. The polarity of –OH bonds in alcohols _____ their polarity in water, and they both can form _____ with each other. This bonding makes short-chain or low molecular-mass alcohols _____ in water. As the hydrocarbon portion of the alcohol increases, however, the water solubility _____ and the molecule moves into the realm of _____. Phenols also can form _____ with water owing to the even greater _____ of their –OH groups. Small ethers have a _____ area that can form a hydrogen bond with water, but larger ethers are _____. Compuonds having any of these three functional groups undergo combustion reactions that produce _____. Alcohols are dehydrated in the presence of _____ to produce _____, depending on reaction conditions. At higher temperatures, _____ are the major products and, at relatively lower temperatures, _____ are the major products. The chemical reactivity of the three different classes of alcohols is _____. Reaction with oxidizing agents such as _____ and _____ provides a means for distinguishing between alcohols and ethers _____. Permanganate or dichromate do not oxidize _____ alcohols but do oxidize _____ alcohols. Primary alcohols produce _____ on oxidation, and secondary alcohols produce _____. Thiols are reversibly oxidized to _____ which are readily _____ back to thiols. Thiols also produce _____ when reacted with heavy metals.

Answers

The introduction of oxygen into a hydrocarbon ring produces new functional groups. Oxygen may be singly or doubly bonded to carbon. When oxygen is singly bonded to either one or two carbons, three functional groups are possible: alcohols, ethers, and phenols. Replacing the oxygen with a sulfur produces the thiol family. Alcohols, characterized by an –OH group, can be classified according to the type of carbon to which the hydroxyl group is bonded. Saturated carbons bonded to one other carbon are primary, and alcohols with the –OH group bonded to them are called primary alcohols. In secondary alcohols the –OH groups are attached to secondary carbons, and in tertiary alcohols the –OH groups are attached to tertiary carbons. There are no quaternary alcohols, because that would imply that carbon was bonded to five atoms, which is not possible because carbon is tetravalent. Alcohols, phenols, and ethers are structurally similar to water because the –OH of the alcohol resembles the –OH of water. The hydrogen on water has been replaced with an aliphatic or aromatic carbon-containing group, given the symbol R. The polarity of –OH bonds in alcohols is similar to their polarity in water, and both can form hydrogen bonds with each other. This bonding makes short-chain or low molecular-mass alcohols soluble in water. As the hydrocarbon portion of the alcohol increases, however, the water solubility decreases, and the molecule moves into the realm of nonpolarity. Phenols also can form hydrogen bonds with water owing to the even greater polarity of their –OH groups. Small ethers have a polar C–O–C area that can form a hydrogen bond with water, but larger ethers are nonpolar. Compunds having any of these three functional groups undergo combustion reactions that produce carbon dioxide and water. Alcohols are dehydrated in the presence of acid to produce alkenes or ethers, depending on reaction conditions. At higher temperatures, alkenes are the major products and, at relatively lower temperatures, ethers are the major products. The chemical reactivity of the three different classes of alcohols is different. Reaction with oxidizing agents such as permanganate and dichromate provides a means for distinguishing between alcohols and ethers because certain alcohols will be oxidized but ethers will not. Permanganate or dichromate do not oxidize tertiary alcohols but do oxidize primary and secondary alcohols. Primary alcohols produce aldehydes and carboxylic acids on oxidation, and secondary alcohols produce ketones. Thiols are reversibly oxidized to disulfides which are readily reduced back to thiols. Thiols also produce insoluble salts when reacted with heavy metals.

TEST YOURSELF

1. Write the general formula for:
   (a) a primary alcohol (b) a secondary alcohol (c) a tertiary alcohol
   (d) an ortho-substituted phenol (e) an ether with an aliphatic group and an aromatic group
   (f) an aliphatic thiol (g) an aliphatic disulfide
2. Draw a reaction map for alcohols. Include combustion, dehydration, and oxidation reactions.
3. Explain why the following molecules are suitable for their typical use:

Molecule	Use
1,2,3-Propanetriol (glycerol)	Moistening agent
2-Propanol	Rubbing alcohol to treat fever
Diethyl ether	Solvent
Ethanethiol	Added to natural gas to detect leaks

4. Draw the structures for the following molecules and arrange them in order of increasing boiling point: a three-carbon primary alcohol; a three-carbon ether; a three-carbon alkane. Explain your answer.
5. Draw the structures of the following molecules and arrange them in order of increasing water solubility: a three-carbon, primary alcohol; a three-carbon ether; a three-carbon diol; a three-carbon triol; a three-carbon alkane. Explain your answer.
6. Methanol is toxic, sometimes lethal, if ingested. Methanol is oxidized with the help of enzymes. Draw the structure of methanol and follow it through oxidation to explain why methanol ingestion can be accompanied by a life-threatening lowering of blood pH.
7. Compare the oxidation products of methanol and ethanol.
8. The application of diethyl ether to the skin causes dryness. Do you think that the dryness is caused by water evaporating from the skin or by loss of skin oils?

Answers

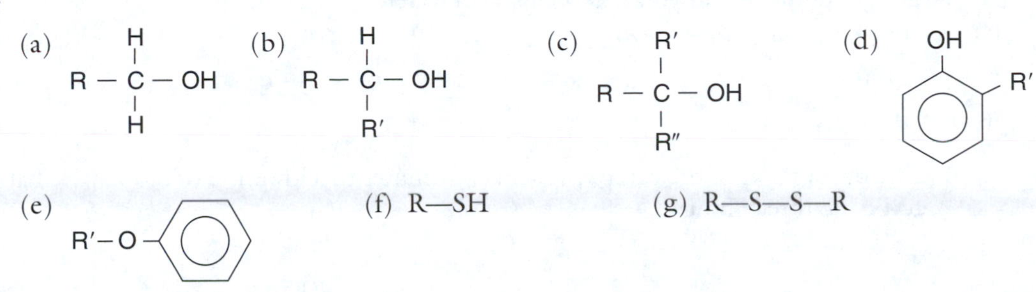

(e) R′–O–(benzene ring) (f) R—SH (g) R—S—S—R

2.

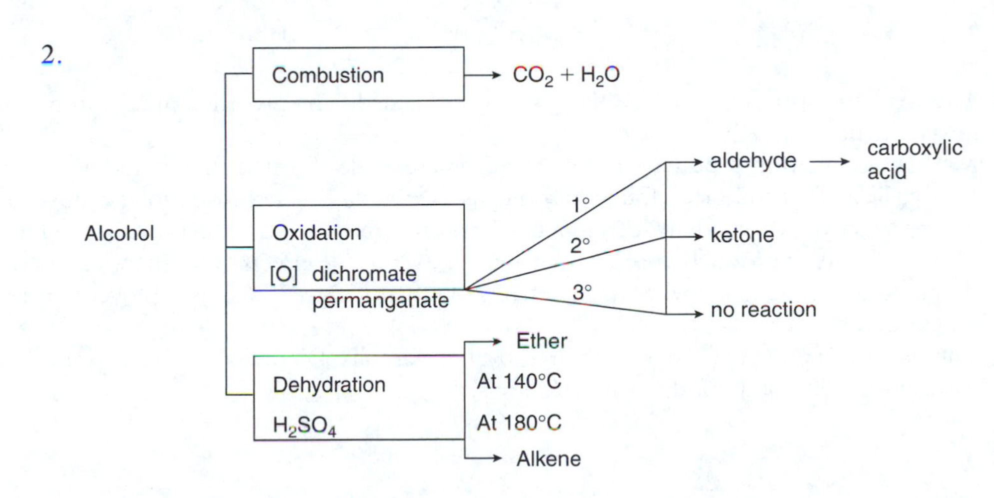

3. (a) Glycerol, with three –OH groups, can form many hydrogen bonds with water. It can attract and hold water to a greater degree and is soluble owing to the –OH : C ratio (3 C: 3–OH).
   (b) 2-Propanol evaporates rapidly from the skin. The evaporation requires overcoming secondary forces in the alcohol to move it from the liquid to the gas phase. The energy to do this comes from the heat produced by the body; so, as the alcohol evaporates, the heat is "carried away."
   (c) Diethyl ether is chemically inert, which makes it an ideal solvent.
   (d) Thiols have strong, unpleasant odors that are easily detected at very low concentrations.
4. As the magnitude of secondary forces between molecules in the liquid phase increases, the boiling point increases. The alkane, with only weak London forces, has the lowest boiling point. The ether has a slightly higher boiling point owing to the presence of stronger dipole forces between molecules. Ether molecules, however, cannot form strong attractions with other ether molecules; so the boiling point is still low, though relatively higher than that of

the alkane. The alcohol has a much higher boiling point than that of the alkane or the ether because of hydrogen bonds between molecules.

```
    H   H   H            H   H        H            H   H   H
    |   |   |            |   |        |            |   |   |
H — C — C — C — H  <  H — C — C — O — C — H  <  H — C — C — C — OH
    |   |   |            |   |        |            |   |   |
    H   H   H            H   H        H            H   H   H
```

5. Water solubility depends on the ability of a molecule to form hydrogen bonds with water, which requires the presence of polar bonds. It also requires that the polar part of a molecule dominate the nonpolar part. The alkane, with no polar bonds, cannot form hydrogen bonds and is soluble. The ether can form a hydrogen bond and has a relatively small nonpolar part; it is more soluble than the alkane but less soluble than the alcohol. The primary alcohol is soluble, but the diol and triol are more soluble owing to their abilities to form greater numbers of hydrogen bonds with water.

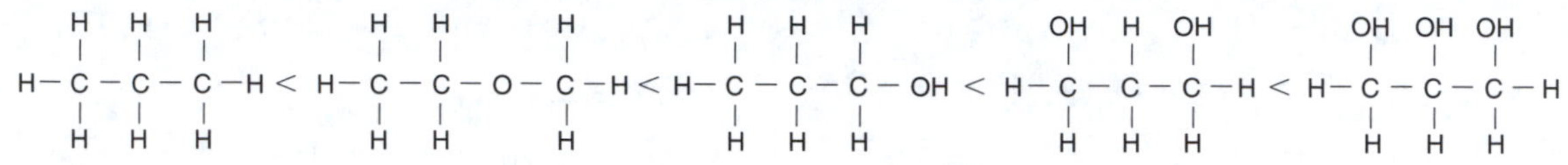

6. The production of the carboxylic acid contributes to a lowering of blood pH that is enough to be life threatening.

```
    H                   O                  O
    |                   ‖                  ‖
H — C — OH   —[O]→      C      —[O]→       C
    |                  / \                / \
    H                 H   H              H   OH
```

The oxidation product of CH_3OH is a carboxylic acid. The presence of acid in blood will lower the blood pH.

7. Methanol oxidation products are a one-carbon aldehyde (formaldehyde) and a one carbon carboxylic acid (formic acid; found in some insect stings). Ethanol oxidation produces a two-carbon aldehyde (acetaldehyde) and a two-carbon carboxylic acid (acetic acid; used in vinegar). Both sets of products are harmful to the body, but the effects of methanol ingestion are dramatic and immediate, whereas the effects of ethanol ingestion show up over longer periods of time.
8. The dryness is most likely due to the removal of skin oils. Diethyl ether is a solvent that dissolves nonpolar compounds such as skin oils.

chapter 14

Aldehydes and Ketones

OUTLINE

14.1 Structure of Aldehydes and Ketones

A. Aldehydes and ketones contain a carbonyl group.
   1. carbonyl structure is C=O
   2. C atom of C=O is called carbonyl carbon
   3. in aldehydes, the C atom of the carbonyl group is bonded to another carbon atom and to a hydrogen atom
   4. in ketones, the C atom of the carbonyl group is bonded to two other carbon atoms
   5. carbonyl structure is produced when an alcohol loses both an H atom on an –OH and an H atom on an adjacent C atom
   6. carbonyl C and O atoms are both sp^2 hybridized (Figure 14.1); produces trigonal geometry around C=O bond
   7. carbonyl structure contains strong C–O sigma bond and weaker C–O pi bond

B. When an atom other than C or H is attached to a carbonyl carbon, other organic families are produced.
   1. carbon : hydrogen ratio is lowered by presence of a double bond (Table 14.1)
   2. aldehydes and ketones have same C : H ratio as that of alkenes: $C_nH_{2n}O$

C. Aldehydes and ketones are isomers of one another.
   1. for aldehydes, there are only constitutional isomers based on carbon skeleton; the C=O is always on the last carbon atom in the chain
   2. for ketones, both constitutional isomers based on carbon skeleton and based on placement of the C=O are possible

Key Terms aldehyde, carbonyl, heteroatom, ketone

Example 14.1 Classification of molecules based on functional groups
Problem 14.1 Classification of molecules based on functional groups
Example 14.2 Drawing isomers of aldehydes and ketones
Problem 14.2 Drawing isomers of aldehydes and ketones

14.2 Naming Aldehydes and Ketones

A. Aldehydes and ketones are named in a manner similar to that of alkanes

IUPAC Rules for Naming Aldehydes and Ketones

1. Identify the longest continuous carbon chain that contains the carbonyl group.
2. The chain is numbered starting at the end closest to the C=O.
   a. For aldehydes, C1 is always the C=O carbon.
   b. For ketones, the number of the carbonyl C is put before the alkanone name.
3. The ending of the alkane is changed to -al if C=O is an aldehyde or -one if C=O is a ketone.
4. Substituents are located with a number and identified; the names are prefixes, preceded by the carbon location number, in front of the base name.
5. Cyclic compounds that have C=O are cycloalkanones.
   a. C number 1 is always the C=O carbon.
   b. Substituents are numbered to give the lowest possible numbering.
6. Benzene ring that contains an aldehyde is called benzaldehyde.

B. Some aldehydes and ketones have common names that are used (Table 14.3).
   1. Simple aldehydes have common prefixes attached to the -aldehyde suffix

Prefix	Number of C's in chain
form-	1
acet-	2
propion-	3
butyr-	4

   2. A ketone is named by identifying the R group attached to the C=O and placing its name before the word ketone; used only when R group is simple

Key Term cycloalkanone

Example 14.3 Naming aldehydes and ketones by using IUPAC rules
Problem 14.3 Naming aldehydes and ketones by using IUPAC rules
Example 14.4 Naming aldehydes and ketones by their common names
Problem 14.4 Naming aldehydes and ketones by their common names

14.3 Physical Properties of Aldehydes and Ketones

A. Properties of short-chain aldehydes and ketones are dominated by the properties of the carbonyl group.
   1. carbonyl group is highly polarized
      a. carbon is partially positive
      b. oxygen is partially negative
   2. carbonyl group has resonance hybrid forms that are a blend of ionic and covalent
   3. carbonyl group polarity results in strong dipole–dipole secondary forces (Figure 14.2)

B. Physical properties of aldehydes and ketones are between those of alcohols and alkanes (Table 14.4).
   1. boiling and melting points are significantly higher than those of alkanes of similar size (Table 14.4); dipole attractions are stronger than London forces
   2. boiling points and melting points are significantly lower than those of similar alcohols (Table 14.4); dipole attractions are weaker than hydrogen bonds
   3. solubility in water is similar to that of alcohols; aldehydes and ketones can form hydrogen bonds with water

C. As fraction of molecule that is C=O becomes smaller with increasing molecular mass, nonpolar properties dominate and differences in boiling points among alkanes, alcohols, aldehydes, and ketones diminish.

Key Terms dipole–dipole force, resonance hybrid

Example 14.5 describing hydrogen bonding in aldehydes and ketones

Problem 14.5 Predicting relative values of physical properties of organic molecules

14.4 Oxidation of Aldehydes and Ketones

A. Aldehydes and ketones undergo combustion to produce carbon dioxide and water.
B. Ketones do not undergo oxidation.
C. Aldehydes undergo oxidation to produce carboxylic acids; selective oxidation with permanganate and dichromate occurs.
D. Tollens's and Benedict's tests distinguish aldehydes from ketones and from primary and secondary alcohols and alkenes.
  1. Tollens's reagent produces a silver mirror in the presence of an aldehyde
    a. contains silver ion in ammonia
    b. requires basic conditions
  2. Benedict's test produces a red-orange precipitate in the presence of α-hydroxy aldehydes and α-hydroxy ketones
    a. blue solution of Cu^{2+} ion with citrate
    b. requires basic conditions
    c. α refers to C next to C=O carbon; hydroxy means that it has an –OH (hydroxyl group)
    d. this test can be used to detect glucose and some other sugars

Key Terms α-hydroxy aldehyde, α-hydroxy ketone, Benedict's reagent, Tollens's reagent

Example 14.6 Identification of reactants that will undergo oxidation and prediction of product
Problem 14.6 Identification of reactants that will undergo oxidation and prediction of product
Example 14.7 Identification of reactants that will react with Tollens's and Benedict's reagents
Problem 14.7 Identification of reactants that will react with Tollens's and Benedict's reagents

14.5 Reduction of Aldehydes and Ketones

A. Aldehydes and ketones can be reduced to alcohols.
  1. aldehydes are reduced to primary alcohols
  2. ketones are reduced to secondary alcohols
  3. reduction adds one hydrogen to the C and one hydrogen to the O of the carbonyl group
  4. process is the reverse of oxidation of alcohols to aldehydes and ketones (Section 13.7)
B. Catalytic hydrogenation
  1. molecular hydrogen, H_2, is reducing agent
  2. requires Ni or Pt metal catalysts
  3. pi bond of C=O breaks, and H is added to both C and O
  4. Also reduces C=C
C. Hydride reduction
  1. metal hydride such as $LiAlH_4$ is reducing agent
  2. requires water as second reagent after metal hydride has reacted with aldehyde or ketone
  3. if water and metal hydride added simultaneously, metal hydride reacts with water so will not reduce C=O
  4. will not reduce C=C
D. Hydride reduction in living cells uses NADH as reducing agent in two-step process catalyzed by the enzyme alcohol dehydrogenase.
  1. first step is addition of H: (from NADH) to C of C=O; breaks double bond; produces negative charge on carbonyl O
  2. second step is addition of H^+ (from H_2O) to negative O to produce –OH group

Key Terms catalytic hydrogenation, catalytic reduction, hydride reduction, NADH

Example 14.8 Identification of reactants that will undergo reduction and prediction of product
Problem 14.8 Identification of reactants that will undergo reduction and prediction of product

14.6 Hemiacetal and Acetal Formation by Reaction with Alcohol

A. Aldehydes and ketones react with alcohols in a two-step process to produce hemiacetals and acetals.

1. requires acid catalyst
2. first step produces the hemiacetal
   a. alcohol (R–OH) adds –O–R to the carbonyl C and adds H to the carbonyl O
   b. two different oxygens bonded to the same carbon, –OH and –OR
3. second step produces the acetal
   a. –OH group is removed from former carbonyl C (H^+ reacts with it to produce H_2O), and it is replaced with –O–R
   b. two different oxygens bonded to same carbon; –OR and –OR
4. hemiacetals are generally unstable; acetals are stable

B Intramolecular reactions can form stable cyclic hemiacetals.
1. formed when one molecule has both alcohol and carbonyl groups
2. H^+ is taken from alcohol group and attached to carbonyl oxygen

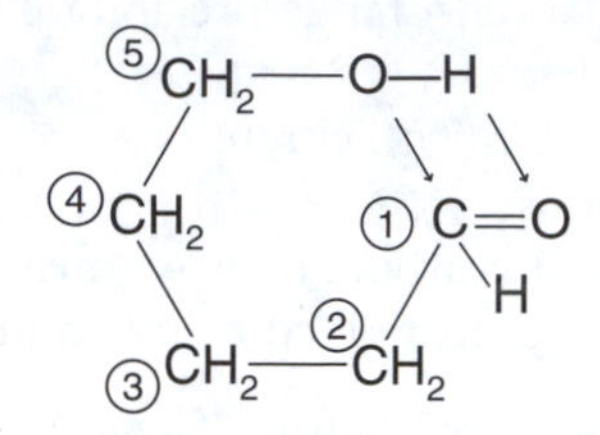

3. O from alcohol group is joined to carbonyl carbon, forming ring

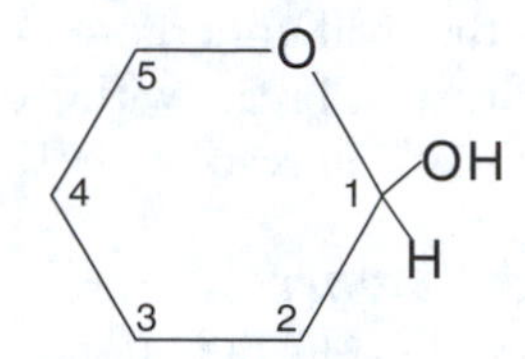

4. five- and six-membered rings are especially stable
5. these stable rings are important in carbohydrate chemistry

C. Hydrolysis reaction with water cleaves acetal to produce aldehyde or ketone and alcohol.
1. reaction is reverse of acetal formation
2. requires excess water and acid catalyst

Key Terms acetal, hemiacetal, hemiketal, hydrolysis, intermolecular reaction, intramolecular reaction, ketal

Example 14.9 Drawing hemiacetal and acetal structures
Problem 14.9 Drawing hemiacetal and acetal structures
Example 14.10 Drawing cyclic hemiacetals formed in intramolecular reactions
Problem 14.10 Drawing cyclic hemiacetals formed in intramolecular reactions
Example 14.11 Identification of structures as hemiacetals or acetals
Problem 14.11 Identification of structures as hemiacetals or acetals
Example 14.12 Prediction of products of hemiacetal and acetal hydrolysis
Problem 14.12 Prediction of products of hemiacetal and acetal hydrolysis

ARE YOU ABLE TO... AND WORKED TEXT PROBLEMS

Objectives Section 14.1 Structure of Aldehydes and Ketones

Are you able to...
- discuss the structure of the carbonyl group (Figure 14.1, Exercises 14.7, 14.8, 14.29)
- discuss the polarity of the carbonyl group
- distinguish between aldehydes and ketones (Table 14.1, Ex. 14.1, Prob. 14.1)
- discuss the molecular formulas of aldehydes and ketones relative to other organic families (Table 14.2, Exercises 14.1, 14.2)

- identify isomers of aldehydes and ketones (Exercises 14.6, 14.32)
- draw isomers of aldehydes and ketones (Ex. 14.2, Prob. 14.2, Exercises 14.3, 14.4, 14.5, 14.31)

Problem 14.1 Classify each of the following compounds as an aldehyde, a ketone, a carboxylic acid, an ester, or an amide.

(a) Ester; the carbonyl carbon atom is bonded to a C atom and to an O atom
(b) Aldehyde; the carbonyl carbon atom is bonded to a C atom and to an H atom
(c) Ketone; the carbonyl carbon atom is bonded to a C atom and to a C atom (follow the bonds on either side of the carbonyl C)
(d) Amide; the carbonyl carbon atom is bonded to a C atom and to an N atom

Problem 14.2 Draw the structural formula for each aldehyde and ketone with the molecular formula $C_5H_{10}O$.

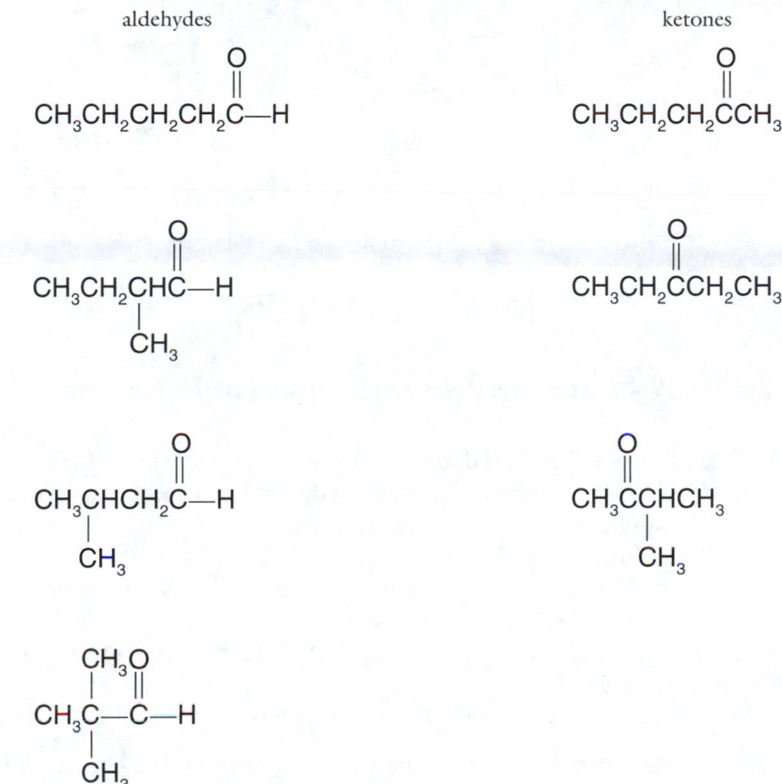

Objectives Section 14.2 Naming Aldehydes and Ketones

Are you able to...

- name aldehydes and ketones by using IUPAC rules (Table 14.3, Exs. 14.3, 14.4, Probs. 14.3, 14.4, Exercises 14.11, 14.12, 14.34)
- draw structures of aldehydes and ketones from their IUPAC names (Exercises 14.9, 14.10, 14.33)
- name aldehydes and ketones by their common names

Problem 14.3 Name the following compounds by using IUPAC rules.

(a) 2,5-Dimethyl-3-hexanone

$CH_3CH—C—CH_2CHCH_3$

methyl→ CH_3 O CH_3 ←methyl

↑ketone

(b) 4,4-Dimethylpentanal

methyl→ CH_3 O

$CH_3CCH_2CH_2—CH$

CH_3 ↑aldehyde

↑methyl

(c) 3-Chlorocyclopentanone

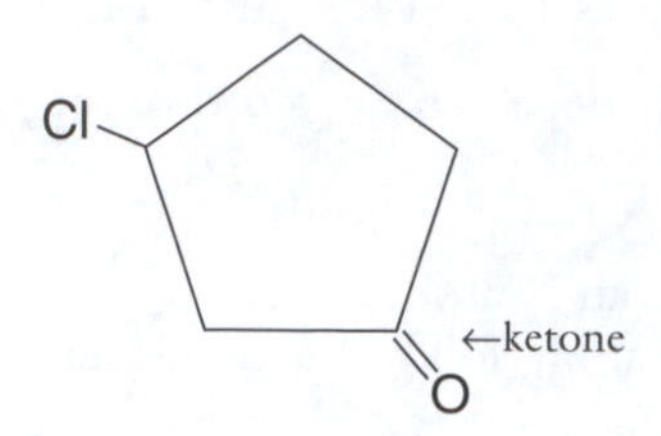

(d) 2-Bromo-4-methylbenzaldehyde

Problem 14.4 Give the common name for each of the following ketones.

(a) *t*-Butylcyclohexyl ketone

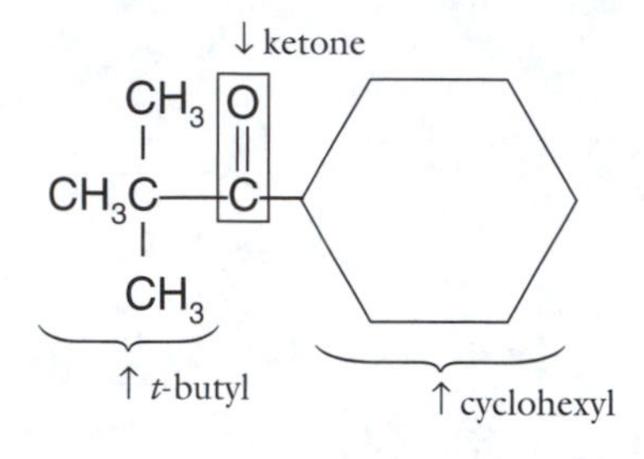

(b) *m*-Chlorophenylpropyl ketone

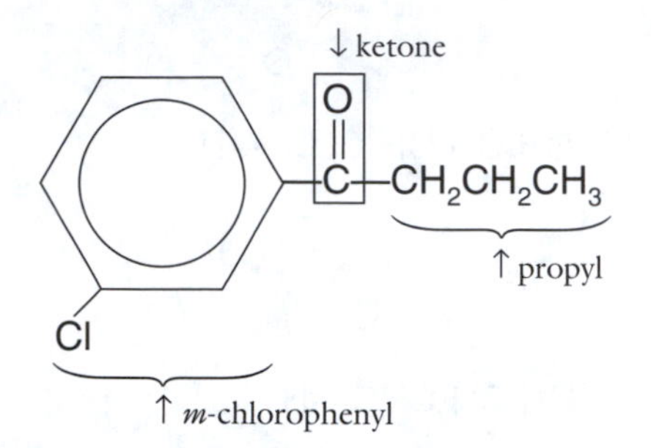

Objectives Section 14.3 Physical Properties of Aldehydes and Ketones

Are you able to...

- explain how the properties of carbonyl groups determine the properties of small aldehydes and ketones
- discuss the formation of dipole–dipole attractions in aldehydes and ketones (Figure 14.2)
- discuss how melting and boiling points of aldehydes and ketones compare with those of alcohols, ethers, and alkanes (Ex. 14.5, Prob. 14.5, Exercises 14.13, 14.35)
- discuss the relative trends in melting and boiling points of aldehydes and ketones on the basis of secondary forces (Table 14.4, Exercise 14.36)
- explain the solubility of aldehydes and ketones relative to that of alcohols and alkanes (Ex. 14.5, Prob. 14.5, Exercises 14.13, 14.35, 14.37)
- describe how aldehydes and ketones form hydrogen bonds with water (Exercises 14.4, 14.38)

Problem 14.5 Indicate which compound in each of the following pairs has the higher value of the specified property. Indicate if values are similar. Explain your reasoning.

(a) Hexanal has a higher boiling point than that of butanal. Boiling points increase with increasing molecular mass within a family because the longer chain provides greater surfaces over which London forces operate.

(b) 1-Propanal is more soluble in water than is pentanal. The solubility of a molecule depends on the presence of polar groups. A rule of thumb is that one polar group makes three carbons water soluble. The ratio is correct for 1-propanal, but nonpolar forces begin to dominate in pentanal.

(c) 2-Propanol has the higher boiling point because the magnitude of the strength of the hydrogen bonds between molecules requires more energy to separate them.

(d) Butanal and butanol have about the same solubility in water because both can form hydrogen bonds to water to almost the same extent.

Objectives Section 14.4 Oxidation of Aldehydes and Ketones

Are you able to...

- predict the products of the combustion of aldehydes and ketones
- predict the products of the oxidation of aldehydes (Ex. 14.6, Prob. 14.6, Exercise 14.15, 14.44, 14.45, 14.46, 14.49, 14.50, Expand Your Knowledge 14.53)
- write reactions for the oxidation of aldehydes (Ex. 14.6, Prob. 14.6, Exercise 14.15)

- identify the reaction conditions under which oxidation of aldehydes takes place
- identify the reaction conditions used for Benedict's and Tollens's tests
- explain the utility and limitations of Benedict's and Tollens's tests (Expand Your Knowledge 14.51)
- predict the products of reactions with Benedict's and Tollens's reagents (Ex. 14.7, Prob. 14.7, Exercises 14.16, 14.17)
- identify unknowns based upon the results of oxidation, Benedict's and Tollens's tests (Exercises 14.18, 14.19, 14.20, 14.39, 14.40, 14.48)

Problem 14.6 Which of the following compounds undergo mild oxidation with dichromate or permanganate? Show the product if oxidation takes place.
(a) Cyclopentanone is a ketone; no reaction takes place.
(b) Pentanal is an aldehyde that is oxidized to the carboxylic acid:

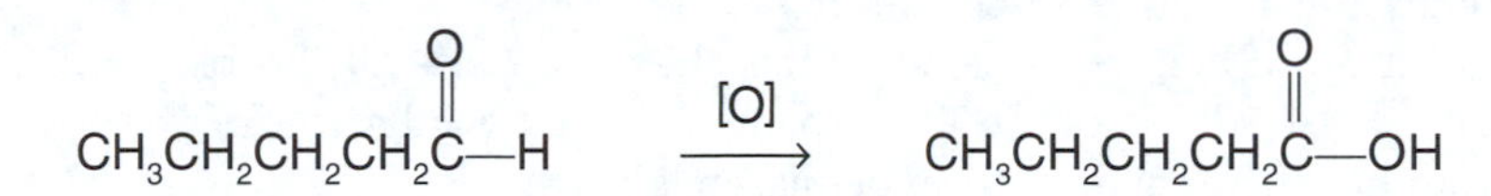

Problem 14.7 For each of the following compounds, indicate whether it gives a negative or positive test with both Tollens's and Benedict's reagents.
(a) α-Hydroxy ketone; gives positive Benedict's test and negative Tollens's test.
(b) Aldehyde with no α-hydroxy group; gives a positive Tollens's test and a negative Benedict's test.
(c) Ketone; gives negatives for both Tollens's and Benedict's tests.
(d) α-Hydroxy aldehyde; gives positives for both Tollens's and Benedict's tests.

Objectives Section 14.5 Reduction of Aldehydes and Ketones

Are you able to...
- predict the products of reduction reactions of aldehydes and ketones (Exercises 14.24, 14.41, Expand Your Knowledge 14.54)
- explain the relation between the reduction of aldehydes and ketones and the oxidation of alcohols
- explain the process of catalytic hydrogenation
- identify the reaction conditions under which catalytic hydrogenation takes place
- predict the products of the catalytic reduction of aldehydes and ketones (Ex. 14.8, Prob. 14.8, Exercise 14.22)
- explain the process of hydride reduction (Exercise 14.23)
- identify the reaction conditions under which hydride reduction takes place
- predict the products of hydride reduction of aldehydes and ketones (Ex. 14.8, Prob. 14.18, Exercises 14.21, 14.47)
- explain how hydride reduction takes place with NADH in living cells

Problem 14.8 Write an equation for each of the following reactions.
(a) Reduction of cyclohexanone by using $LiAlH_4$ produces cyclohexanol:

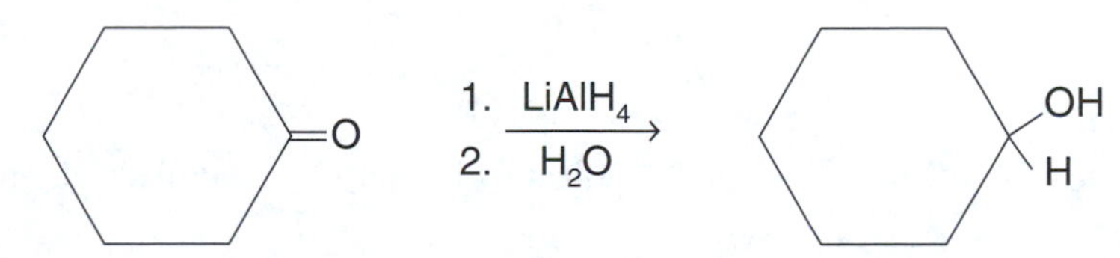

(b) Catalytic reduction of benzaldehyde produces benzene with alcohol group attached:

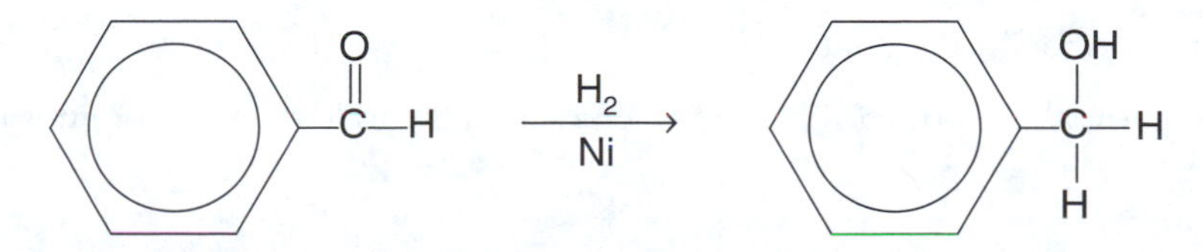

(c) Reduction of 2-pentanone with NADH produces 2-pentanol:

$$CH_3CH_2CH_2\overset{O}{\overset{\|}{C}}CH_3 \xrightarrow[2.\ H_2O]{1.\ NADH} CH_3CH_2CH_2\overset{OH}{\overset{|}{C}}HCH_3$$

Objectives Section 14.6 Hemiacetal and Acetal Formation by Reaction with Alcohol

Are you able to...

- identify hemiacetals and acetals (Exercise 14.26)
- describe the formation of hemiacetals and acetals (Exs. 14.9, 14.10, Probs. 14.9, 14.10, Exercises 14.25, 14.27, 14.42, 14.43, 14.44, 14.45, Expand Your Knowledge 14.55)
- explain how intramolecular reactions form hemiacetals
- identify the products of the hydrolysis of acetals (Exercise 14.28)

Problem 14.9 Show the structures of the hemiacetal and the acetal formed from the reaction of butanal with excess methanol.

The first step attaches an H to the carbonyl oxygen and the group $-OCH_3$ to the carbonyl carbon; this is the hemiacetal.

$$CH_3CH_2CH_2\overset{O}{\overset{\|}{C}}—H + CH_3OH \xrightarrow{H^+} —CH_3CH_2CH_2\underset{OCH_3}{\underset{|}{\overset{OH}{\overset{|}{C}}}}H$$

The second step replaces $-OH$ on the former carbonyl carbon with $-OCH_3$; this is the acetal.

$$CH_3CH_2CH_2\underset{OCH_3}{\underset{|}{\overset{OH}{\overset{|}{C}}}}H + CH_3OH \xrightarrow{H^+} CH_3CH_2CH_2\underset{OCH_3}{\underset{|}{\overset{OCH_3}{\overset{|}{C}}}}H$$

Problem 14.10 Show the structure of the hemiacetal formed by the intramolecular reaction between the $-OH$ and carbonyl groups of 6-hydroxyl-2-hexanone.

$$CH_3\overset{O}{\overset{\|}{C}}CH_2CH_2CH_2\overset{OH}{\overset{|}{C}}H_2 \longrightarrow$$

O—CH_2 HO C CH_2 CH_3 CH_2—CH_2

Problem 14.11 For each of the compounds, indicate whether it is a hemiacetal, an acetal, or something else.

(a) Acetal; carbon is bonded to two different O's with same R group.
(b) Hemiacetal; carbon is bonded to two different O's; one an –OH and one an –OR.
(c) Neither.
(d) Acetal; carbon is bonded to two oxygens.

Problem 14.12 Show the product(s) formed when each of the following compounds undergoes acid-catalyzed hydrolysis.

(a) Each $-OCH_2CH_3$ group is removed and two molecules of CH_3CH_2OH are formed.

$$CH_3CH_2O{-}\underset{\displaystyle CH_2C_6H_5}{\underset{|}{CH}}{-}OCH_2CH_3 \xrightarrow[H_2O]{H^+} 2\ CH_3CH_2OH + \underset{\displaystyle CH_2C_6H_5}{\underset{|}{\overset{\displaystyle O}{\overset{\|}{C}}}}{-}H$$

(b) One molecule of CH_3OH is formed.

$$\text{(2-methoxytetrahydrofuran)} \xrightarrow[H_2O]{H^+} HO{-}CH_2{-}CH_2{-}CH_2{-}\underset{}{\overset{\displaystyle H}{\overset{|}{C}}}{=}O + CH_3OH$$

MAKING CONNECTIONS

Construct concept maps or write paragraphs describing the relation among the members of each of the following groups.

(a) carbonyl, pi bond, sp^2 hybridization
(b) carbonyl, aldehyde, ketone
(c) carbonyl, polarity, water solubility
(d) carbonyl, dipole attractions, physical properties
(e) dipole attractions, boiling point, melting point
(f) boiling point, alkane, alcohol, aldehyde, ketone
(g) isomers, aldehyde, ketone, molecular formula
(h) oxidation, aldehyde, ketone
(i) reduction, aldehyde, ketone
(j) hydrogen, reduction, aldehyde, ketone
(k) NADH, reduction, aldehyde, ketone
(l) NADH, reduction, cell
(m) alcohol, aldehyde, hemiacetal, acetal
(n) intramolecular hemiacetal formation, cyclic structure, aldehyde, alcohol
(o) hydrolysis, acetal, alcohol, aldehyde

FILL-INS

Aldehydes and ketones are similar in that they both contain the _____. They differ in the _____. In aldehydes, the carbonyl carbon is bonded to a _____. In ketones, the carbonyl carbon is bonded to _____. In ketones, the two _____ may be part of a cyclic structure. The presence of the carbonyl group imparts some _____ to aldehydes and ketones, which is manifested in the higher boiling and melting points of small-chain aldehydes and ketones relative to similar alkanes. The values are still considerably _____ than those of similar alcohols because the _____ between aldehyde and ketone molecules are not as strong as the _____ in alcohols. As the molecular masses of these molecules increase, however, the _____ character of the hydrocarbon chain dominates and the differences between them diminish. The polar carbonyl group can form bonds with water molecules, so small aldehydes and ketones are _____ as similar alcohols are. The chemistry of the carbonyl group determines the _____ of aldehydes and ketones. They can be burned in _____ reactions and produce carbon dioxide and water. The oxidation of aldehydes adds an _____ atom to the carbonyl carbon. This creates a _____. Ketones do not undergo oxidation, because there is _____ on the carbonyl carbon that can participate in the reaction. Both aldehydes and ketones can be reduced to _____. Aldehydes are reduced to _____ and ketones are reduced to _____. Reacting an aldehyde or a ketone with an alcohol produces, first, an unstable _____, which is converted into the more stable _____. _____ hemiacetal formation occurs in molecules such as glucose that contain both an _____ and an _____. The product is a _____ with a heteroatom, _____, incorporated into the ring.

Answers

Aldehydes and ketones are similar in that they both contain the carbonyl group, –C=O. They differ in the atoms to which the carbon of the carbonyl is attached. In aldehydes, the carbonyl carbon is bonded to a carbon and to a hydrogen. In ketones, the carbonyl carbon is bonded to two different carbon atoms. In ketones, the two different carbon atoms may be part of a cyclic structure. The presence of the carbonyl group imparts some polarity to aldehydes and ketones, which is manifested in the higher boiling and melting points of small-chain aldehydes and ketones relative to similar alkanes. The values are still considerably lower than those of similar alcohols because the dipole attractions between aldehyde and ketone molecules are not as strong as the hydrogen bonds in alcohols. As the molecular masses of these molecules increase, however, the nonpolar character of the hydrocarbon chain dominates and the differences between them diminish. The polar carbonyl group can form bonds with water molecules, so small aldehydes and ketones are about as soluble in water as similar alcohols are. The chemistry of the carbonyl group determines the reactivity of aldehydes and ketones. They can be burned in combustion reactions and produce carbon dioxide and water. The oxidation of aldehydes adds an additional oxygen atom to the carbonyl carbon. This creates a carboxylic acid. Ketones do not undergo oxidation, because there is no hydrogen on the carbonyl carbon that can participate in the reaction. Both aldehydes and ketones can be reduced to alcohols. Aldehydes are reduced to primary alcohols and ketones are reduced to secondary alcohols. Reacting an aldehyde or a ketone with an alcohol produces, first, an unstable hemiacetal, which is converted into the more stable acetal. Intramolecular hemiacetal formation occurs in molecules such as glucose that contain both an aldehyde and an alcohol group. The product is a cyclic structure with a heteroatom, O, incorporated into the ring.

TEST YOURSELF

1. Explain how the structures of aldehydes and ketones are related.
2. Write the general formula for:
   (a) the formation of a hemiacetal and an acetal by addition of an alcohol to an aldehyde.
   (b) the formation of an aldehyde and an alcohol by hydrolysis of an acetal.
   (c) the formation of a ketal by addition of an alcohol to a ketone.
   (d) the hydrogenation of a ketone.
   (e) the hydrogenation of an aldehyde.
3. Draw a reaction map for aldehydes and ketones. Include combustion, oxidation, reduction, and acetal formation reactions.
4. Explain the differences in boiling points for the following compounds. Compare the molecular masses of the compounds.

Compound	Boiling point (°C)
Butane	0
Propanone	56
1-Propanol	97

5. Draw the structures for the molecules in Question 4 and arrange them in order of increasing water solubility. Explain your answer.
6. Explain why hydrogen bonds do not form between aldehyde molecules and ketone molecules.

7. Glucose is a simple sugar that is used by the body as an energy source. It forms a cyclic hemiacetal through an intramolecular reaction. Show which groups on the glucose molecule participate in the reaction.

$$\mathrm{H{-}\overset{\overset{\large O}{\|}}{C}{-}\underset{\underset{\large OH}{|}}{\overset{\overset{\large H}{|}}{C}}{-}\underset{\underset{\large H}{|}}{\overset{\overset{\large OH}{|}}{C}}{-}\underset{\underset{\large OH}{|}}{\overset{\overset{\large H}{|}}{C}}{-}\underset{\underset{\large OH}{|}}{\overset{\overset{\large H}{|}}{C}}{-}\underset{\underset{\large OH}{|}}{\overset{\overset{\large H}{|}}{C}}{-}H}$$

8. Draw the generic structures of:
   (a) an aldehyde that gives a positive Benedict's test.
   (b) a ketone that gives a positive Benedict's test.
9. How could you distinguish between a ketone, an aldehyde, and an α-hydroxy aldehyde?

Answers

1. Both aldehydes and ketones contain a carbonyl group. The difference is onto which atoms the carbon of the carbonyl is bonded. In an aldehyde, it is bonded to another carbon and to hydrogen. In a ketone, it is bonded to two different carbons.
2. Write the general formula for:
   (a) the formation of a hemiacetal and an acetal by addition of an alcohol to an aldehyde

$$\mathrm{R'{-}O{-}H + R{-}\overset{\overset{\large O}{\|}}{C}{-}H \rightleftharpoons R{-}\underset{\underset{\large OR'}{|}}{\overset{\overset{\large OH}{|}}{C}}{-}H}$$

$$\mathrm{R{-}\underset{\underset{\large OR'}{|}}{\overset{\overset{\large OH}{|}}{C}}{-}H + R'O{-}H \rightleftharpoons R{-}\underset{\underset{\large OR'}{|}}{\overset{\overset{\large OR'}{|}}{C}}{-}H + H_2O}$$

   (b) the formation of an aldehyde and an alcohol by hydrolysis of an acetal

$$\mathrm{R{-}\underset{\underset{\large OR'}{|}}{\overset{\overset{\large OR'}{|}}{C}}{-}H + H{-}O{-}H \rightleftharpoons R{-}\overset{\overset{\large O}{\|}}{C}{-}H + 2R'OH}$$

   (c) the formation of a ketal by addition of an alcohol to a ketone

$$\mathrm{R^1{-}OH + \underset{R^2\ \ R^3}{\overset{\overset{\large O}{\|}}{C}} \rightleftharpoons R^2{-}\underset{\underset{\large OR^1}{|}}{\overset{\overset{\large OR^1}{|}}{C}}{-}R^3}$$

   (d) the hydrogenation of a ketone

$$\mathrm{R{-}\overset{\overset{\large O}{\|}}{C}{-}R' \xrightarrow[Pt]{H_2} R{-}\underset{\underset{\large H}{|}}{\overset{\overset{\large O{-}H}{|}}{C}}{-}R'}$$

(e) the hydrogenation of an aldehyde

3.

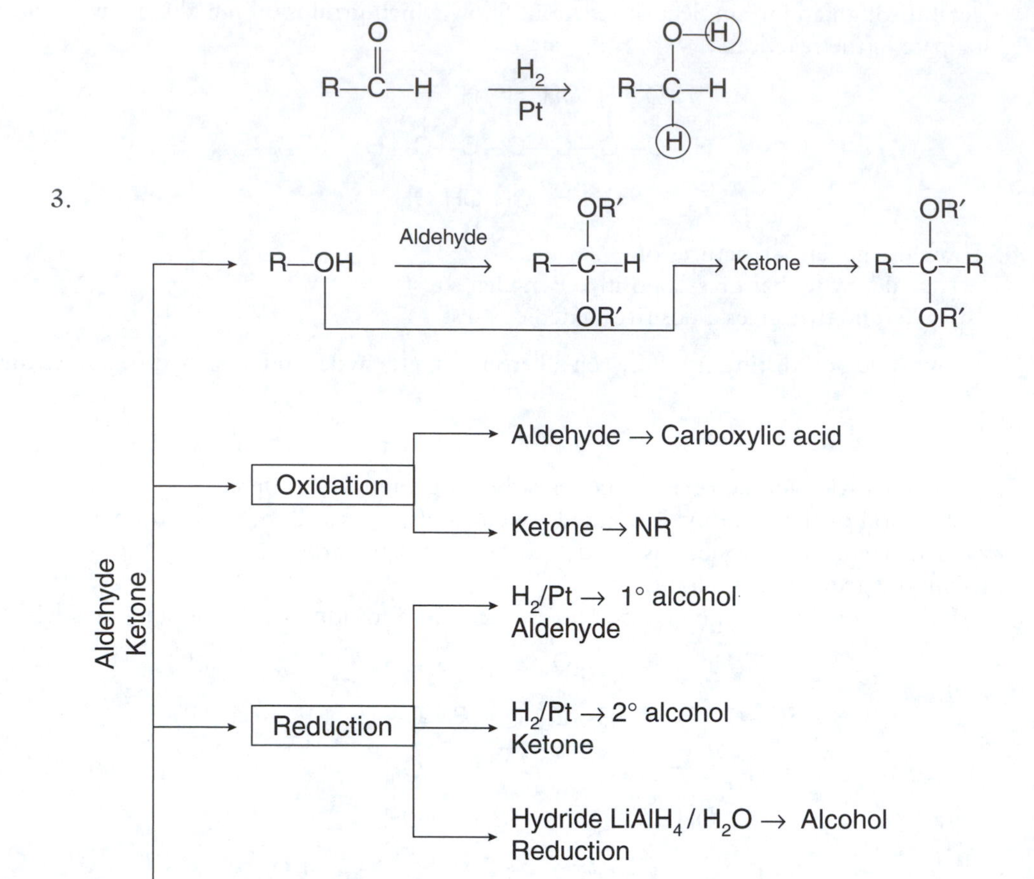

Combustion → Produces → $CO_2 + H_2O$

4. The molecular masses are the same for each of the compounds: 58 g/mol. The boiling point of butane is the lowest, indicating the weakest attractions between molecules. Butane is a hydrocarbon, and only weak London forces operate between molecules, which is highlighted by the fact that butane is a gas at room temperature. Propanone has dipole–dipole interactions between molecules. The strength of these attractions relative to London forces means that it takes more energy to separate these ketones than to separate alkanes of similar weight. Propanone is a liquid at room temperature, emphasizing the strength of the secondary dipole attractions. Propanol is also a liquid at room temperature and has a much higher boiling point than propanone owing to the strong hydrogen bonding between alcohol molecules.
5. As the polarity of a molecule increases, its water solubility increases. The water solubility also increases as the ability to form hydrogen bonds increases. Butane is not polar and cannot form hydrogen bonds with water; it is be insoluble. Propanone has the polar carbonyl group that can form a hydrogen bond with a water molecule; it is soluble. Propanol's secondary forces are hydrogen bonds, just like those of water, and it is soluble in water.

$CH_3CH_2CH_2CH_3$ < $CH_3\overset{O}{\overset{\|}{C}}CH_3$ < $CH_3CH_2CH_2OH$

butane: no polar bonds — propane: polar carbonyl — 1-propanol: polar –O–H forms hydrogen bonds

6. Aldehyde molecules and ketone molecules cannot form hydrogen bonds with each other, because the criterion for hydrogen-bond formation is not met: hydrogen must be bonded to either O, N, or F. The polarity of the carbonyl sets up dipole interactions, not hydrogen bonds.

7. 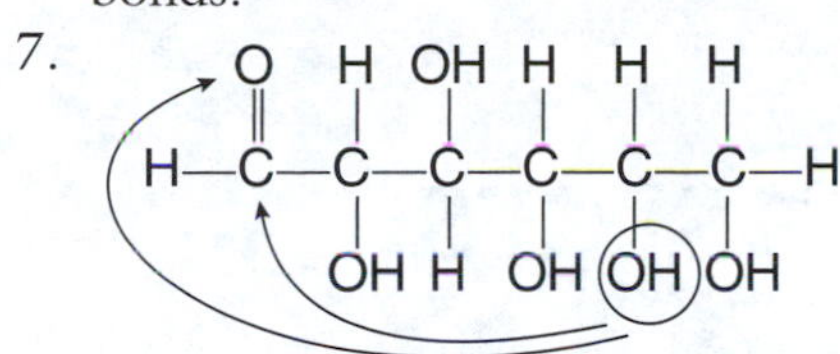

8. (a) An aldehyde that gives a positive Benedict's test is an α-hydroxyaldehyde. The carbonyl carbon is bonded to a carbon atom that is also bonded to an –OH group.

$\overset{OH}{|}\,\overset{O}{\|}$ $CH_2C—H$	$\overset{OH}{|}\,\overset{O}{\|}$ CH_2CCH_3
α–hydroxyaldehyde	α–hydroxyketone

(b) A ketone that gives a positive Benedict's test is an α-hydroxyketone. The carbonyl carbon is bonded to a carbon atom that is also bonded to an –OH group.

9. Benedict's test distinguishes between an aldehyde and an α-hydroxyaldehyde. Tollens's test distinguishes between aldehydes and ketones. Ketones will not react with either reagent.

chapter 15

Carboxylic Acids, Esters, and Other Acid Derivatives

OUTLINE

15.1 Carboxylic Acids and Their Derivatives Compared

A. Carboxylic acids contain a carbonyl group, C=O, in which the carbonyl carbon atom is also bonded to an –OH group.
   1. the carboxyl group

$$\overset{\displaystyle \text{O}}{\underset{\diagup}{\overset{\|}{\text{C}}}}\text{— OH}$$

   2. abbreviated –COOH, $-CO_2H$
   3. carboxylic acids are abbreviated R–COOH or RCO_2H

B. Derivatives of carboxylic acids are (Table 15.1):
   1. esters
      a. carbonyl carbon is bonded to a C atom and to an O atom
      b. esters are formed from reactions of carboxylic acids and alcohols
   2. acid anhydrides
      a. two carbonyl groups are bonded to the same O atom and each is bonded to another C atom
      b. acid anhydrides are formed by a dehydration reaction between two carboxylic acid molecules
   3. acid halides: carbonyl carbon is bonded to a C and to a halogen (X)
   4. amides: carbonyl carbon is bonded to a C and to an N
   5. all four can be synthesized from carboxylic acids and hydrolyzed back to carboxylic acids

C. Isomers of carboxylic acids contain both an alcohol and either a ketone or an aldehyde, but their properties are different.

Key Terms acid halide, amide, anhydride, carboxyl group, carboxylic acid, carboxylic acid derivative, ester

Key Table Table 15.1 –COOH's and their derivatives

Example 15.1 Classification of organic molecules
Problem 15.1 Classification of organic molecules

15.2 Synthesis of Carboxylic Acids

A. Carboxylic acids are synthesized by:
   1. selective oxidation of primary alcohols

2. selective oxidation of aldehydes
3. selective oxidation of an alkyl group on a benzene ring; produces benzoic acid

Key Term selective oxidation

Example 15.2 Writing oxidation equations for the synthesis of carboxylic acids
Problem 15.2 Writing oxidation equations for the synthesis of carboxylic acids

15.3 Naming Carboxylic Acids

A. Carboxylic acids are named by using IUPAC rules (Table 15.2).

IUPAC Rules for Naming Carboxylic Acids

1. Identify the longest chain that contains the carboxyl carbon. The carboxyl carbon is carbon number 1.
2. The ending -oic acid replaces the -e of the alkane name.
3. Substituents are identified and their location designated as with alkanes.
4. If a molecule contains two –COOH groups, it as an alkanedioic acid. The position of each –COOH must be specified and put in front of the acid name.
5. A benzene ring with a –COOH group is called benzoic acid. The ring C to which the –COOH is attached is C1.

B. Common names are often used for carboxylic acids (Table 15.2); prefixes are the same as for common aldehydes.
C. Carboxylic acid molecules that contain other functional groups are called:
1. hydroxycarboxylic acids: contain an additional hydroxyl group, –OH
2. ketocarboxylic acids: contain an additional carbonyl group, –C=O
3. aminocarboxylic acids: contain an amino group, $-NH_2$

Key Terms aminocarboxylic acid, benzoic acid, hydroxycarboxylic acid, ketocarboxylic acid
Key Table Table 15.2 IUPAC and common names of –COOH's

Example 15.3 Naming carboxylic acids by using IUPAC rules
Problem 15.3 Naming carboxylic acids by using IUPAC rules

15.4 Physical Properties of Carboxylic Acids

A. The combination of a doubly bonded O and a hydroxyl group bonded to the same carbon gives carboxylic acids their unique properties.
B. The carboxyl group has two polar groups.
1. carbonyl group is polar

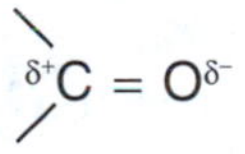

$$>{}^{\delta+}C = O^{\delta-}$$

2. hydroxyl group is polar

$$-{}^{\delta-}O—H^{\delta+}$$

3. the partial positive charge on C has an electron-withdrawing effect that stabilizes the ionized form, the carboxylate ion

C. Two polar groups result in:
1. strong hydrogen bonding between carboxylic acid molecules
   a. causes carboxylic acids to exist as dimers

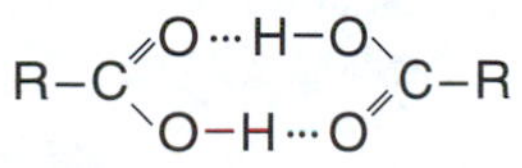

$$R-C\begin{matrix}\nearrow O\cdots H-O \searrow \\ \searrow O-H\cdots O \nearrow\end{matrix}C-R$$

b. causes melting and boiling points to be even higher than those of alcohols (Table 15.3)
2. three hydrogen-bonding sites with water molecules
a. solubility in water is slightly greater than that of alcohols
b. solubility decreases gradually as carbon chain increases

Key Term dimer
Key Table Table 15.3 Comparison of boiling points of compounds from different families

15.5 *Acidity of Carboxylic Acids*

A. Carboxylic acids are weak acids.
1. they undergo ionization in water to form carboxylate ions and hydronium ions

$$R—COOH + H_2O \rightleftharpoons R—COO^- + H_3O^+$$

a. about 1 of every 100 molecules is ionized in water
b. this is an equilibrium reaction
c. ionized form, carboxylates, predominates at physiological pH
2. they react with NaOH to form salt and water

$$R—COOH + NaOH \rightarrow R—COO^- Na^+ + H_2O$$

a. reaction goes to completion; not an equilibrium
b. salt formed is a carboxylate salt
3. they react with $NaHCO_3$ to produce salt and H_2CO_3

$$R—COOH + NaHCO_3 \rightarrow R—COO^- Na^+ + H_2CO_3$$

a. reaction goes to completion; not an equilibrium
b. salt formed is a carboxylate salt
c. H_2CO_3 decomposes to H_2O and CO_2

B. Relative acidities of acidic compounds (Table 15.4):
1. benzoic acid > acetic acid (2 C's) > phenol > ethanol = water
2. carboxylic acids will turn blue litmus paper red

Key Terms carboxylate ion, pH
Key Table Table 15.4 Relative acidities of organic compounds

Example 15.4 Writing ionization reactions of carboxylic acids and identifying conditions under which they ionize
Problem 15.4 Writing ionization reactions of carboxylic acids and identifying conditions under which they ionize

15.6 *Carboxylate Salts*

A. Carboxylate salt is formed from the reaction between a carboxylic acid and a strong base.

$$R—COOH + NaOH \rightarrow R—COO^- Na^+ + H_2O$$

1. carboxylate ion, $R–COO^-$ is the anion
2. base contributes the cation, M^+, or the ammonium ion, NH_4^+
3. carboxylate ions are weak bases
a. they can accept protons
b. they undergo acid-base reactions with water and strong acids
c. Carboxylate salt solutions turn red litmus paper blue

B. Carboxylate salts are ionic compounds.
1. have significantly higher melting and boiling points than those of the corresponding carboxylic acids

2. are solids at room temperature
3. are water soluble to a greater degree than the corresponding carboxylic acids

C. Carboxylate salts are named as follows:
1. positive ion is named first
2. the -ic ending of the parent carboxylic acid is changed to -ate for the carboxylate ion: cation carboxylate

Key Term carboxylate salt

Example 15.5 Naming carboxylate salts by using IUPAC rules
Problem 15.5 Naming carboxylate salts using IUPAC rules
Example 15.6 Predicting relative melting points of organic compounds
Problem 15.6 Predicting relative water solubility of organic compounds

15.7 Soaps and Their Cleaning Action

A. Soaps are carboxylate salts.
1. they contain long-chain carboxylate ions derived from fatty acids
2. they typically have from 12 to 20 aliphatic carbons in various unbranched mixtures

B. Soaps are produced from fats and oils in a process called saponification.

C. Soap molecules have both hydrophilic and hydrophobic areas (Box 15.3).
1. they are amphipathic
2. the hydrophobic part is the long hydrocarbon chain of the carboxylate salt; dirt and grease are nonpolar and the hydrophobic part can interact with them
3. the hydrophilic part is the $-COO^{-}Na^{+}$; water is polar and the hydrophilic part interacts with water

Key Terms amphipathic, fatty acid, hard water, hydrophilic, hydrophobic, saponification

Key Figure and Box
Figure 15.1 The cleaning action of soap
Box 15.3 Hard Water and Detergents

15.8 Esters from Carboxylic Acids and Alcohols

A. Esters are produced by reacting a carboxylic acid with an alcohol.
1. the process is called esterification
   a. –OH from carboxyl group combines with H from hydroxyl group of alcohol to produce water
   b. carbonyl C of the carboxyl fragment joins the hydroxyl O of the alcohol fragment to produce the ester

$$R-\overset{O}{\overset{\|}{C}}-OH \quad H-OR' \rightleftharpoons R-\overset{O}{\overset{\|}{C}}-O-R' + H_2O$$

↑ from carboxylic acid ↑ from alcohol

   c. this process is reversible and can be represented as an equilibrium
2. esterification requires an acid catalyst

B. Thioesters are produced by esterification of carboxylic acids and thiols.

Key Terms ester, esterification, thioester

Example 15.7 Predicting products and writing esterification reactions for carboxylic acids and alcohols
Problem 15.7 Predicting products and writing esterification reactions for carboxylic acids and alcohols

15.9 Names and Physical Properties of Esters

A. Esters are named as follows:
   1. the group that came from the alcohol is named first as a separate word
      a. this fragment, –O–R, is called an alkoxy or aryloxy group
      b. name it as if it were an alkyl substituent by looking at the R group
   2. the group that came from the carboxylic acid is named second
      a. this fragment is called the acyl group
      b. change the ending of the parent carboxylic acid from -oic to –ate: alcohol alkyl carboxylate

B. Physical properties of esters are due to the lower polarity of the molecule.
   1. slight polarity of O and C bonds allows hydrogen bonding with water; solubility is about equal to that of aldehydes and ketones
   2. polarity and secondary forces are lower than in aldehydes and ketones but higher than in alkanes; melting points and boiling points are relatively lower

Key Terms acyl, alkoxy, aryloxy

Example 15.8 Naming esters by using IUPAC rules
Problem 15.8 Naming esters by using IUPAC rules

15.10 Polyester Synthesis

A. Esterification can be used to produce polymers; called condensation polymerization because water is produced as a by-product

B. Reactants with two functional groups are monomers.
   1. they are called bifunctional reagents
   2. dicarboxylic acid and diol can combine to form a polyester; reaction is acid catalyzed

$$HO{-}\overset{\overset{O}{\Vert}}{C}{-}R{-}\overset{\overset{O}{\Vert}}{C}{-}O\boxed{H + HO}{-}R{-}OH \longrightarrow \left[{-}\overset{\overset{O}{\Vert}}{C}{-}R{-}\overset{\overset{O}{\Vert}}{C}{-}O{-}R{-}O{-}\right]_n$$

   3. a reactive functional group must be at each end of the molecule for the polymerization to continue

Key Terms bifunctional reactants, condensation polymerization, polyester, repeat unit

Example 15.9 Writing equations and predicting products for polyesterification reactions
Problem 15.9 Writing equations and predicting products for polyesterification reactions and esterification reactions

15.11 Hydrolysis of Esters

A. Esters can be broken down to component alcohol and carboxylic acid by hydrolysis
   1. hydrolysis is the reverse of esterification
   2. water is used to split a molecule, with the H- going to one part and the –OH to another part

$$R{-}\overset{\overset{O}{\Vert}}{C}{-}O{-}R' + H{-}OH \rightleftharpoons R{-}\overset{\overset{O}{\Vert}}{C}{-}\textcircled{OH} + \textcircled{H}{-}O{-}R'$$

b. Hydrolysis can take place in acidic or basic conditions
   1. acid hydrolysis requires a strong acid such as H_2SO_4

$$R{-}\overset{\overset{O}{\Vert}}{C}{-}O{-}R' + H{-}OH \overset{H^+}{\rightleftharpoons} R{-}\overset{\overset{O}{\Vert}}{C}{-}OH + H{-}OR'$$

      a. acid is a catalyst, not a reactant

b. products are the carboxylic acid and the alcohol
c. this is an equilibrium reaction

2. base hydrolysis requires a strong base such as NaOH

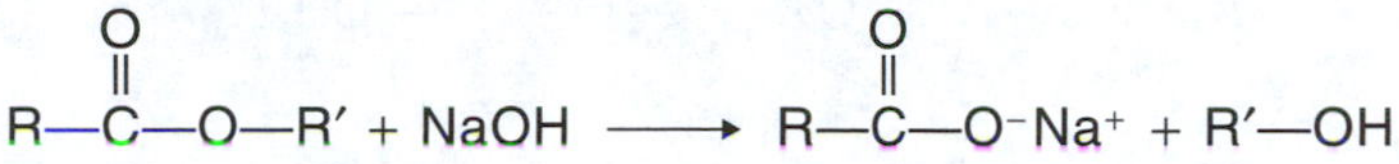

$$R{-}\overset{\overset{\displaystyle O}{\|}}{C}{-}O{-}R' + NaOH \longrightarrow R{-}\overset{\overset{\displaystyle O}{\|}}{C}{-}O^{-}Na^{+} + R'{-}OH$$

a. base is a reactant and is consumed in reaction
b. products are carboxylate salt and alcohol
c. base hydrolysis is not an equilibrium reaction
d. reaction is also called saponification

Key Terms acid hydrolysis, base hydrolysis, saponification

Example 15.10 Identifying conditions under which esters react
Problem 15.10 Writing equations and predicting products for reactions of esters

15.12 Carboxylic Acid Anhydrides and Halides

A. An acid anhydride is formed by dehydration reactions between two molecules of carboxylic acid

B. Carboxylic acid anhydrides have the generic formula $R{-}\overset{\overset{\displaystyle O}{\|}}{C}{-}O{-}\overset{\overset{\displaystyle O}{\|}}{C}{-}R$.

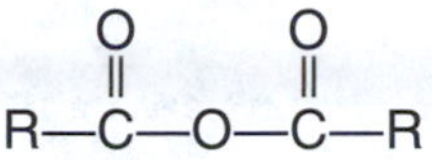

C. Acid halides have the generic formula $R{-}\overset{\overset{\displaystyle O}{\|}}{C}{-}Cl$.

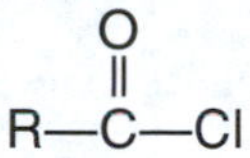

D. Both acid anhydrides and acid halides are very reactive toward water.

E. Acid anhydrides and acid halides act as acyl transfer agents; they transfer the RCO– (acyl) group to the O of an alcohol or phenol.

F. Carboxylic acid anhydrides are named by dropping the acid from the parent carboxylic acid and replacing it with anhydride.

G. Acid halides are named by changing the -ic acid of the parent carboxylic acid to -yl halide.

Key Terms acid anhydride, acid halides, acyl halides, acyl transfer agents, acyl transfer reactions

Example 15.11 Naming acid anhydrides and acid halides
Problem 15.11 Drawing structures from the names of acid anhydrides and acid halides
Example 15.12 Writing reactions and predicting products of acyl transfer reactions
Problem 15.12 Writing reactions and predicting products of acyl transfer reactions

15.13 Phosphoric Acids and Their Derivatives

A. The phosphoric acid family consists of:

1. phosphoric acid

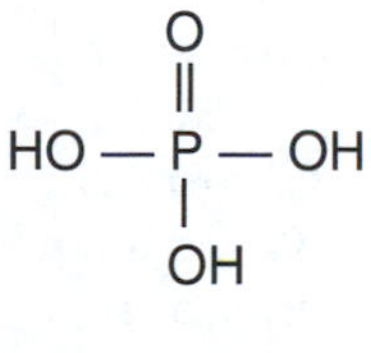

$$HO{-}\underset{\underset{\displaystyle OH}{|}}{\overset{\overset{\displaystyle O}{\|}}{P}}{-}OH$$

2. diphosphoric acid, an anhydride

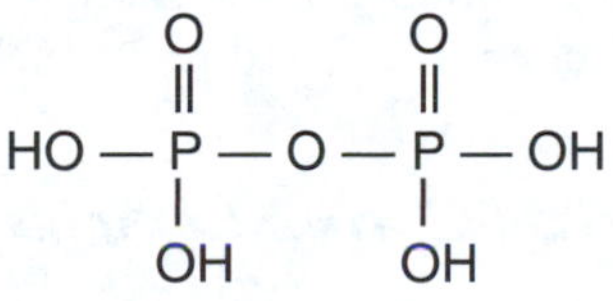

$$HO{-}\underset{\underset{\displaystyle OH}{|}}{\overset{\overset{\displaystyle O}{\|}}{P}}{-}O{-}\underset{\underset{\displaystyle OH}{|}}{\overset{\overset{\displaystyle O}{\|}}{P}}{-}OH$$

3. triphosphoric acid, an anhydride

```
       O        O        O
       ||       ||       ||
HO — P — O — P — O — P — OH
       |        |        |
       OH       OH       OH
```

B. The dehydration between two phosphoric acid molecules produces diphosphoric acid.
C. The dehydration of between phosphoric acid and diphosphoric acid produces triphosphoric acid.
D. The HO–P=O groups are analogous to the HO–C=O in carboxylic acids; they are stronger acids than –COOH.
E. HO–P=O undergoes esterification with alcohols and phenols.
   1. reaction is analogous to carboxylic acid esterification
   2. products are phosphate esters

```
               O                        O
               ||                       ||
R—O[H + HO]—P—OH  ———→  [R—O]—P—OH
               |                        |
               OH                       OH
```

F. Phosphate esters are named by giving the alcohol fragment its alkyl name first followed by the phosphoric acid name.

Key Terms diphosphoric acid, phosphate ester, phosphoric acid, phosphoric acid ester, triphosphoric acid

Example 15.13 Writing reactions and predicting products for reactions of phosphoric acid with strong base
Problem 15.13 Writing reactions and predicting products for reactions of phosphoric acid with strong base
Example 15.14 Writing equations and predicting products of phosphate ester formation
Problem 15.14 Writing equations and predicting products of phosphate ester formation

ARE YOU ABLE TO... AND WORKED TEXT PROBLEMS

Objectives Section 15.1 Carboxylic Acids and Their Derivatives Compared

Are you able to...
- describe the structure of carboxylic acids
- identify derivatives of carboxylic acids (Table 15.1, Ex. 15.1, Prob. 15.1, Exercise 15.1)
- identify the carboxyl group
- describe the structures of derivatives of carboxylic acids
- recognize isomers of carboxylic acids (Exercises 15.2, 15.3, 15.4, 15.52)

Problem 15.1 Classify each of the following compounds as a carboxylic acid, ester, amide, anhydride, or something else.
1. Ester; the carbonyl C is bonded to a C and to an O
2. Aldehyde; the carbonyl C is bonded to a C and to an H
3. Amide; the carbonyl C is bonded to a C and to an N
4. Acid halide; the carbonyl carbon is bonded to a C and to a halide, Br
5. Acid anhydride; two carbonyl C's are bonded to the same O
6. Carboxylic acid; carbonyl C is attached to a C and to an –OH

Objectives Section 15.2 Synthesis of Carboxylic Acids

Are you able to...
- explain how carboxylic acids are synthesized (Expand Your Knowledge 15.68)

- write equations for the synthesis of carboxylic acids (Ex. 15.2, Prob. 15.2, Exercises 15.5, 15.6)
- predict the products of carboxylic and synthesis reactions (Ex. 15.2, Prob. 15.2, Exercises 15.5, 15.6)

Problem 15.2 Write the equation for the selective oxidation of each of the following compounds:

(a) Ethylbenzene is an alkyl benzene. Selective oxidation yields benzoic acid, CO_2, and H_2O.

$$C_6H_5{-}CH_2CH_3 \xrightarrow[Cr_2O_7^{2-}]{MnO_4^-} C_6H_5{-}\overset{O}{\overset{\|}{C}}{-}OH + CO_2 + H_2O$$

(b) Ethanol is a primary alcohol that is oxidized to ethanoic acid.

$$H{-}\underset{H}{\overset{H}{C}}{-}\underset{H}{\overset{H}{C}}{-}OH \xrightarrow[\text{or } Cr_2O_7^{2-}]{MnO_4^-} H{-}\underset{H}{\overset{H}{C}}{-}\overset{O}{\overset{\|}{C}}{-}OH$$

(c) Acetaldehyde is oxidized to ethanoic acid.

$$H{-}\underset{H}{\overset{H}{C}}{-}\overset{O}{\overset{\|}{C}}{-}H \xrightarrow[\text{or } Cr_2O_7^{2-}]{MnO_4^-} H{-}\underset{H}{\overset{H}{C}}{-}\overset{O}{\overset{\|}{C}}{-}OH$$

(d) 2-Propanone is a ketone that is not oxidized.

Objectives Section 15.3 Naming Carboxylic Acids

Are you able to...

- name carboxylic acids by using IUPAC rules (Table 15.2, Ex. 15.3, Prob. 15.3, Exercise 15.7)
- name carboxylic acids by using common names (Table 15.2)
- draw the structures of carboxylic acids from their IUPAC names (Exercises 15.8, 15.49,15.50, 15.52, 15.53)

Problem 15.3 Name each of the following compounds by using IUPAC rules.

(a) 5-Methyl-2-heptanoic acid
(b) 3-*t*-Butylbenzoic acid
(c) 2-Methyl-pentanedioic acid

Objectives Section 15.4 Physical Properties of Carboxylic Acids

Are you able to...

- describe the polarity of the carboxyl group
- explain how the properties of the carboxylic acids compare with those of alcohols, aldehydes, and ketones (Exercises 15.11, 15.12)
- describe the effect of the carboxyl group on secondary forces in carboxylic acids (Exercises 15.11, 15.12)
- describe carboxylic acid dimers (Exercise 15.9)
- describe the way in which carboxylic acids interact with water (Exercise 15.10)

Objectives Section 15.5 Acidity of Carboxylic Acids

Are you able to...

- explain how the acidic properties of carboxylic acids arise (Exercises 15.13, 15.14, 15.18)
- compare the acidity of carboxylic acids with that of other organic compounds (Table 15.4, Exercises 15.14, 15.18)
- write the equation for the ionization of a carboxylic acid in water (Ex. 15.4, Prob. 15.4, Exercise 15.17)

- write the equation for the reaction of a carboxylic acid and a strong base (Ex. 15.4, Prob. 15.4, Exercises 15.15, 15.16)
- predict the products of the reaction of a carboxylic acid and a strong base (Ex. 15.4, Prob. 15.4)
- identify the conditions under which carboxylic acids react (Ex. 15.4, Prob. 15.4, Exercises 15.15, 15.17)
- use acidic properties of carboxylic acids to identify an unknown (Exercises 15.19, 15.20, 15.57, 15.58)

Problem 15.4 Indicate whether a reaction takes place under the following conditions. If a reaction takes place, write the appropriate equation.

(a) Propanoic acid ionizes in water to produce the ion $CH_3CH_2COO^-$ and H_3O^+ in an equilibrium reaction.

$$CH_3CH_2\overset{\overset{O}{\|}}{C}{-}OH + H_2O \rightleftharpoons CH_3CH_2\overset{\overset{O}{\|}}{C}{-}O^- + H_3O^+$$

(b) Propanoic acid and aqueous NaOH (strong base) produce the carboxylate salt $CH_3CH_2COO^-\ Na^+$ and H_2O; reaction goes to completion.

$$CH_3CH_2\overset{\overset{O}{\|}}{C}{-}OH + NaOH \longrightarrow CH_3CH_2\overset{\overset{O}{\|}}{C}{-}O^-Na^+ + H_2O$$

(c) Propanoic acid and aqueous $NaHCO_3$ (weaker base than NaOH, strong enough to react with –COOH) produces the carboxylate salt $CH_3CH_2COO^-\ Na^+$ and H_2CO_3; reaction goes to completion.

$$CH_3CH_2\overset{\overset{O}{\|}}{C}{-}OH + NaHCO_3 \longrightarrow CH_3CH_2\overset{\overset{O}{\|}}{C}{-}O^-Na^+ + H_2O + CO_2$$

(d) Phenol and aqueous $NaHCO_3$; phenol is a weak acid; no reaction takes place.

Objectives Section 15.6 Carboxylate Salts

Are you able to...

- identify the carboxylate salt formed by reaction of a carboxylic acid and a strong base
- write the formulas for carboxylate salts formed by reaction of carboxylic acids with a strong base
- name carboxylate salts by using IUPAC rules (Ex. 15.5, Prob. 15.5, Exercise 15.22)
- identify the structure of a carboxylate salt from its name (Exercise 15.21)
- discuss the basic properties of carboxylate salts (Exercise 15.24)
- discuss the ionic properties of carboxylate salts (Exercise 15.23)
- explain the differences in physical properties between carboxylic acids and the corresponding carboxylate salts (Ex. 15.6, Prob. 15.6, Exercise 15.23)
- use the properties of carboxylate salts to identify an unknown (Exercises 15.25, 15.26)

Problem 15.5 Name the following carboxylate salts:

(a) Sodium 3-methylbutanoate (b) Aluminum ethanoate

Problem 15.6 Place the following compounds in order of increasing water solubility.

$CH_3(CH_2)_{14}COOK$ is a salt; $CH_3(CH_2)_{14}COOH$ is a carboxylic acid; $CH_3(CH_2)_{14}OH$ is an alcohol. The salt is an ionic compound and is soluble in water. The carboxylic acid and alcohol both have

16 carbons and are not water soluble, despite the presence of –COOH and –OH groups. Generally, the carboxylic acid is more soluble than the alcohol because it forms hydrogen bonds with water more extensively than an alcohol does.

$$CH_3(CH_2)_{14}OH \cong CH_3(CH_2)_{14}COOH < CH_3(CH_2)_{14}COOK$$

Objectives Section 15.7 Soaps and Their Cleaning Action

Are you able to...
- describe the structure of carboxylate salts in soaps (Exercise 15.27)
- explain the process of saponification
- discuss the amphipathic nature of soaps (Exercise 15.28)
- explain the cleaning action of soaps (Box 15.3, Exercise 15.28)

Objectives Section 15.8 Esters from Carboxylic Acids and Alcohols

Are you able to...
- explain how esters are produced by reaction of carboxylic acids and alcohols
- write equations for esterification reactions (Ex. 15.7, Prob. 15.7, Exercises 15.29, 15.30, 15.59, 15.61)
- identify the reaction conditions under which esterification takes place (Ex. 15.7, Prob. 15.7)
- predict the products of esterification reactions (Ex. 15.7, Prob. 15.7, Exercises 15.29, 15.30, 15.51, 15.59)
- identify the carboxylic acid and alcohol from which an ester is derived (Exercises 15.31, 15.32)
- describe thioesters

Problem 15.7 Give the equation for ester formation between butanoic acid and phenol.

The –OH group from butanoic acid will combine with the H from the –OH on phenol to form H_2O. The butanoate ion will combine with the O from the –OH on phenol.

$$CH_3CH_2CH_2C(=O){-}OH + HO{-}C_6H_5 \xrightarrow[-H_2O]{H^+} CH_3CH_2CH_2C(=O){-}O{-}C_6H_5 + H_2O$$

Objectives Section 15.9 Names and Physical Properties of Esters

Are you able to...
- name esters by using IUPAC rules (Ex. 15.8, Prob. 15.58, Exercise 15.34)
- draw ester structures from their IUPAC names (Exercise 15.32)
- identify the alkoxy part of an ester (Ex. 15.8, Prob. 15.8)
- identify the acyl part of an ester (Ex. 15.58, Prob. 15.8)
- discuss the physical properties of esters relative to other organic compounds (Exercise 15.35)
- discuss the nature of secondary forces in esters (Exercise 15.36)

Problem15.8 Give the IUPAC names of the following compounds.

(a) *p*-Ethylphenyl 2-methylbutanoate
(b) Pentyl methanoate
(c) Methyl hexanoate

Objectives Section 15.10 Polyester Synthesis

Are you able to...
- describe the process of condensation polymerization
- identify bifunctional reagents that can be monomers in polymerization (Exercise 15.37)
- write equations for condensation polymerization (Ex. 15.9, Prob. 15.9)
- predict the polyester formed by condensation polymerization (Ex. 15.9, Prob. 15.9, Exercise 15.38)

Problem 15.9 Each of the following pairs of reactants undergoes esterification. Which of the pairs produces a polymer? Write the equation for any reaction(s) that produce(s) a polymer.

Polyesterification occurs when a molecule with two alcohols, a diol, and a carboxylic acid with two carboxyl groups, a dicarboxylic acid, react.

(a) A dicarboxylic acid and an alcohol; no polyesterification
(b) A carboxylic acid and a diol; no polyesterification
(c) A dicarboxylic acid and a diol; polyesterification takes place. The product formed is:

$$H{-}O{-}CH_2CH_2{-}OH \quad HO{-}\overset{O}{\overset{\|}{C}}CH_2CH_2CH_2CH_2\overset{O}{\overset{\|}{C}}{-}OH \xrightarrow{H_2O}$$

$$\left[O{-}CH_2CH_2{-}O{-}\overset{O}{\overset{\|}{C}}{-}CH_2CH_2CH_2CH_2\overset{O}{\overset{\|}{C}}{-}O\right]_n$$

Objectives Section 15.11 Hydrolysis of Esters

Are you able to...
- describe the process of hydrolysis of esters
- describe the relation between hydrolysis and esterification
- identify the conditions under which hydrolysis takes place (Ex. 15.10)
- write equations for acid and base hydrolysis of esters (Prob. 15.10, Exercises 15.39, 15.40, 15.60, 15.61)
- predict the products of acid and base hydrolysis of esters (Prob. 15.10, Exercises 15.39, 15.40, 15.60, 15.61, 15.62, 15.63)
- compare acid hydrolysis and base hydrolysis of esters

Problem 15.10 Write the equation for each of the following reactions.

(a) Acidic hydrolysis of phenyl butanoate produces phenol and butanoic acid.

$$CH_3CH_2CH_2\overset{O}{\overset{\|}{C}}{-}O{-}C_6H_5 \xrightarrow[H_2O]{H^+} CH_3CH_2CH_2\overset{O}{\overset{\|}{C}}{-}OH \quad HO{-}C_6H_5$$

(b) Saponification of propyl propanoate with KOH produces potassium propanoate and propanol.

$$CH_3CH_2CH_2{-}O{-}\overset{O}{\overset{\|}{C}}CH_2CH_3 + KOH \longrightarrow CH_3CH_2\overset{O}{\overset{\|}{C}}{-}O^-K^+ + CH_3CH_2CH_2OH$$

Objectives Section 15.12 Carboxylic Acid Anhydrides and Halides

Are you able to...
- describe the formation of anhydrides from carboxylic acids
- describe the formation of acid halides from carboxylic acids
- describe the reactivity of anhydrides and acid halides in water (Exercises 15.43, 15.44)

- describe the properties of an acyl transfer agent
- explain what is meant by acyl transfer
- write equations for acyl transfer reactions (Ex. 15.12, Prob. 15.12)
- predict the products of acyl transfer reactions (Ex. 15.12, Prob. 15.12)
- name acid anhydrides and acyl halides (Ex. 15.11, Prob. 15.11, Exercise 15.42)
- draw the structures of acid anhydrides and acyl halides from their names (Exercise 15.41)

Problem 15.11 Give the structure of each of the following compounds:

(a) Butanoyl chloride

$$CH_3CH_2CH_2C(=O)—Cl$$

(b) Acetic anhydride

$$CH_3—C(=O)—O—C(=O)—CH_3$$

Problem 15.12 Write the equation for the acyl transfer reaction between benzoyl chloride and methanol.

$$C_6H_5—C(=O)—Cl + HO—CH_3 \longrightarrow C_6H_5—C(=O)—O—CH_3 + HCl$$

Objectives Section 15.13 Phosphoric Acids and Their Derivatives

Are you able to...

- identify the members of the phosphoric acid family
- draw the structures of members of the phosphoric acid family
- compare the structures of di- and triphosphoric acid with that of an acid anhydride
- discuss the formation of di- and triphosphoric acid by dehydration reaction
- show how carboxylic acid groups and HO–P=O groups are similar
- describe how the dehydration between two phosphoric acid molecules produces diphosphoric acid
- describe the acidic properties of phosphoric acids (Exercises 15.47, 15.48)
- write equations for reactions of phosphoric acids with a strong base (Ex. 15.13, Prob. 15.13)
- predict the products of the reaction of phosphoric acids with a strong base (Ex. 15.13, Prob. 15.13)
- write esterification reactions for phosphoric acids (Ex. 15.14, Prob. 15.14, Exercise 15.45)
- predict the products of esterification of phosphoric acids and alcohols (Ex. 15.14, Prob. 15.14, Exercise 15.45)
- name phosphate esters
- draw structures of phosphate esters from their names (Exercise 15.46)

Problem 15.13 Show the final product after diphosphoric acid reacts with excess KOH.

Each –OH group can react with KOH to produce the K^+ salt; there are four –OH groups on diphosphoric acid.

$$HO—P(=O)(OH)—O—P(=O)(OH)—OH \xrightarrow{KOH} K^+\ {}^-O—P(=O)(O^-K^+)—O—P(=O)(O^-K^+)—O^-K^+$$

Problem 15.14 Write the equation for esterification of two of the –OH groups in phosphoric acid with methanol. Name the ester.

Dimethyl phosphate is the ester formed.

$$HO-\overset{\overset{\displaystyle O}{\|}}{\underset{\underset{\displaystyle OH}{|}}{P}}-OH + 2CH_3OH \longrightarrow CH_3-O-\overset{\overset{\displaystyle O}{\|}}{\underset{\underset{\displaystyle OH}{|}}{P}}-O-CH_3 + 2H_2O$$

MAKING CONNECTIONS

Construct a concept map or write a paragraph describing the relation among the members of each of the following groups:

(a) carboxyl group, carbonyl, alcohol
(b) carboxylic acid, ketone, aldehyde, alcohol
(c) carboxylic acid, ester, amide, anhydride
(d) alcohol, ketone, aldehyde, carboxylic acid, oxidation
(e) carbonyl, carboxyl, alcohol, melting point, boiling point
(f) carbonyl, carboxyl, alcohol, water solubility
(g) alkane, long-chain carboxylic acid, short-chain carboxylic acid, water solubility
(h) carboxylic acid, dimer, hydrogen bond, physical properties
(i) carboxyl group, acidity, carboxylate ion, basicity
(j) carboxylic acid, ionization, water, acid solution, basic solution
(k) carboxylic acid, strong acid, strong base, water, equilibrium
(l) melting point, boiling point, water solubility, carboxylic acid, carboxylate salt
(m) carboxylate salt, soap, amphipathic molecule, water
(n) saponification, hydrolysis, acid, base
(o) ester, alcohol, carboxylic acid, esterification, hydrolysis
(p) thioester, ester, polyester
(q) alkoxy group, acyl group, ester, carboxylic acid, alcohol
(r) physical properties, ester, carboxylic acid, alcohol, secondary forces
(s) polymerization, esterification, monomer, polymer
(t) acid halide, acid anhydride, carboxylic acid, phosphoric acid
(u) acyl transfer reaction, alcohol, acid halide

FILL-INS

Carboxylic acids are characterized by two functional groups, _____ and _____ bonded to the _____. Together, they are called the _____. The unique properties of the carboxyl group include the ability _____ with water or another carboxylic acid. The result is that carboxylic acids exist as _____. Carboxylic acids are also _____ water soluble than comparable alcohols, aldehydes, and ketones. This effect can be seen when looking at the solubilities of comparable alcohols because _____. Six-carbon alcohols are not water soluble, but six-carbon carboxylic acids are. The strength of the secondary forces also contributes to the relatively _____ melting and boiling points of carboxylic acids compared with those of alcohols, aldehydes, and ketones. Soluble carboxylic acids are weak acids and _____ in water. As the pH of an aqueous solution is increased, the number of ionized carboxylic acid molecules _____. The ionization of a carboxylic acid in water produces _____. Carboxylic acids react with bases to produce _____ and water. The _____ of carboxylate salts differ from those of the parent carboxylic acids. Carboxylate salts are more _____ and are _____, which makes them solids at room temperature. Carboxylic acid derivatives include _____. Along with carboxylic acids, these derivatives react with alcohols to produce _____. _____ and _____ react to produce polyesters. Polyesters are _____ polymers that form when the elements of water, H_2O, are abstracted from _____, and they are bonded in the process. Esters can be converted back into carboxylic acids and alcohols by _____ under acidic conditions. In basic solution, the process is called _____ and it produces a _____ and alcohol. Inorganic acids such as phosphoric acid also can form _____ by reaction with _____.

Answers

Carboxylic acids are characterized by two functional groups, a carbonyl and an –OH bonded to the carbonyl carbon. Together, they are called the carboxyl group. The unique properties of the carboxyl group include the ability to form three hydrogen bonds with water or another carboxylic acid. The result is that carboxylic acids exist as dimers, two molecules joined together by hydrogen bonds. Carboxylic acids are also much more water soluble than comparable alcohols, aldehydes, and ketones. This effect can be seen when looking at the solubilities of comparable alcohols because water insolubility in alcohols starts with smaller carbon chains. Six-carbon alcohols are not water soluble, but six-carbon carboxylic acids are. The strength of the secondary forces also contributes to the relatively high melting and boiling points of carboxylic acids compared with those of alcohols, aldehydes, and ketones. Soluble carboxylic acids are weak acids and ionize slightly in water. As the pH of an aqueous solution is increased, the number of ionized carboxylic acid molecules increases. The ionization of a carboxylic acid in water produces a carboxylate anion, $RCOO^-$, and a hydronium ion, H_3O^+. Carboxylic acids react with bases to produce carboxylate salts and water. The properties of carboxylate salts differ from those of the parent carboxylic acids. Carboxylate salts are more water soluble and are ionic compounds, which makes them solids at room temperature. Carboxylic acid derivatives include acid anhydrides and acid halides. Along with carboxylic acids, these derivatives react with alcohols to produce esters. Dicarboxylic acids and diols react to produce polyesters. Polyesters are condensation polymers that form when the elements of water, H_2O, are abstracted from two monomers, and they are bonded in the process. Esters can be converted back into carboxylic acids and alcohols by hydrolysis under acidic conditions. In basic solution, the process is called saponification and it produces a carboxylate salt and alcohol. Inorganic acids such as phosphoric acid also can form esters by reaction with alcohols.

TEST YOURSELF

1. How does the structure of a carboxyl group differ from that of an alcohol or an aldehyde?
2. Draw a reaction map for carboxylic acids; include ionization in water, acid and base esterification, reaction with alcohol, and acid and base hydrolysis of ester.
3. Draw the dimer that forms from two molecules of propionic acid.
4. Lactic acid (below) is produced in the muscles during exercise; it is also a component of sour milk. What form does lactic acid take at physiological pH, 7.4?

$$CH_3-\underset{\displaystyle OH}{\underset{|}{CH}}-\underset{\displaystyle O}{\underset{\|}{C}}-O-H$$

5. The solubilities in water are given below for three carboxylic acids and three alcohols. Explain the trends.

Compound	H_2O solubility (g/100 mL)
1-Butanol	7.9
1-Pentanol	2.2
1-Hexanol	0.75
Butyric acid	Infinitely soluble
Pentanoic acid	5.1
Hexanoic acid	1.0

6. How do the products of ester hydrolysis and saponification compare?
7. Why is an amphipathic molecule necessary for the cleansing action of soap?
8. Why is a carboxylate ion formed instead of a carboxylic acid when an ester is cleaved under basic conditions?

9. A medicine that is designed to be injected and that has a carboxyl group is usually prepared as its carboxylate salt. Why do you think this is so?

Answers

1. The carboxyl group is a combination of both a carbonyl and an alcohol. The electron-withdrawing effect of the carbonyl O creates a partial positive charge on the carbon that has an electron-withdrawing effect on the O in the –OH group. This partial charge makes the carboxyl group even more polar than the carbonyl group, which is itself more polar than the hydroxyl group owing to the electron-withdrawing effect of the doubly bonded O.
2.

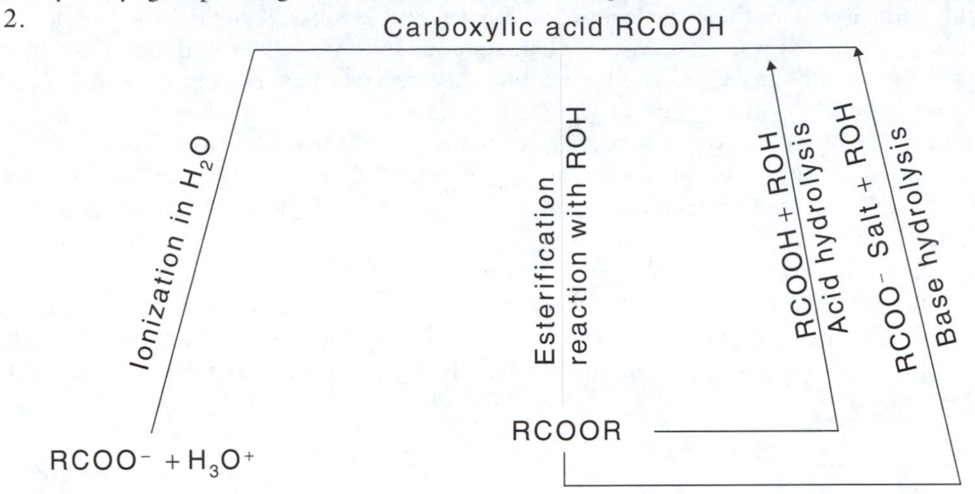

3. Propionic acid has the formula CH_3CH_2COOH:

$$CH_3CH_2{-}C\begin{matrix} \diagup O{-}H\cdots O\diagdown \\ \diagdown O\cdots H{-}O\diagup \end{matrix}C{-}CH_2CH_3$$

4. At physiological pH, the ionic form of lactic acid predominates.

$$CH_3{-}\overset{\overset{\displaystyle OH}{|}}{C}{-}\overset{\overset{\displaystyle O}{\|}}{C}{-}O^-$$

5. Alcohols and carboxylic acids of similar molecular mass have different solubilities. Longer-chain carboxylic acids have solubilities similar to those of shorter-chain alcohols because of the presence of the carboxyl group, which can form extensive hydrogen bonds with water. As the carbon chain gets longer, the solubility difference between alcohols and carboxylic acids diminishes: the difference between four-carbon alcohols and carboxylic acids is close to 8 g/100 mL, but the difference between six-carbon alcohols and carboxylic acids is 0.25 g/100 mL. This difference in solubility shows that the increasing nonpolarity of each molecule dominates the secondary forces.
6. Hydrolysis of an ester requires an acid catalyst and produces the carboxylic acid and the alcohol when the components of water add to the ester as the ester bond is cleaved. Saponification is base hydrolysis in which the carboxylate salt and the alcohol are formed.
7. To clean, a soap must attract and hold dirt and grease molecules, which are largely nonpolar. To carry the dirt away, the molecule must also be attracted to the water molecules that surround it. To do both, a soap molecule needs both a hydrophilic and a hydrophobic part.
8. Under basic conditions, the carboxylic acid, which is a weak acid, donates its proton to the strongly basic anion, $CH_3{-}O^-$, which is formed when the ester bond is cleaved.
9. The solubility of the carboxylate salt allows aqueous solutions of higher concentrations to be prepared. The higher concentrations allow for faster absorption after injection.

chapter 16

Amines and Amides

OUTLINE

16.1 Amines and Amides Compared

A. Amines are organic derivatives of ammonia in which the hydrogens are replaced by organic groups.

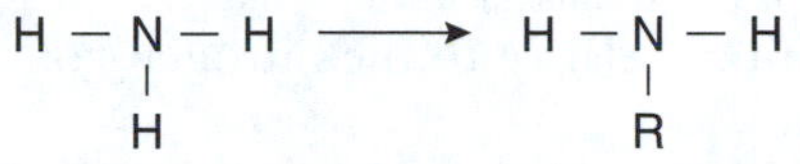

```
H — N — H  ———→  H — N — H
    |                |
    H                R
```

B. Amides are organic derivatives of ammonia in which one hydrogen has been replaced by a carbonyl group.

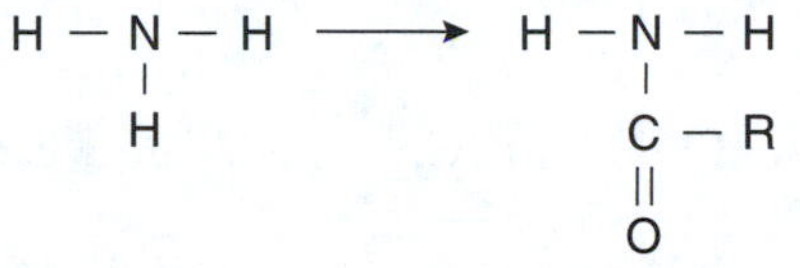

```
H — N — H  ———→  H — N — H
    |                |
    H                C — R
                     ||
                     O
```

C. Nitrogen in amines is trivalent with a single lone pair of electrons.

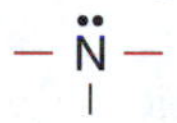

```
  ••
— N —
  |
```

Key Terms amide, amide group, amine

16.2 Classifying Amines

A. Amines are classified on the basis of the number of carbons to which the nitrogen is bonded.
   1. primary amines: nitrogen is bonded to one carbon atom

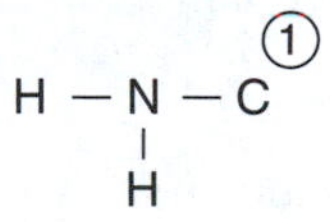

```
           (1)
H — N — C
    |
    H
```

   2. secondary amines: nitrogen is bonded to two different carbon atoms

```
(2)        (1)
C — N — C
    |
    H
```

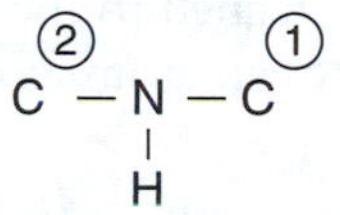

   3. tertiary amines: nitrogen is bonded to three different carbon atoms

```
(3)        (1)
C — N — C
    |
    C(2)
```

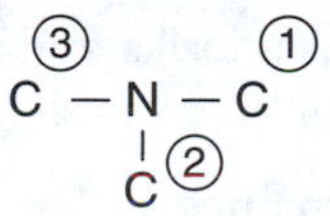

B. Aromatic amines have a benzene or an aromatic group bonded directly to the nitrogen; also called aryl amines.

C. Aliphatic amines do not have a benzene or an aromatic group bonded directly to the nitrogen.
D. Constitutional isomers of amines arise owing to:
   1. varied arrangements of carbon skeleton
   2. varied placement of nitrogen on carbon skeleton

Key Terms aliphatic amine, aromatic amine, primary amine, secondary amine, tertiary amine

Example 16.1 Identification and classification of amines and amides
Problem 16.1 Identification and classification of amines and amides
Example 16.2 Drawing constitutional isomers of amines
Problem 16.2 Drawing constitutional isomers of amines

16.3 Naming Amines

A. Amines are named by using the *Chemical Abstracts* (CA) system.
B. Simple amines are named by alkyl names of groups attached to nitrogen.
   1. -amine suffix is given
   2. used for
      a. unbranched alkyl groups methyl- (1 C) through decyl- (10 C)
      b. three- and four-carbon branched alkyl groups
      c. unsubstituted cycloalkyl groups
C. Benzene with $–NH_2$ ($C_6H_5NH_2$) is called aniline.
   1. substituted anilines are named as -anilines
   2. substituents are identified relative to the nitrogen-containing carbon by using ortho-, meta-, or para- prefixes
   3. substituents on nitrogen are given an *N* prefix
   4. CA uses benzenamine for this group

CA Rules for Naming Amines

1. Primary amines are named by identifying the longest carbon chain that has the nitrogen attached.
2. The alkane ending is changed from -e to -amine. The number of the carbon to which nitrogen is attached is designated before the alkanamine name. Carbon number 1 is the carbon nearest the nitrogen.
3. Substituents are numbered and identified as with other organic families.
4. Compounds with two amine groups are called alkanediamines.
5. Cyclic compounds that are not aromatic and contain an attached nitrogen are called cycloalkanamines.
6. Secondary and tertiary amines are named by identifying the group attached to the nitrogen that has the longest carbon chain. The names of the other groups attached to the nitrogen are given the prefix N- and the alkyl name.
   a. $–NH_2$ is called amino.
   b. –NHR is called N-alkylamino.
   c. $–NHR_2$ is called N,N-dialkylamino.
   d. Oxygen-containing functional groups have preference over the N-containing group and are named as an amino group attached to the base oxygen-containing group.
7. Heterocyclic amines contain a nitrogen atom as part of a ring structure and have common names (Table 16.1).

D. Alkaloids are physiologically active amines produced by plants; alkaloids are usually heterocycles.

Key Terms alkaloid, alkaneamine, CA rules, cycloalkaneamine, heterocyclic amine, nitrogen heterocyclic
Key Table Table 16.1 Common names of heterocyclic structures

Example 16.3 Naming amines by using CA names
Problem 16.3 Naming amines by using CA names

16.4 *Physical Properties of Amines*

A. Primary, secondary, and tertiary amines have different physical properties.
  1. differences are due to the difference in type and magnitude of secondary forces (Fig. 16.1)
  2. primary and secondary amines form hydrogen bonds between molecules
  3. tertiary amines form weak attractions owing to C–N bond but cannot form hydrogen bonds

B. Melting points and boiling points (Table 16.2).
  1. of primary and secondary amines are comparable to those of aldehydes and ketones
  2. of tertiary amines are comparable to those of ethers and hydrocarbons; they are significantly lower than those of primary and secondary amines
  3. hydrogen bonds between primary and secondary amines require relatively more energy to separate
  4. N is not as electronegative as O in alcohol, so amine hydrogen bonds are weaker than alcohol hydrogen bonds but similar to dipole interactions in aldehydes and ketones
  5. tertiary amines cannot form hydrogen bonds and have only weak interactions owing to slightly polar C–N bonds that are similar to London forces in hydrocarbons and slightly polar interactions in ethers

C. Small molecules of all three classes of amines are water soluble.
  1. solubility is similar to that of aldehydes and ketones but less than that of alcohols
  2. primary and secondary amines form hydrogen bonds with water (Fig. 16.2)
  3. tertiary amines form limited hydrogen bonds between the *N* of the tertiary amine and an H of water (Fig. 16.3)
  4. one- to four-carbon amines are water soluble
  5. water solubility is insignificant when the amine contains more than seven carbons

Key Figures and Table

Figure 16.1	Hydrogen bonding in secondary amines
Figure 16.2	Hydrogen bonding between water and a secondary amine
Figure 16.3	Hydrogen bonding between water and a tertiary amine
Table 16.2	Comparison of boiling points of amines and compounds from different families

16.5 *Basicity of Amines*

A. Nitrogen lone pair can accept a proton.

B. Ammonia, NH_3, is converted into ammonium ion, NH_4^+, when it accepts a proton.

C. Amines are converted into cations:
  1. primary amine + proton → RNH_3^+
  2. secondary amine + proton → $R_2NH_2^+$
  3. tertiary amine + proton → R_3NH^+

D. Amines react as bases with water.
  1. $RH_2N: + H_2O \rightleftharpoons RH_3N^+ + OH^-$
  2. this is an equilibrium reaction
  3. about 7 of 100 amine molecules are ionized in water

E. Amines are stronger bases than water, alcohols, or ethers (Table 16.3).
  1. oxygen is more electronegative than nitrogen and holds its electrons more tightly
  2. oxygen-containing organic molecules accept protons to a lesser extent than does nitrogen in amines
  3. amines turn red litmus paper blue just like carboxylate salts
  4. arylamines are weaker bases than aliphatic amines; electron-withdrawing effect of benzene ring causes nitrogen electrons to be less available for bonding

F. Reaction with acid forms amine salt:

$$RH_2N: + HCl \rightarrow RH_3N^+ \; Cl^-$$

  1. reaction goes to completion
  2. amine is called a free base to distinguish it from the amine salt
  3. amines are almost completely ionized at physiological pH

Key Terms amine salt, ammonium ion
Key Table Table 16.3 Relative basicities of amines

Example 16.4 Writing dissociation reactions and acid-base reactions for amines
Problem 16.4 Writing dissociation reactions and acid-base reactions for amines

16.6 Amine Salts

A. Amine salts are named as follows:
   1. cation is named first
      a. aromatic and heterocyclic amines have amine name -e changed to -ium
      b. aliphatic amines are named by changing -amine ending to -ammonium
   2. anion is named second
B. Properties of amine salts:
   1. amine salts are ionic compounds
      a. have stronger secondary forces than do covalent compounds
      b. melting and boiling points are higher than those of corresponding amines
      c. amine salts are solids at room temperature
      d. amine salts are more water soluble than corresponding amines
      e. amine salt solubilities increase in environments of neutral or acidic pH because amine is in salt form
C. Quaternary ammonium salts:
   1. amine salts with four R groups are called quaternary ammonium salts; $R_4N^+Cl^-$
      a. quaternary ammonium salts are more resistant to pH effects
         i. N is not bonded to an H atom that can be removed by a base
         ii. N does not have nonbonded pair of electrons to react with a proton

Key Terms ammonium, quaternary ammonium salt

Example 16.5 Naming amine salts
Problem 16.5 Naming amine salts
Example 16.6 Predicting relative magnitude of melting point
Problem 16.6 Predicting relative water solubilities

16.7 Classifying Amides

A. Amides are characterized by a carbonyl group bonded directly to the nitrogen.

```
        O
        ||
  — N — C — R
    |
```

B. Just like amines, amides are classified according to the number of R groups bonded to the nitrogen.

```
         O                  O                  O
         ||                 ||                 ||
H — N — C — R     R′ — N — C — R     R′ — N — C — R
    |                  |                  |
    H                  H                  R″
   (1°)               (2°)               (3°)
```

Key Terms primary amide, secondary amide, tertiary amide

Example 16.7 Identification and classification of amides
Problem 16.7 Identification and classification of amides

16.8 Synthesis of Amides

A. Amides are synthesized by reaction of a carboxylic acid with ammonia or with a primary or secondary amine.
   1. must be carried out at relatively high temperatures

2. products are determined by temperature and class of amine reactant
   a. carboxylic acid + ammonia → primary amide + H_2O
   b. carboxylic acid + primary amine → secondary amide + H_2O
   c. carboxylic acid + secondary amine → tertiary amide + H_2O
3. reaction is similar to esterification
   a. –COOH is substituted when dehydration occurs with amine; -OH from carboxylic acid and H- from amine form water
   b. amine N is bonded to carbonyl C
   c. this is an acyl-transfer reaction with –COOH as acyl transfer agent
4. this is a reversible reaction
5. tertiary amines cannot form amides, because there is no H to participate in dehydration
6. at temperatures between 20°C and 50°C, an acid base reaction produces an ammonium carboxylate salt; amide formation requires temperatures above 100°C
7. acid halides and anhydrides form amides faster than –COOH and proceed to completion

Key Term amide

Example 16.8 Recognizing conditions under which amide formation takes place
Problem 16.8 Recognizing conditions under which amide formation takes place
Example 16.9 Writing reactions for amide synthesis from an acid halide
Problem 16.9 Writing reactions for amide synthesis from an acid halide

16.9 Polyamide Synthesis

A. Polyamides are produced by polymerization of bifunctional carboxylic acids and bifunctional amines.

$$HO-\overset{\overset{\Large O}{||}}{C}-R-\overset{\overset{\Large O}{||}}{C}-OH$$

Bifunctional carboxylic acid

$$H_2N-R'-NH_2$$

Bifunctional amine

1. bifunctional amines are called diamines
2. reaction is a condensation polymerization; water is produced as a by-product
3. Proteins are polyamides; called polypeptides

Key Term polyamide

16.10 Naming Amides

A. IUPAC rules are used to name amides.

IUPAC Rules for Naming Amides

1. Carboxylic acid is base name.
   a. -oic acid ending is changed to -amide.
   b. Acyl part is derived from carboxylic acid.
2. Names of substituents attached to nitrogen are placed first with the use of an *N* for each prefix. Amino part is derived from amine.

Example 16.10 Naming amides
Problem 16.10 Naming amides

16.11 Physical and Basicity Properties of Amides

A. Amides have the strongest secondary forces of covalent organic compounds (Table 16.4).
   1. produces higher melting and boiling points than those of carboxylic acids
   2. amide structure is resonance hybrid that includes a dipolar ion
      a. nitrogen bears positive charge
      b. carbonyl oxygen bears negative charge

3. primary and secondary amides also participate in hydrogen bonding, but tertiary amides do not; primary and secondary amides have higher melting and boiling points than tertiary amides (Table 16.4)

B. Amides are slightly more soluble than carboxylic acids in water; amides participate in both hydrogen bonding and dipole interactions with water.

C. Amides are about as basic as ethers and alcohols.
   1. amides are weaker bases than amines
   2. electron-withdrawing effect of carbonyl group on nitrogen atom binds nonbonding electron pair more tightly and these electrons are less able to accept a proton

Key Term dipolar ion

Key Table Table 16.4 Comparison of carboxylic acids and amines

16.12 Hydrolysis of Amides

A. Hydrolysis of bond between N and carbonyl C can take place under certain conditions
   1. the bond between N and the carbonyl C is called an amide bond
   2. amide bonds are resistant to hydrolysis at neutral pH

B. Hydrolysis of amides can take place in the presence of a strong acid or a strong base.
   1. acid hydrolysis of amides is similar to acid hydrolysis of esters
      a. reaction produces the corresponding carboxylic acid and amine
      b. addition of strong acid can convert the amine into the amine salt
   2. basic hydrolysis of amides is similar to basic hydrolysis of esters
      a. reaction produces the corresponding carboxylic acid and amine
      b. addition of strong base converts carboxylic acid into carboxylate salt
   3. acid and base are reactants, not catalysts

Key Term amide bond

Example 16.11 Predicting products and identifying conditions under which hydrolysis of amides takes place
Problem 16.11 Predicting products and identifying conditions under which hydrolysis of amides takes place

ARE YOU ABLE TO ... AND WORKED TEXT PROBLEMS

Objectives Section 16.1 Amines and Amides Compared

Are you able to...

- describe the relation between amine molecules and ammonia
- identify amines and amides (Ex. 16.1, Prob. 16.1)
- describe the bonding capacity of nitrogen in amines and amides

Objectives Section 16.2 Classifying Amines

Are you able to...

- classify amines as primary, secondary, or tertiary (Ex. 16.1, Prob. 16.1, Exercises 16.3, 16.4, 16.50)
- describe the structural features of aromatic amine
- describe the structural features of aliphatic amines
- identify the structural features that produce constitutional isomers of amines (Ex. 16.2, Prob. 16.2, Exercises 16.46, 16.59)
- draw constitutional isomers of amines (Ex. 16.2, Prob. 16.2, Exercises 16.1, 16.2, Exercise 16.45)

Problem 16.1 Classify each of the following compounds.

1. Primary aliphatic amine; N is bonded to one C on an alkane chain
2. Tertiary aliphatic amine; N is bonded to an aliphatic ring and two methyl groups
3. Primary amine; N is bonded to one C; N is not bonded directly to aromatic group and so it is aliphatic
4. Secondary aliphatic amine; following the bonds from N on either side brings you to a carbon
5. Amide; a carbonyl C is bonded directly to N

6. The molecule contains a ketone and a tertiary amine; the N is bonded to 3 R groups
7. Primary aromatic amine; N is bonded directly to a C in a disubstituted benzene ring
8. Tertiary amine; the N is bonded to three alkyl groups

Problem 16.2 Draw the constitutional isomers of $C_4H_{11}N$.

The isomers can vary in both the arrangement of the carbon skeleton and the placement of the amine group. There are four different arrangements of the carbon skeleton, with different placements of the N producing three primary amines and a secondary amine. There are three different ways in which various R groups can be placed on *N*, producing two secondary amines and one tertiary amine.

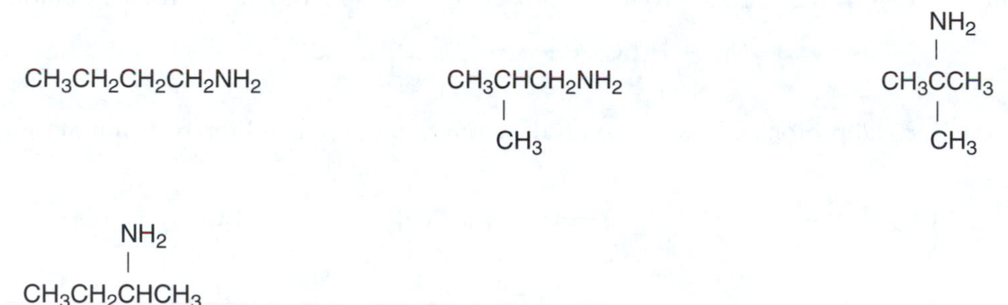

$CH_3CH_2 - N - CH_2CH_3$ (H on N)

$CH_3 - N - CH_2CH_2CH_3$ (H on N)

$CH_3 - N - CH_3$ (CH_2CH_3 on N)

Objectives Section 16.3 Naming Amines

Are you able to...

- name amines by using the CA system (Ex. 16.3, Prob. 16.3, Exercises 16.7, 16.8)
- identify heterocyclic amines and name them by using common names (Table 16.1, Ex. 16.3, Prob. 16.3)
- draw structures of amines from their CA names (Exercises 16.5, 16.6, 16.45)
- explain what alkaloids are

Problem 16.3 Give the common or CA name, whichever is appropriate, for each of the following compounds.

(a) Tertiary amine; name longest chain attached to N as base; *N,N*-ethylmethyl-2-butanamine
(b) Primary amine; 1,2-butanediamine
(c) Tertiary amine; 3-ethyl-*N,N*,-dimethylcyclopentamine
(d) A substituted aniline; *N*-ethyl-*m*-isopropyl-*p*-methylaniline
(e) Primary amine; 4-ethyl-2,6-dimethyl-2-heptanamine

Objectives Section 16.4 Physical Properties of Amines

Are you able to...

- explain the basis for the difference in physical properties between primary, secondary, and tertiary amines (Figure 16.1)
- discuss the nature of secondary forces in primary, secondary, and tertiary amines (Exercise 16.9)
- describe how primary, secondary, and tertiary amines interact with water (Figures 16.2, 16.3, Exercise 16.10)
- discuss the trends in physical properties of primary, secondary, and tertiary amines relative to other organic families (Exercises 16.11, 16.12, 16.51)

Objectives Section 16.5 Basicity of Amines

Are you able to...

- explain the origin of basicity in amines
- discuss the strength of amine basicity relative to other bases (Table 16.3, Exercises 16.15, 16.16)
- identify the products of protonation of primary, secondary, and tertiary amines
- discuss the reactivity of amines toward water (Ex. 16.4, Prob. 16.4, Exercises 16.14, 16.20)

- write reactions for the formation of amine salts by reaction of amines and acids (Ex. 16.4, Prob. 16.4, Exercises 16.13, 16.14, 16.19)
- predict the products of the reaction of an amine and an acid (Exercises 16.13, 16.14, 16.19, Expand Your Knowledge 16.60)
- identify an unknown compound by using the properties of basic amines (Exercises 16.17, 16.18, 16.53, 16.54, 16.55, Expand Your Knowledge 16.62, 16.64)

Problem 16.4 Write the equation for each of the following reactions.

(a) Aniline will react with H_2SO_4 to produce protonated aniline and the hydrogen sulfate anion, HSO_4^-.

$$2C_6H_5NH_2 + H_2SO_4 \longrightarrow [C_6H_5NH_3^+]_2SO_4^{2-}$$

(b) Diethylamine will protonate in water to produce the amine anion and the hydronium ion.

$$CH_3CH_2-\underset{\displaystyle H}{\overset{}{N}}-CH_2CH_3 \rightleftharpoons CH_3CH_2-\underset{\displaystyle H}{\overset{\displaystyle H}{N^+}}-CH_2CH_3 + OH^-$$

Objective Section 16.6 Amine Salts

Are you able to...

- name amine salts (Ex. 16.5, Prob. 16.5, Exercise 16.22)
- draw the structure of amine salts from their names (Exercises 16.21, 16.46)
- compare the properties of amine salts with the parent amines and with other organic families (Ex. 16.6, Prob.16.6, Exercises 16.23, 16.51)
- discuss the nature of secondary forces in amine salts
- write acid-base reactions for ammonium salts (Exercise 16.24)
- predict the products of acid-base reactions of ammonium salts (Exercise 16.24)
- describe the formation of quaternary ammonium salts
- describe the properties of quaternary ammonium salts
- compare the properties of quaternary ammonium salts with other amine salts
- identify an unknown by using the properties of amine salts (Exercises 16.25, 16.26)

Problem 16.5 Draw the structures of the following amine salts.

(a) Cyclohexylammonium chloride

The positive ion is cyclohexylammonium and the negative ion is chloride. Cyclohexyl is from cyclohexane, which is a six-membered cyclic alkane. The N is bonded to only one alkyl group, so it also bonded to three hydrogens.

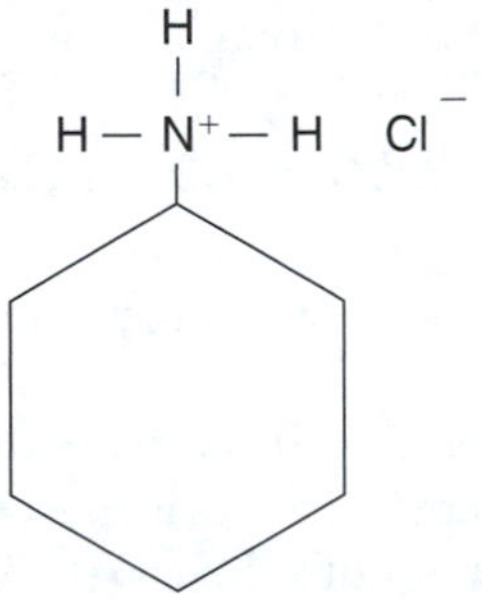

$$H-\overset{\displaystyle H}{N^+}-H \quad Cl^-$$
(N bonded to a cyclohexane ring)

(b) *N,N*-Dimethyl-1-butanammonium hydrogen sulfate

$$CH_3CH_2CH_2CH_2\underset{\displaystyle CH_3}{\overset{\displaystyle H}{N^+}}-CH_3 \quad HSO_4^-$$

(c) Pyridinium bromide (Table 16.1)

Pyridinium, the positive ion, is derived from pyridine, which is a six-membered heterocyclic amine with alternating double bonds. The protonated form has a proton attached to the N in the ring. Bromide is the negative ion.

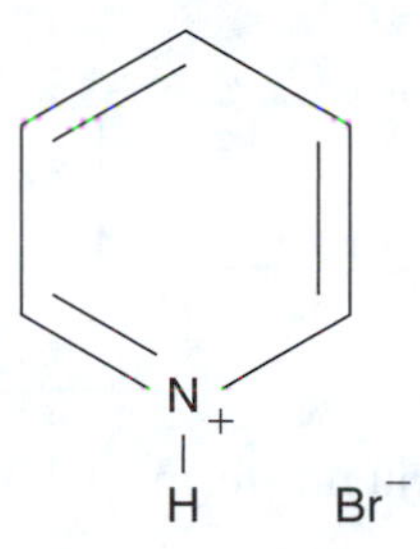

Problem 16.6 Place the following compounds in order of increasing water solubility.

$CH_3(CH_2)_6NH_2$ is a 7-carbon primary amine; $CH_3(CH_2)_{14}NH_3{}^+Cl^-$ is a 15-carbon ionic salt; $CH_3(CH_2)_{14}NH_2$ is a 15-carbon primary amine. The 15-carbon and 7-carbon amines are both insoluble in water. In general, the longer the chain, the less soluble the amine. The 15-carbon ionic salt is water soluble owing to the extensive interaction of the ionic portion with water.

$$CH_3(CH_2)_{14}NH_2 < CH_3(CH_2)_6NH_2 \ll CH_3(CH_2)_{14}NH_3{}^+Cl$$

Objectives Section 16.7 Classification of Amides

Are you able to...

- identify amides (Ex. 16.7, Prob. 16.7, Exercise 16.27)
- classify amides as primary, secondary, or tertiary (Ex. 16.8, Prob. 16.8)
- draw structures of constitutional isomers of amides (Exercises 16.28, 16.59)

Problem 16.7 Which of the following compounds are amides? Classify them as primary, secondary, or tertiary amides.

1. Tertiary amide; N is bonded to carbonyl and two other alkyl groups; N is not bonded to any H's
2. Secondary amide; N is bonded to carbonyl and aromatic ring and to one H
3. Not an amide; contains both an aldehyde and an amine, but the N is not directly bonded to the carbonyl C
4. Tertiary amide; N is bonded to a carbonyl group and to two other alkyl groups; N is not bonded to any H's

Objectives Section 16.8 Synthesis of Amides

Are you able to...

- describe the formation of amides by the reaction of carboxylic acids and ammonia
- describe the formation of amides by the reaction of carboxylic acids with primary and secondary amines
- describe the formation of amides by reaction of an acid halide and amines (Exercise 16.58)
- discuss the reaction conditions under which amide formation takes place (Ex. 16.8, Prob. 16.8, Exercises 16.29, 16.30, 16.58)
- write reactions for the formation of an amide (Ex. 16.9, Prob. 16.9, Exercises 16.29, 16.30)
- predict the products of amide synthesis (Exs. 16.8, 16.9, Probs. 16.8, 16.9, Exercises 16.29, 16.30, 16.56, 16.58)
- compare amide formation and esterification
- identify the parent carboxylic acid and amine of an amide (Exercises 16.31, 16.32)

Problem 16.8 Complete each of the following reactions. If no reaction takes place, write NR.

(a) benzoic acid and methylamine at 20–50°C will not form an amide; an acid-base reaction takes place.

$$C_6H_5C(=O){-}OH + H{-}N(H){-}CH_3 \longrightarrow C_6H_5C(=O){-}O^- + H{-}N^+(H)(H){-}CH_3$$

(b) Acetic acid and pryidine above 100°C will not form an amide, because pyridine is a tertiary amine; an acid-base reaction takes place.

$$CH_3C(=O){-}OH + C_5H_5N \xrightarrow{>100°C} CH_3C(=O){-}O^- + C_5H_5N^+{-}H$$

(c) Benzoic acid and dimethylamine above 100°C forms an amide.

$$C_6H_5C(=O){-}OH + H_3C{-}N(H){-}CH_3 \xrightarrow{>100°C} C_6H_5C(=O){-}N(CH_3)_2 + H_2O$$

(OH from the acid and H from the amine form H_2O)

Problem 16.9 Give the equation for amide formation between propanoic anhydride and dimethylamine.

Cleavage of –C—O—C– bonds produces the three-carbon carboxylic acid propanoic acid and the secondary amide *N,N*-dimethyl propamide.

$$CH_3CH_2C(=O){-}O{-}C(=O)CH_2CH_3 + H_3C{-}N(H){-}CH_3 \longrightarrow CH_3CH_2C(=O){-}OH + CH_3CH_2C(=O){-}N(CH_3){-}CH_3$$

Objectives Section 16.9 Polyamide Synthesis

Are you able to...

- discuss the formation of polyamides by condensation polymerization
- identify the reactants in polyamide reactions (Exercises 16.33, 16.57, 16.58)
- predict the repeating unit formed in polyamide reactions (Exercises 16.34, 16.57)
- compare polyester formation to polyamide formation

Objective Section 16.10 Naming Amides

Are you able to...

- name amides by using IUPAC rules (Ex. 16.10, Prob. 16.10, Exercises 16.37, 16.38)
- draw amide structures from their IUPAC names (Exercises 16.35, 16.36, 16.46)
- identify the acyl- and amino- parts of an amide

Problem 16.10 Name the following compounds.

(a) *N*-Butyl-*N*-methylpropanamide; N is attached to a methyl group, a butyl group, and a C3 carbonyl group from propanoic acid

(b) *N*-Phenylbenzamide; N is attached to a carbonyl carbon on a benzene ring from benzoic acid; the other group is a benzene ring named as an alkyl substituent, phenyl

(c) 2-Aminopropanal; not an amide; carbonyl C is not bonded directly to N

(d) *N*-Benzyl-*N*-methylisobutyramide; N is attached to a carbonyl carbon from isobutyric acid; the other groups to which N is attached are a benzyl group, $CH_2C_6H_5$, and a methyl group

Objectives Section 16.11 Physical and Basicity Properties of Amides

Are you able to...

- discuss the nature of secondary forces in amides (Exercise 16.40)
- describe the structure of a dipolar ion
- explain the relative magnitudes of melting and boiling points of amides compared with other organic families (Table 16.4, Exercises 16.39, 16.51, 16.52)
- explain the relative water solubility of amides compared with other organic families (Exercises 16.39, 16.51, 16.52)
- discuss the basicity of amides relative to other organic families (Exercises 16.41, 16.42)
- identify unknowns by using the properties of amides (Exercises 16.53, 16.54, 16.55, Expand Your Knowledge 16.62)

Objectives Section 16.12 Hydrolysis of Amides

Are you able to...

- discuss the hydrolysis of amides
- identify amide bonds
- explain the conditions under which hydrolysis of amide bonds takes place (Ex. 16.11, Prob. 16.11, Exercises 16.44, 16.58)
- write reactions for the hydrolysis of an amide (Ex. 16.11, Prob. 16.11, Exercises 16.43, 16.44)
- predict the product of acidic or basic hydrolysis of amides (Ex. 16.11, Prob. 16.11, Exercises 16.43, 16.44, 16.58, Expand Your Knowledge 16.64)

Problem 16.11 Complete the following equations. Indicate if no reaction takes place.

(a) The amide bond is cleaved under acidic conditions to produce aniline which is protonated by acid and forms the amine salt anilinium chloride. Butanoic acid also is produced.

$$CH_3CH_2CH_2C(=O){-}NH{-}C_6H_5 \xrightarrow[H_2O]{HCl} CH_3CH_2CH_2C(=O){-}OH + H{-}\overset{+}{N}H_2{-}C_6H_5Cl^-$$

Cleave Here (C–N bond); –H Here; –OH Here

(b) The amide bond is cleaved under basic conditions to produce the amine 1-propamine and the carboxylic acid, which is ionized by base to form the carboxylate salt sodium propionate.

$$CH_3CH_2C(=O){-}NH{-}CH_2CH_2CH_3 \xrightarrow[H_2O]{NaOH} CH_3CH_2C(=O){-}O^-\ Na^+ + H_2NCH_2CH_2CH_3$$

Cleave Here (C–N bond)

MAKING CONNECTIONS

Construct concept maps or write a paragraph describing the relation among the members of each of the following groups:

(a) amine, amide, ammonia
(b) carbonyl, amide, carboxylic acid, ester
(c) amine, salt, acid, base
(d) amine, carboxylic acid, acid, water
(e) primary amine, secondary amine, tertiary amine, quaternary ammonium salt
(f) amine, amide, secondary forces, physical properties,
(g) amine, amide, water solubility, secondary forces
(h) amine, amide, basicity
(i) water, solubility, ammonium salt, amine
(j) hydrolysis, amide, ester
(k) esterification, amide synthesis, polyamide synthesis
(l) polyester, polyamide, monomer
(m) polarity, hydroxyl group, carbonyl group, carboxyl group, amide group
(n) amide linkage, ester linkage, hydrolysis
(o) amines, amides, biological molecules, physiological activity

FILL-INS

The amine and amide families can be looked at as derivatives of _____. When an aliphatic or aromatic group replaces a hydrogen, an _____ is produced. The nitrogen may have _____ hydrogen atoms replaced by R groups. Classification of amines is based on the _____ nitrogen is bonded: in primary amines N is bonded to one R group; in secondary amines N is bonded to two R groups; in tertiary amines N is _____. In _____ N is bonded to aliphatic groups, whereas in an _____ there is at least one benzene ring bonded directly to the nitrogen. A nitrogen bonded to _____ characterizes amides. They also are classified as primary, secondary, or tertiary, depending on the _____. Amines and amides play very important roles in biological systems. Physiologically, amines and amides are found in _____. Alkaloids are amines that have _____ and are produced by _____. Opium _____ such as heroin and codeine are powerful _____ but are addictive. Caffeine is an amide that is a _____. Amine salts are important components of many medicines because they are more _____ than amines and can be administered as _____. In the stomach, free amines are converted into _____ owing to the _____ environment, which makes it easy for amines in oral medications to be _____. Amides are synthesized by reaction of a _____, or _____ with _____ or a _____. The competing _____ reaction dominates at low temperatures if one of the reactants is a carboxylic acid. Amide formation is a _____reaction that is similar to esterification in that the _____. Hydrolysis of amides proceeds under _____ or _____ conditions. In acid, the amine is _____ and the _____ is produced. In base, the carboxylic acid is _____ and the _____ is produced. Polyamides are analogous to polyesters and are formed by _____ reaction. The basicity of amines and amides is _____. Amines are _____, but amides are _____. The nonbonded electrons on nitrogen are _____ in amides than in amines because of the _____ of the carbonyl group on the amide nitrogen.

Answers

The amine and amide families can be looked at as derivatives of ammonia, NH_3. When an aliphatic or aromatic group replaces a hydrogen, an amine is produced. The nitrogen may have one, two, or three hydrogen atoms replaced by R groups. Classification of amines is based on the number of carbons to which the nitrogen is bonded: in primary amines N is bonded to one R group, in secondary amines N is bonded to two R groups; in tertiary amines N is bonded to three R groups. In aliphatic amines N is bonded to aliphatic groups, whereas in an aromatic amine there is at least one benzene ring bonded directly to the nitrogen. A nitrogen bonded to the carbon of a carbonyl group characterizes amides. They also are classified as primary, secondary, or tertiary, depending on the number of carbons to which the nitrogen is bonded. Amines and amides play very important roles in biological systems. Physiologically, amines and amides are found in proteins, nucleic acids, hormones, and vitamins. Alkaloids are amines that have physiological effects and are produced by plants. Opium alkaloids such as heroin and codeine are powerful pain relievers but are addictive. Caffeine is an amide that is a mild stimulant. Amine salts are important components of many medicines because they are more water soluble than amines and can be administered as aqueous solutions. In the stomach, free amines are converted into amine salts owing to the highly acidic environment, which makes it easy for amines in oral medications to be absorbed into the bloodstream. Amides are synthesized by reaction of a carboxylic acid, acid halide, or acid anhydride with ammonia or a primary or secondary amine. The competing acid-base reaction dominates at low temperatures if one of the reactants is a carboxylic acid. Amide formation is a reversible reaction that is similar to esterification in that the carboxylic acid undergoes substitution with dehydration between the carboxylic acid and the other reagent. Hydrolysis of amides proceeds under strongly acidic or strongly basic conditions. In acid, the amine is protonated and the amine salt is produced. In base, the carboxylic acid is ionized and the carboxylate salt is produced. Polyamides are analogous to polyesters and are formed by the same type of condensation polymerization reaction. The basicity of amines and amides is different. Amines are weak bases, but amides are about as basic as water. The nonbonded electrons on nitrogen are less available in amides than in amines because of the electron-withdrawing effect of the carbonyl group on the amide nitrogen.

TEST YOURSELF

1. Draw a reaction map for amines and amides; include protonation of amines in water, reaction of amines in acid, amide synthesis, polyamide synthesis, acid and basic hydrolysis of amides.
2. Show how each of the following molecules interacts with water: a primary alcohol, a carboxylic acid, a primary amine, an amide, and a primary amine salt.
3. For the molecules in Question 2, show how each molecule interacts with another molecule of the same type.
4. On the basis of your drawings in Question 2, what can you say about the water solubility of each molecule?
5. On the basis of your drawings in Question 3, what can you say about the strength of secondary forces of each molecule?
6. Why must amine medications that are injected be in salt form, but amine medications taken orally do not have to be?
7. What organic families are isomers for the molecular formula C_4H_9NO?
8. Given the following data, for the same number of moles of each compound, how many molecules of every 100 of each sample are ionized?

Compound	Percent ionization
NaOH	100
CH_3NH_2	6.3
$C_6H_5NH_2$	0.0062
H_2O	0.0001

9. Which of the compounds in Question 8 is the strongest base? Which is the weakest base? Explain your answer. How do the acidities of all compounds in Question 8 compare?

Answers

1.

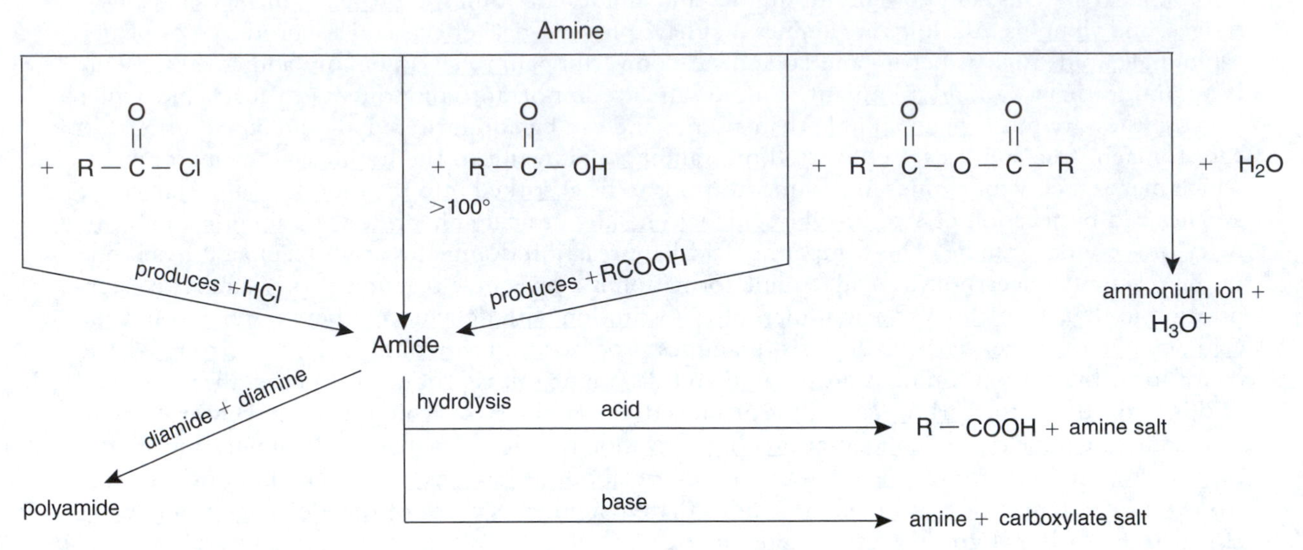

2.

δ+ H δ− O H δ+

H bonds

1° alcohol

1° amine

Carboxylic acid

Amide

Salt

3.

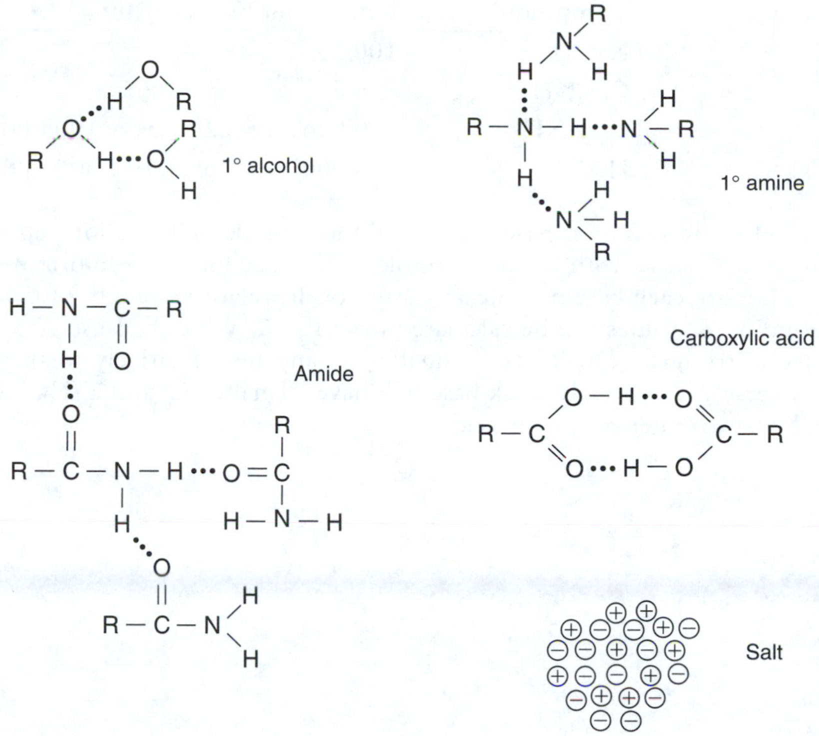

1° alcohol

1° amine

Amide

Carboxylic acid

Salt

4. Primary alcohols and primary amines each form two hydrogen bonds with water; carboxylic acids and amides each form two hydrogen bonds and an additional dipolar attraction between the carbonyl oxygen and hydrogen on water; the amine salt is ionic and is hydrated by water molecules attracted to the charged parts of the molecule. The ionic salt is most soluble; the carboxylic acid and amide are next; the primary alcohol and amine are last.
5. Alcohols and primary amines can each form hydrogen bonds with other molecules like themselves. The hydrogen bonds in alcohols are a bit stronger owing to the greater polarity of the –O–H bond relative to the –N-H bond. Carboxylic acids have a hydrogen bond and a dipole attraction between two molecules and exist as dimers. Amides have three dipolar attractions between every two molecules. Amine salts are ionic compounds and are solids.
6. When medications are injected, they are immediately introduced into an aqueous environment from which they must be absorbed. The solubility of an amine is increased when it is in its salt form and absorption can occur more quickly and to a greater degree. The salt form allows aqueous solutions of higher concentrations to be prepared. Medications taken orally will first encounter the acidic environment of the stomach. In this case, the amine will be converted into salt because of the acidity of stomach fluid and will be rendered more soluble and easily absorbed as a result.
7. Alcohols, ketones, amines, and amides can all be constructed from the molecular formula C_4H_9NO.

8.

Compound	Number of molecules ionized out of every 100
NaOH	100
CH_3NH_2	6.3
$C_6H_5NH_2$	0.0062, or 6.2 of every 100,000
H_2O	0.0001, or 1 of every 1,000,000

9. NaOH is the strongest base because all the formula units are ionized in water. Water is the weakest base, with the fewest molecules ionized in an equilibrium dissociation. The K_b values of each base provide a measure of the relative strength of the base. The corresponding K_a values can be calculated from the K_b values because $K_a \times K_b = K_w$. A relatively strong base will have a smaller K_a and be a relatively weaker conjugate acid. Conversely, a relatively weak base will have a smaller K_b and a relatively larger K_a and will be a stronger conjugate acid.

chapter 17

Stereoisomerism

OUTLINE

17.1 Review of Isomerism

A. There are different classes of isomers (Fig. 17.1); all are different compounds that have the same molecular formula.

B. Structural isomers
  1. differ in the way that atoms are connected to one another
     a. different carbon skeletons
     b. different placement of functional group on various carbon skeletons
     c. different placement of functional group on the same carbon skeleton

C. Stereoisomers
  1. differ in ways in which atoms are oriented in space relative to one another
     a. enantiomers are nonsuperimposable mirror images of one another
     b. diastereomers are nonsuperimposable on one another but are not mirror images
        i. cis-trans isomers are diastereomers
        ii. diastereomers contain stereocenter carbons

Key Terms configuration, conformation, connectivity, constitutional isomer, diastereomer, enantiomer, geometric stereoisomer, stereocenter, stereoisomer, structural isomer, tetrahedral stereocenter, trigonal stereocenter

Key Figure Figure 17.1 Different types of isomers

17.2 Enantiomers

A. Two compounds are enantiomers if (Figs. 17.5, 17.8)
  1. they contain a tetrahedral stereocenter; stereocenter is a carbon bonded to four different substituents
  2. they are mirror images of each other
  3. they are not superimposable on each other (Fig. 17.7)

B. Enantiomers are chiral.
  1. chiral compounds have handedness (Figs. 17.2, 17.3, 17.4)
  2. achiral compounds do not have handedness

C. Enantiomers can be represented in different ways.
  1. wedge-bond structures show bonds extending in front of the plane of the paper as solid wedges and bonds extending behind the plane of the paper as dashed wedges (Fig. 17.6)
  2. Fischer projections look like crosses, but horizontal lines are understood to extend in front of the plane of the paper and vertical lines are understood to extend behind the plane of the paper (Fig. 17.6)

Key Terms achiral, chiral, chirality, enantiomer, Fischer projection, handedness, mirror image, mirror-image condition, nonsuperimposable condition, tetrahedral stereocenter, wedge-bond

Key Figures

Figure17.2	Chiral objects
Figure 17.3	Chirality of hands
Figure 17.4	Achiral objects
Figure 17.5	Enantiomers of 2-butanol
Figure 17.6	Alternate representations of 2-butanol enantiomers
Figure 17.7	Mirror-image molecules of 2-methyl-2-butanol
Figure 17.8	Flowchart for determining if a structural formula represents a single compound or a pair of enantiomers

Example 17.1	Determining chirality in compounds and drawing structural formulas of enantiomers
Problem 17.1	Determining chirality in compounds and drawing structural formulas of enantiomers

17.3 Interpreting Structural Formulas of Enantiomers

A. Molecular models are the best way to unambiguously represent molecules.

B. If two-dimensional representations are used, one representation is designated as the orientation with which the other will be compared (Fig. 17.9).
   1. this normalizes the observation angle
   2. if Fischer projections are used, any two structures can be compared by rotating the non-designated molecule by 180° in the plane of the paper
   3. rotations are limited to 180° and must be kept in the plane of the paper

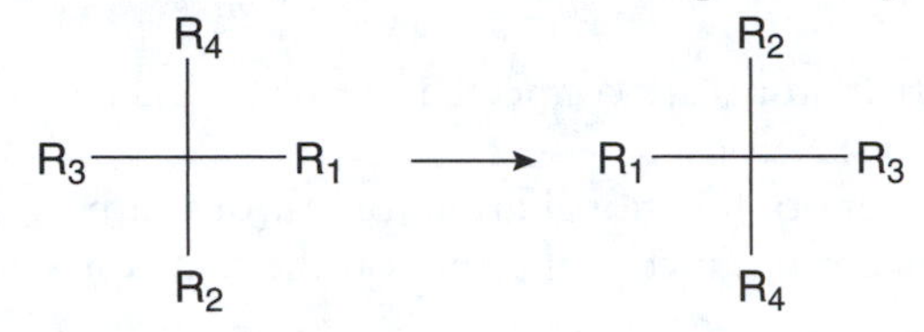

C. To determine if the representations are enantiomers or the same compound (Fig. 17.10),
   1. are the two drawings superimposable?
      a. If yes, they are the same compound
      b. If no, are they mirror images? If yes, they are a pair of enantiomers
   2. If no, rotate one drawing 180° in plane of paper and go back to the first step.

Key Figures

Figure 17.9	Different representations of one pair of enantiomers of 1-bromo-1-chloroethane
Figure 17.10	Procedure for determining if two structural drawings represent the same compound or a pair of enantiomers

Example 17.2	Determining whether two structural drawings represent a pair of enantiomers by using or the same compound
Problem 17.2	Determining whether two structural drawings represent a pair of enantiomers by using or the same compound

17.4 Nomenclature for Enantiomers

A. Enantiomers are distinguished by the use of one of two systems.
   1. (R) / (S) system designates the order in which ranked substituents fall relative to each other when viewing the molecule in a specified orientation (Box 7.1)
   2. D/L system can be used under the following circumstances:
      a. tetrahedral stereocenter has a hydrogen substituent
      b. tetrahedral stereocenter also has a heteroatom substituent (–OH, $-NH_2$) bonded to it
      c. the tetrahedral stereocenter has the carbons of two different R groups bonded to it

B. Fischer projections drawn from a designated perspective can be used to identify D/L isomers.
   1. R substituents are placed on vertical line (extend back behind plane)
      a. carbon (attached to tetrahedral stereocenter) with fewest hydrogens is at the top
      b. carbon (attached to tetrahedral stereocenter) with most hydrogens is at the bottom

2. H and heteroatom substituents are on horizontal line (extend in front of plane)
   a. if heteroatom is bonded on left side, enantiomer is L-
   b. if heteroatom is bonded on right, enantiomer is D-
3. figure may be rotated 180° in plane of paper to arrive at this orientation

Key Terms D/L system, (R) / (S) system
Key Box 7.1 The R/S nomenclature system for enantiomers

Example 17.3 Identifying D- and L-enantiomers
Problem 17.3 Identifying D- and L-enantiomers

17.5 Properties of Enantiomers

A. Chiral compounds and achiral compounds are different in two ways:
   1. physical property of optical activity differs
      a. chiral compounds rotate the plane of polarized light and are optically active
      b. the D- and L-enantiomers rotate polarized light to exactly the same degree in opposite directions
         i. clockwise is dextrorotatory, (+)
         ii. counterclockwise is levorotatory, (−)
         iii. (+) and (−) do not correspond to D and L respectively
      c. for standard solution, degree to which polarized light is rotated is called the specific rotation
         i. degree of rotation is measured by using a polarimeter (Fig. 17.11)
         ii. specific rotation, [α], is related to measured rotation and both the concentration of the solution and the length of the sample through which light must pass

$$[\alpha] = \frac{\alpha}{CL}$$

← this is measured by polarimeter
← these are given; C is concentration in g/mL and L is path length in decimeters

      d. equimolar mixtures of D/L enantiomers cancel each other and no rotation is seen; such mixtures are called racemic mixtures
      e. achiral compounds do not rotate the plane of polarized light and are not optically active
   2. chemical property of chemical recognition differs (Fig. 17.12)
      a. chiral compounds usually have different chemical activity toward chiral molecules
         i. think of this property as being analogous to trying to fit a glove: right hands fit right-hand gloves but not left-hand gloves; gloves are chiral, and so are hands
         ii. active sites of many enzymes are chiral and will bind with only one enantiomer (Fig. 17.12)
      b. achiral compounds usually have the same chemical reactivity toward chiral molecules
B. Other physical properties of pairs of enantiomers are the same: melting point, boiling point, solubility, density.

Key Terms active site, analyzer, chiral discrimination, chiral recognition, dextrorotatory, levorotatory, optical activity, optically active, optically inactive, plane-polarized light, polarimeter, polarizer, racemate, racemic mixture, specific rotation

Key Figures
Figure 17.11 Polarimeter measurement of optical activity
Figure 17.12 Chiral recognition of enantiomers by chiral enzyme

Example 17.4 Calculation of specific rotation from observed rotation and concentration information
Problem 17.4 Calculation of observed rotation from specific rotation and concentration information

17.6 Compounds Containing Two or More Tetrahedral Stereocenters

A. As the number of tetrahedral stereocenters increases, the number of possible stereoisomers increases.
   1. 2^n is maximum possible where n is the number of stereocenters
   2. many biological molecules contain numerous stereocenters

B. When two tetrahedral stereocenters have different sets of substituents, four isomers that are all diastereomers exist.
   1. exist as diastereomers and enantiomers
      a. diastereomers are stereoisomers that are neither superimposable nor mirror images
      b. there are two pairs of enantiomers

C. When two tetrahedral stereocenters have the same set of substituents, there are three possible stereoisomers that are diastereomers of each other.
   1. a pair of enantiomers
   2. a meso compound

Key Terms diastereomer, meso compound

Example 17.5 Determining stereoisomers of compounds with two tetrahedral stereocenters
Problem 17.5 Determining stereoisomers of compounds with two tetrahedral stereocenters

17.7 Cyclic Compounds Containing Tetrahedral Stereocenters

A. When cyclic compounds have tetrahedral stereocenters, a pair of enantiomers can exist (Fig. 17.13).

B. To be a tetrahedral stereocenter, a ring carbon must meet the following criteria:
   1. it must be attached to two different nonring substituents
   2. the ring must not be symmetric with respect to the carbon in criterion 1

Key Figure
Figure 17.13 Mirror images of cyclic molecules

Example 17.6 Determining stereoisomers of cyclic compounds
Problem 17.6 Determining stereoisomers of cyclic compounds

ARE YOU ABLE TO... AND WORKED TEXT PROBLEMS

Objectives Section 17.1 Review of Isomerism

Are you able to...

- distinguish between different classes of isomers (Exercise 17.37)
- classify isomers as structural or stereoisomers
- explain the features of structural isomers
- draw constitutional isomers of a molecular formula (Exercises 17.1, 17.2, 17.3, 17.4)
- explain the features of stereoisomers
- draw stereoisomers of a molecular formula (Exercises 17.5, 17.6)
- explain the relation between enantiomers and diastereomers
- define stereocenter
- explain how connectivity, configuration, and conformation differ

Objectives Section 17.2 Enantiomers

Are you able to...

- determine if two compounds are enantiomers (Ex. 17.1, Prob. 17.1, Exercises 17.9, 17.10)
- identify tetrahedral carbon stereocenters (Exercises 17.9, 17.10, 17.31, 17.32, 17.39)
- list the criteria for compounds to be enantiomers (Ex. 17.1, Prob. 17.1)
- identify chiral objects (Exercises 17.7, 17.8)
- explain chirality and chiral compounds
- represent chiral compounds by using Fischer projections (Ex. 17.1, Prob. 17.1)
- interpret wedge-bond and Fischer representations of structures (Ex. 17.1, Prob. 17.1, Exercise 17.37)

Problem 17.1 Which of the following compounds can exist as a pair of enantiomers and which cannot? If enantiomers are possible, draw structural formulas to represent the enantiomers.

(a) 3-Hydroxypropanal contains an aldehyde at the number 1 carbon and an alcohol at the number 3 carbon. There are no tetrahedral carbons bonded to four different substituents. Carbon 1 is not tetrahedral; carbon 2 is bonded to two hydrogens; carbon 3 is bonded to two hydrogens.

(b) 2-Hydroxypropanal contains one chiral carbon. Carbon number 2 is bonded to a $-CH_3$, an $-OH$, an $-H$, and an aldehyde, $-CHO$. This compound exists as a pair of enantiomers.

```
    H   OH  O              CHO              CHO
    |   |   ||              |                |
H — C — C*— C — H     H — C — OH       HO — C — H
    |   |                   |                |
    H   H                  CH3              CH3

                            I                II
```

Objectives Section 17.3 Interpreting Structural Formulas of Enantiomers

Are you able to...

- represent structural formulas by using Fischer drawings
- compare Fischer representations to determine the relation between them (Ex. 17.2, Prob. 17.2, Exercises 17.11, 17.12, 17.37)
- orient Fischer drawings to compare them (Ex. 17.2, Prob. 17.2, Exercises 17.11, 17.12)

Problem 17.2 Indicate whether each of the following pairs of structural drawings represents a pair of enantiomers or the same molecule.

(a) As drawn, structures 1 and 2 are not superimposable and they do not appear to be mirror images. Rotation of the vertical group in structure 1 by 180° in the plane of the paper produces a structure that is a nonsuperimposable mirror image of structure 2. These structures are a pair of enantiomers.

(b) Structures 3 and 4 are not superimposable and they are not mirror images. Rotation of the vertical group in structure 3 by 180° in the plane of the paper produces a structure that is superimposable on structure 4. These structures represent the same compound.

Objectives Section 17.4 Nomenclature for Enantiomers

Are you able to...

- identify the two systems used to distinguish enantiomers
- explain the circumstances in which D/L designations can be assigned
- designate Fischer projections of enantiomers as D or L (Ex. 17.3, Prob. 17.3, Exercises 17.13, 17.14)
- draw Fischer projections of enantiomers from their names and D/L designation (Exercises 17.19, 17.20)

Problem 17.3 Name the following enantiomer of alanine as D- or L-.

The structure must be oriented with the R substituents on the chiral carbon oriented vertically, the substituent with the carbon with the fewest hydrogens ($-COOH$) pointing upward and the substituent with the carbon with the most hydrogens ($-CH_3$) pointing downward. The heteroatom substituent ($-NH_2$) is on the left, so the enantiomer is L-alanine.

Objectives Section 17.5 Properties of Enantiomers

Are you able to...

- explain the ways in which chiral compounds differ from achiral compounds (Exercises 17.15, 17.16)
- explain the way in which chiral compounds differ from each other (Expand Your Knowledge 17.40)
- explain how enantiomers are similar to each other (Exercises 17.21, 17.22)
- discuss the property of optical activity
- describe the relation between optical activity and D/L designation (Exercise 17.18)
- explain how optical activity is measured
- explain the terms dextrorotatory and levorotatory

- describe the design and function of a polarimeter
- calculate the specific rotation of a compound (Ex. 17.4, Exercises 17.17, 17.23, 17.24)
- use the specific rotation of a compound to calculate the observed rotation of a solution (Prob. 17.4, Exercises 17.25, 17.26)
- explain how a racemic mixture affects optical activity (Exercises 17.15, 17.16, Expand Your Knowledge 17.41)
- discuss the chemical property of chemical recognition (Exercise 17.38, Expand Your Knowledge 17.43)
- explain the relation between chiral compounds and chemical recognition

Problem 17.4 D-Glucose has a specific rotation of +53°. Calculate the rotation of a solution of D-glucose at a concentration of 30 g/L when measured in a 15-cm-long sample cell.

First make sure that all units for the measurements given are in the appropriate units to use the specific rotation equation: concentration is in g/L and should be in g/mL and cm should be changed to dm: 30 g glucose/L ⇒ 30 g glucose/1000 mL, or 0.030 g glucose/mL; 15 cm ⇒ 1.5 dm.

The specific rotation, +53°, is given and you are asked to find the observed rotation for the sample. The form of the equation is:

$$\text{this is given} \rightarrow [\alpha] = \frac{\alpha}{CL} \quad \begin{array}{l} \leftarrow \text{this needs to be calculated} \\ \leftarrow \text{these are given} \end{array}$$

Rearrange the equation to solve for α: $C \times L \times [\alpha] = \alpha$

$(0.030 \text{ g/mL}) \times (1.5 \text{ dm}) \times (+53°) = +2.4°$

Objectives Section 17.6 Compounds Containing Two or More Tetrahedral Stereocenters

Are you able to...

- relate the number of tetrahedral stereocenters to the number of possible stereoisomers (Exercises 17.31, 17.32)
- determine the isomers that exist when two tetrahedral stereocenters have different sets of substituents (Ex. 17.5, Exercises 17.27, 17.28)
- determine the isomers that exist when two tetrahedral stereocenters have the same set of substituents (Prob. 17.5, Exercises 17.27, 17.28)
- identify meso compounds (Prob. 17.5, Exercises 17.27, 17.28)
- explain the relation between enantiomers and diastereomers
- draw stereoisomers of a Fischer projection (Exercises 17.29, 17.30)

Problem 17.5 Draw structural formulas of all stereoisomers of tartaric acid, 2,3-dihydroxy-1,4-butanedioic acid. Which stereoisomers are optically active? Which are enantiomers and which are diastereomers?

There are two stereocenters and both have the same set of substituents attached to the tetrahedral carbon, so a meso compound and one pair of enantiomers produce three diastereomers. The enantiomers are optically active.

```
     COOH            COOH              COOH              COOH
      |               |                 |                 |
 HO— C*—H        H— C*— OH         H— C*— OH        HO— C*— H
      |               |                 |                 |
 HO— C*—H        H— C*— OH        HO— C*— H          H— C*— OH
      |               |                 |                 |
     COOH            COOH              COOH              COOH

             Meso                              Enantiomers
```

Objectives Section 17.7 Cyclic Compounds Containing Tetrahedral Stereocenters

Are you able to...

- determine if a cyclic compound has a tetrahedral carbon stereocenter (Ex. 17.6, Prob. 17.6, Exercises 17.33, 17.34, 17.35, 17.36)

- draw the isomers of a cyclic compound with a tetrahedral stereocenter (Ex. 17.6, Prob. 17.6, Exercises 17.33, 17.34)
- describe the relation between the isomers of a cyclic compound with a tetrahedral stereocenter (Ex. 17.6, Prob. 17.6)

Problem 17.6 Draw the stereoisomers of (a) chlorocyclohexane and (b) 1-chloro-3-methylcyclohexane. Which stereoisomers are optically active? Which are enantiomers and which are diastereomers?

(a) Chlorocyclohexane has a tetrahedral carbon with two different nonring substituents. They are superimposable and the ring is symmetric. There are no carbon stereocenters and the compound is not optically active. Chlorocyclohexane exists as a single compound.

(b) 1-Chloro-3-methylcyclohexane has two tetrahedral carbons with two different nonring substituents. The ring is asymmetric. There exist a pair of nonsuperimposable cis isomers and a pair of nonsuperimposable trans isomers. Each member of one enantiomeric pair is also a diastereomer of the two members of the other enantiomeric pair. All four compounds are optically active.

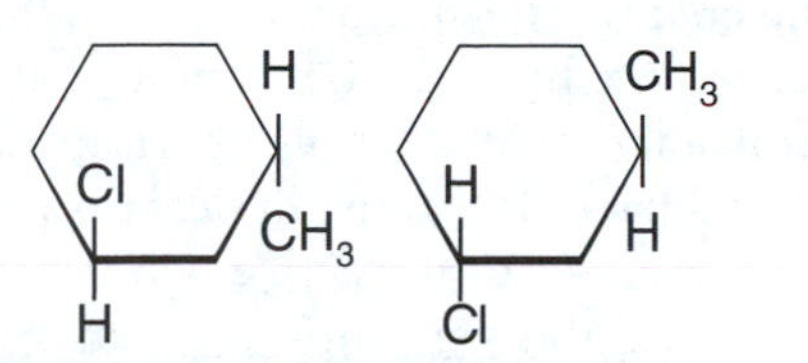

pair of trans enantiomers

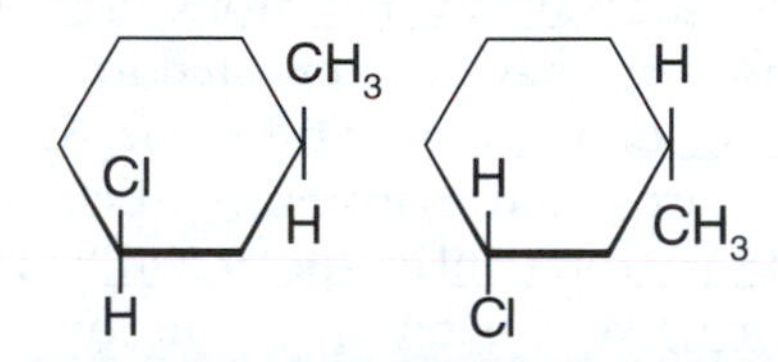

pair of cis enantiomers

MAKING CONNECTIONS

Draw a concept map or write a paragraph describing the relation among the members of each of the following groups:

(a) isomer, stereoisomer, structural isomer, enantiomer, diastereomer
(b) connectivity, conformation, configuration
(c) molecular formula, enantiomer, diastereomer
(d) mirror image, enantiomer, diastereomer, constitutional isomer
(e) Fischer projection, wedge-bond figure, enantiomer
(f) chiral carbon, enantiomer, stereocenter
(g) chiral carbon, handedness, nonsuperimposable mirror image
(h) D/L, enantiomer, optical activity, (+)/(−)
(i) D/L, Fischer projection, enantiomer
(j) physical properties, chemical properties, enantiomers, constitutional isomers
(k) specific rotation, measured rotation, enantiomer
(l) specific rotation, levorotatory, dextrorotatory, polarized light
(m) polarimeter, specific rotation, polarized light, measured rotation
(n) specific rotation, measured rotation, solution concentration, path length
(o) tetrahedral stereocenters, stereoisomers, meso compounds
(p) possible isomers, tetrahedral stereocenters with two different substituents, tetrahedral stereocenters with two identical substituents, cyclic compounds

FILL-INS

_____ and _____ are the elements that distinguish constitutional isomers from stereoisomers. The _____ defines different structural isomers. Stereoisomers may be connected in the same way, but _____ is different. Stereoisomers that are nonsuperimposable mirror images of one another are called _____. Enantiomers have _____ physical properties except for _____. They are optically _____, whereas achiral compounds are optically _____. Chemically, enantiomers may exhibit _____ behavior. This property is especially important when chiral molecules in the body interact with _____, such as enzymes. Many biological molecules have specific orientations in space and interact only with molecules that _____. For example, enzymes have _____ that have a "handedness" that requires a molecule to _____ in order for the enzyme to do its job. The differences in biological activity between enantiomers can produce results that vary from _____ to _____. Many drugs are synthesized as _____ in which one of the enantiomers produces a beneficial response and one produces a negative response. Molecular shape also plays a role in the interaction between receptors for _____ and molecules in food. Diastereomers are _____ but are not _____. Diastereomers usually have _____ physical properties. The existence of stereoisomers of a molecule can be determined by examining the _____ to carbons in the molecule. A carbon atom that is _____ bonded to _____ substituents is a _____ and is characteristic of _____ compounds. Fischer projections are a means of representing _____ molecules in _____ space. The protocol for drawing Fischer projections allows for the _____ to determine their relation.

Answers

Connectivity and conformation are the elements that distinguish constitutional isomers from stereoisomers. The way in which atoms are connected to one another defines different structural isomers. Stereoisomers may be connected in the same way, but the orientation of atoms relative to one another in three dimensions is different. Stereoisomers that are nonsuperimposable mirror images of one another are called enantiomers. Enantiomers have the same physical properties except for the way in which they rotate plane-polarized light. They are optically active, whereas achiral compounds are optically inactive. Chemically, enantiomers may exhibit very different behavior. This property is especially important when chiral molecules in the body interact with other chiral molecules, such as enzymes. Many biological molecules have specific orientations in space and interact only with molecules that complement them. For example, enzymes have active sites that have a "handedness" that requires a molecule to fit the site in order for the enzyme to do its job. The differences in biological activity between enantiomers can produce results that vary from negligible to deadly. Many drugs are synthesized as racemic mixtures in which one of the enantiomers produces a beneficial response and one produces a negative response. Molecular shape also plays a role in the interaction between receptors for smell and taste and molecules in food. Diastereomers are stereoisomers but are not enantiomers. Diastereomers usually have different physical properties. The existence of stereoisomers of a molecule can be determined by examining the number and identity of substituents bonded to carbons in the molecule. A carbon atom that is tetrahedrally bonded to four different substituents is a stereocenter and is characteristic of chiral compounds. Fischer projections are a means of representing three-dimensional molecules in two-dimensional space. The protocol for drawing Fischer projections allows for the comparison of different isomers to determine their relation.

TEST YOURSELF

1. Explain what the following Fischer projections tell you about the three-dimensional aspects of the molecules that they represent:

(a)
CHO
H—OH
CH_2OH

(b)
CHO
H—OH
H—OH
CH_2OH

2. Draw a ball-and-stick representation by using wedge bonds for the following molecule:

$$\begin{array}{c} CH_2OH \\ H - \!\!\!+\!\!\! - OH \\ CH_2OH \end{array}$$

3. What physical property would distinguish (−)-carvone, spearmint oil, from (+)-carvone, caraway seed oil? How would this distinction change if you measure the specific rotation of a racemic mixture of (−)-carvone and (+)-carvone?
4. The compound known as amphetamine is a mixture of the following isomers. Dexedrine is the commercial name given the compound on the left and Benzedrine is a mixture of both isomers. Dexedrine alone is a more powerful central-nervous-system stimulant than Benzedrine. What type of a mixture is Benzedrine?

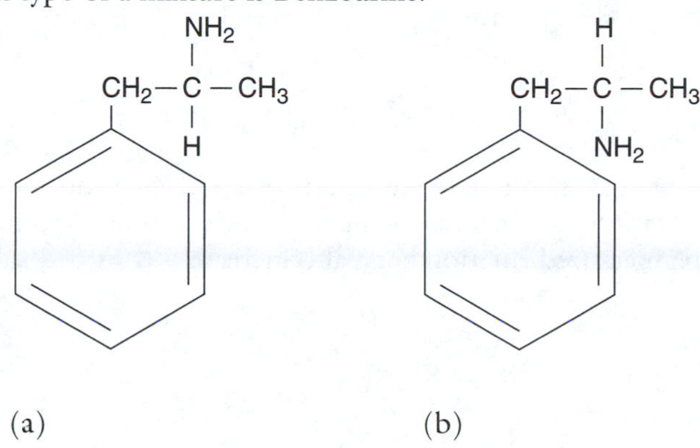

5. How can you explain the difference in physiological activity of the two isomers in Question 4? How can you explain the similarity in physiological activity of the two isomers?
6. Assign D and L designations to the isomers in Question 4.
7. The relation between the specific rotation of a molecule in solution and the measured rotation is given through the following equation: $[\alpha] = \alpha / C\,L$
   What changes in the specific rotation, $[\alpha]$, will result from changing either the concentration, C, or the pathlength, L? What changes in the measured rotation, α, will result from changing either the concentration, C, or the pathlength, L?

Answers

1. Structure (a) is based on a three-carbon chain. There is an aldehyde at C1 and two alcohols at C2 and C3. C2 is a chiral carbon, so this molecule exists as a pair of enantiomers. Structure (b) is based on a four-carbon chain. There is an aldehyde at C1 and three alcohols at C2, C3, and C4. There are two chiral carbons, so this molecule exists as enantiomers.
2. 

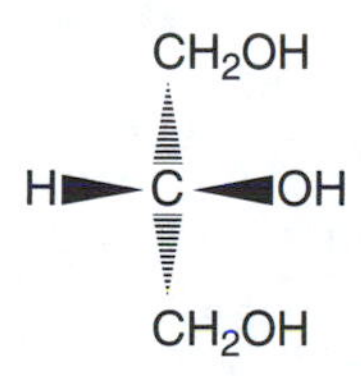

3. The only physical property that differs is the way in which each enantiomer of carvone rotates polarized light: they each rotate polarized light to the same degree, but in opposite direc-

tions. A racemic mixture is not optically active, because the rotations cancel each other.

4. Benzedrine is a racemic mixture.
5. Both the differences and the similarities in physiological activity can be traced to the shapes of the molecules. The fact that they exhibit similar activities implies that they both interact with the same molecules in the body. The difference may be due to the extent of the interaction caused by different binding abilities because of the subtle shape differences.
6. Reorient the structures so that they follow Fischer projection protocol. Structure (a) has the heteroatom group ($-NH_2$) on the right; it is the D-isomer and, in fact, is called dextroamphetamine. Structure (b) has the heteroatom group ($-NH_2$) on the left; it is the L-isomer.

CH_2–C_6H_5 / H — C — NH_2 / CH_3

(a)

CH_2–C_6H_5 / H_2N — C — H / CH_3

(b)

7. The specific rotation is an inherent property of a molecule and does not change. In order for the specific rotation to remain unchanged, if the solution concentration or path length are increased, the measured rotation must also increase.

chapter 18
Carbohydrates

OUTLINE

Introduction

A. Biochemistry is the study of chemical activities within individual cells and among organizations of cells.
B. The relation between molecular and cellular structure and function is an important theme in biochemistry.
C. Functions of life include:
   1. extraction of energy and materials from the environment
   2. use of materials and energy to maintain the structure of the organism and to carry out essential processes
   3. reproduction
D. There is a remarkable similarity in biomolecules and biochemical processes taking part in the functions of life.
E. Biomolecules can be studied with respect to
   1. their structures
      a. the components
      b. organization of components
      c. relation between molecular structure and physiological function
   2. transmission of information
      a. molecular basis for storage of genetic information
      b. molecular basis for transmission of genetic information
      c. molecular basis for translation of genetic information
   3. metabolism: all physical and chemical processes by which an organism extracts, transforms, and uses material and energy from its environment
F. Both inorganic and organic compounds are important to sustain life.
   1. inorganic compounds include ions and small molecules
      a. an inorganic compound may be a tiny part of the structure of an organic molecule but may be vital to the proper functioning of that organic molecule
      b. physiological functions may depend on differences in concentrations of inorganic ions
   2. organic compounds include
      a. carbohydrates: provide energy and structural component
      b. lipids: form cell membranes, provide energy, act as hormones
      c. proteins: structural material, enzymes, transport oxygen, immune system function, act as hormones
      d. nucleic acids: store, transmit hereditary information, direct and control protein synthesis

Key Terms biochemistry, biomolecule, carbohydrate, lipid, metabolism, nucleic acid, protein

18.1 *Introduction to Carbohydrates*

A. Carbohydrates are the most abundant family of organic compounds found in nature.
   1. also called saccharides
   2. -ose ending used for smaller saccharides
   3. larger polysaccharides are given common names

B. Carbohydrates perform a variety of functions.
   1. provide and store energy
   2. are precursors for synthesis of other biomolecules
   3. play a role in cell recognition functions and processes
   4. are components of nucleic acids

C. Carbohydrate structures:
   1. every carbohydrate has hydroxyl groups; they are polyhydroxy molecules
   2. every carbohydrate has either an aldehyde or ketone
      a. if it is an aldehyde, it is called a polyhydroxyaldehyde
      b. if it is a ketone, it is called a polyhydroxyketone
   3. monosaccharides
      a. simple saccharides that play important biological roles
      b. are monomers for larger saccharides
   4. disaccharides: contain two monosaccharide units
   5. oligosaccharides
      a. a saccharide that is larger than a monosaccharide but smaller than a polysaccharide
      b. can be categorized by using prefixes: di-, tri- tetra-, and so forth
   6. polysaccharides
      a. contain many monosaccharide units
      b. are structural and energy-storage molecules

Key Terms carbohydrate, disaccharide, monomer, monosaccharide, oligosaccharide, polyhydroxyaldehyde, polyhydroxyketone, polysaccharide, repeat unit, residue, saccharide, sugar, tetrasaccharide, trisaccharide

18.2 *Monosaccharides*

A. Monosaccharide name denotes classification:
   1. the prefixes for the number of carbons in the chain and the carbonyl functional group are joined to an -ose ending (Figs. 18.1, 18.2)

      EXAMPLE: aldopentose ketohexose

   2. the prefix for the type of carbonyl in the molecule

      aldose ⇒ ald- denotes aldehyde (Fig. 18.1)
      ketose ⇒ ket- denotes ketone (Fig. 18.2)

   3. the prefix for the number of carbons in the chain

      triose ⇒ tri- denotes three carbons pentose ⇒ pent- denotes five carbons
      tetrose ⇒ tetr- denotes four carbons hexose ⇒ hex- denotes six carbons

   4. the prefix D- or L- identifies the enantiomer
      a. biochemical systems use D sugars exclusively
      b. if no D/L designation, assume D sugar
      c. D/L is determined by placement of –OH at tetrahedral stereocenter farthest from the carbonyl carbon

B. Important monosaccharides:
   1. hexoses
      a. D-glucose: called blood sugar or dextrose; used as structural component and energy source
      b. D-fructose: called levulose; found bonded to glucose in fruit and table sugars
      c. D-galactose: found bonded to glucose in mammalian milk

2. pentoses
   a. D-ribose
      i. component of deoxyribonucleic acid (DNA) as D-ribose and 2-deoxy-D-ribose
      ii. deoxy sugars have one fewer oxygen atom than corresponding saccharide
   b. D-xylose: component of plant polysaccharides
3. trioses: both are important in metabolic processes
   a. D-glyceraldehyde
   b. dihydroxyacetone

Key Terms aldose, blood sugar, D-galactose, D-glucose, D-glyceraldehyde, D-ribose, D-xylose, deoxysaccharide, deoxysugar, dextrose, dihydroxyacetone, fructose, hexose, ketose, lactose, levulose, pentose, tetrose, triose

Key Figures
Figure 18.1 D-Aldoses
Figure 18.2 D-Ketoses

Example 18.1 Classification of monosaccharides according to number of carbons and identity of carbonyl group
Problem 18.1 Classification of monosaccharides according to number of carbons and identity of carbonyl group
Example 18.2 Recognizing structural relations between saccharides
Problem 18.2 Recognizing structural relations between saccharides

18.3 Cyclic Hemiacetal Structures

A. Hexoses and pentoses exist primarily as cyclic hemiacetals.
   1. six-membered cyclic hemiacetals are called pyranose rings (Fig. 18.3)
   2. five-membered cyclic hemiacetals are called furanose rings (Fig. 18.4)
B. The open-chain structure shown by Fischer projection can be converted into the hemiacetal structure, called a Haworth projection, as follows:
   1. turn the Fischer projection sideways with C1 closest to you
   2. bring the C5 –OH group close to the carbonyl group
   3. add the –OH to the carbonyl group
   4. break the –OH bond
   5. bond the H of the –OH to the carbonyl O
   6. bond the O of the –OH to the carbonyl carbon
C. The formation of a hemiacetal creates a new stereocenter called the hemiacetal carbon.
   1. the hemiacetal carbon is the one with two different O– groups attached to it: an –OH and an –OR
   2. the position of the –OH on the hemiacetal carbon relative to the CH_2OH group on C5 produces anomers
      a. anomers are diastereomers that differ in their configuration at the hemiacetal carbon
      b. if the –OH and $–CH_2OH$ groups are trans to one another, it is an α-anomer
      c. if the –OH and $–CH_2OH$ groups are cis to one another, it is a β-anomer
   3. –OH groups that were on the right in the Fischer projection are positioned downward from the ring
   4. –OH groups that were on the left in the Fischer projection are positioned upward from the ring

Key Terms anomer, cyclic hemiacetal, furanose ring, Haworth projection, Haworth structure, hemiacetal carbon, open-chain structure, pyranose ring

Key Figures
Figure 18.3 Hemiacetal formation in D-glucose
Figure 18.4 Hemiacetals in an equilibrium mixture of D-fructose

Example 18.3 Drawing cyclic structure of an aldohexose
Problem 18.3 Drawing the cyclic structure of a ketohexose

18.4 Chemical and Physical Properties of Monosaccharides

A. The presence of multiple –OH groups permits extensive hydrogen bonding
   1. between monosaccharides; they are solids at room temperature
   2. between monosaccharides and water; they are very soluble in water
   3. concentrated solutions are viscous liquids
   4. most have a sweet taste (Table 18.1)

B. Solutions of monosaccharides are equilibrium systems.
   1. open-chain, α-hemiacetal, and β-hemiacetal forms interchange rapidly; rings open, close, reopen
   2. equilibrium produces mutarotation in aqueous solutions
      a. each form contributes to the observed optical activity of the solution at equilibrium
      b. at equilibrium, a measurable optical activity results from the equilibrium percentages of each component
      c. each anomer has its own specific rotation that can be measured in its pure form

C. Aldehydes and hydroxyketones can be oxidized with mild oxidizing agents such as Benedict's reagent.
   1. the reaction takes place between the aldehyde of the open-chain form and the oxidizing agent
   2. the hemiacetal is not oxidized
   3. aldehyde is oxidized to –COOH with corresponding Benedict's reagent color change from blue to red owing to precipitate formation
   4. α-hydroxy ketoses are converted into aldoses
   5. the oxidation is quantitative and can be used to correlate concentrations with observed color changes; forms the basis for using dipsticks to test for glucose in urine
   6. sugars that are oxidized under these conditions are called reducing sugars

D. An acetal forms when a monosaccharide (in hemiacetal form) and an –OH from another molecule undergo a dehydration reaction.
   1. acid catalyst is required in the laboratory, but enzymes catalyze the reaction in living systems
   2. reaction takes place preferentially at the hemiacetal –OH group
   3. creates an acetal carbon attached to two –OR groups

E. Carbohydrate acetals are called glycosides.
   1. glycosidic linkages include the acetal C and the two –OR groups
   2. glycoside linkages can be α or β, depending on which hemiacetal, α or β, takes part in reaction
   3. glycosides are not reducing sugars
   4. glycosides do not undergo mutarotation
   5. glycosides can be hydrolyzed with water in the presence of acid or enzyme to produce the saccharide and alcohol

F. Other monosaccharide derivatives:
   1. phosphate esters can be formed by reaction of an –OH on a saccharide with phosphoric acid
   2. acidic sugars have a carboxyl group, –COOH, at the position on the ring where a CH_2OH was located
   3. aminosugars have an amino group, $-NH_2$, at the position on the ring where an –OH was located

Key Terms acidic sugar, aminosugar, Benedict's reagent, glycoside, glucose oxidase test, glycosidic linkage, mutarotation, phosphate ester, reducing sugar

Key Table Table 18.1 Sweetness of various compounds relative to sucrose

Example 18.4 Predicting products of monosaccharide reactions with methanol

Problem 18.4 Predicting products of monosaccharide oxidation with Benedict's reagent

18.5 Disaccharides

A. Two monosaccharide units joined by a glycosidic linkage form a disaccharide; visualize link as a dehydration between the hemiacetal –OH of one monosaccharide and an –OH of another monosaccharide.
B. Disaccharides may be produced by the joining of two monosaccharides together or by the breakdown of larger polysaccharides.
C. Disaccharides are characterized according to the following features:
  1. the monosaccharide units
  2. the nature of the glycosidic linkage; that is, is it an α- or a β-linkage?
  3. the atoms through which the monosaccharides are linked
    a. are they both linked through hemiacetal –OH groups?
    b. is one linked through a hemiacetal –OH and the other through an alcohol –OH? if yes, which –OH group participates?
  c. notation used looks like this: α $(1 \rightarrow 4)$
    i. α: indicates the configuration of the glycosidic linkage between the two monosaccharides (THIS IS NOT THE SAME as the α that refers to the anomer)
    ii. 1: indicates the position of the hemiacetal carbon in the same monosaccharide
      C1 in aldoses
      C2 in ketoses
    iii. 4: indicates the position of the carbon on the other monosaccharide to which it is attached
D. Important disaccharides:
  1. maltose (malt sugar or corn sugar)
    a. consists of two glucose units linked by a glycosidic bond
    b. glycosidic bond in maltose is characterized by an α $(1 \rightarrow 4)$ linkage
      i. C1 on first glucose is joined to C4 of second glucose
      ii. configuration at the linkage on C1 in the glycosidic linkage is α
    c. maltose has a free hemiacetal on one of the two glucose units; so it can open to the aldehyde and is a reducing sugar
    d. produced by the partial hydrolysis of starch by the enzyme amylase in the mouth and small intestine
    e. maltose is hydrolyzed to two glucose monosaccharides by the enzyme maltase
  2. lactose (milk sugar)
    a. consists of a galactose unit and a glucose unit linked by a glycosidic bond
    b. glycosidic bond in lactose is characterized by a β $(1 \rightarrow 4)$ linkage
      i. C1 on galactose is joined to C4 on glucose
      ii. configuration at the linkage on C1 is β
    c. lactose has a free hemiacetal on the glucose unit; so it can open to the aldehyde and is a reducing sugar
    d. lactose is found in mammalian milk and provides energy for nursing babies
    e. lack of the enzyme lactase, which cleaves lactose, is a hallmark of lactose intolerance
  3. sucrose (table sugar and fruit sugar)
    a. consists of a glucose unit and a fructose unit linked by a glycosidic bond
    b. glycosidic bond in sucrose is characterized by α,β $(1 \rightarrow 2)$ linkage
      i. C1 on glucose is joined to C2 on fructose
      ii. linkage is a to C1 on glucose
      iii. linkage is b to C2 on fructose
    c. the anomeric positions in both monosaccharides are part of the glycosidic linkage, and there are no hemiacetals to open either ring; it is not a reducing sugar
  4. Cellobiose
    a. consists of two glucose units linked by a glycosidic bond
    b. glycosidic bond in cellobiose is characterized by a β $(1 \rightarrow 4)$ linkage

i. C1 on one glucose is joined to C4 on the other glucose unit
ii. Linkage is β at C1
c. cellobiose has a free hemiacetal on one glucose unit and can open to form the aldehyde; it is a reducing sugar
d. cellobiose is produced by partial hydrolysis of cellulose by the enzyme cellulase
e. cellobiose is hydrolyzed by the enzyme cellobiase

E. Carbohydrates must be digested (hydrolyzed) by enzymes to monosaccharides in order to be absorbed.

Key Terms cellobiase, cellobiose, cellulase, corn sugar, lactose, maltase, maltose, malt sugar, milk sucrose, sugar

Key Figures
Figure 18.5 Structure of maltose
Figure 18.6 Structure of cellobiose
Figure 18.7 Structure of lactose
Figure 18.8 Structure of sucrose

Example 18.5 Drawing disaccharide structures
Problem 18.5 Determining the monosaccharides formed when a disaccharide is hydrolyzed

18.6 Polysaccharides

A. Polysaccharides contain hundreds or thousands of monosaccharide units.
B. Polysaccharides perform different functions as determined by their structures.
C. Polysaccharide structures differ from one another by:
1. the identity of the constituent monosaccharide units
2. the characteristics of the –OH group(s) that participates in linking the monosaccharide
3. the type of glycosidic linkages when one of the –OH groups is a hemiacetal –OH
4. the extent or absence of branching in the polysaccharide

D. Important polysaccharides:
1. starch
a. starch is a storage molecule for D-glucose in plant cells
b. it can be subdivided into two types:
i. amylose: an unbranched chain of glucose units connected by $\alpha(1 \rightarrow 4)$ linkages (Fig. 18.9); tends to form helical structures that are easily solvated (Fig. 18.14)
ii. amylopectin: contains randomly branched chains of glucose units connected by $\alpha(1 \rightarrow 4)$ linkages in the chain and $\alpha(1 \rightarrow 6)$ linkages at the branch points (Figs. 18.10, 18.11)
2. glycogen
a. glycogen is a storage molecule for D-glucose in animals
b. contains randomly branched chains of glucose units connected by $\alpha(1 \rightarrow 4)$ linkages in the chain and $\alpha(1 \rightarrow 6)$ linkages at the branch points
c. more highly branched than amylopectin (Fig. 18.11)
3. cellulose
a. structural polysaccharide found in plants
b. straight chains of glucose are linked by $\beta\ (1 \rightarrow 4)$ glycosidic linkages (Fig. 18.12)
i. forms straight chains joined to one another by hydrogen bonds, conferring strength (Fig. 18.13)
ii. makes cellulose insoluble in water
c. $\beta(1 \rightarrow 4)$ linkages have a different shape than $\alpha(1 \rightarrow 4)$ linkages

E. Digestion of starches:
1. amylose, amylopectin, and glycogen are digested (hydrolyzed) by human enzymes
2. cellulose is not digested by humans because they do not have the enzyme cellulase, which cleaves the $\beta(1 \rightarrow 4)$ linkages between glucose units

F. Polysaccharides play a role in cell recognition as glycolipids and glycoproteins.
1. glycolipids have a sugar part and a lipid part

2. glycoproteins have a sugar part and a protein part
   a. lipid or protein is integrated into cell membrane and sugar is on cell surface
   b. attractive forces between sugar or protein part and other molecules play a role in the recognition by cells of foreign proteins and molecules and of host proteins and molecules
3. cell recognition processes depend on
   a. complementary secondary forces between cell receptors and other molecules
   b. complementary fit of shapes between cell receptors and other molecules
   c. ability of the cell to distinguish foreign molecules from host molecules and to discriminate between host molecules
4. processes that require cell recognition include fertilization, blood transfusion (Table 18.2), cell growth, and infection of cells by bacteria and viruses

Key Terms amylopectin, amylose, antibody, antigen, cell recognition, dextrin, dextrinase, glycogen, glycogenesis, glycogenolysis, glycolipid, glycoprotein, nutritional polysaccharide, polysaccharide, storage polysaccharide, structural polysaccharide

Key Figures and Table

Figure 18.9 Structure of amylose
Figure 18.10 Structure of amylopectin
Figure 18.11 Representation of branching in amylopectin and glycogen
Figure 18.12 Structure of cellulose
Figure 18.13 Chain conformations of cellulose
Figure 18.14 Helical conformation of amylose
Table 18.2 ABO blood-typing system

18.7 Photosynthesis

A. Photosynthesis is the process by which organisms synthesize carbohydrates from carbon dioxide, water, and sunlight.
   1. photosynthesis converts carbon in CO_2 (inorganic) into carbon in carbohydrates (organic) while releasing oxygen; called carbon fixation
   2. process is carried out by chlorophyll in plants
      a. $n\,CO_2 + n\,H_2O + \text{sunlight} \rightarrow (CH_2O)_n + n\,O_2$
      (CH_2O) = monosaccharides with $n = 3$–6
      b. chlorophyll absorbs energy from sunlight; the energy is used to drive reactions that make saccharides
      c. energy stored in saccharides is retrieved when they are metabolized

Key Terms carbon cycle, carbon fixation, oxygen cycle, photosynthesis
Key Figure Figure 18.15 Photosynthesis

ARE YOU ABLE TO... AND WORKED TEXT PROBLEMS

Objectives Introduction

Are you able to. . .
- define biochemistry
- list the functions associated with living systems
- identify biomolecules
- explain which aspects of biomolecules are important in understanding biochemistry

Objectives Section 18.1 Introduction to Carbohydrates

Are you able to. . .
- define carbohydrate
- identify a carbohydrate by its name

- list the roles played by carbohydrates
- describe the structural features common to all carbohydrates (Exercise 18.47)
- describe the structural features that distinguish one carbohydrate from another
- compare monosaccharide, disaccharide, oligosaccharide, and polysaccharide (Exercise 18.48)

Objectives Section 18.2 Monosaccharides

Are you able to. . .

- define the terms associated with monosaccharide nomenclature (Exercises 18.1, 18.2)
- classify monosaccharides by the type of carbonyl in the molecule (Ex. 18.1, Prob. 18.1)
- classify monosaccharides by the number of carbons in the chain (Ex. 18.1, Prob. 18.1)
- combine both classifications to identify a monosaccharide (Ex. 18.1, Prob. 18.1)
- identify D/L monosaccharides from Fischer projections (Exercises 18.50, 18.51)
- identify tetrahedral stereocenters in monosaccharides (Exercises 18.21, 18.22)
- list important hexoses, pentoses, and trioses (Exercise 18.53)
- recognize the structural relations between saccharides (Ex. 18.2, Prob. 18.2, Exercises 18.3, 18.4, 18.5, 18.6, 18.7, 18.8, 18.52, 18.55)

Problem 18.1 Classify each of the following monosaccharides to indicate both the type of carbonyl group and the number of carbons present

(1) Aldotriose; aldose and triose (three-carbon chain)
(2) Ketohexose; ketose and hexose (six-carbon chain)
(3) Aldopentose; aldose and pentose (five-carbon chain)
(4) Aldotetrose; aldose and tetrose (four-carbon chain)

Problem 18.2 For each of the following pairs of compounds, indicate whether the pair consists of different compounds that are (1) constitutional isomers or (2) stereoisomers that are enantiomers or (3) stereoisomers that are diasteromers or (4) not isomers.

(a) Stereoisomers that are diastereomers
(b) Aldopentose and ketohexose; not isomers
(c) Enantiomers; D/L forms of same isomer
(d) Aldotetrose and ketotetrose; constitutional isomers
(e) Stereoisomers that are diastereomers

Objectives Section 18.3 Cyclic Hemiacetal Structures

Are you able to. . .

- discuss the formation of pyranose and furanose rings from hexoses and pentoses (Ex. 18.3, Prob. 18.3, Exercises 18.11, 18.12)
- explain how a hemiacetal is formed
- discuss the use of Haworth and Fischer projections for monosaccharide structures (Exercises 18.9, 18.10)
- convert an open-chain Fischer projection into a cyclic Haworth projection (Ex. 18.3, Prob. 18.3, Exercises 18.15, 18.16, 18.17, 18.18, 18.56, 18.62)
- identify α- and β-anomers of cyclic hemiacetals (Exercises 18.13, 18.14)

Problem 18.3 Draw the cyclic structure of β-D-sorbose by reference to Figures 18.2 and 18.4.

Figure 18.2 gives the open-chain structure of D-sorbose and Figure 18.4 shows the formation of a hemiacetal from a ketose. In a ketose, the carbonyl carbon, C2, joins to the O of the –OH group on C5; the carbonyl O becomes the –OH at C2; it points upward in the β form.

CH_2OH
C=O
OH
HO
OH
CH_2OH
D-sorbose

CH_3OH
O OH
OH H
H CH_2OH
H OH
β-D-sorbose

Objectives Section 18.4 Chemical and Physical Properties of Monosaccharides

Are you able to. . .

- explain the influence that the many monosaccharide hydroxyl groups have on the properties of monosaccharides
- discuss the nature of secondary forces between monosaccharides
- discuss monosaccharide solubility in water
- identify the equilibrium components of monosaccharide solutions
- discuss mutarotation in equilibrium solutions of monosaccharides (Exercise 18.20)
- discuss the oxidation reactions of monosaccharides with Benedict's reagent (Exercises 18.19, 18.25)
- identify reducing sugars (Exercises 18.27, 18.28, 18.54)
- explain the features required for a sugar to be a reducing sugar
- discuss acetal formation between hemiacetal monosaccharides and –OH groups (Exercises 18.23, 18.24)
- predict products of acetal formation (Ex. 18.4, Prob. 18.4, Exercises 18.23, 18.24)
- explain the relation between glycosides and acetals
- compare glycosides with hemiacetal monosaccharides
- predict the product of the hydrolysis of a glycoside
- describe a glycosidic link
- identify glycosidic linkages in Haworth projections
- identify monosaccharide derivatives (Exercise 18.26)

Problem 18.4 Show the product(s) formed when Benedict's reagent oxidizes D-fructose.

D-fructose is an α-hydroxyketone and is converted into an aldohexose by oxidation with Benedict's reagent. The product is D-glucose. D-glucose reacts with Benedict's reagent to form

```
        O⁻
        |
        C=O
        |—OH
   HO—|
        |—OH
        |—OH
        |
      CH₂OH
```

Objectives Section 18.5 Disaccharides

Are you able to. . .

- identify the features that characterize disaccharides
- describe the glycosidic linkage features in disaccharides
- explain the notation used to describe glycosidic linkages (Exercises 18.29, 18.30)
- explain the origin of disaccharides
- identify common disaccharides (Exercise 18.53)
- identify the monosaccharide units in maltose, lactose, sucrose, and cellobiose
- identify the glycosidic linkages in maltose, lactose, sucrose, and cellobiose
- discuss the reducing ability of maltose, lactose, sucrose, and cellobiose and other disaccharides (Ex. 18.5, Prob. 18.5, Exercises 18.37, 18.38)
- identify sources of maltose, lactose, sucrose, and cellobiose
- list functions of maltose, lactose, sucrose, and cellobiose
- discuss the digestion of disaccharides by enzymes
- draw disaccharide structure from component monosaccharides (Ex. 18.5, Exercises 18.33, 18.34, 18.35, 18.36, 18.63)
- draw the monosaccharide products of disaccharide hydrolysis (Prob. 18.5, Exercises 18.31, 18.32)

Problem 18.5 What monosaccharides are formed when the following disaccharide is hydrolyzed? Is the disaccharide a reducing sugar? Does it undergo mutarotation? Characterize the linkage between monosaccharide residues as α or β.

For the disaccharide to be a reducing sugar, the disaccharide must be in equilibrium with an open-chain structure that is an aldehyde or α-hydroxyketone. The hemiacetal carbon of the five-membered ring is free to open and therefore undergoes mutarotation. It will form an α-hydroxyketone, which is a reducing sugar. The monosaccharides produced are D-fructose and D-mannose.

Objectives Section 18.6 Polysaccharides

Are you able to. . .

- define polysaccharide
- distinguish polysaccharides from other classes of saccharides
- list some functions of polysaccharides
- draw polysaccharides from their monosaccharide units (Exercises 18.43, 18.44, 18.64)
- identify the structural features that distinguish polysaccharides from each other
- identify the monosaccharide units in amylose, amylopectin, glycogen, and cellulose
- identify the glycosidic linkages in amylose, amylopectin, glycogen, and cellulose
- identify sources of amylose, amylopectin, glycogen, and cellulose
- compare the structures of amylose, amylopectin, glycogen, and cellulose (Exercises 18.39, 18.40)
- compare the functions of amylose, amylopectin, glycogen, and cellulose (Exercises 18.39, 18.40, 18.57)
- compare the properties of amylose, amylopectin, glycogen, and cellulose
- compare the hydrolysis and digestion of amylose, amylopectin, glycogen, and cellulose (Exercises 18.41, 18.42, 18.58, 18.59)
- explain the process of cell recognition
- discuss the role that polysaccharides play in cell recognition
- describe glycolipids and glycoproteins

Objectives Section 18.7 Photosynthesis

Are you able to. . .

- describe the process of photosynthesis (Fig. 18.15)
- write a balanced equation for a photosynthetic reaction (Exercises 18.65)
- compare the energy quality of carbon in carbon dioxide with that of carbon in carbohydrates
- identify immediate and ultimate sources of organic compounds for animals and plants (Exercises 18.45, 18.46)

MAKING CONNECTIONS

Construct a concept map or write a paragraph describing the relation among the members of each of the following groups:

(a) biochemistry, biomolecules, metabolism
(b) biomolecules, structure, biological function
(c) carbohydrate, protein, lipid, nucleic acid
(d) monomer, polymer, biomolecules
(e) carbonyl, aldose, ketose, monosaccharide, hydroxyl
(f) Fischer projection, Haworth projection, monosaccharide, isomer
(g) monosaccharide, disaccharide, oligosaccharide, polysaccharide (Exercise 18.48)
(h) glucose, galactose, fructose, ribose
(i) energy, structure, chemical messenger, carbohydrate
(j) hexose, pentose, triose
(k) pyranose ring, furanose ring, hexose, pentose
(l) hemiacetal, aldose, ketose
(m) α-anomer, β-anomer, cyclic hemiacetal
(n) mutarotation, open-chain monosaccharide, cyclic hemiacetal
(o) reducing sugars, glucose, galactose, fructose, aldose, ketose
(p) glycoside, acetal, monosaccharide
(q) hydrolysis, glycoside, dehydration
(r) aminosugar, acidic sugar, phosphate ester
(s) glycosidic link, enzyme, hydrolysis
(t) maltose, lactose, sucrose, cellobiose

(u) amylose, amylopectin, glycogen, cellulose
(v) cell recognition, glycolipid, glycoprotein
(w) photosynthesis, carbohydrate, carbon dioxide

FILL-INS

Carbohydrates are characterized by the presence of two types of functional groups: a carbonyl as either an _____ and multiple _____ groups. Called _____, or sugars, carbohydrates are classified in several ways. One classification scheme is based on the identity of the _____ and the number of _____. The other classification is based on the number of _____ in a carbohydrate. Carbohydrates are important _____ and _____ components. Larger polysaccharides such as _____ are storage molecules for glucose monomers. They differ in the extent to which _____. The end glucose units are _____ as needed for energy. The _____ nature of glycogen allows enzymes to access many end units when needed. The macrostructure of digestible starch is _____ owing to _____ between hydroxyl units along chains. Cellulose in plants is a _____ polysaccharide that is tough and fibrous owing to _____ between sheets of glucose chains. Although cellulose is made of _____, humans cannot digest cellulose because the bonds joining glucose units are of a type that cannot _____. Termites, on the other hand, have symbiotic microorganisms that have the _____ and readily digest cellulose for use as an energy source. The structures of monosaccharides can be represented by _____ Fischer projections or _____ Haworth projections.

Most monosaccharide molecules exist in the _____, which is produced when an _____ takes place. The linking of two monosaccharides to form a _____ is a _____ reaction, and the bond that forms is called a _____. If one of the disaccharide rings is able to open, it can react with _____. They are called _____ sugars.

Answers

Carbohydrates are characterized by the presence of two types of functional groups: a carbonyl as either an aldehyde or a ketone and multiple hydroxyl groups. Called saccharides, or sugars, carbohydrates are classified in several ways. One classification scheme is based on the identity of the carbonyl-containing group and the number of carbons in the chain. The other classification is based on the number of monomer units in a carbohydrate. Carbohydrates are important energy sources and structural components. Larger polysaccharides such as amylose, amylopectin, and glycogen are storage molecules for glucose monomers. They differ in the extent to which glucose side chains branch off from longer chains. The end glucose units are enzymatically cleaved as needed for energy. The highly branched nature of glycogen allows enzymes to access many end units when needed. The macrostructure of digestible starch is helical owing to very weak hydrogen bonding between hydroxyl units along chains. Cellulose in plants is a structural polysaccharide that is tough and fibrous owing to extensive hydrogen bonding between sheets of glucose chains. Although cellulose is made of glucose, humans cannot digest cellulose because the bonds joining glucose units are of a type that cannot be cleaved by human enzymes. Termites, on the other hand, have symbiotic microorganisms that have the enzyme cellulase and readily digest cellulose for use as an energy source. The structures of monosaccharides can be represented by open-chain Fischer projections or cyclic Haworth projections. Most monosaccharide molecules exist in the cyclic form, which is produced when an intramolecular hemiacetal reaction takes place. The linking of two monosaccharides to form a disaccharide is a dehydration reaction, and the bond that forms is called a glycosidic bond. If one of the disaccharide rings is able to open, it can react with mild oxidizing agents such as Benedict's reagent. They are called reducing sugars.

TEST YOURSELF

1. Identify the functional group and the number of carbons in the following monosaccharides:
   (a) a ketopentose (b) an aldotriose (c) a ketohexose (d) an aldohexose
2. What property of carbohydrates makes them suitable for playing a role in cell-recognition processes?
3. Glucose chains in cellulose form extensive hydrogen bonds between them. Why is cellulose insoluble in water?
4. If you let a saltine cracker sit in your mouth without chewing, you begin to taste sweetness. What is happening to the cracker?

5. Glycogen is a highly branched polysaccharide. Why is this structure an efficient way of storing glucose for energy needs?
6. Lactose intolerance results when the enzyme that cleaves lactose is missing or diminished. Can absorption of lactose occur in the absence of the enzyme? Explain your answer.
7. Compare the structural formulas of the disaccharides sucrose, lactose, and maltose. How are they similar? How are they different? Are any of the disaccharides reducing sugars?
8. Examine the structural formulas of glucose, fructose, and galactose. What are the molecular formulas of these monosaccharides? How are they related to one another? Are any of the monosaccharides reducing sugars?
9. How do the molecular formulas of the disaccharides in Question 7 compare with the molecular formulas of Question 8? Explain any differences.
10. Invert sugar, which is obtained by the hydrolysis of sucrose to fructose and glucose, has a relative sweetness of 1.58 compared with that of sucrose. Why is invert sugar sweeter than the original sucrose?

Answers

1. (a) five carbons, ketone (b) three carbons, aldehyde
   (c) six carbons, ketone (d) six carbons, aldehyde
2. The presence of many polar hydroxyl groups makes monosaccharides areas of negative and positive charge to which other polar or charged substances can be attracted.
3. The hydroxyl groups ineract to establish the strong three-dimensional structure characteristic of cellulose. Thus the high molecular weight and lack of hydroxyl groups to form hydrogen bonds with water make cellulose insoluble. Most of the hydroxyl groups in cellulose participate in the formation of intramolecular hydrogen bonds and are not available to interact with water.
4. The starch in the cracker is hydrolyzed by the enzyme amylase in your mouth. Specifically, maltose and short-chain polysaccharides are produced.
5. The compact form permits the storage of many glucose units in a small amount of space. The extensive branching presents a large surface area over which enzyme activity can take place. Each end is a glucose unit that can be cleaved by enzymes when energy is needed.
6. Lactose is a disaccharide. Disaccharides are too large for absorption to occur. They must be broken down into smaller monosaccharides.

7.

	Molecular formula	**Monosaccharide units**	**Glycosidic bonds**	**Reducing sugar**
Lactose	$C_{12}H_{22}O_{11}$	glucose + galactose	β (C1 galactose → C4 glucose)	Yes
Maltose	$C_{12}H_{22}O_{11}$	glucose + glucose	α (C1 glucose → C4 glucose)	Yes
Sucrose	$C_{12}H_{22}O_{11}$	glucose + fructose	α, β (C1 glucose → C2 fructose)	No

All are isomers of one another.

8.

	Molecular formula	**Reducing sugar**
Glucose	$C_6H_{12}O_6$	Yes
Galactose	$C_6H_{12}O_6$	Yes
Fructose	$C_6H_{12}O_6$	Yes

All are isomers of one another.

9. The disaccharides differ from the monosaccharides by two hydrogens and an oxygen (H_2O). H_2O is lost when a glycosidic bond is formed between two monosaccharides (dehydration).
10. The combined sweetness of individual fructose and glucose molecules results in a greater sweetness than that of the original disaccharide. The relation between the response of receptors for sweetness and different molecular structures is a reasonable explanation for the differences.

chapter 19

Lipids

OUTLINE

Introduction

A. The family of molecules called lipids is defined on the basis of a property, insolubility in water, rather than a specific structural component.
B. Lipids are soluble in nonpolar or low-polarity solvents.
C. Lipids have a variety of structural components.
D. Lipid function is related to lipid structure.
E. Lipids perform a variety of functions.
   1. energy storage
      a. lipids provide more energy than does glycogen
      b. carbon atoms in triglycerides contain more hydrogen and less oxygen than do those in glycogen
      c. carbon in triglycerides is more reduced, or less oxidized, than carbon in monosaccharides
   2. membrane components; some lipids are able to form nonpolar layers that separate two aqueous solutions
   3. chemical messengers
      a. hormones are synthesized and transported in the bloodstream to the site of action
      b. eicosanoids are local hormones that act in the same tissue in which they are synthesized
   4. digestive roles as bile salts; bile salts act as soaps and aid in fat digestion
   5. vitamins

Key Term lipid

19.1 Classifying Lipids

A. Lipids can be divided into hydrolyzable and nonhydrolyzable lipids depending on whether or not they undergo hydrolytic cleavage in the presence of acid, base, or digestive enzymes (Fig. 19.1)
   1. all hydrolyzable lipids contain at least one ester group; hydrolyzable lipids are called saponifiable
   2. some hydrolyzable lipids contain
      a. an amide
      b. a phosphate
      c. an acetal
   3. hydrolyzable lipids include
      a. triacylglycerols
      b. waxes
      c. glycerophospholipids
      d. sphingolipids

4. nonhydrolyzable lipids do not contain ester, amide, phosphate, or acetal groups; nonhydrolyzable lipids are called nonsaponifiable
5. nonhydrolyzable lipids include
   a. steroids
   b. eicosanoids
   c. fat-soluble vitamins

Key Terms amphipathic, hydrolyzable lipid, nonhydrolyzable lipid, nonsaponifiable, saponifiable
Key Figure Figure 19.1 Classification of lipids

19.2 Fatty Acids

A. A fatty acid is a carboxylic acid with a long hydrocarbon chain.
   1. the carboxyl part is called the "head"; the head is polar, or hydrophilic
   2. the hydrocarbon chain is called the "tail"; the tail is nonpolar, or hydrophobic
   3. the number of carbons in the chain is an even number in plant and animal fatty acids
   4. the chains are generally not branched in plant and animal fatty acids
   5. fatty acids are not usually found in uncombined free acid form

B. Fatty acids may be saturated or unsaturated (Table 19.1).
   1. saturated fatty acids have no carbon-carbon double bonds (C=C)
      a. all carbons (except carbonyl) have four single bonds
      b. there is free rotation around single bonds
   2. unsaturated fatty acids have one or more carbon-carbon double bonds
      a. if one, the fatty acid is monounsaturated
      b. if more than one, the fatty acid is polyunsaturated
      c. rotation around double bonds is restricted
   3. in plant and animal fatty acids, any double bonds are usually cis

C. Unsaturated fatty acids have an omega number, ω.
   1. indicates the position of the first carbon in the first double bond counting from the carbon farthest from the carboxyl group; ω-carbon is the carbon farthest from the hydroxyl group
   2. assignment of ω numbers is opposite of IUPAC system, which identifies the carboxyl group C as C number 1

D. The physical properties of fatty acids are related to the number of carbons in the hydrocarbon tail and the presence of double bonds (Fig. 19.2).
   1. the number of carbons in the hydrocarbon chain
      a. even small fatty acids have low water solubility
         i. the carboxyl group is polar and can form hydrogen bonds
         ii. the hydrocarbon tail is nonpolar and its properties dominate the molecule
      b. water solubility decreases even more as the length of the tail increases
   2. the presence of double bonds
      a. the number of double bonds is important
      b. the configuration (cis or trans) around double bonds is important

E. Melting points of fatty acids are directly related to the length of the hydrocarbon tail and the absence or presence of carbon-carbon double bonds (Fig. 19.2).
   1. the melting point increases when the tail length increases and decreases when the tail length decreases; longer tails have larger surface areas over which London forces act
   2. the melting point decreases as the number of carbon-carbon double bonds increases
      a. cis double bonds lower the melting point more than trans double bonds do
      b. multiple double bonds lower the melting point more than one double bond does
      c. double bonds introduce kinks into the molecule that change the molecular shape and disrupt the surface over which London forces act; cis double bonds bend the molecule more than trans double bonds do

Key Terms essential fatty acids, fatty acid, monounsaturated, nonessential fatty acids, omega number, polyunsaturated, saturated fatty acid, unsaturated fatty acid

Key Figure and Table
Figure 19.2 Molecular shapes of saturated and unsaturated fatty acids
Table 19.1 Common saturated and unsaturated fatty acids

Example 19.1 Recognizing naturally occurring fatty acids
Problem 19.1 Recognizing naturally occurring fatty acids

19.3 Structure and Physical Properties of Triacylglycerols

A. Animal fats and plant oils in our diets are triacylglycerols (Table 19.2).
  1. also called triglycerides
  2. fats and oils are mixtures of different triglycerides
    a. fats are solids at room temperature and generally come from animal sources
      i. triglycerides from animal sources (except fish) contain a higher percentage of saturated fatty acids than do triglycerides from plants
      ii. fewer double bonds and a higher degree of saturation allow stronger, more extensive secondary interactions
    b. oils are liquids at room temperature and generally come from plant sources
      i. triglycerides from plants (except in palm and coconut oils) contain a higher percentage of unsaturated fatty acids than do most triglycerides from animals
      ii. more double bonds and a higher degree of unsaturation weaken the strength of secondary interactions

B. Triacylglycerols are triesters of glycerol.
  1. glycerol backbone is esterified to three fatty acids

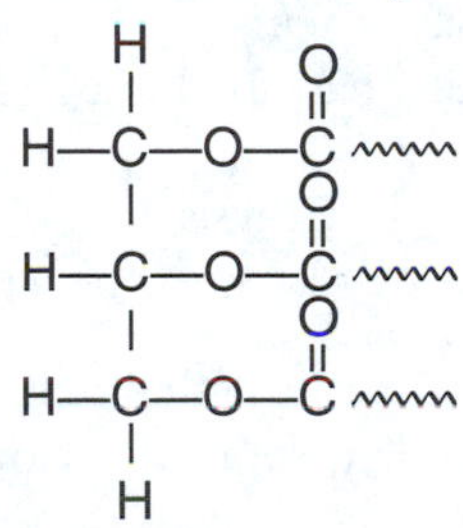

  2. the fatty acids may be the same or different
    a. simple triglycerides have the same three fatty acid tails
    b. mixed triglycerides have different fatty acid tails

Key Terms animal fat, mixed triacylglycerol, plant oil, saturated fat, simple triacylglycerol, triacylglycerol, triglyceride, unsaturated fat
Key Table Table 19.2 Fatty acid composition of triacylglycerols

Example 19.2 Drawing the structures of triacylglycerols
Problem 19.2 Drawing the structures of triacylglycerols

19.4 Chemical Reactions of Triacylglycerols

A. Triglycerides undergo hydrolysis, catalytic hydrogenation, and rancidity reactions.

B. Hydrolysis:
  1. under acidic or basic conditions or in the presence of digestive enzymes, the ester bonds can be cleaved to produce glycerol, a triol, and the corresponding fatty acids
  2. if this is done in base, it is called saponification; saponification produces glycerol and fatty acid salts instead of fatty acids
  3. digestion of dietary triglycerides is hydrolysis
    a. lipases (enzymes) break down fats with the help of bile salts
    b. products of hydrolysis include monoacyl- and diacylglycerides, fatty acids, and glycerol that diffuse across intestinal cell membrane

c. intestinal cells repackage components in a chylomicron
   i. chylomicron is an amphipathic particle that contains lipids and proteins (it is a lipoprotein)
   ii. chylomicrons transport lipids through aqueous lymphatic system and bloodstream
d. cells take components from chylomicron and use them for energy
e. if triglycerides are not used immediately, they are stored in adipose cells (fat cells)

C. Catalytic hydrogenation:
1. hydrogen is added to the carbons of double bonds in the presence of a Ni or Pt catalyst
   a. the process changes unsaturated triglycerides into saturated triglycerides
   b. the process converts liquid oils into solid fats
   c. hydrogenation can be partial (hydrogen added to some double bonds only) or total (hydrogen added to all double bonds)
      i. in partial hydrogenation, some remaining double bonds in cis configuration are changed to trans configuration
      ii. the process is called isomerization
      iii. partial hydrogenation produces trans fatty acids with properties similar to those of saturated fatty acids

D. Rancidity:
1. triglycerides can be oxidized by air or bacteria causing them to develop unpleasant odors and flavors; in this state, the oxidized material is called rancid
2. oxidation by air
   a. breaks alkene double bonds
   b. each carbon of the double bond is converted into a carboxylic acid
      i. one has a single carboxyl group
      ii. the other is a dicarboxylic acid
   c. products are small, volatile, bad smelling
3. oxidation by bacteria
   a. breaks the ester groups by hydrolysis
   b. produces carboxylic acids that proceed to be oxidized by air

Key Terms adipocyte, adipose cell, catalytic hydrogenation, chylomicron, fat cell, hardening, isomerization, lipase, rancidity, saponification

Example 19.3 Writing equations for the hydrolysis of triacylglycerols
Problem 19.3 Writing equations for the hydrolysis of triacylglycerols
Example 19.4 Writing equations for hydrogenation of triacylglycerols
Example 19.4 Writing equations for hydrogenation of triacylglycerols
Example 19.5 Writing equations for the oxidation of fatty acids in air (rancidity reactions)
Problem 19.5 Writing equations for the oxidation of fatty acids in air (rancidity reactions)

19.5 Waxes

A. Most waxes are mixtures of esters of fatty acids and long-chain alcohols.
1. the alcohol usually has only one –OH group
2. both the fatty acid and the alcohol usually have an even number of carbons, 14–36
3. both the fatty acid and the alcohol are usually straight chains

B. Waxes play protective roles.
1. coatings on plants keep water in and prevent excessive water loss
2. coatings on plants keep parasites out
3. coatings on feathers waterproof them
4. bee secretions are used for structural purposes
5. some organisms use waxes for energy storage

Key Term wax

19.6 Amphipathic Hydrolyzable Lipids

A. Amphipathic hydrolyzable lipids can be divided into two broad categories (Fig. 19.3):
   1. phosphoglycerides
   2. sphingolipids

B. Amphipathic structure has both polar and nonpolar parts.
   1. these molecules can create nonpolar barriers in aqueous environments
   2. they can be used to form cell membranes that keep two different aqueous environments separated

C. Glycerophospholipids (Fig. 19.3):
   1. structural components are
      a. glycerol-derived backbone
      b. two esterified fatty acid chains
      c. a phosphodiester from phosphoric acid
      d. a group esterified to the phosphate group
   2. dehydration of components initially forms phosphatidic acid; further esterification with an alcohol replaces the acidic –H on the phosphate group and produces the phosphodiester (Table 19.3)
      a. this group is from an alcohol and determines the subcategory of the molecule
      b. it can be an amine alcohol such as ethanolamine, choline, or serine
         i. if ethanolamine, it is called a cephalin
         ii. if choline, it is called a lecithin
      c. it can be from a saccharide such as inositol

D. Sphingolipids (Fig. 19.3):
   1. structural components are
      a. sphingosine-derived backbone (an unsaturated, 16-C amino alcohol)
      b. an amide group at C2
      c. either a phosphate group or a saccharide unit at C1
         i. if it is a phosphate, the molecule is a sphingophospholipid; the same alcohols as those in glycerophospholipids are esterified to phosphate groups (Table 19.3)
         ii. if it is a saccharide, the molecule is a sphingoglycolipid; saccharide is joined through an acetal link; saccharide may be a mono- or oligosaccharide

E. Both sphingoglycolipids and glycerophospholipids are found in cell membranes; percentages and identities vary according to the function of the membrane.

Key Terms cephalin, glycerophospholipid, lecithin, phospatidyl, phosphatic acid, phosphoglyceride, phospholipid, sphingoglycolipid, sphingolipid, sphingophospholipid

Key Figure and Table
Figure 19.3 Structural comparison of different hydrolyzable lipids
Table 19.3 Alcohols in glycerophospholipids

Example 19.6 Drawing the structures of glycerophospholipids
Problem 19.6 Drawing the structures of glycerophospholipids
Example 19.7 Drawing the structures of sphingolipids
Problem 19.7 Drawing the structures of sphingolipids

19.7 Steroids: Cholesterol, Steroid Hormones, and Bile Salts

A. Steroids are characterized by the steroid ring structure (Fig. 19.4):

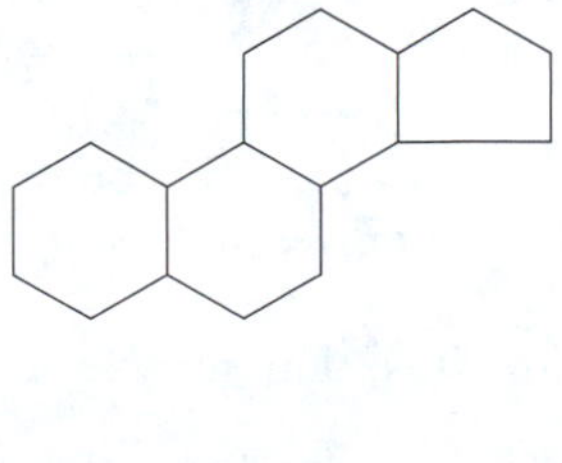

1. cholesterol is a steroid
    a. it is a component of cell membranes
    b. it is a precursor to steroid hormones
    c. it is a precursor to bile salts
2. some hormones and bile salts are steroids

B. Steroid hormones regulate a variety of cell functions.
1. adrenocortical hormones are produced by adrenal glands at the top of the kidneys
2. cortisol regulates response to short-term stress by
    a. directing cells to increase metabolism of fats and proteins for energy
    b. blocking immune system (anti-inflammatory and antiallergy)
3. aldosterone regulates electrolyte concentrations; stimulates Na^+, K^+, and water reabsorption by kidneys
4. sex hormones are classified as
    a. female
        i. estrogens
        ii. progestins
    b. male: androgens
    c. sex hormones regulate the development of primary and secondary sexual characteristics (Boxes 19.2, 19.3)

C. Bile salts:
1. bile salt has a carboxylate group attached to a steroid ring
    a. bile salts are amphipathic
    b. the carboxylate group is hydrophilic
    c. the steroid ring is hydrophobic
2. bile salts aid in lipid digestion by acting as an emulsifier of lipids
    a. bile salts break large lipid globules into smaller ones much as soap breaks up dirt molecules
    b. the smaller droplets are more accessible to digestive lipases
3. bile salts are synthesized from cholesterol in the liver and stored in the gall bladder as part of bile
4. bile salts are also needed for efficient absorption of fat-soluble vitamins

Key Terms adrenocortical hormone, androgen, bile, bile acid, bile salt, cholesterol, estrogen, hormone, progestin, sex hormone, steroid

Key Figure and Boxes

Figure 19.4 Steroids
Box 19.2 Menstrual cycle and contraceptive drugs
Box 19.3 Anabolic steroids

19.8 Eicosanoids

A. Eicosanoids are nonhydrolyzable lipids derived from arachidonic acid; arachidonic acid is a polyunsaturated (four double bonds) 20-C fatty acid (Figure 19.5).

B. Eicosanoids can be divided into three groups that are derivatives of the arachidonic structure:
1. leukotrienes
    a. basic arachidonic structure is maintained
    b. amine, amide, hydroxyl, and thiol functional groups are added as parts of substituents on the main structure
2. prostaglandins: cyclopentane is incorporated into the 20-C chain of arachidonic acid
3. thromboxanes: six-membered cyclic ether is incorporated into the 20-C chain of arachidonic acid

C. Eicosanoids act as local hormones and act in the tissue within which they are synthesized; they are not transported in the bloodstream.

D. Eicosanoids play a part in:
1. inflammatory responses in joints, skin, muscle, and eyes

2. the production of pain and fever in disease and injury
3. blood-pressure regulation
4. labor induction
5. regulation of wake–sleep cycles
6. allergic and asthmatic reactions

Key Terms eicosanoid, leukotriene, local mediator, prostaglandin, thromboxane
Key Figure Figure 19.5 Eicosanoids

19.9 Fat-Soluble Vitamins

A. Vitamins are classified as water soluble or fat soluble.
B. Water-soluble vitamins contain enough polar groups to form hydrogen bonds with water.
   1. Vitamins B and C are water soluble
   2. water-soluble vitamins can be excreted in urine and do not build to toxic levels easily
C. Fat-soluble vitamins are nonpolar, nonhydrolyzable lipids (Table 19.4).
   1. Vitamins A, D, E, and K are fat soluble
   2. fat-soluble vitamins may be stored in fatty tissue and can build up to toxic levels

Key Terms fat-soluble vitamin, vitamin, water-soluble vitamin
Key Table Table 19.4 Fat-soluble vitamins

19.10 Biological Membranes

A. The organization of cells in organisms requires that environments inside and outside cells be separated.
B. Biological membranes maintain the separation and regulate the flow of material into and out of cells.
C. Functions of biological membranes:
   1. they regulate the movement of materials from the environment into a cell and of cellular products from inside the cell to the outer environment
      a. membrane proteins act as transfer sites
      b. small nonpolar molecules can diffuse through the membrane
   2. they play a key role in cell-recognition processes and act as information-transfer sites
      a. membrane proteins may interact with other molecules
      b. glycolipids and glycoproteins act as receptor sites
   3. they maintain concentration gradients on either side of the membrane; protein pumps use energy to maintain concentration gradients
D. Structure of biological membranes:
   1. the fluid-mosaic model of biological membranes describes membrane structure (Fig. 19.6)

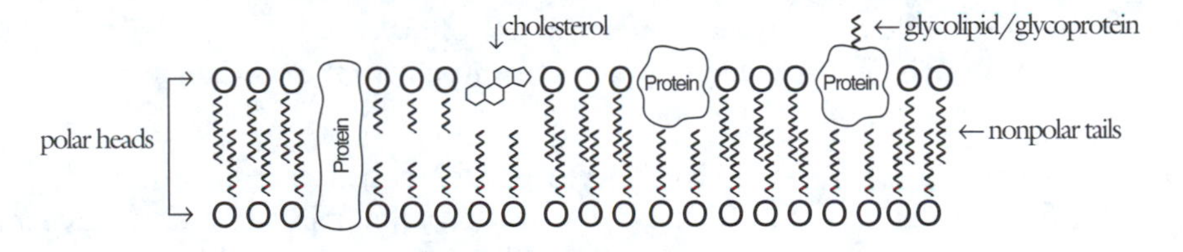

      a. the membrane contains amphipathic lipids
         i. the polar heads face inward and outward to the aqueous environments and form the inner and outer surfaces of the membrane
         ii. the nonpolar tails form the layer between the polar heads
         iii. lipids may have carbohydrates attached and are called glycolipids
         iv. lipid aggregation is stereospecific
      b. the membrane contains proteins
         i. if they span the entire membrane, the proteins are integral; integral proteins have both hydrophobic and hydrophilic areas

   ii. if they are embedded in one side of the membrane, the proteins are peripheral; peripheral proteins associate with the polar heads of the lipid layer
   iii. proteins diffuse laterally through the fluidlike matrix of the membrane
   iv. proteins may have carbohydrates attached and are called glycoproteins
  c. the identities and percentages of individual components vary according to the function of the membrane
  d. the fluidity of the matrix depends on:
   i. temperature: lower temperatures result in decreased motion of lipid tails
   ii. percentage of cholesterol in membrane: high levels make membrane more rigid
   iii. identity of fatty acid tails: shorter, unsaturated tails increase fluidity
 2. membrane construction is based on:
  a. secondary attractive forces between areas of similar polarity
   i. attractive forces between hydrophobic tails are maximized
   ii. attractive forces between hydrophilic heads and aqueous environments are maximized
  b. repulsive forces between areas of opposite polarity; structure prevents interactions between hydrophobic tails and aqueous environments

E. Methods of transport through biological membranes:
 1. simple transport is diffusion down a concentration gradient (Fig. 19.7)
  a. small, uncharged molecules move from areas of higher concentration to areas of lower concentration
  b. molecules that diffuse are O_2, N_2, H_2O (although current research indicates that water molecules may get help from proteins), and urea
 2. facilitated transport is transport down a concentration gradient with help from transport proteins (Fig. 19.7)
  a. proteins called transporters or permeases assist transport
  b. molecules that are transported in this way include glucose, chloride ions, and bicarbonate ions
 3. active transport moves something against the concentration gradient and requires energy (Fig. 19.7)
  a. proteins act as pumps to move materials from areas of lower concentration to areas of higher concentration
  b. the process is coupled to an energy-producing metabolic reaction in order to supply the energy needed
 4. exocytosis (Fig. 19.8)
  a. the movement of large molecules out of the cell is called exocytosis
  b. materials are packaged in vesicles that fuse with the outer membrane and release contents outside the cell
 5. endocytosis (Fig. 19.8)
  a. the intake of large molecules into the cell is called endocytosis
  b. materials are packaged in vesicles that fuse with the outer membrane, allowing the contents to be released inside the cell

Key Terms active transport, channel, endocytosis, exocytosis, extrinsic protein, facilitated transport, fluid-mosaic model, gate, head, intrinsic protein, lipid bilayer, Na^+/K^+ pump, organelle, permease, pump, simple transport, tail, transporter, vesicle

Key Figures

Figure 19.6 Fluid-mosaic model of cell membrane
Figure 19.7 Transport across biological membranes
Figure 19.8 Exocytosis and endocytosis

ARE YOU ABLE TO... AND WORKED TEXT PROBLEMS

Objectives Introduction

Are you able to. . .

- identify the property that defines a lipid
- identify some lipid groups
- describe the variety of functions performed by lipids

Objectives Section 19.1 Classifying Lipids

Are you able to. . .

- categorize lipids as hydrolyzable or nonhydrolyzable (Exercises 19.29, 19.39, 19.71)
- identify the structural features of hydrolyzable lipids
- give examples of hydrolyzable and nonhydrolyzable lipids

Objectives Section 19.2 Fatty Acids

Are you able to. . .

- identify fatty acids and their derivatives
- describe the structural features of fatty acids
- describe the features that are common in plant and animal fatty acids (naturally occurring fatty acids) (Ex. 19.1, Prob. 19.1, Exercises 19.1, 19.2)
- distinguish between saturated and unsaturated fatty acids (Exercises 19.3, 19.4)
- relate the physical properties of fatty acids to their structures (Exercises 19.5, 19.6)
- explain the term omega number
- identify the omega carbon in a fatty acid (Exercises 19.4, 19.65)
- discuss the reactions that fatty acids undergo
- describe the roles played by fatty acids in plants and animals
- give examples of fatty acids
- distinguish between essential and nonessential fatty acids (Exercise 19.64)

Problem 19.1 Which of the following carboxylic acids are present in animal and plant lipids?

The criteria for a fatty acid to be of plant or animal origin are that it have an even number of carbons and be linear, or unbranched.

(1) This fatty acid has 16 carbons and no branches; it is palmitic acid and is naturally occurring.

(2) This fatty acid has an even number of carbons (16), but it is branched; it is not present in animals or plants.

(3) This fatty acid has an even number of carbons (16) and is unsaturated with a cis double bond; it is palmitoleic acid and is naturally occurring.

Objectives Section 19.3 Structure and Physical Properties of Triacylglycerols

Are you able to. . .

- describe the structures of triglycerides
- compare fats and oils
- relate the structures of fatty acid chains to the properties of triglycerides (Exercises 19.9, 19.10)
- relate the structure of triglycerides to their function
- draw the structure of a triglyceride from the components (Ex. 19.2, Prob. 19.2, Exercises 19.7, 19.8)
- identify the original molecules from which a triglyceride is derived
- describe the chemical reactions that produce a triglyceride

Problem 19.2 Draw the structure of the triacyglycerol containing equimolar amounts of lauric, palmitoleic, and linoleic acids.

$$
\begin{array}{l}
\quad\;\; \mathrm{H} \qquad\;\; \mathrm{O} \\
\quad\;\; | \qquad\;\;\; \| \\
\mathrm{H{-}C{-}O{-}C{-}(CH_2)_{10}{-}CH_3} \\
\quad\;\; | \qquad\;\;\; \mathrm{O} \\
\quad\;\; | \qquad\;\;\; \| \\
\mathrm{H{-}C{-}O{-}C(CH_2)_7CH{=}CH(CH_2)_5CH_3} \\
\quad\;\; | \qquad\;\;\; \mathrm{O} \\
\quad\;\; | \qquad\;\;\; \| \\
\mathrm{H{-}C{-}O{-}C{-}(CH_2)_6CH_2CH{=}CH(CH_2)_4CH_3} \\
\quad\;\; | \\
\quad\;\; \mathrm{H}
\end{array}
$$

Objectives Section 19.4 Chemical Reactions of Triacylglycerols

Are you able to. . .

- predict the products of the hydrolysis of a triglyceride (Ex. 19.3, Prob. 19.3, Exercises 19.11, 19.12, 19.66)
- identify the conditions under which the hydrolysis of triglycerides takes place (Exercises 19.11, 19.12)
- relate hydrolysis and saponification (Prob. 19.3)
- relate hydrolysis and digestion (Exercises 19.12, 19.69)
- identify the original triglyceride from hydrolysis products
- predict the products of hydrogenation of a triglyceride (Ex. 19.4, Prob. 19.4, Exercises 19.11, 19.66)
- predict the products of saponification of a triglyceride (Exercise 19.66)
- explain how hydrogenation changes the properties of a triglyceride (Exercise 19.15)
- predict the products of the air or bacterial oxidation of a triglyceride (Ex. 19.5, Prob. 19.5, Exercises 19.13, 19.14)
- explain rancidity (Exercise 19.76)
- describe the processes of digestion, absorption, and transport of triglycerides in the body

Problem 19.3 Write the equation for the saponification with NaOH of the triacylglycerol containing equimolar amounts of myristic, stearic, and palmitoleic acids.

$$
\begin{array}{l}
\mathrm{H{-}C{-}O{-}C(=O){-}(CH_2)_{12}{-}CH_3} \\
\quad\;\; | \\
\mathrm{H{-}C{-}O{-}C(=O){-}(CH_2)_{16}{-}CH_3} \\
\quad\;\; | \\
\mathrm{H{-}C{-}O{-}C(=O){-}(CH_2)_7{-}CH{=}CH{-}(CH_2)_5{-}CH_3}
\end{array}
\;(\text{with H at top and bottom C})\; + 3\,\mathrm{NaOH} \longrightarrow
\begin{array}{l}
\mathrm{Na^+\ {}^-O{-}C(=O){-}(CH_2)_{12}{-}CH_3} \\
\mathrm{Na^+\ {}^-O{-}C(=O){-}(CH_2)_{16}{-}CH_3} \\
\mathrm{Na^+\ {}^-O{-}C(=O){-}(CH_2)_7{-}CH{=}CH{-}(CH_2)_5{-}CH_3}
\end{array}
+
\begin{array}{l}
\quad\;\; \mathrm{H} \\
\mathrm{H{-}C{-}OH} \\
\mathrm{H{-}C{-}OH} \\
\mathrm{H{-}C{-}OH} \\
\quad\;\; \mathrm{H}
\end{array}
$$

Problem 19.4 Write the equation for the Pt-catalyzed hydrogenation of the triacylglycerol that contains equimolar amounts of palmitic, oleic, and linoleic acid.

The hydrogenation adds H to each of the double bonds in the fatty acid tails. The resultant triacylglycerol, which is fully saturated, contains two palmitic acid tails and a stearic acid tail.

$$
\begin{array}{l}
\mathrm{H{-}C{-}O{-}C(=O){-}(CH_2)_{14}{-}CH_3} \\
\quad\;\; | \\
\mathrm{H{-}C{-}O{-}C(=O){-}(CH_2)_7{-}CH{=}CH{-}(CH_2)_7{-}CH_3} \\
\quad\;\; | \\
\mathrm{H{-}C{-}O{-}C(=O){-}(CH_2)_6{-}(CH_2CH{=}CH)_2{-}(CH_2)_4{-}CH_3}
\end{array}
+ 3\,\mathrm{H_2} \xrightarrow{\mathrm{Pt}}
\begin{array}{l}
\mathrm{H{-}C{-}O{-}C(=O){-}(CH_2)_{14}{-}CH_3} \\
\quad\;\; | \\
\mathrm{H{-}C{-}O{-}C(=O){-}(CH_2)_{16}{-}CH_3} \\
\quad\;\; | \\
\mathrm{H{-}C{-}O{-}C(=O){-}(CH_2)_{16}{-}CH_3}
\end{array}
$$

(In each structure, the top and bottom glycerol carbons also carry an H.)

Problem 19.5 Write the equation for the air oxidation of oleic acid.

$$CH_3{-}(CH_2)_7{-}CH{=}CH{-}(CH_2)_7{-}COOH$$

A molecule of oleic acid is cleaved at the double bond by oxidation in air. The products are a monocarboxylic acid of nine carbons (nonanoic acid) and a dicarboxylic acid of nine carbons (nonanedioic acid).

$$CH_3(CH_2)_7CH{=}CH(CH_2)_7{-}COOH \rightarrow CH_3(CH_2)_7\overset{O}{\overset{\|}{C}}{-}OH + HO\overset{O}{\overset{\|}{C}}{-}(CH_2)_7{-}\overset{O}{\overset{\|}{C}}{-}OH$$

Objectives Section 19.5 Waxes

Are you able to. . .

- describe the structural features of waxes (Exercises 19.17, 19.18)
- describe some roles played by waxes (Exercise 19.67)
- relate the structures of waxes to their functions

Objectives Section 19.6 Amphipathic Hydrolyzable Lipids

Are you able to. . .

- describe amphipathic structures
- classify hydrolyzable lipids as phosphoglycerides or sphingolipids (Fig. 19.3)
- describe the structural features of phosphoglycerides (Exercises 19.19, 19.70)
- list functions of phosphoglycerides
- relate the structures of phosphoglycerides to their functions
- describe the structural features of sphingolipids (Exercises 19.19, 19.70)
- list functions of sphingolipids
- relate the structures of sphingolipids to their functions
- draw structures of phosphoglycerides and sphingolipids from the components (Exercises 19.21, 19.22)
- describe the formation of phosphoglycerides and sphingolipids
- describe the original structure of a phosphoglyceride or a sphingolipid from its hydrolysis products (Exs. 19.6, 19.7, Probs. 19.6, 19.7, Exercises 19.21, 19.22)
- predict the products of reactions of a phospholipid or a sphingolipid (Exercises 19.27, 19.28, 19.68, 19.74)
- identify amine alcohols (Table 19.3)
- classify a phosphoglyceride according to the amine alcohol that it contains

Problem 19.6 Draw the glycerophospholipid whose hydrolysis yields equimolar amounts of glycerol, stearic acid, linoleic acid, phosphoric acid, and choline.

The original molecule had fatty acid ester groups at two carbons of the glycerol backbone. The third group was a phosphodiester with choline, phosphatidylcholine.

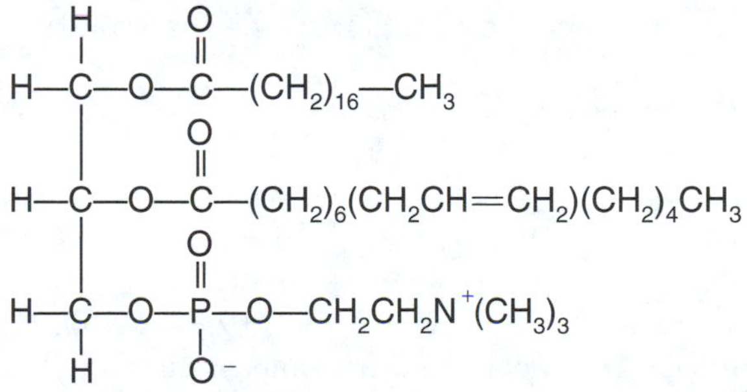

$$\begin{array}{l}
H{-}\overset{H}{\overset{|}{C}}{-}O{-}\overset{O}{\overset{\|}{C}}{-}(CH_2)_{16}{-}CH_3 \\
\quad | \\
H{-}C{-}O{-}\overset{O}{\overset{\|}{C}}{-}(CH_2)_6(CH_2CH{=}CH_2)(CH_2)_4CH_3 \\
\quad | \\
H{-}\underset{H}{\underset{|}{C}}{-}O{-}\underset{O^-}{\underset{|}{\overset{O}{\overset{\|}{P}}}}{-}O{-}CH_2CH_2N^+(CH_3)_3
\end{array}$$

Problem 19.7 Draw the sphingolipid whose hydrolysis yields equimolar amounts of sphingosine, oleic acid, and β-D-glucose.

The original molecule was a sphingoglycolipid (glucose tells you that it was a glycolipid). The backbone was a sphingosine with an amide at the number two carbon that had the oleic acid fatty tail and glucose at the first carbon.

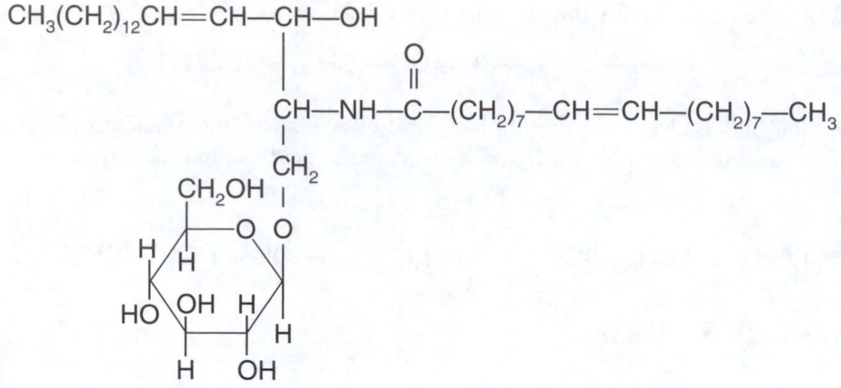

```
CH3(CH2)12CH=CH—CH—OH
                |        O
                |        ||
                CH—NH—C—(CH2)7—CH=CH—(CH2)7—CH3
                |
                CH2
        CH2OH   |
          |——O  O
       H /H    \ |
        /       \|
    HO \OH  H   /H
        \______/
        H    OH
```

Objectives Section 19.7 Steroids: Cholesterol, Steroid Hormones, and Bile Salts

Are you able to. . .

- identify steroids
- describe the structural feature that defines a steroid (Exercises 19.30, 19.36, 19.50, 19.70)
- list some steroid functions (Exercises 19.33, 19.34, 19.35, 19.78)
- give examples of steroids
- identify bile salts
- identify the structural features of bile salts (Exercise 19.37)
- describe the properties of bile salts
- describe the functions of bile salts (Exercise 19.38)
- relate the structures of bile salts to their functions

Objectives Section 19.8 Eicosanoids

Are you able to. . .

- identify eicosanoids
- identify the structural features of eicosanoids (Exercise 19.40)
- classify eicosanoids as leukotrienes, prostaglandins, or thromboxanes (Exercises 19.41, 19.42)
- list some functions of eicosanoids (Exercises 19.43, 19.44)

Objectives Section 19.9 Fat-Soluble Vitamins

Are you able to. . .

- define vitamin (Exercises 19.45, 19.46)
- compare fat-soluble and water-soluble vitamins (Exercises 19.47, 19.48)
- identify fat-soluble vitamins (Table 19.4)
- describe the roles played by fat-soluble vitamins (Table 19.4, Exercise 19.49)

Objectives Section 19.10 Biological Membranes

Are you able to. . .

- describe a membrane (Exercises 19.51, 19.80)
- describe a lipid bilayer
- list the functions performed by membranes (Exercise 19.52)
- identify the components of membranes (Exercises 19.53, 19.54, 19.60, 19.72, 19.81)
- discuss the roles of lipids, proteins, and carbohydrates in membrane function (Exercises 19.53, 19.54, 19.60)
- relate the function of a membrane to its structure (Exercise 19.81)
- describe the fluid-mosaic model of a membrane (Exercise 19.80)
- describe different transport mechanisms (Exercises 19.57, 19.58, 19.59, 19.61, 19.62, 19.81, 19.82)
- compare diffusion, facilitated transport, active transport, endocytosis, and exocytosis (Exercises 19.57, 19.58, 19.59, 19.61, 19.62, 19.81, 19.82)
- discuss the role played by membranes and their components in transport processes
- identify the factors that affect membrane fluidity (Exercises 19.55, 19.56)
- describe the forces that underlie the construction of a membrane

MAKING CONNECTIONS

Draw a concept map or write a paragraph describing the relation between the members of each of the following groups:

(a) hydrolyzable lipid, nonhydrolyzable lipid, ester group
(b) saponification, hydrolysis, hydrolyzable lipid
(c) wax, fat, oil, phospholipid, sphingolipid
(d) steroid, eicosanoid, vitamin A
(e) fatty acid, triglyceride, glycerol
(f) fatty acid, saturated, unsaturated, fat, oil
(g) saturation, melting point, boiling point, cis double bond
(h) triglyceride, hydrolysis, fatty acid
(i) triglyceride, bacteria, carboxylic acid
(j) fat, oil, catalytic hydrogenation
(k) wax, fatty acid, fatty alcohol, protective coat
(l) glycolipid, lipoprotein, cell recognition
(m) cholesterol, bile salt, estrogen
(n) bile salt, amphipathic, emulsification
(o) lipid bilayer, membrane transport, diffusion
(p) fluid-mosaic model, peripheral protein, integral protein, transport
(q) membrane fluidity, temperature, cholesterol, fatty acid chain
(r) simple transport, facilitated transport, active transport
(s) sphingolipid, phospholipid, cholesterol, cell membrane

FILL-INS

Lipids are biomolecules that are not characterized by a _____. A biomolecule is classified as a lipid if it is _____ and not _____. The property of nonpolarity implies that there are many more nonpolar than polar _____ in lipids. _____, a triol, form the basis for triglycerides, more commonly known as _____. They are dietary lipids that we ingest when eating _____ products. The differences in properties between fats and oils can be attributed _____ that make up the _____ fatty acid components of the triglyceride. The presence of many double bonds, indicating a polyunsaturated triglyceride, reduces the ability of molecules _____. This reduced ability translates into _____ melting and boiling points for these molecules. Polyunsaturated triglycerides tend to be _____ at room temperature. Longer carbon chains and fewer double bonds allow for _____ between molecules and result in relatively _____ melting and boiling points. These molecules tend to be _____, _____ at room temperature. Triglycerides can be cleaved by _____, _____ catalytically in the laboratory, or _____ by air and bacterial action. The _____ of lipids makes them unsuitable for transport in aqueous environments such as blood. After digestion and absorption, they must be _____ to be moved around the body. _____, lipoproteins that have both _____ parts, do this job. Triglycerides are more _____ molecules for energy than is glycogen and are deposited and stored in _____ if they are not immediately needed. Other classes of lipids include _____, which are amphipathic hydrolyzable lipids. Both of these molecules are important in cell membranes owing to their _____ nature. Cell membranes are _____ that consist of an inner _____ matrix sandwiched between two _____ surfaces. They regulate the movement of _____ the cell and its environment. The _____ of a membrane describes the nonpolar matrix as fluid. Within this matrix move _____ proteins, and spanning the matrix are _____ proteins. _____, containing a carbohydrate attached to a protein, and membrane proteins are vital participants in _____. They translate chemical information from the cellular environment into directions that _____, just to name a few. _____, such as estrogens, are synthesized in and secreted by endocrine glands and _____ to target tissues where they _____ a variety of cell functions. _____ act as hormones, but are _____.

Answers

Lipids are biomolecules that are not characterized by a monomeric subunit or specific functional group. A biomolecule is classified as a lipid if it is soluble in nonpolar solvents and not soluble in water. The property of nonpolarity implies that there are many more nonpolar than polar structural features in lipids. Fatty acids and glycerol, a triol, form the basis for triglycerides, more commonly known as fats and oils. They are dietary lipids that we ingest when eating plant or animal products. The differences in properties between fats and oils can be attributed to the identity of the hydrocarbon tails that make up the three fatty acid components of the triglyceride. The presence of many double bonds, indicating a polyunsaturated triglyceride, reduces the ability of molecules to interact strongly with one another. This reduced ability translates into lower melting and boiling points for these molecules. Polyunsaturated triglycerides tend to be oils, or liquids, at room temperature. Longer carbon chains and fewer double bonds allow for more extensive interactions between molecules and result in relatively higher melting and boiling points. These molecules tend to be fats, or solids, at room temperature. Triglycerides can be cleaved by digestive enzymes, hydrogenated catalytically in the laboratory, or oxidized by air and bacterial action. The nonpolarity of lipids makes them unsuitable for transport in aqueous environments such as blood. After digestion and absorption, they must be repackaged to be moved around the body. Chylomicrons, lipoproteins that have both hydrophilic and hydrophobic parts, do this job. Triglycerides are more efficient storage molecules for energy than is glycogen and are deposited and stored in fat cells if they are not immediately needed. Other classes of lipids include sphingolipids and phosphoglycerides, which are amphipathic hydrolyzable lipids. Both of these molecules are important in cell membranes owing to their amphipathic nature. Cell membranes are lipid bilayers that consist of an inner nonpolar matrix sandwiched between two polar surfaces. They regulate the movement of material between the cell and its environment. The fluid-mosaic model of a membrane describes the nonpolar matrix as fluid. Within this matrix move peripheral proteins, and spanning the matrix are integral proteins. Glycoproteins, containing a carbohydrate attached to a protein, and membrane proteins are vital participants in cell recognition and information transfer. They translate chemical information from the cellular environment into directions that mediate cell processes, regulate hormone responses, and activate immune system processes, just to name a few. Hormones, such as estrogens, are synthesized in and secreted by endocrine glands and transported in the blood to target tissues where they regulate a variety of cell functions. Eicosanoids act as hormones but are synthesized in the same tissues in which they act.

TEST YOURSELF

1. Match the lipid on the left with a description on the right.

(a) Wax	(1) plays a role in inflammatory responses
(b) Triglyceride	(2) is found in myelin sheaths
(c) Eicosanoid	(3) waterproofs feathers
(d) Sphingolipid	(4) is an important component of cell membranes
(e) Phospholipid	(5) emulsifies fat droplets in the small intestine
(f) Steroid	(6) plays a key role in vision
(g) Bile salt	(7) adds rigidity to cell membranes
(h) Vitamin A	(8) provides protection and energy storage
(i) Cholesterol	(9) stimulates the synthesis of protein

2. Match the lipid on the left with a structural feature on the right.

(a) Wax	(1) has three esterified fatty acids
(b) Triglyceride	(2) has three fused six-membered rings and a five-membered ring
(c) Eicosanoid	(3) is made from the esterification of a fatty alcohol and a fatty acid
(d) Sphingolipid	(4) has an ionized carboxylate head
(e) Phospholipid	(5) has a phosphate diester and two fatty acids esterified to a glycerol backbone
(f) Steroid	
(g) Bile salt	(6) is a steroid derivative
(h) Cholesterol	(7) has a sphingosine backbone that is esterified to a fatty acid and a sugar
	(8) is based on arachidonic acid

3. Draw a reaction map for the hydrolysis (as digestion), saponification, hydrogenation, and air oxidation of a generic, mixed triglyceride with a monounsaturated and a diunsaturated fatty acid tail.
4. Lecithins are used as emulsifying agents in many foods. Draw a lecithin and explain why the structural features result in properties that make it an appropriate choice for this task.

5. Gallstones can sometimes block the flow of bile from the gallbladder to the duodenum, the point from which it is released into the intestine. What effect does this blockage have on the digestion of fats?
6. Draw the structure of a generic fatty acid. How will a group of similar fatty acids arrange themselves in an aqueous environment? Draw the possible arrangements.
7. Kidney dialysis machines must mimic the function of biological membranes. If the major activity through dialysis tubing is simple diffusion, what properties does the membrane need to have?

Answers

1. (a) 3; (b) 8; (c) 1; (d) 4; (e) 4; (f) 9; (g) 5; (h) 6; (i) 7
2. (a) 3; (b) 1; (c) 8; (d) 7; (e) 5; (f) 2; (g) 4; (h) 6
3.

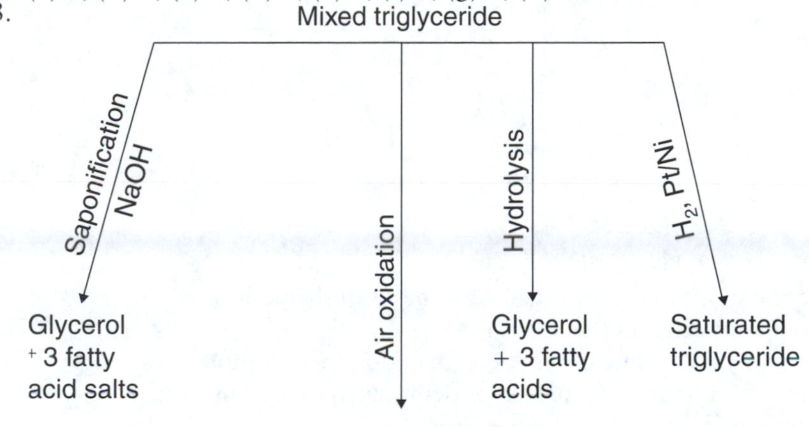

4. A lecithin is a phosphoglyceride that contains the amino acid choline. The polar nature of the phosphate-amino alcohol group coupled with the nonpolar fatty acid tails makes this an amphipathic molecule. Emulsifiers have hydrophobic and hydrophilic areas that keep food from separating (for example, peanut butter oil, or oil and water in mayonnaise). Lecithins are able to keep hydrophobic molecules suspended in aqueous matrices from which they would otherwise separate because they can interact with both polar and nonpolar components.
5. Bile salts are essential to the digestion of fats. Like emulsifiers, they are able to interact with fats and the aqueous environments in the body. In the intestine, bile salts act as emulsifiers and break large fat globules into smaller fat droplets. Large fat globules have a limited surface area over which digestive enzymes can act. Small fat droplets provide more surface area and easier access. Without bile salts, fats cannot be digested normally.
6. The polar ends will always arrange to be in contact with the aqueous environment, and the nonpolar ends will always arrange to minimize contact with the aqueous environment.

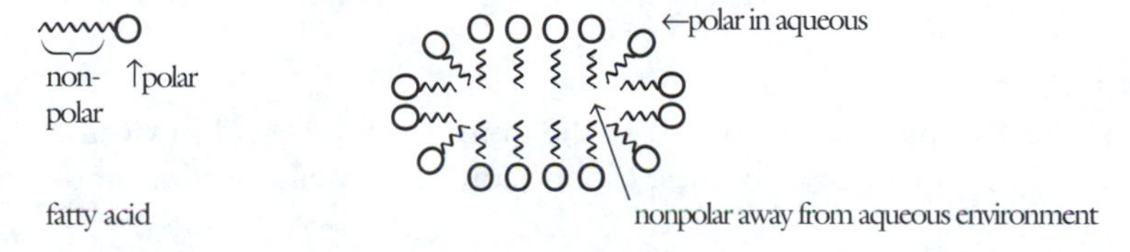

7. Simple diffusion requires that a concentration gradient be maintained on either side of the membrane. The membrane must have openings large enough for the diffusing species to move through. It also must prevent entrance or exit by certain molecules. The openings must be small enough to be selective. The membrane must also have a polarity that does not repel the material that needs to diffuse out.

chapter 20
Proteins

OUTLINE

Introduction

A. Proteins are the workhorses of the body.
   1. they are the most abundant organic chemicals in the body
   2. they build and maintain the organism
   3. they have a greater range of functions than do other biomolecules
   4. the biological function of a protein is determined by its amino acid sequence

B. Protein functions include:
   1. enzyme activity
      a. enzymes catalyze biological reactions in the body
      b. examples include proteins that catalyze the synthesis and utilization of carbohydrates, fats, and proteins
   2. transport
      a. transport proteins bind and carry specific molecules or ions throughout the body
      b. examples include proteins that carry oxygen, copper, and iron
   3. regulation
      a. regulatory proteins control cellular activity
      b. examples include proteins that regulate nerve transmission, gene expression, and glucose metabolism
   4. structural components
      a. structural proteins give shape, strength, and protection to an organism
      b. examples include parts of bone, tendon, hair, nails, horn, and feathers
   5. contraction
      a. contractile proteins provide the ability to move or change shape
      b. examples include proteins participating in muscle contraction and in flagella and cilia movement
   6. protection
      a. protective proteins are active in the immune system and in wound healing
      b. examples include antibodies, blood-clotting proteins, toxins, and venoms
   7. storage
      a. storage proteins act as reservoirs of nutrients and nitrogen
      b. examples include proteins in egg whites, mammalian milk, and seed

C. Structure of proteins
   1. proteins are polymers of α-amino acids
   2. proteins are also called polypeptides

Key Terms catalytic protein, contractile protein, enzyme, polypeptide, protective protein, regulatory protein, storage protein, structural protein, transport protein

20.1 α-Amino Acids

A. Every α-amino acid contains:
   1. a central tetrahedral carbon, called the α-carbon; all α-carbons are stereocenters, except that in glycine
   2. an amine end, $-NH_2$
      a. this end is a base because it can accept a proton
      b. all α-amino groups are primary amines except in proline, in which the α-amino group is a secondary amine
   3. a carboxylic acid end, –COOH; this end is an acid because it can donate a proton
   4. an R group
      a. the R group determines the identity and the properties of the individual amino acid
      b. R groups can be classified by the following characteristics:
         i. nonpolar neutral: contains a neutral hydrophobic group
         ii. polar neutral: contains a neutral hydrophilic group
         iii. polar acidic: contains an acidic hydrophilic group
         iv. polar basic: contains a basic hydrophilic group

B. Most polypeptides of all species of plants and animals are constructed from combinations of 20 different α-amino acids (Table 20.1).
   1. mammals can synthesize only half of them
   2. mammals must obtain 10 essential α-amino acids from their diets

C. α–Amino acids are given common names (Table 20.1).
   1. each common name has a three-letter abbreviation
   2. one-letter abbreviations are used by biochemists
   3. α-amino acid (except glycine) exists as a pair of enantiomers (Fig. 20.1)
      a. designation is determined by the position of the $-NH_2$ group in the same way as that for monosaccharides (Fig. 20.1)
      b. D-amino acids are rare in nature; most plant and animal amino acids are L-α-amino acids

Key Terms acidic protein, basic protein, neutral protein

Key Figure and Table
Figure 20.1 D- and L-enantiomers
Table 20.1 α-Amino acids

20.2 Zwitterionic Structure of α-Amino Acids

A. α-Amino acids exist as dipolar ions called zwitterions.
   1. zwitterions form by an intramolecular acid-base reaction between the acidic carboxyl group and the basic amino group

$$H_2N\text{—}\underset{|\atop R}{CH}\text{—}COO(H) \rightleftharpoons H_3N^+\text{—}\underset{|\atop R}{CH}\text{—}COO^-$$

   2. the carboxylate ion is the negative end with a 1 – charge
   3. the protonated amine is the positive end with a 1+ charge
   4. zwitterions are neutral overall

B. The physical properties of amino acids are due to the existence of zwitterions.
   1. zwitterions are responsible for strong secondary forces between amino acids
   2. amino acids have higher melting points than those of comparable biomolecules
   3. amino acids are water soluble, owing to the polarity of the zwitterions

a. water solubility of amino acids varies with pH
b. amino acids are least soluble at the isoelectric point because the opposite charges of zwitterions are attracted to one another and they associate with themselves

4. amino acids are zwitterions in the solid state

C. α-Amino acids are amphoteric, acting as both proton donors and proton acceptors.

1. $-NH_2$ and basic R groups are proton acceptors
2. –COOH and acidic R groups are proton donors
3. in aqueous solution, zwitterions are in equilibrium with cation and anion forms

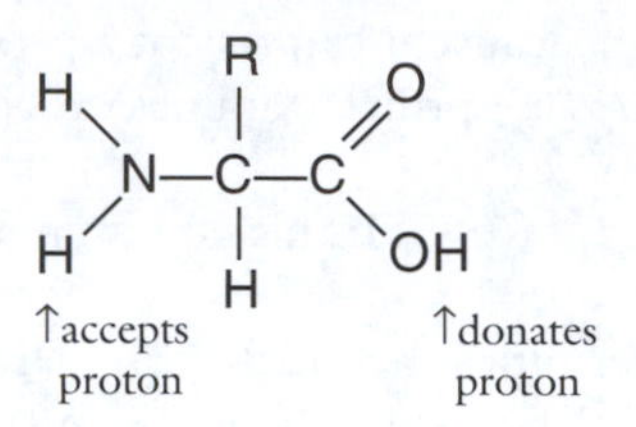

4. the identity of the amino acid and the pH of the solution determine which forms exist
   a. at the pH that is the isoelectric point (pI), almost all ions are zwitterions; there is no net electrical charge in this form
   b. at pHs below the isoelectric point, the number of cations increases and the number of zwitterions decreases
   c. at pHs above the isoelectric point, the number of anions increases and the number of zwitterions decreases

D. The acid-base behavior is different for different types of R groups.

1. neutral polar and nonpolar R groups: the amino acids act as buffers
   a. if pH of solution is within 2 pH units above or below the pI, most molecules are still zwitterions
   b. pIs for neutral amino acids are between 5.05 and 6.30
   c. cation forms dominate at pHs below 1
   d. anion forms dominate at pHs above 12
   e. carboxyl groups are negatively charged at physiological pH (7.35 for blood; 6.8–7.1 for most cells)
   f. amino groups are positively charged at physiological pH
   g. neutral amino acids exist as zwitterions at physiological pH
2. acidic R groups produce acidic amino acids
   a. acidic amino acid pIs are below those of neutral amino acids
   b. the acidic carboxyl group on the side chain must be protonated for zwitterion to exist
   c. all acidic side groups are negatively charged at physiological pH
   d. acidic amino acids have an overall charge of 1− at physiological pH
3. basic R groups produce basic amino acids
   a. basic amino acid pIs are above those of neutral amino acids
   b. the basic amino group on the side chain must be unprotonated for zwitterion to exist
   c. all basic side groups are positively charged at physiological pH
   d. basic amino acids have an overall charge of 1+ at physiological pH

E. Mixtures of amino acids can be separated by electrophoresis in an applied electric field (Fig. 20.2).

1. amino acids are separated on the basis of their overall charges
   a. amino acids migrate at different rates in an electric field
   b. the rate of migration is related to the overall charge on the molecule
   c. there is no migration at pH = pI because there is no net charge on the molecule
   d. at physiological pH, electrophoresis can distinguish between neutral, acidic, and basic amino acids
      i. neutral amino acids do not migrate
      ii. acidic amino acids have a 1− charge and migrate toward the positive electrode (anode)
      iii. basic amino acids have a 1+ charge and migrate toward the negative electrode (cathode)

Key Terms amphoteric, dipolar ion, electrophoresis, isoelectric point, paper electrophoresis, zwitterion
Key Figure Figure 20.2 Electrophoresis

Example 20.1 Determination of the ionic structure of α-amino acids in solutions of different pH
Problem 20.1 Determination of the ionic structure of α-amino acids in solutions of different pH

20.3 Peptides

A. Peptide structure:
1. peptides are polyamides formed from reactions between amino acids
    a. the reaction is analogous to a dehydration reaction between the carboxyl end of one amino acid and the amino end of a second amino acid
    b. each amino acid becomes a residue, or monomer
    c. the bond formed is called a peptide bond
    d. peptides can be given prefixes such as di-, tri-, and so forth, as high as deca (10), depending on the number of residues
    e. peptides containing between 1 and 20 residues are commonly called oligopeptides
    f. peptides that are larger are called polypeptides

B. Peptide isomers and nomenclature
1. peptides have constitutional isomers based on the different sequences in which residues can link
    a. every peptide isomer has an amino end called the N-terminal amino group
    b. every peptide isomer has a carboxyl end called the C-terminal carboxyl group
    c. convention places the N-terminal residue at the left and the C-terminal residue at the right when the peptide is drawn
2. peptides are named by beginning at the N-terminal end and moving to the C-terminal end
    a. the suffix (-ine) of the amino acid names are replaced by a -yl suffix, except tryptophan
    b. the C-terminal residue retains its common name
3. three-letter abbreviations separated by hyphens are generally used to designate peptide sequences
4. the number of isomers possible for a group of amino acids can be calculated by using $n!$ (n factorial)

EXAMPLE: $n!$ for $n = 5$ is $5 \times 4 \times 3 \times 2 \times 1$

C. Peptide bond structure
1. the peptide bond has properties of double bonds owing to resonance between carbonyl C and electrons on N
2. there is no free rotation around the C–N bond
    b. bond length is shorter than that of a single bond
    c. O=C–N all lie in the same plane
3. peptide segments are trans to one another because cis-relations would cause interference between groups of atoms
4. peptide bonds affect three-dimensional structures and functions of polypeptides

D. Peptide properties:
1. a peptide has an isoelectric point at which it has an overall zero charge and does not migrate in electrophoresis
    a. if the peptide contains only neutral amino acid residues or equal numbers of acidic and basic residues, pI is in the same range as that for neutral amino acids: pH 5.05–6.30
    b. if the peptide contains more acidic than basic amino acid residues, pI is below pH 5.05–6.30
    c. if the peptide contains more basic than acidic amino acid residues, pI is above pH 5.05–6.30
2. the solubilities and electrophoretic properties of peptides are pH dependent
    a. a peptide is at its minimum solubility when pH = pI
    b. peptide solubility in water increases above and below the pI
    c. electrophoresis can separate peptides on the basis of their charges

Key Terms amino acid residue, C-terminal, N-terminal, oligopeptide, peptide, peptide bond, peptide group, polypeptide

Example 20.2 Drawing and naming peptides
Problem 20.2 Drawing and naming peptides
Example 20.3 Calculating the number of constitutional isomers of peptides
Problem 20.3 Calculating the number of constitutional isomers of peptides
Example 20.4 Writing peptide structures at different pH values
Problem 20.4 Writing peptide structures at different pH values

20.4 Chemical Reactions of Peptides

A. Peptides are cleaved by hydrolysis.
   1. peptide formation and peptide hydrolysis are the reverse of one another; amino acid + amino acid ⇌ peptide + H_2O
   2. hydrolysis is carried out enzymatically in the body or in acidic or base solution in the laboratory

B. Sulfhydryl groups (–SH) on cysteine residues can undergo oxidation and reduction reactions.
   1. oxidation joins two S atoms to form a disulfide bridge, creating a cystine residue
   2. reduction breaks disulfide bond to form sulfhydryl groups
   3. the formation of disulfide bridges is important in the overall structure of a protein
   4. disulfide bonds are not cleaved by hydrolysis or digestion
   5. disulfide bonds between residues are designated by joining the three-letter abbreviations of the linked residues with –S–S–

Key Terms disulfide bridge, sulfhydryl group

20.5 Three-Dimensional Structure of Proteins

A. The three-dimensional structure of a protein determines its biological function.

B. The three-dimensional structure is the result of several layers of organization and is the most stable conformation, as determined by the bonds in the amino acid sequence and the interactions between side groups (Fig. 20.3).
   1. the amino acid sequence is the primary structure; this structure determines what interactions can take place between side groups and peptides
   2. local conformations are called secondary structure
      a. local folding by rotation about single bonds can occur
         i. folding usually occurs in an effort to shield hydrophobic groups from aqueous interactions
         ii. this folding is called the hydrophobic effect and sets the stage for hydrophobic interactions between side groups
      b. there are common structural patterns that repeat in proteins
         i. α-helices (Fig. 20.6): look like a coiled spring stabilized by hydrogen bonds between –C=O and $–NH_2$ groups on different residues in the same chain; side groups protrude outward from chain; can accommodate larger side groups than can β-pleated sheets, but proline is a helix breaker
         ii. β-pleated sheets (Fig. 20.7): hydrogen bonding between adjacent chains results in pleated sheets; favored for peptides with small side groups
         iii. β-bends: abrupt directional changes in the chain tend to connect adjacent intramolecular segments of β-pleated sheets
         iv. loops: irregularly ordered segments
         v. different patterns may be repeated in a single protein or the one pattern may dominate
   3. the three-dimensional relations between areas of different secondary structures determine the tertiary structure
      a. in many proteins, this level of structure produces the biologically active form
      b. tertiary structure is dominated by side-group interactions (Figs. 20.4, 20.5): hydrophobic interactions, hydrogen bonds, salt bridges, disulfide links

4. when two or more polypeptide molecules need to be associated for the protein to be biologically active, the protein possesses quaternary structure.

C. The factors that influence the most stable conformation include:
1. most proteins fold to shield hydrophobic groups from contact with water
   a. neutral nonpolar side groups are buried within the protein structure
   b. this is called a hydrophobic effect
2. hydrogen bonds can exist between the –C=O of one peptide and the –NH group of another
   a. hydrogen bonding between peptides dominates the secondary structure of proteins; this produces α-helices, β-pleated sheets, β-turns, and loops
   b. hydrogen bonding between peptide groups works to shield hydrophobic groups from water
3. side groups can interact with one another in various ways
   a. as a result of the hydrophobic effect, nonpolar neutral side groups are placed near one another and participate in hydrophobic interactions
   b. neutral polar residues placed near each other can participate in hydrogen bonding
4. polar side groups can interact with water
   a. proteins that perform functions that require solubility usually have many polar groups on the surface
   b. proteins that perform functions that require insolubility have few polar groups on the surface
5. disulfide bonds formed between cystine residues are the only covalent interactions that stabilize protein structure (Fig. 20.5); disulfide bonds can form between residues on the same peptide sequence or between different sequences

D. Biologically active peptides and proteins differ in the number of amino acid residues constituting the molecule, but the actual lines are blurred (Table 20.2).
1. proteins generally have more than 50 amino acid residues
2. peptides range from a few to 50 amino acid residues

E. Proteins can be classified in a variety of ways.
1. they can be classified by structure
   a. fibrous proteins are water-insoluble proteins that serve structural and contractile functions
      i. α-keratin is the main structural component of hair, horn, nails, skin, and wool
      ii. collagen is the major stress-bearing component of connective tissues
      iii. β-keratin is found in feathers
      iv. silk fibroin is the protein in silk
   b. globular proteins are able to move in aqueous environments and do most of the metabolic work
      i. myoglobin picks up oxygen in muscle cells
      ii. hemoglobin carries oxygen in red blood cells
      iii. lysozyme is an enzyme that hydrolyzes bacterial cell walls
2. They can be classified by composition (Table 20.3)
   a. simple proteins are peptides
   b. conjugated proteins contain nonpolypeptide molecules
      i. polypeptide part is called the apoprotein
      ii. nonpolypeptide part is called the prosthetic group
      iii. prosthetic groups may be ions or small molecules
3. they can be classified by function (see Introduction: B)

Key Terms α-helix, apoprotein, β-bend, β-pleated sheet, β-turn, conjugated protein, helix breaker, hydrophobic effect, prosthetic group, salt-bridge, simple protein

Key Figures and Tables

Figure 20.3 Conformations about single bond in polypeptide chain
Figure 20.4 Noncovalent tertiary interactions between side groups
Figure 20.5 Disulfide bridges

20.6 Fibrous Proteins

A. Fibrous proteins are water insoluble and perform structural functions.
   1. α-keratin is the main structural component of hair, horn, nails, skin, and wool (Figs. 20.8, 20.9, 20.10)
      a. hair is composed of triple helices that coil into protofibrils that are bundled together to form micro- and macrofibrils
      b. number of disulfide bridges from cysteine units determines whether α-keratins are hard or soft
   2. collagen is the major stress-bearing component of connective tissues (Fig. 20.11); three collagen polypeptides coil into a triple helix, or tropocollagen
   3. β-keratin: found in bird feathers and reptile scales
   4. silk fibroin: protein in silk

Key Terms α-keratin, β-keratin, collagen, cross-link, fibrous protein, macrofibril, microfibril, protofibril, silk fibroin, triple helix, tropocollagen

Key Figures

20.7 Globular Proteins

A. Globular proteins perform functions in aqueous environments.
   1. myoglobin (Fig. 20.12)
      a. myoglobin stores oxygen in muscle tissue
      b. a single polypeptide chain with a heme group binds oxygen at the Fe^{2+} site
      c. oxygen binds more tightly to myoglobin than to hemoglobin
      d. all polar residues (except two histidines) are on the protein surface
   2. hemoglobin (Fig. 20.13)
      a. hemoglobin carries oxygen from the lungs to tissue and brings carbon dioxide waste from tissue to the lungs
      b. four polypeptide chains that each have a heme group bind oxygen at the Fe^{2+} site carbon monoxide binds more strongly to Fe^{2+} than does oxygen and can cause death
      c. lysozyme (Fig. 20.14): an enzyme that disposes of dead bacteria

Key Terms globular protein, heme, hemoglobin, lysozyme, myoglobin

Key Figures

20.8 Mutations: Sickle-Cell Hemoglobin

A. Genetic mutations are alterations in DNA that change the primary structure of a protein.
   1. substitution of an amino acid can occur
      a. if the substitute has size, shape, and polarity similar to those of the original, the effect on protein function may be minimal
      b. if the substitute is very different in size, shape, or polarity, the effect may alter the biological function of the protein

B. Sickle-cell hemoglobin: sickled hemoglobin and normal hemoglobin differ by one amino acid residue (Figs. 20.15, 20.16); a neutral valine replaces a polar glutamic acid, and the molecule cannot pick up oxygen

Key Terms genetic mutation, sickle-cell anemia, sickle-cell hemoglobin, sickle-cell trait

Key Figures
Figure 20.15 Electron micrograph of normal and sickled red blood cells
Figure 20.16 Formation of rodlike structures by aggregation of deoxygenated sickle-cell hemoglobin

20.9 Denaturation

A. The biological function of a protein is determined by its structure; the native conformation is the biologically active conformation of the protein.
B. The loss of native conformation results in loss of biological function; this loss is called denaturation.
C. Factors that affect the native conformation and can denature a protein include
   1. temperature: increases in temperature disrupt noncovalent attractions by increasing thermal motion of molecules
   2. ultraviolet light and ionizing radiation: can cause chemical reactions in the chains that change the native conformation
   3. mechanical energy: violent mixing or whipping can disrupt noncovalent interactions
   4. changes in pH: can disrupt salt-bridge interactions by donating or abstracting protons from side groups and changing the charges in those areas
   5. organic chemicals: soaps, detergents, alcohol, and urea can disrupt side-chain interactions
   6. salts of heavy metals: Pb^{2+}, Hg^{2+}, and Ag^{+} react with sulfhydryl groups to form metal–disulfide bridges that prevent the formation of native conformations
   7. oxidizing and reducing agents: oxidizing agents form disulfide bridges and reducing agents cleave disulfide bridges

Key Terms denaturation, denatured protein, native conformation, native protein

ARE YOU ABLE TO... AND WORKED TEXT PROBLEMS

Objectives Introduction

Are you able to . . .
- describe the range of activities in which proteins take part
- list protein functions
- classify a protein on the basis of its function
- describe the structure of proteins

Objectives Section 20.1 α-Amino Acids

Are you able to . . .
- list the structural components of amino acids
- describe the configuration and conformation of an amino acid
- classify amino acids as α, β, γ (Exercise 20.1)
- describe the properties of an amino acid
- relate amino acid properties to amino acid structure (Exercise 20.70)
- explain the significance of R groups
- classify R groups
- define essential amino acids
- identify amino acids from their abbreviated names and R groups (Exercises 20.3, 20.4)
- give the abbreviation for a common amino acid
- identify L- and D-enantiomers of amino acids (Exercises 20.2, 20.69)

Objectives Section 20.2 Zwitterionic Structure of α-Amino Acids

Are you able to . . .

- describe the structure of a zwitterion
- discuss the properties of a zwitterion
- identify zwitterions
- relate the physical structures of zwitterions to their properties
- describe the nature of secondary forces between zwitterions
- explain why amino acids are amphoteric (Exercise 20.5)
- identify equilibrium products of a zwitterion in water
- discuss the pH dependency of zwitterion structure
- discuss the effect of pH on zwitterion properties (Exercises 20.7, 20.8)
- predict the zwitterion species in solutions of varying pH
- draw the structure of anionic, cationic, and neutral zwitterions (Ex. 20.1, Prob. 20.1, Exercises 20.7, 20.8, 20.71, 20.72)
- define isoelectric point (Exercise 20.6)
- explain the differences in the relative isoelectric points of acidic, basic, and neutral amino acids
- discuss the acid-base behavior of different R groups
- explain why amino acids can be separated by electrophoresis (Exercises 20.7, 20.8)
- predict amino acid structures at physiological pH (Exercise 20.71)

Problem 20.1 Show the predominant structure of phenylalanine at its isoelectric point (5.96), at physiological pH (7), at acidic pH < 1 and at basic pH > 12.

Phenylalanine has a neutral nonpolar side group, so only the amino and carboxylic acid ends of the molecule are affected by pH changes. At pH below 1, all proton-accepting sites will have protons; $-NH_3^+$. At pH above 12, all acidic sites have lost their protons; $-COO^-$. At physiological pH (within 2 pH units of pI) and at pI, the molecule is in the zwitterion form.

$H_3N^+—CH(CH_2C_6H_5)—COOH$ $H_3N^+—CH(CH_2C_6H_5)—COO^-$ $H_2N—CH(CH_2C_6H_5)—COO^-$

pH < 1   physiological pH   pH > 12

Objectives Section 20.3 Peptides

Are you able to . . .

- describe the structure of peptides
- classify peptides on the basis of the number of amino acid residues that they contain
- discuss the nature of the peptide bond
- describe the configuration of atoms around peptide bonds
- explain how peptide bonds are formed
- name peptides (Ex. 20.2, Prob. 20.2, Exercise 20.17)
- draw the structure of a peptide sequence from the individual amino acids (Ex. 20.2, Prob. 20.2, Exercises 20.74, 20.77, 20.78)
- draw the structures of peptides at various pH values (Ex. 20.4, Prob. 20.4, Exercises 20.9, 20.10, 20.11, 20.12, 20.13, 20.14, 20.18, 20.19, 20.20, Expand Your Knowledge 20.96)
- predict the number of isomers possible for a peptide sequence (Ex. 20.3, Prob. 20.3, Exercises 20.15, 20.16, 20.73, 20.75)
- relate peptide properties to the properties of amino acid residues (Expand Your Knowledge 20.97)
- compare the isoelectric points of peptides and amino acids
- compare the properties of peptides and amino acids
- predict electrophoretic behavior of peptides (Exercises 20.72, 20.79, 20.80, 20.82, 20.85, 20.86, 20.87, 20.88)

- interpret electrophoresis results for peptides (Exercises 20.86, 20.87)

Problem 20.2 Show the formation of Ala-Lys-Phe from the individual α-amino acids by using structural formulas. Give the full name for this tripeptide. Show all amino and carboxyl groups in their ionized form.

The tripeptide is from alanine-lysine-phenylalanine. The N-terminal residue is alanine and the C-terminal residue is phenylalanine. The full name is alanyllysylphenylalanine and the structure is:

$$H_3\overset{+}{N}-CH(CH_3)-C(=O)-O^- + H_3\overset{+}{N}-CH((CH_2)_4\overset{+}{N}H_3)-C(=O)-O^- + H_3\overset{+}{N}-CH(CH_2C_6H_5)-C(=O)-O^-$$

dehydration dehydration

$$H_3\overset{+}{N}-CH(CH_3)-C(=O)-NH-CH(CH_2CH_2CH_2CH_2\overset{+}{N}H_3)-C(=O)-NH-CH(CH_2C_6H_5)-C(=O)-O^- + 2H_2O$$

Problem 20.3 How many different constitutional isomers are possible for tripeptides containing glutamic acid, isoleucine, and lysine? Give the abbreviated names for the different amino acid sequences.

There are three different amino acids, so the number of isomers comes from using the formula $3! = 3 \times 2 \times 1 = 6$ isomers. The isomers are:

Glu-Ile-Lys Lys-Ile-Glu
Glu-Lys-Ile Lys-Glu-Ile
Ile-Glu-Lys Ile-Lys-Glu

Problem 20.4 Show the structures of the following peptides at physiological pH. (a) Ala-Lys-Ala; (b) Asp-Lys-Asp. Does each peptide migrate in electrophoresis at physiological pH? If it migrates, to which electrode? Is the pI value for each peptide on the acidic or basic side of the pI values for polypeptides containing only neutral amino acid residues?

Residue	R-group	Property
Ala	$-CH_3$	neutral nonpolar
Lys	$-(CH_2)_4-NH_2$	basic
Asp	$-(CH_2)COOH$	acidic

(a) Ala-Lys-Ala

Two neutral nonpolar groups and a basic group; there is one extra basic group; the pI is higher than 5.05–6.30; it is on the basic side; at physiological pH, the peptide has an overall 1+ charge and migrates toward the negative electrode (cathode).

$$H_3\overset{+}{N}{-}CH(CH_3){-}\overset{O}{\overset{\|}{C}}{-}NH{-}CH(CH_2CH_2CH_2CH_2\overset{+}{N}H_3){-}\overset{O}{\overset{\|}{C}}{-}NH{-}CH(CH_3){-}\overset{O}{\overset{\|}{C}}{-}O^-$$

(b) Asp-Lys-Asp

Two acidic groups and a basic group; there is an extra acidic group; the pI is lower than 5.05–6.30; it is on the acidic side; at physiological pH, the peptide has an overall 1– charge and migrates toward the positive anode (cathode).

$$\overset{+}{N}H_3{-}CH(CH_2COO^-){-}\overset{O}{\overset{\|}{C}}{-}NH{-}CH(CH_2CH_2CH_2CH_2\overset{+}{N}H_3){-}\overset{O}{\overset{\|}{C}}{-}NH{-}CH(CH_2COO^-){-}COO^-$$

Objectives Section 20.4 Chemical Reactions of Peptides

Are you able to . . .

- predict the products of hydrolysis of a peptide (Exercises 20.21, 20.22, 20.24, 20.25, 20.26)
- draw the structure of a peptide from the hydrolysis products (Exercise 20.23)
- explain the relation between peptide formation and hydrolysis
- discuss the oxidation and reduction of sulfhydryl groups (Exercises 20.25, 20.26, 20.83)
- describe disulfide bonds

Objectives Section 20.5 Three-Dimensional Structure of Proteins

Are you able to . . .

- discuss the relation between protein structure and the protein's biological function
- describe the levels of structural organization in a protein (Exercises 20.37, 20.38, 20.90, 20.91)
- relate the primary, secondary, and tertiary structures to one another and to the overall shape of the protein (Exercises 20.29, 20.30, 20.31, 20.32, 20.38)
- describe the specific structural features of each level (Exercises 20.29, 20.30, 20.31, 20.32)
- describe the factors that contribute to the final conformation of a protein (Exercises 20.29, 20.30, 20.31, 20.32, 20.35, 20.36)
- explain how each of the preceding factors afftects protein structure (Exercises 20.29, 20.30, 20.31, 20.32)
- analyze the R groups in an amino acid sequence to determine what types of interactions can take place (Exercises 20.33, 20.34)
- classify proteins by structure, function, or composition
- distinguish between simple and conjugated proteins (Exercise 20.27)

Objectives Section 20.6 Fibrous Proteins

Are you able to . . .

- describe the features and functions of fibrous proteins (Exercises 20.39, 20.41, Expand Your Knowledge 20.100)
- relate the structural features of fibrous proteins to their properties and functions (Exercises 20.40, 20.42, 20.89, 20.90)
- give examples of fibrous proteins (Exercises 20.43, 20.44)
- discuss the specific structural features of common fibrous proteins (Expand Your Knowledge 20.98)

Objectives Section 20.7 Globular Proteins

Are you able to . . .

- describe the features and functions of globular proteins (Exercises 20.52, 20.53, 20.54)
- relate the structural features of globular proteins to their properties and functions (Exercises 20.45, 20.88, 20.89, 20.92)
- give examples of globular proteins (Exercises 20.46, 20.47, 20.48, 20.49, 20.50, 20.51)
- discuss the specific structural features of globular proteins (Exercises 20.46, 20.47, 20.48, 20.49, 20.50, 20.51, 20.88)

Objectives Section 20.8 Mutations: Sickle-Cell Hemoglobin

Are you able to . . .

- define genetic mutation (Exercise 20.55)
- describe the ways in which genetic mutations can altar protein shape (Exercises 20.57, 20.58, 20.59, 20.60, 20.93, Expand Your Knowledge 20.99)
- give examples of genetic mutations altering biological activity (Exercise 20.56)

Objectives Section 20.9 Denaturation

Are you able to . . .

- define native protein and native conformation (Exercise 20.61)
- compare denaturation and digestion (Exercise 20.63)
- explain how denaturation affects the native conformation of a protein (Exercise 20.62)
- list the factors that can cause denaturation (Exercises 20.64, 20.65, 20.66, 20.67, 20.68)
- describe how each factor can cause denaturation of proteins (Exercise 20.94)

MAKING CONNECTIONS

Draw a concept map or write a paragraph describing the relation between the members of each of the following groups:

(a) carbohydrates, lipids, proteins, biological function
(b) metalloprotein, enzyme, metal ion
(c) glycoprotein, saccharides, cell recognition, antibody
(d) amino acid, R group, amino, carboxyl
(e) amino acid, zwitterion, pH
(f) zwitterion, cation, anion, equilibrium
(g) amino acid, peptide, isomer
(h) peptide bond, conformation, single bond
(i) electrophoresis, zwitterion, neutral amino acid
(j) apoprotein, prosthetic group, conjugated protein
(k) peptide bond, digestion, hydrolysis
(l) hydrophobicity, hydrophilicity, stearic interactions, side-group interactions, stable conformation
(m) α-helix, β-pleated sheet, β-turn, loop
(n) sulfhydryl group, oxidation, reduction
(o) fibrous protein, globular protein, protein function
(p) fibrous protein, globular protein, R groups, secondary structure

(q) genetic mutation, denaturation, primary sequence, biological function
(r) biological function, native conformation, physiological conditions
(s) denaturation, primary structure, secondary structure, tertiary structure, quaternary structure

FILL-INS

Proteins perform many more functions than do _____. The _____ makes them more versatile than other biomolecules. Comprising sequences of _____, the primary structure of a protein lays down the foundation from which _____ of structure are derived. Hydrogen bonding between backbone groups on the same or adjacent peptide chains produces _____ structural features such as _____. The assembly of the secondary structure based on side-group interactions directs the formation of _____. Factors such as hydrophobicity, steric accommodation, and hydrophilicity contribute to the final _____ adopted by the protein. The biologically active conformation, called the _____, may have a fourth level of structure if _____. This _____ is the overall assemblage of two or more polypeptides, each with its own tertiary structure. Any change in shape of the native conformation that results in disruption of the protein's biological activity is called _____. Factors that can cause this change include _____. Changes in the native conformation can also be caused when the _____ is altered by genetic mutations.

Answers

Proteins perform many more functions than do <u>carbohydrates or lipids</u>. The <u>complexity of their three-dimensional structures</u> makes them more versatile than other biomolecules. Comprising sequences of <u>amino acids</u>, the primary structure of a protein lays down the foundation from which <u>other levels</u> of structure are derived. Hydrogen bonding between backbone groups on the same or adjacent peptide chains produces <u>secondary</u> structural features such as <u>α-helices and β-pleated sheets</u>. The assembly of the secondary structure based on side-group interactions directs the formation of <u>the tertiary structure</u>. Factors such as hydrophobicity, steric accommodation, and hydrophilicity contribute to the final <u>stable conformation</u> adopted by the protein. The biologically active conformation, called the <u>native conformation</u>, may have a fourth level of structure if <u>an association of separate polypeptides is required for biological activity</u>. This <u>quaternary structure</u> is the overall assemblage of two or more polypeptides, each with its own tertiary structure. Any change in shape of the native conformation that results in disruption of the protein's biological activity is called <u>denaturation</u>. Factors that can cause this change include <u>temperature increases, pH changes, exposure to UV light, whipping, heavy-metal salts, and organic substances</u>. Changes in the native conformation can also be caused when the <u>primary sequence</u> is altered by genetic mutations.

TEST YOURSELF

1. Table 20.3 lists some examples of conjugated proteins. Read the descriptions and classify each protein according to its function (that is, catalytic, transport, regulatory, structural, contractile, storage).
2. The following figure represents a protein in its native conformation. Redraw the figure to show what it looks like (a) after denaturation and (b) after digestion.

3. Explain how the following interactions can be disrupted:
   (a) Salt bridge (b) Disulfide bond (c) Hydrophobic attraction (d) Hydrogen bond

4. Compare an α-helix and a β-pleated sheet. How are they the same? How are they different?
5. Where would you expect to find the following amino acid residues when they are part of a globular protein? Explain your answer?
   (a) Glycine (b) Leucine (c) Serine (d) Phenylalanine
6. Explain why a very high fever is dangerous.
7. Explain why acidosis and alkalosis, blood pH values below and above the normal range, are dangerous.
8. Explain why milk is sometimes given as a treatment for heavy-metal poisoning.
9. In light of what you learned about protein structure, why is it important to maintain an adequate intake of minerals such as iron and zinc?

Answers

1. Immunoglobulin: protective protein; heme: storage protein; chlyomicron: transport protein; metal ions: catalytic proteins; nucleoprotein: regulatory protein; casein: storage protein.
2. After denaturation, the strand is uncoiled but not broken. Secondary and tertiary structure is gone, but the primary structure remains intact. After digestion, the strand is broken into individual amino acids and the primary structure is destroyed.

Individual amino acids

(a) ~~~ Uncoiled 1° intact

(b) — — — — — 1° disrupted

3. (a) Salt bridges are ionic attractions that can be disrupted by the neutralization of one of the charged species or by insertion of a charged species that attracts one or both members of the original pair.
   (b) Disulfide bonds can be disrupted by chemical reduction, which cleaves them.
   (c) Hydrophobic attractions can be disrupted by the presence of a hydrophobic substance with which hydrophobic residues can interact.
   (d) Hydrogen bonds can be disrupted by the presence of acids or bases or another substance that can form dipolar or hydrogen bonds with one or both members of the original pair.
4. An α-helix and a β-pleated sheet are similar in that both are local conformations that result from hydrogen bonding within peptide groups of the backbone and not between the side groups. They are different in that the spiral of an α-helix is held in place by weak hydrogen bonds between members on the same chain. β-Pleated sheets, on the other hand, are held together by stronger hydrogen bonds between adjacent chains, which results in a linear configuration. Extensive α-helix structure is found in water-soluble proteins, and extensive β-pleated sheet structure is found in proteins that are not soluble in water.
5. Glycine, leucine, and phenylalanine are all neutral nonpolar residues that would most likely be folded inside a protein, away from aqueous environments. Serine is a neutral polar residue and would most likely be on the outside of the protein and taking part in hydrogen bonding with water.
6. High fevers are dangerous because the additional thermal energy can cause denaturation of proteins and disrupt normal metabolic activity.
7. Acidosis and alkalosis are dangerous because both conditions can change protein shape by disrupting hydrogen bonding and salt bridges in proteins. This disruption causes denaturation, which leads to disruption of metabolic activities.
8. Milk is sometimes given as a temporary treatment for heavy-metal poisoning because heavy metals bind to the protein in milk. This binding changes the solubility of the protein, precipitating it out of solution and temporarily removing the heavy metal from metabolic activity until the stomach can be cleared by pumping.
9. Minerals such as iron and zinc are prosthetic groups in metalloproteins. They provide catalytic centers for enzymatic activity and stabilize structure through interactions with polar or charged species.

chapter 21

Nucleic Acids

OUTLINE

Introduction

A. DNA and RNA direct and carry out growth and reproduction.
B. Chromosomes contain DNA that directs synthesis of proteins from the nucleus.
   1. human cells contain 23 pairs of chromosomes
   2. chromosomes contain genes
   3. genes carry the information for synthesizing polypeptides
C. RNA directs synthesis of proteins from ribosomes.

Key Terms chromosome, deoxyribonucleic acid (DNA), diploid, gene, haploid, nucleic acid, ribonucleic acid (RNA)

21.1 Nucleotides

A. Nucleotides are monomeric units of nucleic acids.
B. All nucleotides contain (Fig. 21.1):
   1. a heterocyclic base
      a. pyrimidine bases are six-membered heterocycles (Fig. 21.2)
         i. uracil , U (only in RNA)
         ii. thymine, T (only in DNA)
         iii. cytosine, C
      b. purine bases are fused five- and six-membered heterocycles (Fig. 21.2)
         i. adenine, A
         ii. guanine, G
   2. a sugar (Fig. 21.2)
      a. D-ribose in RNA
      b. D-deoxyribose in DNA
      c. carbons in sugars are numbered with primed numbers (for example, 5′) to distinguish them from base ring positions
   3. a phosphate; derived from phosphoric acid
C. Nucleotides are named according to the identity of the sugar and base.
   1. purine base + sugar: replace -ine ending of base with -osine
   2. pyrimidine base + sugar: first three letters of base name (first two for uracil) followed by -idine
   3. if sugar is D-deoxyribose, prefix deoxy- is used
   4. 5′-monophosphate indicates phosphate group at C5′ of sugar

Key Terms deoxyribonucleotide, nucleotide, purine, pyrimidine, ribonucleotide

Key Figures

Figure 21.1 Relation of nucleotide components and nucleic acids

Figure 21.2 Individual components of nucleotides

Problem 21.1 Drawing the structure of a nucleotide from its components and naming it
Problem 21.2 Determining the components of a nucleotide from its abbreviated name and providing the full name

21.2 Nucleic Acid Formation from Nucleotides

A. Nucleotides are joined by phosphodiesters to become nucleic acids (Fig. 21.3).
  1. products are those of a dehydration of –OH at C5′ of one and –OH of C3′ of the other, joined by a phosphodiester
  2. RNA from ribonucleotides and DNA from deoxyribonucleotides
  3. by convention, the sequence of nucleic acids is read from the top to bottom or from right to left
    a. designated 5′ → 3′ direction
    b. 5′-end has a phosphate at C5′ that is attached to only one pentose ring
    c. 3′-end has a pentose ring with an unreacted –OH group at C3′
    d. the P–OH groups are ionized at physiological pH

B. Nucleotide sequences are abbreviated to represent only the base sequence on the sugar-phosphate chain (Fig. 21.4)
    a. use the one-letter abbreviations for the bases
    b. RNA is inferred from the presence of U instead of T in the sequence
    c. nucleotide sequence determines α-amino acid sequence in a polypeptide

Key Terms 3′-end, 5′-end, phosphodiester

Key Figures
Figure 21.3 Synthesis of dCMP-dAMP-dTMP
Figure 21.4 Different representations of a nucleotide sequence

Problem 21.3 Drawing the complete structure of a nucleotide sequence from the abbreviated name

21.3 Three-Dimensional Structure of Nucleic Acids

A. DNA:
  1. primary structure is sequence of bases along sugar-phosphate chain
  2. secondary structure is a right-handed helix formed by hydrogen bonds and hydrophobic interactions between bases: base-pairing (Figs. 21.5, 21.6)
  3. DNA double helix: complementary strands that run in opposite directions
    a. A–T and C–G pairs create similar dimensions throughout helix (Fig. 21.6)
    b. phosphate outside, heterocyclic bases inside
  4. large molecule compaction due to folding around nucleosome cores
  5. Levels of structure: base pairs ⇒ double helix ⇒ nucleosome ⇒ chromatin fiber ⇒ chromosome

B. RNA:
  1. single strand of nucleotides containing ribose as the sugar and uracil instead of thymine
  2. several types, each with different structure
    a. transfer RNA (tRNA) (Fig. 21.7)
      i. secondary and tertiary structure results from intramolecular base-pairing in short sequences
      ii. acceptor stem at 3′-end
      iii. anticodon at loop farthest from acceptor stem
    b. messenger RNA (mRNA)
    c. ribosomal RNA (rRNA)

Key Terms acceptor stem, anticodon, base-pairing, chromatin fiber, complementary strands, DNA double helix, DNA strands, duplex, genome, histone, loops, nucleosome, nucleosome core, transfer RNA

Key Figures
Figure 21.5 Different representations and views of a DNA double helix
Figure 21.6 Complementary base pairs A-T and G-C
Figure 21.7 Levels of DNA atructure from molecular to chromatin fiber
Figure 21.8 Representations of structure of a tRNA molecule

Example 21.1 Determining the base sequence in a DNA strand from the base sequence in its complementary strand

Problem 21.4 Determining the base sequence in a DNA strand from the base sequence in its complementary strand

21.4 *Information Flow from DNA to RNA to Polypeptide*

A. The sequence of genetic information flow is (Figure 21.9):
   1. genetic information is stored in DNA as the nucleotide sequence
   2. information is replicated as complementary DNA strand (replication)
   3. information is transcribed from the complementary DNA strand to RNA (transcription)
   4. information is translated from RNA to polypeptide (translation)

Key Terms messenger RNA, replication, ribosomal RNA, transcription, transfer RNA, translation

Key Figures Figure 21.9 Overview of molecular genetics

21.5 *Replication*

A. Overview of replication (Fig. 21.10): DNA strand unwinds ⇒ forms template parent strands ⇒ DNA polymerase catalyzes reaction that produces new DNA strand ⇒ daughter DNA duplex produced
B. Key points:
   1. short spans of unwound DNA, called replication bubbles, have replication forks at each end
   2. template strands of original DNA are used to synthesize complementary strands of DNA
   3. DNA polymerase catalyzes addition of nucleotides to growing DNA strand
   4. new DNA grows in $3' \rightarrow 5'$ direction
   5. Each new daughter DNA molecule consists of a strand from the parent DNA and a new strand

Key Terms daughter, DNA ligase, DNA polymerase, lagging strand, leading strand, Okazaki fragments, parent, replication bubble, replication fork, semiconservative, strand, template strand

Key Figure Figure 21.10 Replication of DNA duplex

Example 21.2 Determining the base sequences in parent and daughter DNA strands

Problem 21.5 Determining the base sequences in parent and daughter DNA strands

21.6 *Transcription*

A. Overview of transcription (Figs. 21.11, 21.12): DNA template strand ($5'$–$3'$) ⇒ ($3'$–$5'$) rRNA, tRNA, mRNA ⇒ invert to read in $5' \rightarrow 3'$ direction ⇒ splice out introns ⇒ final RNA
B. Base-pairing to DNA template determines RNA sequence.
   1. RNA has same base sequence as that of the DNA nontemplate strand
   2. uracil replaces thymine in RNA
C. Key points:
   1. RNA polymerase controls transcription
   2. initiation site on base sequence has start signal
   3. termination site on base sequence has stop signal
   4. RNA is synthesized as a single strand
   5. splicing deletes introns and retains exons (Fig. 21.12)

Key Terms base modification, end capping, exon, initiation site, intron, nontemplate strand, posttranscriptional processing, primary transcript RNA, RNA polymerase, splicing, termination site, transcription bubble

Key Figures

Figure 21.1 Transcription of DNA to form RNA

Figure 21.12 Posttranscriptional processing of ptRNA into mRNA by splicing

Example 21.3 Determining the base sequences in nontemplate DNA and RNA strands from a sequence in DNA template strand

Problem 21.6 Determining the base sequences in nontemplate DNA and RNA strands from a sequence in DNA template strand

21.7 Translation

A. Overview:

mRNA codon ⇒ tRNA anticodon ⇒ amino acid ⇒ polypeptide synthesis outside the nucleus

B. Key points:
   1. mRNA carries codons (5′ → 3′) for base triplets that each specify one amino acid
   2. tRNA carries the anticodon (3′ → 5′) for one specific amino acid (Fig. 21.13)

C. The genetic code is the complete list of mRNA codons and the amino acids that they specify (Table 21.1).
   1. degenerate sequences are different but code for the same amino acid
   2. some sequences are stop codons that terminate polypeptide synthesis
   3. one sequence is an initiation codon that starts polypeptide synthesis (Fig. 21.14)

D. Polypeptide is processed into its final functional form.
   1. initiation amino acid is usually cleaved
   2. polypeptides are folded into active conformations
   3. amino acid residues may be modified
   4. if quaternary structure is required for active conformation, different polypeptides must be brought together and assembled

Key Terms acyl transferase, aminoacyl site, aminoacyl-tRNA synthetase, base triplet, codon, genetic code, genetic message, peptidyl site, peptidyl transferase, posttranslational processing, repressor proteins

Key Figures and Table

Figure 21.13 Picking up an α-amino acid at the 3′-end of tRNA
Figure 21.14 Synthesis of a polypeptide
Table 21.1 Genetic code: codon assignments

Problem 21.7 Determining tRNA anticodon and amino acid from codon sequence
Example 21.4 Determining the polypeptide structure encoded by a template DNA strand
Problem 21.8 Determining the polypeptide structure encoded by a template DNA strand

21.8 Mutations

A. Mutations are errors in the base sequence of a gene.
   1. substitution, or point, mutations arise when one base substitutes for another
   2. frameshift mutations arise when a base is inserted or deleted from a normal sequence

B. Silent mutations do not affect the functioning of an organism.
   1. mutations can result in a degenerate sequence
   2. replacement by an amino acid of similar size, shape, charge, and polarity may not change the three-dimensional structure of the protein
   3. replacement by a dissimilar amino acid but in a region of the protein that is not important for its function
   4. there may be multiple copies of genes that still function if some are mutated

C. Mutations can have negative effects (Table 21.2).
   1. replacement by dissimilar amino acid in a region that is important for biological function may destabilize structure
   2. if the mutation converts a codon into a stop codon, it may stop the function
   3. frameshift mutations in exons shift the entire sequence of codons
   4. mutant DNA may produce toxic polypeptides or proteins

Key Terms carcinogen, frameshift mutation, genetic diseases, germ cell, mutagen, mutation, point mutation, silent mutation, somatic cell, spontaneous mutation, substitution mutation

Key Table Table 21.2 Hereditary diseases

Example 21.5 Determining the effect of a mutation on the amino acid sequence of a polypeptide
Problem 21.9 Determining the effect of a mutation on the amino acid sequence of a polypeptide

21.9 *Antibiotics*

A. Antibiotics are chemicals that fight infections by bacteria, mold, and yeast.
B. Antibiotics interfere with the genetic functioning of the microorganism (Table 21.3).
   1. must be specific for microorganism
   2. must have minimal effect on protein synthesis in host
C. Prolonged use of antibiotics fosters production of mutant strains and may have negative effects on host, or lead to antibiotic-resistant strains.

Key Term antibiotic
Key Table Table 21.3 Antibiotic inhibition of protein synthesis in bacteria

21.10 *Viruses*

A. Viruses are infectious, parasitic particles of DNA or RNA encapsulated by a protein coat.
B. DNA viruses:
   1. DNA viruses contain only DNA
   2. they enter host nucleus and integrate themselves into the host genome
   3. the host treats the modified DNA as its own
C. RNA viruses:
   1. RNA viruses contain only RNA; viral RNA is produced in host cell by using enzyme, RNA replicase, produced by host
   2. RNA retroviruses synthesize viral DNA from viral RNA (Fig. 21.15)
      a. viral DNA is incorporated into host genome
      b. host produces viral RNA from DNA template
      c. HIV and oncogenic viruses are retroviruses (Fig. 21.15)
D. Vaccines are a preemptive strategy against viral infection.
   1. vaccines contain weakened strains of a virus or its proteins
   2. the presence of a vaccine in the body stimulates the immune system to generate antibodies that recognize the virus's antigens

Key Terms Cap sid, DNA virus, oncogenic virus, retrovirus, reverse transcriptase, RNA replicase, RNA virus, vaccine, virus
Key Figure Figure 21.15 Reproduction of HIV in a T cell

21.11 *Recombinant DNA Technology*

A. Genetic engineering alters the genome of an organism by transplanting DNA from another organism into it.
B. Genetic engineering can contribute to (Table 21.4):
   1. production of therapeutic proteins (Fig. 21.16)
      a. donor DNA is spliced into bacterial DNA plasmids
      b. bacterial replication produces clones of altered DNA plasmid
      c. clones produce wanted proteins that can be separated
   2. transgenic breeding of plants and animals
      a. selective production of organisms whose genes have been permanently altered by recombinant DNA technology
      b. produces organisms with desired traits
      c. reduces the amount of time before organism with desired trait is produced
   3. gene therapy
      a. recombinant DNA technology can be used to modify a person's genome by inserting a gene to correct for genetic diseases
      b. Human Genome Project
         i. identify and locate the genes of the human genome
         ii. data leads to identification of genes responsible for specific genetic diseases
         iii. leads to better treatment of those diseases

4. gene testng
   a. enables prenatal and newborn screening
   b. presymptomatic testing for adult-onset diseases
   c. diagnosis confirmation for symptomatic individuals
5. DNA fingerprinting-based on repeated sequences of base pairs—variable number of tandem repeats

Key Terms chemical shock, clone, donor DNA, gene therapy, genetic engineering, Human Genome Project, plasmid, recombinant DNA, recombinant DNA technology, restriction enzyme, tolerence gene, transgenic breeding, variable number of tandem repeats, vector DNA

Key Figure and Table

Figure 21.16 Formation of recombinant DNA
Table 21.4 Products of recombinant DNA technology

ARE YOU ABLE TO... AND WORKED TEXT PROBLEMS

Objectives Introduction

Are you able to...

- explain the function of DNA and RNA
- explain the relation between DNA, chromosomes, genes, and polypeptides
- explain the relation between RNA, ribosomes, and proteins

Objectives Section 21.1 Nucleotides

Are you able to...

- explain the relation between nucleotides and nucleic acids
- describe the structural components of nucleotides
- distinguish between purine and pyrimidine bases
- provide the names and abbreviations for bases
- identify RNA and DNA by the sugar component
- name nucleotides

Problem 21.1 Draw the structure of the nucleotide consisting of phosphoric acid, deoxyribose, and guanine. Give the full and abbreviated names.

Guanine is a purine base. Guanine + deoxyribose + phosphoric acid → deoxyguanosine-5′-monophosphate, the full name. d-GMP is the abbreviated name. Phosphoric acid joins deoxyribose at C5′ of the deoxyribose ring. Deoxyribose joins guanine at C1′ of the deoxyribose ring and N1 of the guanine ring.

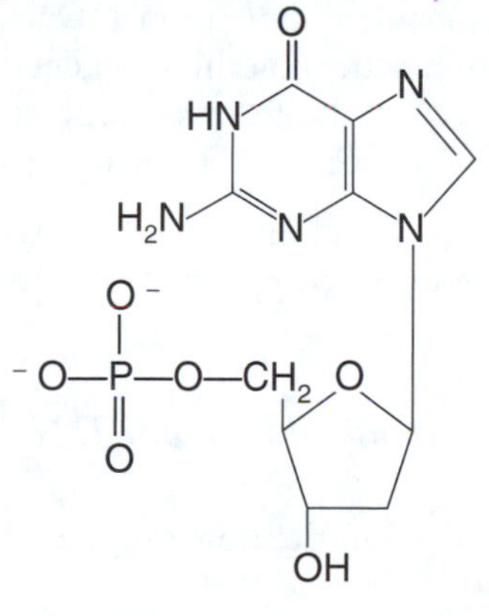

O
HN N
H_2N N N
O^-
$^-O—P—O—CH_2$ O
O
OH

Problem 21.2 What components make up the nucleotide UMP? Give the full name for UMP.

There is no prefix and the base is uracil, so the sugar is ribose. U is uridine, so the base is uracil; MP is a monophosphate. The full name is uridine-5′-monophosphate.

Objectives Section 21.2 Nucleic Acid Formation from Nucleotides

Are you able to...

- explain how nucleic acids are formed from nucleotides
- provide the base sequence for a nucleotide
- distinguish RNA sequences from DNA sequences
- draw the structure of a nucleotide sequence

Problem 21.3 Draw the complete structure of GMP-UMP. Indicate the 5′ → 3′ direction.

By convention, the sequence is left to right, so the 5′-end is GMP and the 3′-end is UMP. GMP contains guanine, ribose, and a phosphate group; UMP contains uracil, ribose, and a phosphate group. The sequence is joined by a phosphodiester group at C3′ on ribose in GMP and to C5′ on ribose in UMP.

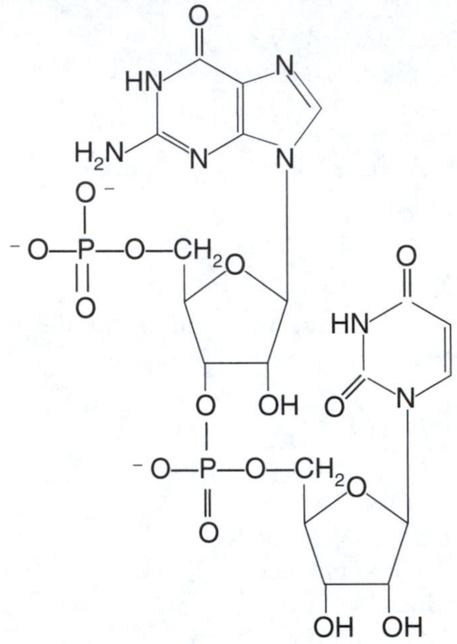

Objectives Section 21.3 Three-Dimensional Structure of Nucleic Acids

Are you able to...

- describe the primary, secondary, and tertiary structure of nucleic acids
- identify the major secondary structural features of nucleic acids
- explain base-pairing and its effect on nucleic acid structure
- explain how a nucleic acid arrives at its three-dimensional structure
- describe the levels of structure from genetic molecules to genetic macrostructure
- determine the base sequence of a DNA strand from its complement
- construct the complementary base sequence for a DNA strand

Problem 21.4 If the base sequence of a section of one DNA strand is 5′-GGCTAT-3′, what is the sequence for the corresponding section of the complementary DNA strand?

3′-CCGATA-5′

Objectives Section 21.4 Information Flow from DNA to RNA to Polypeptide

Are you able to...

- outline the sequence of events involved in the flow of genetic information from DNA to the final polypeptide
- identify the key events as information is translated from DNA into a protein

Objectives Section 21.5 Replication

Are you able to...

- give an overview of the replication process

- identify key events in the replication process
- identify key molecules in the replication process
- determine the base sequence in a DNA parent strand from the daughter strand
- determine the base sequence in a DNA daughter from the parent strand

Problem 21.5 The sequence 5′-ACGTGC-3′ is part of one strand of a parent DNA double helix. What is the corresponding sequence on the complementary strand of the parent DNA double helix? What is the corresponding sequence on the new daughter strand made from this parent strand during replication?

Parent complement → 3′-TGCACG-5′ → daughter 5′-ACGTGC-3′

Objectives Section 21.6 Transcription

Are you able to...

- give an overview of the transcription process
- identify key events in the transcription process
- identify key molecules in the transcription process
- determine the base sequence in an RNA strand from a DNA template
- determine the base sequence in a DNA template from an RNA strand
- determine the base sequence in a nontemplate DNA strand from an RNA sequence
- determine the base sequence in an RNA strand from a nontemplate DNA strand

Problem 21.6 For the sequence 5′-TAAGTCAAC-3′ in a DNA template strand, what is the base sequence for the DNA nontemplate strand? What is the base sequence for the synthesized RNA?

DNA template 5′-TAAGTCAAC-3′; → nontemplate DNA is 3′-ATTCAGTTC-5′; → RNA is 5′-UAAGUCAAG-3′

Objectives Section 21.7 Translation

Are you able to...

- give an overview of the translation process
- identify key events in the translation process
- identify key molecules in the translation process
- determine the base sequence in an amino acid from an RNA strand
- determine the base sequence in an amino acid from a DNA template
- determine the base sequence in an amino acid from a nontemplate DNA strand
- explain the terms degenerate, stop codon, and initiation codon
- distinguish between different types of RNA
- explain how a polypeptide is processed into its final functional form
- determine a tRNA anticodon and amino acid from codon sequence

Problem 21.7 For each of the following codons, what is the complementary tRNA anticodon and what amino acid is specified by the codon?

Codon	tRNA anticodon	Amino acid
(a) UCC	AGG	serine
(b) CAG	GUC	glutamatic acid
(c) AGG	UCC	arginine
(d) GCU	CGA	alanine

Problem 21.8 Sections X, Y, and Z of the following part of a template DNA strand are exons of a gene.

5′- [CAC] CAC [ACCGTA] TGTGGA [CAT] -3′
X Y Z

What are the structures of the primary transcript and mRNA made from this template DNA? What polypeptide sequence will be synthesized?

Primary transcript: 3′- [GUG] GUG [UGGCAU] ACACCU [GUA] -5′
X Y Z

Exons are kept and introns are spliced out; so 3′-GUGUGGCAUGUA-5′

Read from 5′ to 3′: mRNA is: 5′-AUGUACGGUGUG-3′
Codon sequence: AUG-UAC-GGU-GUG
Polypeptide sequence: Met-Tyr-Gly-Val

Objectives Section 21.8 Mutations

Are you able to...

- describe the different types of mutations
- explain the effects that different mutations have on protein function
- determine the effect of a mutation on the amino acid sequence of a polypeptide

Problem 21.9 Imagine the following mutations in the template DNA strand in Problem 21.8: (a) GTA of Y exon is mutated to GAA; (b) GTA of Y exon is mutated to TTA; (c) TGT of the second intron is mutated to AGT. What effect would each of these mutations have on the codon of the resulting mRNA? What effect would they have on the amino acid sequence of the resulting polypeptide?

(a) DNA template: 5′- [CAC] CAC [ACCGTA] TGTGGA [CAT] -3′
(X = CAC, Y = ACCGTA, Z = CAT)

Mutated DNA template: 5′- [CAC] CAC [ACCGAA] TGTGGA [CAT] -3′
(X = CAC, Y = ACCGAA, Z = CAT)

Primary transcript : 3′- [GUG] GUG [UGGCUU] ACACCU [GUA] -5′
(X = GUG, Y = UGGCUU, Z = GUA)

Exons are kept and introns are spliced out so: 3′-GUGUGGCUUGUA-5′
Read from 5′ to 3′: mRNA is 5′-AUGUUCGGUGUG-3′
Codon sequence: AUG-UUC-GGU-GUG
Original polypeptide sequence: Met-Tyr-Gly-Val
Mutated polypeptide sequence: Met-Phe-Gly-Val
Phenylalanine, a large, nonpolar, neutral amino acid has substituted for tyrosine, a large, polar, neutral amino acid.

(b) DNA template: 5′- [CAC] CAC [ACCGTA] TGTGGA [CAT] -3′
(X = CAC, Y = ACCGTA, Z = CAT)

Mutated DNA template: 5′- [CAC] CAC [ACCTTA] TGTGGA [CAT] -3′
(X = CAC, Y = ACCTTA, Z = CAT)

Primary transcript: 3′- [GUG] GUG [UGGAAU] ACACCU [GUA] -5′
(X = GUG, Y = UGGAAU, Z = GUA)

Exons are kept and introns are spliced out so: 3′-GUGUGGAAUGUA-5′
Read from 5′ to 3′: mRNA is 5′-AUGUAAGGUGUG-3′
Codon sequence: AUG-UAA-GGU-GUG
Original polypeptide sequence: Met-Tyr-Gly-Val
Mutated polypeptide sequence: Met-STOP-Gly-Val
A STOP codon has been substituted for tyrosine

(c) Intron mutation probably does not affect the polypeptide sequence.

Objectives Section 21.9 Antibiotics

Are you able to...

- define antibiotic
- explain how antibiotics fight infections
- identify some microorganisms responsible for infections
- describe some of the negative effects of long-term antibiotic use

Objectives Section 21.10 Viruses

Are you able to...

- define virus
- compare DNA and RNA viruses
- explain how DNA and RNA viruses operate in a host
- explain the use of vaccines against viral infections

Objectives Section 21.11 Recombinant DNA Technology

Are you able to...

- define genetic engineering and recombinant DNA technology
- identify some benefits of genetic engineering
- explain how genetic engineering is used to produce therapeutic proteins
- explain how genetic engineering is used in transgenic breeding of plants and animals
- explain how genetic engineering is used in gene therapy
- identify some social and ethical issues associated with genetic engineering

MAKING CONNECTIONS

Draw a concept map or write a paragraph describing the relation between the members of each of the following groups:

(a) monosaccharide, phosphate, purine, pyrimidine
(b) nucleic acid, nucleotide, phosphodiester
(c) nucleic acid, primary structure, secondary structure, base-pairing
(d) base-pairing, hydrogen bond, hydrophobic interaction, stabile structure
(e) DNA, RNA, genetic information, nucleotide
(f) DNA, replication, transcription, translation, RNA
(g) DNA, gene, chromosome, histone
(h) replication, replication fork, Okazaki fragment, replication bubble
(i) messenger RNA, transfer RNA, ribosomal RNA, template DNA
(j) codon, anticodon, amino acid, gene
(k) genetic code, mutation, amino acid
(l) mutagen, mutation, frameshift mutation, point mutation, silent mutation
(m) intron, exon, mRNA
(n) antibiotic, vaccine, gene therapy
(o) antibiotic, bacteria, vaccine, virus
(p) bacteria, plasmid, clone
(q) cloning, gene therapy, transgenic breeding
(r) recombinant DNA technology, enzyme, clone

FILL-INS

Nucleic acids are classified as either _____ or _____ depending on the _____ component of the nucleotides. Both types of nucleic acids are polymers of nucleotides that are composed of a _____ base, a _____, and a _____. The sugars are _____ in RNA and _____ in DNA. The bases _____ are found in all nucleic acids, but _____ is found only in RNA and thymine only in DNA. The sugar-phosphate backbone of nucleic acids is the same for all DNA molecules, but the _____ along the backbone varies. The sequence of bases is the _____ of a DNA molecule, and it is the sequence that carries an individual organism's _____. The secondary structure is a _____ that results from _____ between complementary bases on the two DNA strands. Base-pairing is due to _____ between specific pairs of bases: _____ in DNA or _____ in RNA. Replication of DNA is the process by which an _____ is produced. Replication starts when a DNA double strand _____ at specific points. The exposed bases pair up with _____ of nucleotides and form a new double helix containing one _____ strand and _____ strand. The translation of genetic information from DNA to a functioning protein is carried out in a sequence of steps called _____. Transcription is the process by which information stored in _____ a DNA molecule is passed on to _____ when they are synthesized as _____ strands to a DNA sequence. Messenger RNA segments are spliced together as _____ carrying _____ information for amino acids. Proteins are built from _____ as directed by mRNA. Any change in the base sequence of a DNA molecule is called a _____ and can produce a _____ protein. Some mutations have little or no effect and are called _____. Others, however, such as _____ mutations and some dissimilar _____ can produce _____ proteins or proteins that are _____ to the organism.

Answers

Nucleic acids are classified as either ribonucleic acids (RNA) or deoxyribonucleic acids (DNA), depending on the sugar component of the nucleotides. Both types of nucleic acids are polymers of nucleotides that are composed of a purine or pyrimidine base, a sugar, and a phosphate group. The sugars are D-ribose in RNA and D-deoxyribose in DNA. The bases adenine, guanine, and cytosine are found in all nucleic acids, but uracil is found only in RNA and thymine only in DNA. The sugar-phosphate backbone of nucleic acids is the same for all DNA molecules, but the sequence of bases along the backbone varies. The sequence of bases is the primary structure of a DNA molecule, and it is the sequence that carries an individual organism's genetic profile. The secondary structure is a double helix that results from base-pairing between complementary bases on the two DNA strands. Base pairing is due to hydrogen bonding and hydrophobic interactions between specific pairs of bases: between C and G and between A and T in DNA or A and U in RNA. Replication of DNA is the process by which an exact copy of a DNA molecule is produced. Replication starts when a DNA double strand unwinds at specific points. The exposed bases pair up with complementary bases of nucleotides and form a new double helix containing one original parent strand and a new strand. The translation of genetic information from DNA to a functioning protein is carried out in a sequence of steps called transcription and translation. Transcription is the process by which information stored in the base sequence of a DNA molecule is passed on to single strands of messenger RNA when they are synthesized as complementary strands to a DNA sequence. Messenger RNA segments are spliced together as exons carrying codon information for amino acids. Proteins are built from amino acid sequences as directed by mRNA. Any change in the base sequence of a DNA molecule is called a mutation and can produce a nonfunctioning protein. Some mutations have little or no effect and are called silent mutations. Others, however, such as frameshift mutations and some dissimilar amino acid substitutions can produce nonfunctioning proteins or proteins that are harmful to the organism.

TEST YOURSELF

1. Identify each of the following bases as a purine or a pyrimidine, and identify its complementary base:
   (a) guanine (b) thymine (c) cytosine (d) adenine (e) uracil
2. What are the structural components of the following nucleotides?
   (a) cytidine-5′-monophosphate (b) deoxyadenosine-5′-monophosphate
   (c) d-GMP (d) UMP
3. Compare the functions of mRNA, rRNA, and tRNA.
4. The codons for leucine are CUA, CUC, CUG, CUU, UUA, and UUG. The codon for tryptophan is UGG. Which of them is more susceptible to mutagenic effects?
5. What are the possible effects of genetic mutation on an organism?
6. List some general features of the genetic code.
7. Identify the following statements as true or false. If false, correct the statement:
   (a) Every nucleotide contains a purine base.
   (b) Uracil is a base found only in DNA.
   (c) Each codon is composed of three bases.
   (d) mRNA is made from exons spliced from the precursor RNA sequence.
   (e) Retroviruses are DNA viruses.
   (f) Antibiotics are used to treat viral infections.
   (g) Base-pairing is due to ionic attractions between nucleotide bases.
8. Recombinant DNA technology has produced major benefits. Identify the contributions made by genetic engineering and the social and ethical issues raised by it.

Answers

1. (a) Purine, cytosine; (b) pyrimidine, adenine; (c) pyrimidine, guanine; (d) purine, thymine in DNA and uracil in RNA; (e) pyrimidine, adenine
2. (a) Cytosine, ribose, phosphate; (b) adenosine, deoxyribose, phosphate; (c) guanine, deoxyribose, phosphate; (d) uracil, ribose, phosphate

3. Messenger RNA (mRNA) carries genetic information from DNA in the cell nucleus to the site of protein synthesis in the cytoplasm. Ribosomal RNA (rRNA) is located in the ribosomes and is involved in protein synthesis there. Transfer RNA (tRNA) delivers amino acids to the site of protein synthesis.
4. Tryptophan is more vulnerable to mutagenic effects. Degenerate sequences for leucine make it more mutant resistant. Each codon for leucine has U as a second base, so a mutation there might be more likely to have an effect.
5. Mutation may have no effect; it may make give an organism a survival advantage and have a positive effect; it may have a detrimental effect; it may have a lethal effect.
6. The genetic code consists of codons which are three-letter sequences that code for specific amino acids. There is degeneracy in the code for some amino acids. The start of a sequence is coded and the termination of a sequence is coded. Many organisms contain the same or similar codes.
7. (a) F; nucleotides contain either a purine or a pyrimidine base. (b) F; uracil is found only in RNA. (c) T. (d) T. (e) F; retroviruses are RNA viruses. (f) F; antibiotics are used to treat bacterial infections. Vaccines are administered to confer immunity against viral infections.(g) F; hydrogen bonds and hydrophobic interactions are responsible for base-pairing.
8. Genetic engineering has contributed products that have been used to counteract the effects of genetic diseases. It has been used to create plant and animal lines that carry desirable traits and may be used to help save endangered species. Gene therapy seeks to insert genes that counteract or modify the negative influence of genes that produce genetic diseases. Many issues have been raised as genetic engineering has grown. The use of genetically modified plants and their effects on the environment are currently topics of debate. Research into the effects of pollen from corn that has been modified to produce a natural pesticide indicates that it kills the caterpillars of monarch butterflies. The development of resistant species of microorganisms and pests is a question that needs to be answered. When a person's genome is known, the potential for discrimination by health-care providers and insurance companies arises. The ability of parents to choose an offspring's traits challenges the societal norm.

chapter 22

Enzymes and Metabolism

OUTLINE

22.1 Cell Structure

A. Eukaryotic and prokaryotic cells differ in their internal organization.
   1. eukaryotic cells have organelles that are surrounded by membranes; organelles are subcellular structures surrounded by their own membranes
   2. prokaryotic cells have organelles that are anchored directly to the cell membrane or move about freely

B. Eukaryotic cells have divided compartments and structures that perform specific functions (Fig. 22.1, Table 22.1).
   1. cell membrane encloses the cytosol and cell contents
      a. cytosol is jellylike matrix of the cell
      b. cytoskeleton contains microtubules and microfilaments that give the cell its shape, provide anchoring, and help with movement
      c. cell membranes act as mediators for transporting ions and molecules into and out of the cell
      d. cell membranes act as biomolecule-recognition centers and receptors by interactions of surface molecules
   2. nucleus contains DNA organized in chromosomes
      a. synthesis of ribosomes takes place in nucleolus, which is in the nucleus
      b. DNA synthesis and repair takes place in the nucleus
   3. endoplasmic reticulum
      a. membranous RNA structures at which protein synthesis takes place
      b. lipid synthesis takes place here
      c. the endoplasmic reticulum also participates in transport of macromolecules into and out of the cell
   4. aerobic processes take place in mitochondria
      a. aerobic reactions require oxygen as a reactant
      b. anaerobic reactions do not require oxygen as a reactant
      c. cellular respiration and the oxidation of carbohydrates and lipids takes place in mitochondria
   5. Golgi apparatus is the site of protein modification
      a. smooth membranes
      b. proteins are packaged for transport out of cell or for incorporation into organelles
   6. lysosomes are sites of hydrolysis of carbohydrates, lipids, proteins, and nucleic acids
   7. peroxisomes are sites where direct reaction with oxygen and decomposition of hydrogen peroxide take place
      a. hydrogen peroxide is potentially harmful to cells
      b. detoxification of substances takes place in peroxisomes

Key Terms aerobic, anaerobic, cell membrane, endoplasmic reticulum, eukaryote, Golgi apparatus, lysosome, microfilament, microtubule, nuclear zone, nucleus, organelle, peroxisome, prokaryote

Key Figure and Table

Figure 22.1 Structure of an animal cell

Table 22.1 Animal cell compartments and their major functions

22.2 General Features of Metabolism

A. Metabolism has the following functions in animals:
   1. obtain chemical energy from the degradation of nutrients
   2. create the precursor molecules for and to synthesize carbohydrates, lipids, proteins, nucleic acids, and other cellular molecules
   3. produce or modify the biomolecules necessary for specific specialized functions or produce specialized cells

B. Metabolism takes place through the interaction of catabolism and anabolism.
   1. catabolism is the process by which energy-containing compounds are degraded; complex molecules are converted into a limited number of simpler molecules
   2. anabolism is the process by which biomolecules are synthesized
      a. a few simple building blocks are converted into varied, complex biomolecules
      b. anabolic processes include biosynthesis, active transport, muscle contraction, and the translation and transcription of genetic information

C. Key features of metabolism are regulation and control.
   1. metabolic processes are sequences of reactions that are regulated
   2. energy-producing reactions fuel energy-consuming reactions

Key Terms anabolism, catabolism, metabolism

22.3 Stages of Catabolism

A. Catabolism consists of three major stages (Fig. 22.2)
   1. nutrient molecules are broken down to simpler molecules: monosaccharides, amino acids, and lipids
   2. simple breakdown products are converted into a common breakdown product: acetyl-*S*-CoA
      a. monosaccharides converted first into pyruvate (three carbons), then into acetyl-*S*-CoA (two carbons)
      b. hydrocarbon chains and amino acids converted into acetyl-*S*-CoA
      c. production of acetyl-*S*-CoA is the end step of the second stage
   3. acetyl-*S*-CoA is broken down to carbon dioxide, water, and ammonia; acetyl-*S*-CoA enters the citric acid cycle, where it is oxidized to carbon dioxide and water

Key Terms acetyl-*S*-coenzyme A, citric acid cycle

Key Figure Figure 22.2 Catabolism of the major energy-yielding compounds of cells

22.4 Transformation of Nutrient Chemical Energy into New Forms

A. Chemical energy obtained from nutrients in the second and third stages of catabolism is used to build products that fuel anabolism.
   1. ATP functions as the carrier of energy to the energy-requiring processes of cells
      a. ATP is made in the mitochondria through the electron-transport chain or by substrate-level phosphorylation
      b. ATP links the catabolic and anabolic processes (Fig. 22.3)
      c. ATP is hydrolyzed and its energy is used to fuel anabolic reactions; ATP is hydrolyzed to ADP, adenosine diphosphate, and inorganic phosphate, Pi (Fig. 22.4)
   2. NADH and NAD(P)H provide the reducing power required for cellular biosynthesis (Fig. 22.5)
      a. NAD^+ links metabolic oxidation and metabolic reduction processes (Fig. 22.8)
      b. most metabolic oxidation reactions are dehydrogenation reactions; two hydrogen atoms and two electrons are transferred to an acceptor
      c. NAD(P)H provides electrons and hydrogen atoms in biosynthesis reactions

3. FAD and FMN link metabolic oxidations to metabolic reductions (Figs. 22.6, 22.7)
   a. they accept two hydrogens
   b. FAD is bound to a protein as a cofactor

Key Terms adenosine triphosphate (ATP), dehydrogenation, flavin adenine dinucleotide (FAD), flavin mononucleotide (FMN), NADP, nicotinamide adenine dinucleotide (NADH), nicotinamide adenine dinucleotide phosphate (NADPH), oxidative phosphorylation, substrate-level phosphorylation

Key Figures

Figure 22.3 ATP links catabolism and anabolism
Figure 22.4 Structures of ATP and ADP
Figure 22.5 Structure of NAD^+
Figure 22.6 Structure of FAD
Figure 22.7 Structure of FMN
Figure 22.8 $NADP^+$ links oxidation reactions of catabolism to reductions of anabolism

22.5 Enzymes

A. Enzymes are proteins that are catalysts for biological reactions.
   1. the function of an enzyme depends on its structure
   2. conditions that denature proteins cause enzymes to lose catalytic activity

B. Catalysis takes place at an active site on an enzyme's surface
   1. part of the active site binds the target molecule, called the substrate
      a. binding is specific for substrates of certain structure and polarity; complementarity of structure includes stereochemistry, shape, electrical charge, and hydrophobic-hydrophilic factors
      b. some active sites are specific for a common structural feature and have broad-based activity
      c. some active sites are specific for one molecule
      d. some active sites can be changed by binding to substrate; called inducible fit
   2. part of the active site is catalytic

C. Some enzymes are biologically active only when a cofactor is present (Table 22.2)
   1. the noncatalytic protein (apoenzyme) part combines with a cofactor to form the catalytic enzyme (holoenzyme)
   2. cofactors may be inorganic ions or small organic molecules such as vitamins (Table 22.3)

Key Terms apoenzyme, coenzyme, cofactor, competitive inhibitor, complementarity, holoenzyme, inducible fit, noncompetitive inhibition

Key Tables

Table 22.2 Inorganic cofactors and their enzymes
Table 22.3 Vitamins and corresponding coenzymes

22.6 Enzyme Classification

A. Enzymes can be classified on the basis of their function (Table 22.4).
   1. oxidoreductases catalyze electron-transfer reactions
   2. transferases catalyze the transfer of small groups or atoms between molecules
   3. hydrolases catalyze hydrolysis reactions
   4. lyases catalyze addition or removal of groups to or from double bonds
   5. isomerases catalyze the formation of isomers by rearranging atoms within a molecule
   6. ligases catalyze the formation of C–C, C–S, C–O, and C–N bonds by condensation with the use of ATP

B. Enzyme names usually have an -ase suffix attached to the name of the substrate.

Key Terms hydrolase, isomerase, ligase, lyase, oxidoreductase, transferase

Key Tables Table 22.4 Enzyme classification based on catalyzed reactions

22.7 *Enzyme Activity*

A. Enzyme activity is defined as the number of moles of substrate converted into product per unit time.
   1. the rate of an enzyme-catalyzed reaction reaches a maximum value when the enzyme reaches a saturation point (Fig. 22.9)
      a. there are a limited number of active sites
      b. when all active sites are occupied, the enzyme is operating at maximum capacity
      c. increasing the substrate concentration beyond this point has no effect on the enzyme rate
      d. the ratio of the catalyzed rate to the uncatalyzed rate gives an idea of the effectiveness of enzyme activity

Key Terms enzyme activity, enzyme rate
Key Figures
Figure 22.9 Enzyme reaction rate dependence on substrate concentration
Figure 22.10 Increase and decrease in enzyme activity with pH change.

22.8 *Control of Enzyme Activity*

A. Enzyme activity in cells is regulated in a variety of ways:
   1. covalent modification: the addition or removal of a group to or from the enzyme modifies enzyme activity
   2. feedback inhibition: the presence or absence of a certain concentration of an enzymatic product modifies the enzyme activity

B. Allosteric enzymes are controlled by activators or inhibitors that bind to the enzyme and change its shape.

Key Terms allosteric enzyme, covalent modification, feedback inhibition, regulatory enzyme
Key Figure Figure 22.11 representations of nonegulatory and regulatory enzymes

22.9 *High-Energy Compounds*

A. High-energy compounds yield energy to run chemically unfavorable reactions.
   1. they tend to be intermediates in reactions whose equilibrium constants are small
   2. they tend to have large equilibrium constants in certain reaction

Key Term high-energy compound

ARE YOU ABLE TO...

Objectives Section 22.1 Cell Structure

Are you able to...

- distinguish between eukaryotic and prokaryotic cells (Exercises 22.2, 22.3, 22.9, 22.10)
- define organelle
- identify the major structural compartments in a eukaryotic cell (Exercise 22.4)
- identify the locations of important biomolecules in the cell (Exercise 22.6)
- identify the major organelles in a eukaryotic cell (Exercise 22.8)
- identify the major functions of cell organelles (Exercises 22.5, 22.7)
- compare aerobic and anaerobic processes (Exercise 22.14)

Objectives Section 22.2 General Features of Metabolism

Are you able to...

- define metabolism
- describe the major functions of metabolic processes in animals (Exercise 22.11)
- identify the key features of metabolism
- define catabolism and anabolism (Exercises 22.12, 22.13)
- explain the relation between anabolism and catabolism
- give examples of anabolic and catabolic processes (Exercise 22.14)

Objectives Section 22.3 Stages of Catabolism

Are you able to...
- describe the three major stages of catabolism (Exercises 22.15, 22.16, 22.17, 22.18)
- identify the common product of nutrient degradation in catabolic reactions
- identify the end product of nutrient degradation in catabolic reactions (Exercise 22.18)

Objectives Section 22.4 Transformation of Nutrient Chemical Energy into New Forms

Are you able to...
- explain how chemical energy is obtained in catabolic reactions
- explain how chemical energy is used in anabolic reactions
- describe the role of ATP in energy flow (Exercise 22.19)
- explain how ATP links catabolic and anabolic processes
- explain how ATP is hydrolyzed to ADP (Exercise 22.20)
- explain the synthesis of ATP (Exercise 22.21)
- describe the roles of NADH and NAD(P)H in metabolism (Exercises 22.22, 22.23)
- describe the role of NAD^+ in metabolic oxidation and metabolic reduction processes (Exercises 22.22, 22.23)
- describe the roles of FAD and FMN in metabolic oxidations and metabolic reductions

Objectives Section 22.5 Enzymes

Are you able to...
- define enzyme (Exercise 22.24)
- explain the role of enzymes in metabolism (Exercise 22.24)
- explain the relation between enzyme function and enzyme structure
- describe the activity at an enzyme's active site (Exercise 22.25)
- explain the concept of complementarity (Exercise 22.26)
- describe the features of an active site (Exercises 22.25, 22.27)
- explain the relation between apoenzymes, holoenzymes, and cofactors (Exercise 22.28)

Objectives Section 22.6 Enzyme Classification

Are you able to...
- classify enzymes on the basis of their function
- explain the role of oxidoreductases
- explain the role of transferases
- explain the role of hydrolases
- explain the role of lyases
- identify enzyme substrates from an enzyme name

Objectives Section 22.7 Enzyme Activity

Are you able to...
- explain what is meant by enzyme activity (Exercise 22.30)
- define turnover number (Exercise 22.31)
- describe the molecular situation when an enzyme reaches its point of maximum activity

Objectives Section 22.8 Control of Enzyme Activity

Are you able to...
- explain how enzyme activity in cells can be regulated
- explain covalent modification
- explain feedback inhibition (Exercise 22.32)
- describe the role of regulatory proteins (Exercise 22.33)

Objectives Section 22.9 High-Energy Compounds

Are you able to...

- describe the role of high-energy compounds in metabolic reactions (Exercise 22.35)
- describe the characteristics of high-energy compounds

MAKING CONNECTIONS

Draw a concept map or write a paragraph describing the relation between the members of each of the following groups:

(a) eukaryotic cell, prokaryotic cell, nucleus
(b) eukaryotic cell, organelle, membrane
(c) cell membrane, compartment, cytosol
(d) cell membrane, transport activities, receptor molecules
(e) DNA synthesis, RNA synthesis, nucleus, nucleolus
(f) endoplasmic reticulum, ribosomes, biosynthesis
(g) Golgi apparatus, transport, membrane
(h) peroxisomes, hydrogen peroxide, detoxification
(i) lysosomes, degradation, hydrolysis
(j) metabolism, catabolism, anabolism
(k) metabolism, chemical energy, nutrients
(l) catabolism, nutrients, chemical energy
(m) anabolism, nutrients, chemical energy
(n) acetyl-*S*-CoA, catabolism, anabolism
(o) ATP, ADP, energy
(p) NAD, NADH, reduction
(q) enzyme, protein, biological reaction
(r) electron transfer, cofactor, oxidoreductase
(s) active site, substrate, complementarity
(t) apoenzyme, holoenzyme, cofactor
(u) enzyme activity, turnover number, substrate concentration, product concentration
(v) regulatory enzyme, feedback inhibition, covalent modification

FILL-INS

Eukaryotic and prokaryotic cells differ in the way in which _____ arranged. In prokaryotic cells, organelles _____ or _____. In eukaryotic cells, the organelles are _____. Each compartment and organelle in a eukaryotic cell has a _____. The combined total of all chemical reactions within the cell is called _____. The major functions of metabolism are the _____. Metabolism can be broken broadly into two categories: _____. Catabolic processes _____ until they finally converge at _____. Acetyl-*S*-CoA enters _____ and is oxidized to _____ with the production of _____. The energy captured through catabolism is used to synthesize _____ that fuel anabolism. _____are the molecules that participate in oxidation-reduction reactions of anabolism. _____ carries energy to energy-requiring processes in cells and _____ provide the reducing power for cellular biosynthesis. Enzymes are the _____ that mediate metabolic processes. The biological function of an enzyme is intimately related to its _____, and any factor that _____ affects the biological activity of the enzyme. An enzyme is classified on the basis of the _____ and is usually named by attaching the suffix _____. Some enzymes need cofactors to be _____. _____ are the biologically active forms containing the _____ and the noncatalytic part, the _____. Enzyme activity takes place at the _____ of the molecule. Active sites have _____ and will accept substrates that are _____ to the active site. Some enzymes have active sites that are relatively more flexible and can act on a wide variety of molecules that have _____. Enzyme activity is regulated through _____ that modify the enzyme _____. _____ or _____ molecules that bind to an enzyme and alter its shape modify regulatory enzyme activity.

Answers

Eukaryotic and prokaryotic cells differ in the way in which their internal organelles are arranged. In prokaryotic cells, organelles float freely in the cytosol or are bound to the cell membrane. In eukaryotic cells, the organelles are contained within their own membranes inside the cell cytosol. Each compartment and organelle in a eukaryotic cell has a function. The combined total of all chemical reactions within the cell is called metabolism. The major functions of metabolism are the acquisition of nutrients from the environment and, through the degradation of nutrients, the production of energy and the construction of new cellular components. Metabolism can be broken broadly into two categories: anabolism and catabolism. Catabolic processes degrade nutrients into smaller molecules until they finally converge at a common intermediate, acetyl-*S*-CoA. Acetyl-*S*-CoA enters the citric acid cycle and is oxidized to carbon dioxide and water with the production of energy. The energy captured through catabolism is used to synthesize three molecules that fuel anabolism. ATP, NADH, and NADPH are the molecules that participate in oxidation-reduction reactions of anabolism. ATP carries energy to energy-requiring processes in cells and NADH or NADPH provide the reducing power for cellular biosynthesis. Enzymes are the biological catalysts that mediate metabolic processes. The biological function of an enzyme is intimately related to its shape, and any factor that alters the shape affects the biological activity of the enzyme. An enzyme is classified on the basis of the type of reaction it catalyzes and is usually named by attaching the suffix -ase to the substrate name. Some enzymes need cofactors to be biologically active. Holoenzymes are the biologically active form containing the cofactor and the noncatalytic part, the apoenzyme. Enzyme activity takes place at the active site of the molecule. Active sites have specific shapes and charges and will accept substrates that are complementary to the active site. Some enzymes have active sites that are relatively more flexible and can act on a wide variety of molecules that have a common structural feature. Enzyme activity is regulated through feedback mechanisms that modify the enzyme output of product. Activator or inhibitor molecules that bind to an enzyme and alter its shape modify regulatory enzyme activity.

TEST YOURSELF

1. Determine whether the following statements are true or false. If false, correct the statement.
   (a) Anabolism is the process by which nutrients are degraded and energy is extracted.
   (b) ATP serves as a reduction agent for biosynthesis.
   (c) Anabolic activities include biosynthesis, muscle contraction, and active transport.
   (d) Denaturation agents do not affect enzymes.
   (e) NAD^+ is an oxidizing agent in the citric acid cycle.
2. Identify the three stages in the catabolic extraction of energy from food.
3. Identify the hydrolysis products of proteins, fats, and polysaccharides.
4. Why is ATP called a high-energy compound?
5. Why does an enzyme have a maximum activity rate?

Answers

1. (a) F; catabolism is the process by which nutrients are degraded and energy is extracted.
   (b) F; ATP carries energy to parts of the cell where energy is needed.
   (c) T.
   (d) F; enzymes are proteins and are affected in the same way as other proteins by denaturation agents.
   (e) T.
2. Digestion; production of acetyl-*S*-CoA; citric acid cycle, electron-transport chain, and oxidative phosphorylation.
3. Hydrolysis of proteins produces amino acids; hydrolysis of fats produces fatty acids and glycerol; hydrolysis of polysaccharides produces monosaccharides.
4. ATP is called a high-energy compound because it releases a large amount of free energy on hydrolysis.
5. Enzymes have maximum activity rates because they can transform only as many substrate molecules as they have active sites. When all the active sites have been saturated, the enzyme is operating at maximum capacity.

chapter 23

Carbohydrate Metabolism

OUTLINE

23.1 Glycolysis

A. Glycolysis is a series of ten reactions by which glucose is oxidized to pyruvate (Fig. 23.1).
   1. the net result is the conversion of one glucose molecule into two molecules of pyruvate, 2 ATP, and 2 NADH
   2. all enzymes for glycolysis are present in the cytoplasm
   3. glycolysis can be divided into two major stages
      a. in the first stage, hexoses are converted into glyceraldehyde-3-phosphate
      b. in the second stage, glyceraldehyde-3-phosphate is oxidized and rearranged to produce two high-energy molecules that are used to phosphorylate ADP

Key Term glycolysis
Key Figure Figure 23.1 The glycolytic pathway

23.2 Chemical Transformations in Glycolysis

A. First stage of glycolysis:
   1. glucose + ATP $\rightarrow$ glucose-6-PO_4 + ATP
      a. the enzyme hexokinase attaches a phosphate group to a glucose molecule
      b. this is not an equilibrium reaction
   2. glucose-6-PO_4 $\rightleftharpoons$ fructose-6-PO_4
      a. the enzyme glucose-6-PO_4 isomerase rearranges the glucose atoms into the fructose isomer
      b. this is an equilibrium reaction
   3. fructose-6-PO_4 + ATP $\rightarrow$ fructose-1,6-bis-PO_4 + ATP
      a. the enzyme phosphofructokinase attaches a second phosphate group to the fructose
      b. this is not an equilibrium reaction
   4. fructose-1,6-bis-PO_4 $\rightleftharpoons$ dihydroxyacetone-PO_4 + glyceraldehyde-3-PO_4 (3-PGAL)
      a. the enzyme aldolase catalyzes the cleavage of the six-carbon fructose into 2 three-carbon molecules
      b. dihydroxyacetone-PO_4 is in equilibrium with glyceraldehyde-3-PO_4 and is converted into it as glyceraldehyde-3-PO_4 is continually removed in the next step

B. Second stage of glycolysis:
   5. 3-PGAL + NAD^+ + P_i $\rightleftharpoons$ 1,3-bisphosphoglycerate (1,3-BPG) + NADH + H^+
      a. the enzyme glyceraldehyde-3-PO_4 dehydrogenase catalyzes the oxidation of the aldehyde functional group to a carboxyl group
      b. carboxyl group is esterified with inorganic phosphate forming 1,3-bisphosphoglycerate (1,3-BPG) high-energy; NAD^+ is reduced to NADH

c. this is an equilibrium reaction

6. 1,3-BPG + ADP $\rightleftharpoons$ 3-phosphoglycerate + ATP
   a. the enzyme phosphoglycerate kinase catalyzes the reaction that uses the high-energy 1,3-BPG to form ATP
   b. this is an equilibrium reaction
7. 3-phosphoglycerate $\rightleftharpoons$ 2-phosphoglycerate
   a. the enzyme phosphoglyceromutase catalyzes the rearrangement of the molecule
   b. this is an equilibrium reaction
8. 2-phosphoglycerate $\rightleftharpoons$ phosphoenolpyruvate (PEP)
   a. the enzyme enolase catalyzes the dehydration reaction that produces the phosphorylated enol of pyruvate
   b. this is an equilibrium reaction
9. PEP + ADP $\rightleftharpoons$ pyruvate + ATP
   a. the enzyme pyruvate kinase catalyzes the reaction that uses the high-energy PEP to form ATP
   b. the enol form of pyruvate rearranges to the keto form
   c. this is an equilibrium reaction
10. pyruvate + NADH + H^+ $\rightleftharpoons$ lactate + NAD^+
   a. the enzyme lactate dehydrogenase catalyzes the reduction of pyruvate to lactate
   b. reaction regenerates NAD^+
   c. NAD^+ must be regenerated from reduced NADH for glycolysis to continue
   d. this is an equilibrium reaction

C. Overall reaction of stages 1 and 2 produces 2 ATP.
   Glucose + 2 ADP + 2 P_i $\rightarrow$ 2 lactate + 2 ATP

Key Term glycolysis

23.3 Pentose Phosphate Pathway

A. The pentose phosphate pathway allows the conversion of hexoses and pentoses and the production of NADPH (reducing power)
   1. pentoses are used for nucleic acid synthesis
   2. NADPH is the reduced cofactor
   3. pathway works in combination with glycolysis
B. This pathway supplies the cell with alternative building blocks and oxidation-reducing molecules
   1. it can supply both NADPH and pentoses
   2. it can supply only NADPH when no pentoses are needed
   3. it can supply pentoses when no NADPH is needed
   4. it can supply both ATP (from glycolysis) and NADPH when both are needed

Key Term pentose phosphate pathway

23.4 Formation of Acetyl-S-CoA

A. Pyruvate from glycolysis enters the mitochondria, where it is transformed into acetyl-*S*-CoA.
   1. pyruvate + NAD^+ + CoA-SH $\rightarrow$ acetyl-*S*-CoA + NADH + H^+ + CO_2
   2. acetyl-*S*-CoA is then oxidized to carbon dioxide and water in the citric acid cycle
   3. five cofactors and three enzymes are located within the same complex (Table 23.1, Fig. 23.2)
      a. this minimizes diffusion distances
      b. there are minimal side reactions
      c. control mechanisms are integrated and coordinated
B. Steps in the formation of acetyl-*S*-CoA.
   1. pyruvate + Enz_1-TPP $\rightarrow$ Enz_1-TPP-pyruvate + CO_2
      a. pyruvate replaces hydrogen on the thiamine holoenzyme, producing CO_2 and the enzyme-pyruvate complex
      b. this is not an equilibrium reaction

2. Enz_1-TPP-pyruvate + lipoic acid holoenzyme ⇌ Enz_1-TPP + acetyl thioester with sulfhydryl group
   a. hydroxyethyl group of enzyme-pyruvate complex is transferred to the lipoic acid holoenzyme to form the precursor to acetyl-*S*-CoA
   b. this is an equilibrium reaction
3. acetylthioester precursor + CoA-SH ⇌ reduced lipoic acid + acetyl-*S*-CoA
   a. the thioester is transferred to CoA-SH, forming acetyl-*S*-CoA
   b. this is an equilibrium reaction
4. reduced lipoic acid + Enz_3-FAD ⇌ lipoic acid holoenzyme + Enz_3-$FADH_2$
   a. reducing power in reduced lipoic acid is transferred to FAD in the Enz_3-FAD complex
   b. this is an equilibrium reaction
5. Enz_3-$FADH_2$ + NAD^+ ⇌ NADH + Enz_3-FAD + H^+
   a. reducing power is transferred from Enz_3-$FADH_2$ to NAD^+ to form NADH
   b. this is an equilibrium reaction
   c. NADH yields its electrons to the electron-transport system

C. Overall process: Pyruvate + NAD^+ + CoA-SH → acetyl-*S*-CoA + NADH + H^+ + CO_2

D. The activity of the enzyme pyruvate dehydrogenase is controlled by both product inhibition and covalent modification.
   a. Le Chatelier's principle governs the activity of enzyme 1; when acetyl-*S*-CoA and NADH are present in high concentrations, steps 3 and 5 are driven in the reverse direction
   b. activators and inhibitors that are allosterically controlled covalently modify enzyme 1

Key Terms acetyl-*S*-CoA, CoA-SH

Key Figure and Tables

Figure 23.2	Structures of cofactors
Table 23.1	Enzymes and cofactors
Table 23.2	Allosteric control of enzymes

23.5 *Citric Acid Cycle*

A. The citric acid cycle is a series of reactions in which acetyl-*S*-CoA is oxidized to carbon dioxide, and coenzymes $FADH_2$ and NADH are produced (Fig. 23.3).
   1. two carbons enter the cycle as acetyl-*S*-CoA
   2. two carbons leave as CO_2
      a. acetyl-*S*-CoA + 3 NAD^+ + FAD + GDP + P_i → 2 CO_2 + CoA-SH + 3 NADH + $FADH_2$ + GTP
      b. the carbons that leave as CO_2 are from the preceding cycle
   3. oxaloacetate is regenerated to start the cycle again
   4. FAD and NAD^+ oxidizing agents are supplied by the electron-transport chain
   5. all enzymes are located in mitochondria (Table 23.3)

B. The citric acid cycle is cyclic, unlike the linear sequence of glycolysis.

Key Terms citric acid cycle, Krebs cycle, TCA cycle

Key Figure and Table

Figure 23.3	Outline of the citric acid cycle
Table 23.3	Mitochondrial location of citric acid cycle enzymes

23.6 *Reactions of the Citric Acid Cycle*

A. The citric acid cycle comprises eight reactions.
   1. acetyl-*S*-CoA + oxaloacetate ⇌ citrate + CoA-SH
      a. the enzyme citrate synthase catalyzes the combining of oxaloacetate and acetyl-*S*-CoA to form the six-carbon compound
      b. CoA-SH can participate in another round of reactions
      c. this is an equilibrium reaction

2. citrate ⇌ isocitrate
   a. the enzyme aconitase catalyzes the isomerization
   b. this is an equilibrium reaction
3. isocitrate + NAD^+ → α-ketoglutarate + CO_2 + NADH + H^+
   a. the enzyme isocitrate dehydrogenase catalyzes the reaction that creates a five-carbon α-ketoglutarate
   b. H is donated to NAD^+ to produce the reduced form, NADH
   c. this is not an equilibrium reaction
4. α-ketoglutarate + CoA-SH + NAD^+ → succinyl-*S*-CoA + CO_2 + NADH + H^+
   a. the enzyme α-ketoglutarate dehydrogenase catalyzes the reaction that produces the four-carbon succinyl-*S*-CoA
   b. H is donated to NAD^+ to produce the reduced form, NADH
   c. this is not an equilibrium reaction
5. succinyl-*S*-CoA + GDP + P_i ⇌ succinate + GTP + CoA-SH
   a. the enzyme succinyl-*S*-CoA synthetase catalyzes the reaction that regenerates CoA-SH
   b. GDP is phosphorylated to GTP, an energy-providing species similar to ATP
   c. this is an equilibrium reaction
6. succinate + E-FAD ⇌ fumarate + E-$FADH_2$
   a. the enzyme succinate dehydrogenase catalyzes the reaction that oxidizes succinate and reduces E-$FADH_2$
   b. this is an equilibrium reaction
7. fumarate ⇌ L-malate
   a. the enzyme fumarase catalyzes the reaction that oxidizes fumarate to L-malate
   b. this is an equilibrium reaction
8. L-malate + NAD^+ ⇌ oxaloacetate + NADH + H^+
   a. the enzyme malate dehydrogenase catalyzes the oxidation of L-malate to oxaloacetate
   b. this regenerates oxaloacetate to react with an incoming acetyl-*S*-CoA
   c. NAD^+ is reduced to NADH
   d. this is an equilibrium reaction

Key Term flavoprotein
Key Table Table 23.3 Mitochondrial location of citric acid cycle enzymes

23.7 Replenishment of Cycle Intermediates

A. Some citric acid intermediates are participants in other metabolic processes.
   1. α-ketoglutarate, succinate, and oxaloacetate are converted into amino acids and pulled out of the cycle
   2. these intermediates must be replenished if they are pulled out of the cycle

23.8 Gluconeogenesis

A. Glucose can be synthesized from noncarbohydrate materials; most of this activity takes place in the liver.
B. Lactate, glycerol, and certain amino acids can be used to synthesize pyruvate, which is converted into glucose.
C. The series of reactions that create glucose from noncarbohydrate materials is a linear sequence.
D. Gluconeogenesis employs a different pathway from that of glycolysis.
   1. lactate + NAD^+ ⇌ pyruvate + NADH + H^+
      a. the enzyme lactate dehydrogenase catalyzes the reaction that oxidizes lactate to pyruvate
      b. NAD^+ is reduced to NADH
      c. this is an equilibrium reaction
   2. the gluconeogenesis pathway ultimately converts pyruvate into glucose through an oxaloacetate intermediary (Fig. 23.4)
      a. pyruvate + CO_2 + ATP → oxaloacetate + ADP

b. the reaction that catalyzes the reaction is pyruvate carboxylase, which is active only in the presence of acetyl-*S*-CoA
c. oxaloacetate cannot leave the site of synthesis, the mitochondrion, and must be converted into malate to move into the cytosol, where it is reoxidized to oxaloacetate
d. oxaloacetate is converted into phosphoenolpyruvate, then into fructose, and then into glucose

Key Term gluconeogenesis
Key Figure Figure 23.4 Gluconeogenesis outline

23.9 Glycogenesis

A. Glycogen is the storage form of glucose.
B. Glycogenesis is the synthesis of glycogen from glucose.
   1. it takes place primarily in the liver and muscle cells
   2. it is an anabolic process that binds glucose units to a polysaccharide chain
   3. the process requires energy that is provided by UTP, a high-energy compound
   4. the net process is
      glucose-1-PO_4+ UTP $\rightarrow$ UDP-glucose + PP_i UDP-glucose + $(\text{glucose})_n \rightarrow (\text{glucose})_{n+1}$+ UDP

Key Terms glycogen, glycogenesis, inorganic pyrophosphate

23.10 Glycogenolysis

A. Glycogenolysis takes place in the liver; muscle cells lack an essential enzyme.
B. Glycogenolysis is the breakdown of glycogen to glucose.
   1. glucose can be cleaved from branches or from the chain
      a. $\alpha(1 \rightarrow 4)$ linkages in glycogen chain are catalytically cleaved by glycogen phosphorylase
      b. $\alpha(1 \rightarrow 6)$ linkages at glycogen branch points are catalytically cleaved by a debranching enzyme
   2. this process releases glucose as glucose-1-phosphate
      $(\text{glucose})_n + P_i \rightarrow$ glucose-1-phosphate + $(\text{glucose})_{n-1}$
   3. glucose-1-phosphate is transformed into glucose-6-phosphate, which is transformed by hydrolysis into glucose and P_i
C. The enzyme phosphorylase, which mediates the removal of a chain-end glucose, is activated through a series of steps initiated by the water-soluble hormones glucagon and epinephrine.
   1. glucagon regulates blood-glucose concentrations over long-term intervals
   2. epinephrine is responsive to blood-glucose concentration changes through the cells of the hypothalmus
   3. water-soluble hormones bind to specific receptors on cell membranes
   4. the receptor activates a membrane-bound enzyme when the hormone is bound to the receptor
   5. each enzyme may activate other enzymes and increase the number of catalytic species available (Figs. 23.5, 23.6)

Key Terms epinephrine, first messenger, glucagon, glycogenolysis, hormone, second messenger

Key Figures
Figure 23.5 Enzymatic cascade in glycogenolysis
Figure 23.6 Schematic representation of enzyme cascade

23.11 and 23.12 Electron-Transport Chain and its Enzymes

A. The electron-transport chain is a series of reactions in which protons and electrons from the oxidation of nutrients are used to reduce molecular oxygen to water.
   $O_2 + 4H^+ + 4\,e^- \rightarrow 2\,H_2O$

B. The enzymes that catalyze the reactions are located within the inner membrane of the mitochondrion (Fig. 23.7).
   1. electrons are passed from one electron carrier to another within the membrane and are finally combined with the final electron acceptor, oxygen
   2. electrons are passed in electron-transport chain
      NADH + $H^+ \rightarrow$ flavin mononucleotide $\rightarrow$ Fe-S protein $\rightarrow$ CoQ $\rightarrow$ cytochrome $b \rightarrow$ Fe-S protein $\rightarrow$ cytochrome $c_1 \rightarrow$ cytochrome $c \rightarrow$ cytochrome a, $a_3 \rightarrow O_2$
   3. the electron carriers are arranged in order of increasing affinity for electrons
   4. cytochromes are related proteins that contain an iron group
   5. ATP is synthesized at three points in the electron-transport chain owing to the generation of a proton gradient across the membrane

C. Carbon monoxide or cyanide can stop electron transport by combining with Fe^{2+} and Fe^{3+}, respectively.

Key Terms electron-transport chain, oxidative phosporylation
Key Figure Figure 23.7 Enzyme complexes of the electron-transport chain

23.13 Production of ATP

A. ATP is synthesized at three points in the electron-transport chain, owing to the generation of a proton gradient across the membrane (Fig. 23.10).
   1. this chemiosmotic mechanism produces change through an osmotic gradient
   2. increasing proton concentration difference between inner and outer membranes increases the net positive charge of the intermembrane space
   3. electron transport stops as positive charge grows
   4. protons flow inward through proton-transport channel, and ATP synthase catalyzes the formation of ATP from ADP and P_i, which allows resumption of electron transport; ATP synthase is a mitochondrial enzyme that is also a proton-transport channel

Key Term chemiosmotic mechanism
Key Figures
Figure 23.8 Origins and fate of reducing power
Figure 23.10 Formation of ATP

23.14 Mitochondrial Membrane Selectivity

A. Inner and outer mitochondrial membranes have different permeabilities.
   1. outer membrane is permeable to almost all low-molecular-weight solutes
   2. inner membrane is permeable only to solutes that correspond to specific transport systems
   3. inner membrane is impermeable to H^+, OH^-, K^+, Cl^-

B. Adenine nucleotide and phosphate translocases are two inner membrane-specific transport systems.
   1. adenine nucleotide translocase
      a. allows ADP and phosphate to enter the mitochondria and ATP to leave
      b. will not transport ATP analogues or other nucleotides
   2. phosphate translocase: promotes simultaneous transport of $H_2PO_4^-$ and H^+ into mitochondria
   3. the activities of the two systems are coordinated to bring ADP and P_i into the mitochondria from the cytosol and to send ATP out (Fig. 23.11)

C. Other specific transport systems of the inner mitochondrial membrane include:
   1. pyruvate system, which brings pyruvate into mitochondria from cytosol to participate in citric acid cycle
   2. dicarboxylate system for malate and succinate
   3. tricarboxylate system for citrate and isocitrate
   4. two systems that allow NADH (reducing power) to enter mitochondria from the cytosol

D. NADH cannot pass through the inner mitochondrial membrane and must deliver its reducing power through indirect paths.
   1. the malate-aspartate shuttle delivers reducing equivalents to complex I of electron-transport chain (Fig. 23.12)
   2. glycerol-phosphate shuttle delivers electrons directly to complex II

Key Figures

Figure 23.9 Mitochondrial membrane
Figure 23.11 Coordinated transport of P_i and ADP
Figure 23.12 Malate-aspartate shuttle

23.15 Energy Yield from Carbohydrate Catabolism

A. The net energy yield of carbohydrate catabolism:
   1. glycolysis yields 2 ATP per mole of glucose
   2. glycolysis plus the citric acid cycle yields 36–38 ATP per mole of glucose
   3. if all the energy in glucose were converted into ATP, it would yield 94 mol of ATP per mole of glucose

B. The rate of glycolysis is faster than the rate of mitochondrial oxidative phosphorylation.
   1. the concentration of glycolytic enzymes is higher than that of citric acid enzymes
   2. greater concentration allows anaerobic processes such as intense muscle activity to take place at useful rates

Key Table Table 23.4 ATP yield

ARE YOU ABLE TO...

Objectives Section 23.1 Glycolysis

Are you able to. . .

- describe the process of glycolysis (Exercises 23.1, 23.2)
- identify the products of glycolysis (Exercise 23.3)
- classify glycolysis as an anabolic or catabolic process

Objectives Section 23.2 Chemical Transformations in Glycolysis

Are you able to. . .

- outline the first stage of glycolysis (Exercise 23.2)
- outline the second stage of glycolysis (Exercise 23.11)
- identify the products of glycolysis (Exercises 2.3, 23.7)
- identify reversible and nonreversible steps (Exercise 23.4)
- identify the regulatory points in glycolysis (Exercise 23.5)
- identify oxidation and reduction steps in glycolysis (Exercise 23.6)
- define substrate-level phosphorylation (Exercise 23.8
- explain how NAD^+ is generated in glycolysis (Exercises 23.9, 23.10)
- write the net reaction for glycolysis

Objectives Section 23.3 Pentose Phosphate Pathway

Are you able to. . .

- explain what the pentose phosphate pathway does
- identify the molecular building blocks and the reducing power that they supply

Objectives Section 23.4 Formation of Acetyl-S-CoA

Are you able to. . .

- identify the location of acetyl-*S*-CoA (Exercise 23.13)

- explain the role of oxygen in the process (Exercise 23.12)
- outline the process by which acetyl-*S*-CoA is formed (Exercises 23.14, 23.15)

Objectives Sections 23.5 and 23.6 Citric Acid Cycle and Its Reactions

Are you able to. . .
- outline the steps in the citric acid cycle (Exercises 23.16, 23.17)
- identify the products of the citric acid cycle
- identify reversible and nonreversible steps
- identify the regulatory pints in the citric acid cycle (Exercise 23.22)
- identify oxidation and reduction steps in citric acid cycle
- define oxidative phosphorylation (Exercises 23.18, 23.40)
- compare the reaction sequence in glycolysis with the reaction sequence in the citric acid cycle (Exercise 23.19)
- discuss the enzyme complexes in the citric acid cycle (Exercises 23.20, 23.21)

Objectives Section 23.7 Replenishment of Cycle Intermediates

Are you able to. . .
- identify citric acid cycle intermediates that participate in other metabolic processes
- explain why some citric acid cycle intermediates need to be replenished
- explain how intermediates are replaced (Exercise 23.23)

Objectives Section 23.8 Gluconeogenesis

Are you able to. . .
- describe the process of gluconeogenesis (Exercises 23.25, 23.26, 23.27)
- identify the materials that can participate in gluconeogenesis
- compare the pathways of glycolysis and gluconeogenesis (Exercise 23.24, Expand Your Knowledge 23.50)
- identify the products of gluconeogenesis
- identify the location of gluconeogenesis (Exercise 23.28)

Objectives Section 23.9 Glycogenesis

Are you able to. . .
- describe the process of glycogenesis (Exercises 23.31, 23.32)
- identify the materials that can participate in glycogenesis
- identify the products of glycogenesis (Exercise 23.29)
- identify the location of glycogenesis (Exercise 23.30)

Objectives Section 23.10 Glycogenolysis

Are you able to. . .
- describe the process of glycogenolysis
- identify the materials that can participate in glycogenolysis
- identify the products of glycogenolysis
- identify the location of glycogenolysis (Exercise 23.39)
- describe the role of hormones in the syntheses and degradation of glycogen (Exercises 23.34, 23.35, 23.36, 23.37, 23.38)
- compare glycogenesis and glycogenolysis

Objectives Sections 23.11 and 23.12 Electron-Transport Chain and Its Enzymes

Are you able to. . .
- describe the activities of the electron-transport chain (Exercise 23.4)
- outline the series of events in the electron-transport chain
- identify the products of reactions of the electron-transport chain
- describe the types of proteins and their roles in electron transport (Exercises 23.42, 23.43)

Objectives Section 23.13 Production of ATP

Are you able to. . .

- describe how ATP is synthesized by operation of the electron-transport chain (Exercises 23.44, 23.46, Expand Your Knowledge 23.51)
- define chemiosmotic mechanism

Objectives Section 23.14 Mitochondrial Membrane Selectivity

Are you able to. . .

- compare the permeabilities of the inner and outer mitochondrial membranes (Exercises 23.45, 23.46)
- explain how NADH delivers its reducing power to the electron-transport chain (Expand Your Knowledge 23.53)
- define the specific transport system

Objectives Section 23.15 Energy Yield form Carbohydrate Catabolism

Are you able to. . .

- compare the efficiencies of glycolysis and the citric acid cycle (Exercise 23.48, 23.49, Expand Your Knowledge 23.54)
- explain the difference in the relative efficiencies (Exercises 23.49, 23.49, Expand Your Knowledge 23.54)
- predict the ATP yield per mole of glucose for both glycolysis and the citric acid cycle (Exercise 23.47)

MAKING CONNECTIONS

Draw a concept map or write a paragraph describing the relation between the members of each of the following groups:

(a) glycolysis, citric acid cycle, electron-transport chain
(b) glycolysis, carbohydrate, oxygen
(c) glycolysis, hexose, pentose
(d) glycolysis, glucose, pyruvate
(e) pyruvate, citric acid cycle, carbon dioxide
(f) allosteric enzyme, citric acid cycle, inhibitor, activator
(g) ATP, glycolysis, citric acid cycle, electron-transport chain
(h) substrate-level phosphorylation, oxidative phosphorylation, electron-transport chain
(i) glycogen, glucose, glycogenesis
(j) gluconeogenesis, glucose, amino acid
(k) glycogenolysis, glycogen, glycogenesis
(l) glucagon, epinephrine, glucose
(m) mitochondrial membrane, NADH, enzyme
(n) enzyme, electron-transport chain, cytochrome

FILL-INS

Glycolysis is a _____ that convert one molecule of _____ into two molecules of _____. Two molecules of _____ and two molecules of _____ also are produced. Glycolysis takes place in the _____ and is controlled by _____ in the series. _____ and _____ are both converted into intermediates that enter the glycolytic pathway. In humans, pyruvate can take part in either of two processes: it can enter the _____ and be converted into _____ under _____ conditions or it can be converted into _____ under _____ conditions. The citric acid cycle is an _____ pathway that processes _____ from _____. Carbon leaves as _____, and the hydrogen atoms and electrons leave as _____. Three _____ regulate the citric acid cycle. _____ levels in the cell direct enzyme activity. Electrons from NADH and $FADH_2$ are passed along to _____ in the electron-transport chain. A series of reactions pass the electrons along to proteins that have _____ for electrons. Four of the proteins are _____, which contain _____ that accepts and transfers the electrons. NAD^+ and FAD are regenerated as _____ release their hydrogen and electrons. There are _____ in the electron-transport chain where ADP is converted into ATP. The _____ explains the synthesis of ATP in response to a flow of protons across the inner mitochondrial membrane. The _____ of glucose produces 36–38 ATP compared with the 2 ATP produced by _____. When blood-glucose levels are high, excess glucose is converted into _____ by the process of _____. Glycogen stored in the liver and muscle is broken down into _____ by the process of _____. The hormone _____ stimulates the conversion of glycogen into glucose and _____ blood-sugar levels. _____ is released in response to environmental factors that increase stress and anxiety, and it stimulates the release of glucose from glycogen, thus _____ blood-sugar levels. Noncarbohydrate molecules can be used to generate glucose in the process of _____. When carbohydrate intake is low, _____ can be converted into glucose in the liver.

Answers

Glycolysis is a series of ten reactions that convert one molecule of glucose into two molecules of pyruvate. Two molecules of ATP and two molecules of NADH also are produced. Glycolysis takes place in the cytosol and is controlled by three key regulating enzymes in the series. Hexoses and fructoses are both converted to intermediates that enter the glycolytic pathway. In humans, pyruvate can take part in either of two processes: it can enter the mitochondria and be converted into acetyl-*S*-CoA under aerobic conditions or it can be converted into lactate under anaerobic conditions. The citric acid cycle is an eight-reaction pathway that processes acetyl-*S*-CoA from glycolysis. Carbon leaves as carbon dioxide, and the hydrogen atoms and electrons leave as NADH and $FADH_2$. Three allosteric enzymes regulate the citric acid cycle. ATP levels in the cell direct enzyme activity. Electrons from NADH and $FADH_2$ are passed along to molecular oxygen in the electron-transport chain. A series of reactions pass the electrons along to proteins that have increasing affinity for electrons. Four of the proteins are cytochromes, which contain iron that accepts and transfers the electrons. NAD^+ and FAD are regenerated as NADH and $FADH_2$ release their hydrogen and electrons. There are three places in the electron-transport chain where ADP is converted into ATP. The chemiosmotic mechanism explains the synthesis of ATP in response to a flow of protons across the inner mitochondrial membrane. The complete oxidation of glucose produces 36–38 ATP compared with the 2 ATP produced by glycolysis. When blood-glucose levels are high, excess glucose is converted to glycogen by the process of glycogenesis. Glycogen stored in the liver and muscle is broken down to glucose by the process of glycogenolysis. The hormone glucagon stimulates the conversion of glycogen into glucose and raises blood-sugar levels. Epinephrine is released in response to environmental factors that increase stress and anxiety, and it stimulates the release of glucose from glycogen, thus increasing blood-sugar levels. Noncarbohydrate molecules can be used to generate glucose in the process of gluconeogenesis. When carbohydrate intake is low, lactate, glycerol, and some amino acids can be converted into glucose in the liver.

TEST YOURSELF

1. Choose a process from Column B that best fits the description in Column A

Column A	Column B
(a) A series of reactions by which glucose is oxidized to pyruvate	(1) glycolysis
(b) ADP is converted into ATP when this process is coupled to the electron-transport chain	(2) glycogenolysis
(c) A series of reactions in which acetyl-*S*-CoA is oxidized to carbon dioxide with the production of NADH and $FADH_2$	(3) gluconeogenesis
(d) Glucose is synthesized from noncarbohydrate molecules	(4) glycogenesis
(e) Glycogen is synthesized from glucose	(5) oxidative phosphorylation
(f) Glycogen is broken down into glucose	(6) electron-transport chain
(g) A series of reactions in which electrons and hydrogens from oxidized glucose are used to reduce molecular oxygen to water	(7) citric acid cycle

2. Write the name of the process in which each of the following enzymes is a participant:
   (a) Phosphofructokinase
   (b) Lactic dehydrogenase
   (c) Cytochrome *c*

3. Briefly explain what each of the enzymes in Question 2 does in the process in which it participates.
4. Fill in the blanks with the name of the process that links the two molecules.

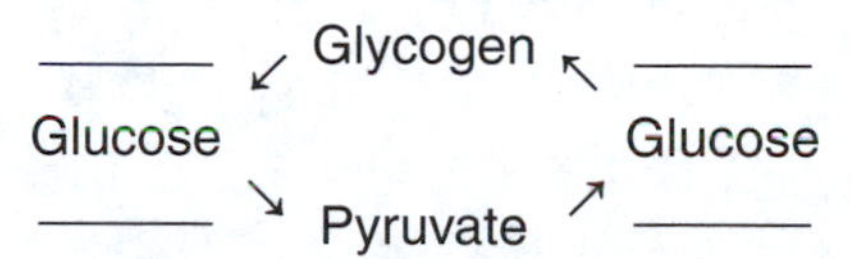

Answers

1. (a) 1; (b) 5; (c) 7; (d) 3; (e) 4; (f) 2; (g) 6.
2. (a) Glycolysis
   (b) Glycolysis
   (c) Electron-transport chain
3. (a) phosphofructokinase; catalyzes phosphorylation of fructose-6-PO_4.
   (b) lactic dehydrogenase; catalyzes the reduction of pyruvate to lactate with the production of NAD^+.
   (c) cytochrome *c*; contains iron centers that undergo reversible oxidation-reduction reactions.
4. 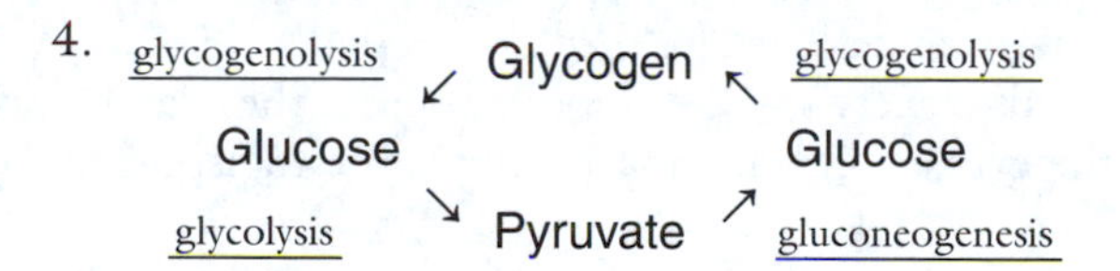

chapter 24
Fatty Acid Metabolism

OUTLINE

24.1 Fatty Acid Mobilization

A. When the body needs fatty acids for energy, hormones from the endocrine system interact with stored triglycerides in adipose tissue in a process called fatty acid mobilization.
   1. epinephrine stimulates the hydrolysis of triacylglycerides to fatty acids and glycerol
   2. fatty acids and glycerol enter the bloodstream
   3. mobilized fatty acids combine with serum albumin, a plasma protein, to form a lipoprotein; the lipoprotein transports fatty acids to where they are needed
      a. fatty acids must be transformed to acetyl-*S*-CoA thioester before they can be oxidized
      b. this process takes place in the cytosol and the molecule must be transported to mitochondria for oxidation
      c. the acetyl-*S*-CoA–fatty acid cannot enter the mitochondria and must be converted into a carnitine derivative to dissolve in the membrane
   4. glycerol is water soluble and is transported by blood to where its needed
      a. glycerol is processed through the glycolytic pathway, where it is transformed into pyruvate through a series of reactions
      b. pyruvate is oxidized in the citric acid cycle or converted into glucose through gluconeogenesis

24.2 Fatty Acid Oxidation

A. Fatty acids must be transformed into lipid-soluble derivatives to enter the mitochondria
B. After being transported into mitochondria, the degradation of the fatty acid proceeds through a β-oxidation process.
   1. β-oxidation is a catabolic process
   2. the process cleaves the fatty acid chain into two-carbon fragments of acetyl-*S*-CoA
   3. the process also produces reduced coenzymes NADH and $FADH_2$, which can enter the electron-transport chain
C. The oxidation process comprises four reactions that oxidize a β-carbon to a ketone.
   1. dehydrogenation of the fatty-acetyl-*S*-CoA produces a double bond between the α- and β-carbons
   2. hydration of the double bond produces an alcohol at the β-carbon
   3. dehydrogenation creates a carbonyl at the β-carbon
   4. the acetyl-*S*-CoA is released and a new fatty acid fragment that is two carbons shorter repeats the process

D. Each pass through the oxidation sequence produces a fatty acid chain that is two carbons shorter than the previous molecule. The last round of β-oxidation produces two molecules of acetyl-*S*-CoA.
   1. the acetyl-*S*-CoA enters the citric acid cycle and is oxidized to carbon dioxide and water
   2. each pair of hydrogens removed from the fatty acid chain shows up as $FADH_2$ and NADH

E. The energy yield from fatty acid oxidation is higher than that from glucose oxidation.

Key Term β-carbon, carnitine

24.3 Ketone Bodies and Cholesterol

A. Ketone bodies are produced from acetyl-*S*-CoA when it is metabolized to acetoacetate, β-hydroxybutyrate, and acetone in the liver.
   1. this happens when a glucose supply is unavailable and the body turns to fatty acids for energy
      a. the level of glycolysis decreases as glucose concentration decreases
      b. the amount of oxaloacetate produced is reduced
      c. as cells turn to gluconeogenesis to make their own glucose, oxaloacetate is diminished further
      d. the citric acid cycle responds to decreased oxaloacetate by decreasing its activity
      e. more acetyl-*S*-CoA is produced by fatty acid oxidation than can be processed by the citric acid cycle
      f. the concentration of acetyl-*S*-CoA increases and the excess is converted in the liver into ketone bodies: acetoacetate, β-hydroxybutyrate, and acetone, which are carried to tissues to be oxidized for energy; this condition is called ketosis

B. Conditions such as untreated diabetes mellitus, fasting, and low-carbohydrate diets can produce ketosis.

Key Terms ketone body, ketosis

24.4 Biosynthesis of Fatty Acids

A. Fatty acid synthesis takes place in the cytosol when excess food is taken in.
   1. concentrations of ATP, NADPH, and acetyl-*S*-CoA are high under these conditions
   2. acetyl-*S*-CoA leaves mitochondria as citrate and is cleaved into acetyl-*S*-CoA and oxaloacetate
   3. fatty acid chains are built up two carbons at a time by the fatty acid synthetase system
      a. the system consists of a single multifunctional polypeptide chain and an acyl carrier protein, ACP, to which the intermediates are attached
      b. the overall reaction for the synthesis of palmitoyl-*S*-CoA from acetyl-*S*-CoA is

$$\text{8 acetyl-}S\text{-CoA} + 7\ ATP^{4-} + 14\ NADPH + 7\ H^+ \rightarrow \text{palmitoyl-}S\text{-CoA} + 14\ NADP^+ + 7\ \text{CoA-SH} + 7\ ADP^{3-} + 7\ P_i^{2-}$$

B. Biosynthesis of fatty acids takes place primarily in the liver.
   1. the liver can lengthen, shorten, saturate, or unsaturate fatty acid chains
   2. the body cannot synthesize polyunsaturated fatty acids
   3. linoleic and linolenic acid in diet can be converted into other polyunsaturated fatty acids

C. Fatty acid synthesis is regulated by the hormones glucagon and insulin.

24.5 Biosynthesis of Triacylglycerols

A. Excess calories are stored as triglycerides in fat cells.

B. Insulin regulates the formation of triglycerides.

C. Triglycerides are formed by the attachment of three fatty acids to a glycerol molecule through condensation reactions.
   1. glyceraldehyde-3-phosphate is converted into glycerol
   2. acetyl-*S*-CoA is converted into a fatty acid
   3. the initial product of reaction between two fatty acyl-*S*-CoA and one molecule of glycerol-3-PO_4 is a diacylglycerol called phosphatidic acid, which is ionized at cellular pH; this molecule is an important intermediate in membrane lipid synthesis

4. phospatidic acid is hydrolyzed to a diacylglycerol, which is combined with another fatty acetyl-*S*-CoA to produce the triglyceride

Key Terms diacylglycerol, phosphatidic acid

24.6 Biosynthesis of Membrane Lipids

A. Polar heads specific to each phospholipid are attached to a diacylglycerol with the use of ATP.
B. Concentrations of membrane lipids are kept constant by balancing synthesis and degradation processes.
C. Degradation of membrane lipids takes place in lysosomes, where they are broken down to water-soluble components that are recycled.
D. In some people, genetic diseases result in the absence of enzymes that degrade membrane lipids (Table 24.1)
   1. lysosomal diseases result in the build up of lipids, which disrupts cell functions
   2. these diseases can cause mental retardation or death

Key Terms lysosomal diseases, lysosome

Key Table Table 24.1 Lysosomal diseases due to faulty degradation of membrane lipids

ARE YOU ABLE TO...

Objectives Section 24.1 Fatty Acid Mobilization

Are you able to. . .

- describe fatty acid mobilization (Exercises 24.4, 24.5)
- describe the conditions under which fatty acid mobilization takes place (Exercise 24.3)
- explain the movement of fatty acids between mitochondria and cytosol (Exercises 24.5, 24.7, 24.8, Expand Knowledge 24.49)
- describe the energy requirements of fatty acid mobilization (Exercise 24.6, Expand Your Knowledge 24.50)
- identify the hormones participating in fatty acid mobilization

Objectives Section 24.2 Fatty Acid Oxidation

Are you able to. . .

- identify the location of fatty acid oxidation
- explain how fatty acid chains are shortened by β-oxidation (Exercises 24.9, 24.10, 24.11, 24.12, 24.13, 24.14, 24.15, 24.16)
- predict the products of fatty acid degradation (Exercises 24.16, 24.43)
- explain the role of acetyl-*S*-CoA in fatty acid oxidation (Exercise 24.45)
- compare the energy yield from fatty acid oxidation with that from glucose oxidation

Objectives Section 24.3 Ketone Bodies and Cholesterol

Are you able to. . .

- describe ketone bodies and how they are used (Exercises 24.17, 24.20)
- describe the relation between ketone bodies and acetyl-*S*-CoA
- describe the conditions under which ketone bodies are produced (Exercises 24.19, 24.44, Expand Your Knowledge 24.48)
- define ketosis (Exercise 24.18)
- explain the role played by HMG-*S*-CoA in lipid metabolism

Objectives Section 24.4 Biosynthesis of Fatty Acids

Are you able to. . .

- explain how fatty acid chains are synthesized (Exercises 24.21, 24.22, 24.25, 24.26, 24.27, 24.28, 24.29, 24.31, 24.32)
- predict the products of fatty acid synthesis
- describe the conditions under which fatty acid synthesis takes place (Exercises 24.21, 24.22, Expand Your Knowledge 24.47)
- explain the role of hormones in fatty acid synthesis (Exercises 24.23, 24.24)
- explain the fatty acid modifications that can be done in the liver

Section 24.5 Biosynthesis of Triacylglycerols

Are you able to. . .

- describe the conditions under which triacylglycerols are produced
- outline the steps by which triacylglycerols are synthesized (Exercises 24.33, 24.34)
- describe the role of hormones in triacylglycerol synthesis
- explain the role of triacylglycerols (Exercise 24.2)
- discuss the energy requirements of triacylglycerol synthesis (Exercise 24.35)

Objectives Section 24.6 Biosynthesis of Membrane Lipids

Are you able to. . .

- outline the steps by which membrane lipids are synthesized (Exercises 24.37, 24.38)
- describe the role of lysosomes in maintaining lipid membrane concentration
- explain the relation between enzymes and lysosomal diseases (Exercises 24.39, 24.40)
- identify some lysosomal diseases (Table 24.1)
- identify some consequences of lysosomal diseases

MAKING CONNECTIONS

Draw a concept map or write a paragraph describing the relation between the members of each of the following groups:

(a) fatty acid mobilization, glycogen, energy
(b) fatty acid mobilization, cytosol, mitochondria
(c) fatty acid mobilization, fatty acid chain, glycerol
(d) fatty acid oxidation, cytosol, mitochondria
(e) β-oxidation, acetyl-*S*-CoA, fatty acid
(f) ketone bodies, cholesterol, fatty acid oxidation, citric acid cycle
(g) fatty acid synthesis, acetyl-*S*-CoA, enzyme
(h) fatty acid chain modification, essential fatty acids, liver
(i) triglyceride, caloric intake, phosphatidic acid
(j) lipid membrane, diacylglycerol, lysosome
(k) lipid membrane, hydrophobic, hydrophilic
(l) mutation, lysosome, lysosomal disease

FILL-INS

When energy is needed above that provided by stored carbohydrates, stored _____ are degraded. Hydrolysis reactions produce _____ from triglycerides. The fatty acids are _____ in the _____ and transported into the _____ as _____ derivatives. Fatty acids are _____ in the mitochondria by _____. This process reduces fatty acid chains by _____ through a _____ sequence of reactions. The acetyl-*S*-CoA formed enters the _____ and is oxidized to carbon dioxide and water. The glycerol produced from fatty acid mobilization enters the _____ after _____ to a glycolytic intermediate. Glycerol is ultimately converted into _____ or oxidized to _____. Excessive fatty acid oxidation _____ the usual metabolic pathways and results in the production of _____: _____. The presence of these molecules can trigger a condition called _____, which is seen in _____ and certain _____ diets. An intermediate in ketone body synthesis is a _____ that is important in the synthesis of _____. Fatty acid synthesis uses _____ as a starting material and takes place in the _____. Energy supplied by _____ is used to join _____ fragments to a growing fatty acid molecule. _____ complex is a multifunctional _____ that mediates _____ activities on different areas of the protein. Triacylglycerols are created in response to _____. An initial _____, is acylated by a third fatty acetyl-*S*-CoA to produce the _____. Membrane lipids have a _____ head and a_____ tail. The hydrophilic head is attached to a _____ in an activated form. Lipid membrane concentration is maintained by a balance of _____ processes. Some genetic diseases impair the ability of _____ to degrade membrane lipids and cause _____ diseases. The consequences of these diseases are severe and include _____.

Answers

When energy is needed above that provided by stored carbohydrates, stored triglycerides are degraded. Hydrolysis reactions produce fatty acids and glycerol from triglycerides. The fatty acids are mobilized in the cytosol and transported into the mitochondria as carnitine derivatives. Fatty acids are oxidized in the mitochondria by β-oxidation. This process reduces fatty acid chains by two carbons through a four-step sequence of reactions. The acetyl-*S*-CoA formed enters the citric acid cycle and is oxidized to carbon dioxide and water. The glycerol produced from fatty acid mobilization enters the glycolytic pathway after modification to a glycolytic intermediate. Glycerol is ultimately converted into glucose or oxidized to carbon dioxide and water. Excessive fatty acid oxidation overloads the usual metabolic pathways and results in the production of ketone bodies: acetoacetate, D-3-hydroxybutyrate, and acetone. The presence of these molecules can trigger a condition called ketosis, which is seen in untreated diabetes mellitus and certain low-carbohydrate diets. An intermediate in ketone body synthesis is a precursor molecule that is important in the synthesis of cholesterol. Fatty acid synthesis uses acetyl-*S*-CoA as a starting material and takes place in the cytosol. Energy supplied by ATP and NADPH is used to join two-carbon fragments to a growing fatty acid molecule. Fatty acid synthetase complex is a multifunctional protein that mediates seven different enzymatic activities on different areas of the protein. Triacylglycerols are created in response to excess caloric intake. An initial diacylglycerol, phosphatidic acid, is acylated by a third fatty acetyl-*S*-CoA to produce the triglyceride. Membrane lipids have a hydrophilic head and a hydrophobic tail. The hydrophilic head is attached to a diacylglycerol in an activated form. Lipid membrane concentration is maintained by a balance of synthetic and degradation processes. Some genetic diseases impair the ability of lysosomes to degrade membrane lipids and cause lysosomal diseases. The consequences of these diseases are severe and include mental retardation and death.

TEST YOURSELF

1. Match each part (a–f) in column A with a term in column B.

Column A	**Column B**
(a) The metabolic pathway by which glycerol is catabolized	(1) β-oxidation
(b) Triglycerides are synthesized using this system	(2) acetyl-*S*-CoA
(c) The metabolic pathway by which fatty acids are catabolized	(3) glycolysis
(d) The metabolic response to excessive fatty acid oxidation	(4) diacylglycerol
(e) The starting material for the biosynthesis of lipid membranes	(5) ketone body production
(f) the starting material for fatty acid synthesis	(6) fatty acid synthase complex

2. Outline the four reactions that take place in one sequence of the oxidation spiral.
3. Where do the following molecules fit in the sequence outlined in Question 2?
   (a) H_2O (b) FAD (c) CoA-SH (d) NAD^+

4. Myristic acid is a 14-carbon fatty acid. When myristic acid is catabolized through β-oxidation, how many times does the spiral reaction sequence takes place? How many molecules of acetyl-*S*-CoA are produced by the complete β-oxidation of this molecule?
5. What modifications can the liver make to a fatty acid chain? What modifications is it unable to make?
6. One mole of the stearic acid, a fatty acid, weighs 284 grams and produces 120 mol of ATP when it is completely catabolized. An equivalent weight of glucose produces 50 mol of ATP. On this basis, how does the efficiency of fats compare with that of glucose as energy-storage molecules?

Answers

1. (a) 3; (b) 6; (c) 1; (d) 5; (e) 4; (f) 2.
2. (a) Dehydrogenation of a fatty acid produces a carbon–carbon double bond.
   (b) The unsaturated molecule is hydrated to produce an alcohol at the β-carbon
   (c) Dehydrogenation of the alcohol produces a ketone at the β-carbon
   (d) The molecule is cleaved to produce acetyl-*S*-CoA and a new fatty-acetyl-*S*-CoA that is two carbons shorter.
3. (a) H_2O is used in the hydration that produces an alcohol.
   (b) FAD is used in the first dehydrogenation that produces a double bond.
   (c) CoA-SH is used in the cleavage of the two-carbon acetyl-*S*-CoA.
   (d) NAD^+ is used in the alcohol dehydrogenation that produces a ketone.
4. Myristic acid enters into six spirals of the β-oxidation sequence and produces seven molecules of acetyl-*S*-CoA.
5. The liver can lengthen, shorten, unsaturate, or add a single double bond to a fatty acid chain. The liver is not able to add multiple double bonds to a fatty acid chain.
6. Fats contain more than twice the energy of carbohydrates, on a weight basis.

chapter 25
Amino Acid Metabolism

OUTLINE

25.1 An Overview of Amino Acid Metabolism

A. Proteins and amino acids are not stored in the body for long periods of time.
B. The metabolic pathways in which amino acids are catabolized require the removal of the amino group from the molecules.
C. The path consists of the following modifications to the abstracted amino group:
   1. it is converted into glutamate, a keto acid, by transamination; glutamate is a common product into which all amino groups from all amino acids are converted
   2. glutamate is converted into a keto acid and ammonia by deamination; this produces α-ketoglutarate and ammonium ion, NH_4^+
   3. NH_4^+, which is toxic, is converted into a nontoxic urea molecule that is excreted in the urine

25.2 Transamination and Oxidative Deamination

A. The degradation processes for unused amino acids include transamination and oxidative deamination (Fig. 25.3).
B. Transamination:
   1. transamination transfers an amino group from an amino acid to an amino group acceptor, an α-keto acid
      a. this produces a new amino acid and a new keto acid
      b. the original amino acid becomes the new keto acid and the carbon skeleton that enters the citric acid cycle
      c. the common product of transamination of various amino acids is usually glutamate
         i. transaminases are enzymes in this pathway that are specific for α-ketoglutarate as the amino group acceptor but take amino groups from a variety of amino acids
         ii. the transaminases in muscle cells are specific for pyruvate, and alanine is the major product of transamination there
      d. keto acids undergo degradation in the citric acid cycle
      e. transaminases use vitamin B_6 in the form of pyridoxal phosphate as a cofactor
   2. the levels of two transaminases in the blood can be used as a diagnostic tool for heart attacks and/or liver disease
      a. tissue damage caused by a heart or liver disease allows cell contents to leak into the blood
      b. the concentrations of the enzymes can be measured to determine the severity of the damage glutamate:pyruvate transaminase (GPT) and glutamate:oxaloacetate transaminase (GOT)

C. Oxidative deamination
1. oxidative deamination converts glutamate from transamination into ketoglutarate and ammonium ion
a. ketoglutarate is regenerated when the amino group is removed from glutamate
b. ammonia is removed from glutamate and converted into ammonium ion
c. oxidative deamination also produces NADH, which enters the electron-transport chain and eventually produces ATP

Key Figure Figure 25.3 Outline of processes by which amino groups are converted into urea

25.3 *Amino Group and Ammonia Transport*

A. Ammonium ions released in oxidative deamination are toxic and must be packaged for transport and excretion.
B. Amino acid degradation products packaged as glutamate cannot pass through the cell membrane and must be repackaged for transport.
C. Ammonium ion is converted into two different nontoxic forms for transport.
1. NH_4^+ combines with glutamate to produce glutamine, which is transported to the liver
a. glutamine, which is uncharged, can pass through the cell membrane
b. this package takes care of both the ammonium ion and the excess glutamate
c. amino groups can be temporarily stored as glutamine for later use
2. NH_4^+ produced in muscle cells is transported as the amino acid alanine through the glucose-alanine cycle, which produces glucose from pyruvate and removes ammonium ion for conversion into urea
a. reductive amination combines NH_4^+ and ketoglutarate to produce glutamate
b. glutamate undergoes transamination with pyruvate to produce alanine, an amino acid, and ketoglutarate
c. alanine can pass through cell membranes and carry ammonia to the liver
d. in the liver, alanine's amino group is transferred to ketoglutarate, forming glutamate and pyruvate
e. oxidative deamination removes NH_4^+ from glutamate or transamination with oxaloacetate produces aspartate
f. pyruvate is converted into glucose by gluconeogenesis, which returns glucose to muscle cells

Key Figures
Figure 25.1 The glucose-alanine cycle
Figure 25.3 Outline of processes by which amino groups are converted into urea

25.4 *Urea Cycle*

A. The urea cycle converts two toxic ammonium ions into a nontoxic form, urea, that can be concentrated for excretion (Fig. 25.1)
1. part of the process takes place in mitochondria, and part of the process takes place in the cytosol
2. two intermediates of the cycle are the amino acids ornithine and citrulline, which are not found in proteins
B. Reactions in mitochondria include the production of carbamoyl phosphate and its subsequent reaction with ornithine to form citrulline.
1. carbamoyl phosphate, the fuel for the urea cycle, is produced by the reaction of NH_4^+ and bicarbonate
2. citrulline leaves the mitochondria and enters the cytosol
C. Reactions in the cytosol:
1. citrulline reacts with aspartate to form argininosuccinate
a. aspartate comes from the transamination between amino acids and oxaloacetate
b. aspartate provides the second ammonium ion

2. argininosuccinate is cleaved to fumarate, a dicarboxylic acid, and arginine, which carries an amino group; fumarate enters the citric acid cycle
3. arginine is cleaved to ornithine and urea
   a. ornithine returns for another cycle
   b. urea moves from the liver into the blood and is excreted by the kidneys
4. the net reaction takes a free ammonium ion and an ammonium from aspartate and packages them into urea; the aspartate becomes fumarate, which is fed into the citric acid cycle

Key Term urea

Key Figures
Figure 25.2 The urea cycle
Figure 25.3 Outline of processes by which amino groups are converted into urea

25.5 *Oxidation of the Carbon Skeleton*

A. Deaminated amino acids form keto acids whose carbon skeletons are catabolized.
B. If the keto acids are the same as those in the glycolytic pathway and citric acid cycle (Fig. 25.4),
   1. they are oxidatively degraded by those pathways
   2. they replenish citric acid cycle intermediates
C. If the keto acids are not the same as those in the glycolytic pathway and citric acid cycle, then they are broken down to acetyl-*S*-CoA or acetoacetyl-*S*-CoA.
   1. these products can enter the citric acid cycle or be converted into ketone bodies
   2. they do not replenish citric acid cycle intermediates
D. Each amino acid follows its own degradation pathway, but a common destination for all is the citric acid cycle and complete oxidation to carbon dioxide and water.
   1. all twenty amino acids are degraded to either pyruvate, acetyl-*S*-CoA, acetoacetyl-*S*-CoA, or other intermediates of the citric acid cycle
      a. pyruvate may be used to generate energy or to synthesize glucose through gluconeogenesis
      b. amino acids that yield derivatives that can be made into glucose are called glucogenic amino acids
      c. amino acids that yield derivatives that are made into acetyl-*S*-CoA or acetoacetyl-*S*-CoA cannot be converted into glucose but can be used to make ketone bodies and fatty acids and are called ketogenic amino acids (Table 25.1)

Key Figure and Table
Figure 25.4 Relation between the catabolism of amino acids and the citric acid cycle
Table 25.1 Glucogenic and ketogenic amino acids

25.6 *Heritable Defects in Amino Acid Metabolism*

A. Defects in enzymes that catalyze steps in amino acid metabolism cause inherited diseases; any degradation step that cannot be catalyzed produces a buildup of undegraded product.
B. Defects in the pathways that metabolize phenylalanine and branched-chain amino acids produce several inherited diseases.
   1. phenylketonuria
      a. phenylalanine accumulates because the reaction that removes the amino group from phenylalanine is defective
      b. high concentrations of phenylalanine allow it to react with pyruvate to form phenylpyruvate, which is not metabolized and accumulates
      c. in young children, the accumulation of phenylpyruvate in nerve cells impairs the normal development of the brain, causing mental retardation

2. alkaptonuria
   a. a phenylalanine intermediate, homogentisate, is not metabolized and is excreted in the urine
   b. reaction with oxygen in the air causes urine to take on a black color
   c. this is a benign genetic defect in that there are no negative physiological consequences
3. maple syrup urine disease
   a. branched-chain amino acids cannot be metabolized and produce acidosis in newborns and young children
   b. mental retardation and death usually result

25.7 Biosynthesis of Amino Acids

A. The human body can synthesize ten of the amino acids that it needs.
   1. the ten amino acids that can be synthesized are called nonessential
   2. those that cannot be synthesized must be obtained from the diet and are called essential

B. Intermediates of both the glycolytic pathway and the citric acid cycle are used as starting materials for synthesis of the nonessential amino acids.
   1. glycolysis intermediates
      a. 3-phosphoglycerate → serine; serine is used to produce cysteine and glycine
      b. pyruvate → alanine
   2. citric acid intermediates
      a. oxaloacetate → aspartate; aspartate is used to produce asparagine
      b. α-ketoglutarate → glutamate; glutamate is used to produce proline, glutamine, and arginine
   3. phenylalanine (an essential amino acid) → tyrosine (the only nonessential amino acid with an aromatic side chain)

ARE YOU ABLE TO...

Objectives Section 25.1 An Overview of Amino Acid Metabolism

Are you able to . . .
- compare the storage of amino acids with that of carbohydrates and lipids (Exercise 25.3)
- compare the overall metabolic pathways of amino acids with those of carbohydrates and lipids
- describe the modifications made to the amino group as it is metabolized
- identify the end products of amino acid catabolism

Objectives Section 25.2 Transamination and Oxidative Deamination

Are you able to . . .
- describe the processes of transamination and oxidative deamination (Exercises 25.1, 25.2, 25.5, 25.6, 25.7, 25.8, Expand Your Knowledge 25.41)
- explain the role of glutamate in transamination and oxidative deamination
- explain the relation between glutamate and enzymes in the pathways that catabolize amino acids
- explain the fate of both the nitrogen group and the carbon skeleton of an amino acid when it is degraded (Exercises 25.10, 25.11, 25.12, 25.13, Expand Your Knowledge 25.41)
- define transaminase and explain how transaminases are used as diagnostic tools

Objectives Section 25.3 Amino Group and Ammonia Transport

Are you able to . . .
- describe why and how ammonium ions and amino acid degradation products are packaged for transport (Exercises 25.11, 25.12, 25.13)

- describe the glucose-alanine cycle (Fig. 25.1)
- describe reductive amination (Expand Your Knowledge 25.40)

Objectives Section 25.4 Urea Cycle

Are you able to . . .

- explain the role of the urea cycle (Fig. 25.2, Exercise 25.35)
- describe the overall processes of the urea cycle (Fig. 25.3)
- describe the processes in the urea cycle that take place in mitochondria (Exercises 25.15, 25.18, 25.19, 25.20)
- describe the processes in the urea cycle that take place in the cytosol (Exercises 25.15, 25.18, 25.19, 25.20)
- identify the amino acids in the urea cycle that are not found in proteins (Exercises 25.16, 25.17, Expand Your Knowledge 25.38)
- discuss the energy costs and production in the urea cycle (Exercises 25.18, 25.19, 25.20, 25.21)
- identify the products of the urea cycle and their fates

Objectives Section 25.5 Oxidation of the Carbon Skeleton

Are you able to . . .

- describe the process by which the amino acid carbon skeleton is oxidized
- explain the relation between amino acid metabolism and the citric acid cycle (Exercise 25.34, Expand Knowledge 25.40)
- explain the relation between amino acid metabolism and the glycolytic pathway (Exercise 25.36)
- explain the relation between amino acid metabolism and ketone bodies
- identify the intermediates into which amino acid carbon skeletons are converted (Exercises 25.26, 25.27)
- identify and explain the difference between glucogenic amino acids and ketogenic amino acids (Exercises 25.22, 25.23, 25.24, 25.25, 25.26, 25.27)

Objectives Section 25.6 Heritable Defects in Amino Acid Metabolism

Are you able to . . .

- explain why enzyme defects can cause negative physiological effects (Exercise 25.29)
- describe the relation between enzymes and genetic mutations (Exercises 25.28, 25.29)
- explain what is meant by a benign genetic defect (Exercise 25.29)
- describe phenylketonuria and its effects (Exercises 25.30, 25.31, Expand Your Knowledge 25.39)
- describe alkaptonuria and its effects (Exercises 25.30, 25.31)
- describe maple syrup urine disease (Exercises 25.30, 25.31)

Objectives Section 25.7 Biosynthesis of Amino Acids

Are you able to . . .

- compare essential and nonessential amino acids (Exercises 25.32, 25.33, Expand Your Knowledge 25.42)
- identify the molecules that are used as starting materials for building amino acids

MAKING CONNECTIONS

Draw a concept map or write a paragraph explaining the relation between the members of each of the following groups:

(a) transamination, oxidative deamination, urea cycle
(b) glutamate, ketoglutarate, ammonium ion, urea

(c) glutamine, glutamate, glucose-alanine cycle, ammonium ion
(d) ornithine, citrulline, arginine
(e) acetyl-*S*-CoA, glucose, amino acid skeleton
(f) glycolysis, citric acid cycle, amino acid catabolism
(g) liver, blood, muscle, ammonium ion
(h) glucogenic amino acid, ketogenic amino acid, catabolism
(i) essential amino acid, nonessential amino acid, diet
(j) reductive amination, glutamate, amino acid
(k) genetic defect, catabolism, amino acid
(l) phenylketonuria, alkaptonuria, maple syrup urine disease
(m) enzyme, genetic disease, amino acid, carbohydrate, lipid

FILL-INS

Proteins and amino acids are not _____ the body, as are carbohydrates and lipids. The major function of amino acids in the body is to provide the _____. Amino acids that are in excess of the body's requirements are _____, and their nitrogen is packaged in _____ for excretion. Their _____ are converted into pyruvate, acetyl-*S*-CoA, acetoacetyl-*S*-CoA, or another of the intermediates in the citric acid cycle and used to produce glucose by _____, or used to produce _____, or converted for storage as _____. The catabolism of nitrogen in amino acids takes place in three stages: _____. Transamination transfers a nitrogen group from one _____ to a _____. Oxidative deamination converts the amino acid into a _____ and _____. In the urea cycle, two _____ are packaged as _____. The carbon skeleton of an amino acid is ultimately _____ to carbon dioxide and water in the _____. _____ amino acids can be converted into glucose, whereas _____ amino acids are converted into acetyl-*S*-CoA. Ketogenic amino acids _____ in the liver. Some amino acids are both _____. Humans can synthesize ten _____ amino acids but cannot synthesize _____ amino acids, which must be provided _____. _____ are the starting materials for amino acids. When amino acid catabolism is impaired by inherited genetic mutations, diseases such as _____disease can result. These diseases are severe and lead to _____.

Answers

Proteins and amino acids are not stored in the body, as are carbohydrates and lipids. The major function of amino acids in the body is to provide the building blocks for proteins. Amino acids that are in excess of the body's requirements are degraded, and their nitrogen is packaged in urea for excretion. Their carbon skeletons are converted into pyruvate, acetyl-*S*-CoA, acetoacetyl-*S*-CoA, or another of the intermediates in the citric acid cycle and used to produce glucose by gluconeogenesis, or used to produce energy, or converted for storage as triglycerides. The catabolism of nitrogen in amino acids takes place in three stages: transamination, oxidative deamination, and the formation of urea. Transamination transfers a nitrogen group from one amino acid to a keto acid. Oxidative deamination converts the amino acid into a keto acid and ammonium ion. In the urea cycle, two ammonium ions are packaged as urea and excreted. The carbon skeleton of an amino acid is ultimately oxidized to carbon dioxide and water in the citric acid cycle. Glucogenic amino acids can be converted into glucose, whereas ketogenic amino acids are converted into acetyl-*S*-CoA. Ketogenic amino acids yield ketone bodies in the liver. Some amino acids are both ketogenic and glucogenic. Humans can synthesize ten nonessential amino acids, but cannot synthesize essential amino acids, which must be provided in the diet. Glycolysis intermediates, citric acid cycle intermediates, and phenylalanine are the starting materials for amino acids. When amino acid catabolism is impaired by inherited genetic mutations, diseases such as phenylketonuria and maple syrup urine disease can result. These diseases are severe and lead to mental retardation and even death.

TEST YOURSELF

1. Match each part (a–k) in column A with a term in column B.

Column A

(a) ammonium ion is transported from muscle cells to the liver as an amino acid
(b) ammonium ion is packaged into a nontoxic, water-soluble form
(c) the removal of an amino group, releasing ammonium ion and ketoglutarate
(d) the enzyme-catalyzed transfer of an amino group to a keto acid
(e) carbon skeleton can be used in the synthesis of glucose
(f) carbon skeleton can be converted into acetyl-*S*-CoA
(g) must be provided in the diet
(h) results from an inability to catabolize phenylpyruvate
(i) results in the production of black-pigmented urine
(j) a common conversion product of a mino acid metabolism
(k) an ammonium ion is incorporated into an amino acid by reaction with α-ketoglutarate and NADPH

Column B

(1) urea cycle
(2) glucose-alanine cycle
(3) alkaptonuria
(4) reductive amination
(5) phenylketonuria
(6) essential amino acid
(7) ketogenic amino acid
(8) glucogenic amino acid
(9) oxidative deamination
(10) transamination
(11) glutamate

2. What biomolecules other than proteins are synthesized from amino acids?
3. Identify the locations in the citric acid cycle into which carbon skeletons from amino acids enter.
4. A person who regularly consumes a high-protein diet must drink more water than usual. Explain.
5. What amino acid degradation intermediate allows excess protein to be converted into fat?
6. What amino acid degradation intermediate allows protein to be converted into glucose?

Answers

1. (a) 2; (b) 1; (c) 9; (d) 10; (e) 8; (f) 7; (g) 6; (h) 5; (i) 3; (j) 11; (k) 4.
2. Purines, pyrimidines, heme, choline, ethanolamine
3. Pyruvate, oxaloacetate, citrate (from acetyl-*S*-CoA), α-ketoglutarate, succinyl CoA, fumarate
4. Water is needed to keep urea in solution at a certain concentration. A person eating a high-protein diet consumes and produces more nitrogen than the body uses. This excess nitrogen is packaged as urea and, without additional water, the blood concentration of urea can increase.
5. Acetyl-*S*-CoA
6. Pyruvate

chapter 26

Nutrition, Nutrient Transport, and Metabolic Regulation

OUTLINE

26.1 Digestive Processes

A. The coordinated control of metabolism relies on the concentrations of nutrients in determining which metabolic pathway a cell chooses.

B. The digestive system starts at the mouth and ends at the large intestine.

C. To be absorbed, nutrients must be degraded into small molecules: amino acids, monosaccharides, fatty acids, glycerol.

D. Each class of nutrients is degraded by specific enzymes (Table 26.1).
  1. enzyme activity may be regulated by hormones
  2. enzyme activity is pH and concentration dependent

E. Proteins:
  1. no hydrolysis of proteins takes place in the mouth
  2. proteins are denatured in the acidic environment of the stomach
     a. the hormone gastrin stimulates the production of HCl and pepsinogen (Fig. 26.1)
     b. the zymogen pepsinogen is converted into the active enzyme pepsin by HCl
     c. activated pepsin catalyzes the conversion of remaining pepsinogen into pepsin
     d. pepsin cleaves large proteins into smaller peptides
  3. in the small intestine, pancreatic and small intestinal enzymes split polypeptides into tripeptides, dipeptides, and free amino acids
     a. entry of acidic stomach contents signals secretion of the hormone secretin; secretin signals the pancreas to secrete bicarbonate into the small intestine, which neutralizes the acidity of the contents
     b. amino acid entry into the small intestine stimulates secretion of enterokinase, an enzyme that converts the zymogen trypsinogen into the active form trypsin
     c. trypsin converts the zymogen chymotrypsinogen into the active form chymotrypsin
     d. both enzymes reduce polypeptides to small peptides
     e. carboxypeptidase and aminopeptidase further hydrolyze these peptides into amino acids, which are absorbed into the intestinal cells and transported through the blood to the liver

F. Carbohydrates:
  1. amylase in the mouth starts breaking down carbohydrates by hydrolysis; amylase is an enzyme secreted by salivary glands
  2. salivary enzymes are denatured in the stomach and become proteins to be degraded; no carbohydrate digestion takes place in the stomach
  3. an amylase produced by the pancreas hydrolyzes carbohydrates to small polysaccharides and disaccharides in the small intestine
     a. enzymes on the surface of intestinal cells hydrolyze the disaccharides to monosaccharides
     b. monosaccharides are absorbed by intestinal cells

4. undigested fiber passes into the large intestine, where bacterial enzymes break it down into fatty acids and gas; short-chain fatty acids are absorbed in the colon and yield energy when metabolized

G. Triglycerides:
1. in the mouth, some solid fats begin to melt as they reach body temperature, and a lipase is secreted by a salivary gland at the base of the tongue
2. some triglycerides are hydrolyzed to diglycerides in the stomach because the lipase secreted in the mouth is acid stable; the churning action of the stomach mechanically mixes fats with water and acid
3. in the intestine, a zymogen called prolipase is converted into an active lipase by enzymes
   a. in conjunction with bile salts from the gall bladder and a protein called colipase, lipase catalyzes the hydrolysis of fatty acids from triglycerides; bile salts emulsify fatty droplets and provide more accessibility to enzymes
   b. mono- and diglycerols are absorbed by intestinal cells and reassembled into triglycerides
   c. these new triglycerides are combined with proteins in chylomicrons, which are transport molecules
   d. chylomicrons move through the lymphatic system and are transported ultimately to the vascular system
   e. lipids absorbed from the blood combine with proteins from the liver to form lipoproteins

Key Terms chylomicron, lipoprotein, zymogen

Key Figure and Table

Figure 26.1 Gastric glands and cells

Table 26.1 Secretions of the human digestive system

26.2 Nutrition

A. The body needs five classes of nutrients in order to function properly (Table 26.2).
1. energy sources (for example, carbohydrates, fats)
2. essential amino acids (those that the body cannot synthesize)
3. essential fatty acids (those that the body cannot synthesize)
4. vitamins
5. minerals

B. The energy content of food is measured by the caloric content (Table 26.3).

C. The basal metabolic rate is a measure of the energy required by a body at rest 12 hours after eating.
1. this energy is the minimum needed for the body to perform most basic functions
2. this value varies according to age, sex, level of daily activity, and body weight (Table 26.4)

D. Proteins:
1. proteins provide amino acids
2. the biological value of a protein is a measure of the types of amino acids present, their proportions, and their biological availability
   a. amino acids must be available in correct amounts
   b. amino acids must be of the correct type
   c. amino acids must be digestible within the amount of time that they are in the digestive system
   d. a biological value of 100 means that (Table 26.5):
      i. a protein provides all the amino acids in the proper proportions
      ii. they are all released on digestion
      iii. they are all absorbed
      iv. a small amount of these proteins will provide enough essential amino acids
   e. a biological value of less than 100 means that:
      i. a protein is not completely digested

ii. a protein is derived from plants and is surrounded by cellulose husks
iii. a protein is deficient in one or more essential amino acids
iv. a very large amount of these proteins must be eaten to obtain enough essential amino acids

f. the chemical score of a protein is obtained on total hydrolysis and comparison of its amino acid composition with that of human milk (Table 26.5)

g. nitrogen balance is a state in which the body maintains the same amount of protein in its tissues from day to day
i. if the body adds protein, the nitrogen balance is positive
ii. if the body loses protein, the nitrogen balance is negative

E. Fatty acids:

1. the body must obtain the essential fatty acids linoleic acid and linolenic acid from the diet; these fatty acids are used to synthesize arachidonic acid, which is a precursor to many lipid molecules
2. the proportion of saturated to polyunsaturated fats in the diet is important
a. diets rich in saturated fats tend to increase cholesterol and low-density lipid concentrations and tend to decreases high-density-lipid concentrations
b. these trends are correlated with an increased incidence of heart disease

F. Vitamins:

1. vitamins are organic nutrients that the body cannot produce in the amounts needed to maintain health (Table 26.6)
a. vitamins assist enzymes in performing metabolic functions
b. vitamins are not broken down to yield energy
c. vitamins are needed in only small amounts
2. some vitamin deficiencies are more common than others
a. deficiencies of water-soluble vitamins—thiamine, riboflavin, niacin, ascorbic acid, and folic acid—are not common in developed countries but are great enough to be life threatening in undeveloped countries; these vitamins are excreted daily and must be replaced on a regular basis
b. deficiencies of fat-soluble vitamins—pyridoxine, pantothenic acid, biotin, vitamin B_{12}, and vitamins A, D, E, and K—are rare
ii. lipid-absorption disorders may cause deficiency in lipid-soluble vitamins
ii. these vitamins are stored in body fat and do not have to be replaced daily

G. Minerals:

1. minerals are inorganic elements that are required in small amounts by the body (Tables 26.7, 26.8)
2. minerals are classified as bulk or trace, depending on the daily amount required by the body

Key Terms basal metabolic rate, biological value, nitrogen, chemical score, intrinsic factor

Key Tables

Table 26.2	Nutrients required by humans
Table 26.3	Energy equivalents of nutrients
Table 26.4	Recommended daily energy allowances
Table 26.5	Chemical scores and biological values of some food proteins
Table 26.6	Vitamin needs of men 23–50 years of age
Table 26.7	Minerals required by humans
Table 26.8	Minerals and their nutritional functions

26.3 Nutrient Transport

A. Composition of blood

1. Blood supplies cells in the body with nutrients and oxygen and carries away waste materials from excretion.

2. Blood can be divided into a solid fraction that contains cells and a liquid fraction called plasma.
   a. plasma is 90% water, with the remaining 10% being 70% plasma proteins, 20% organic metabolites, and 10% inorganic salts (Tables 26.9, 26.10, 26.11)
   b. cell fraction consists of erythrocytes (red blood cells), leukocytes (white blood cells), and platelets (clotting cells)

B. Transport of lipids
1. Particles that carry lipids in the blood must have hydrophobic and hydrophilic parts.
2. A lipoprotein has a hydrophobic lipid core surrounded by a shell of amphipathic lipids and proteins (Fig. 26.2)
   a. lipoproteins are classified on the basis of the relative amounts of lipid and protein that they contain (Table 26.12)
   b. protein is more dense than lipid, so higher protein content produces higher density
      i. chylomicrons are the least dense
      ii. very low density lipoproteins (VLDLs)
      iii. low-density lipoproteins (LDLs)
      iv. high-density lipoproteins (HDLs)
3. Each type of lipoprotein has a specific function.
   a. proteins in lipoproteins solubilize specific lipids in target cells and activate enzymes that hydrolyze and unload lipids from lipoproteins
   b. cell recognition between proteins of the lipoproteins and receptors on cell membranes creates specific binding interactions
   c. chylomicrons transport dietary triglycerides from intestine to adipose tissue and cholesterol to the liver
   d. VLDLs transport cholesterol and triglycerides synthesized in the liver to other tissues; when they unload lipids, they become LDLs
   e. LDLs transport cholesterol from the liver to other cells for the construction of cell membranes and the synthesis of steroids
   f. HDLs are synthesized in the liver and remove cholesterol from cells undergoing degradation

C. Transport of oxygen
1. Oxygen enters the body through the lungs and is carried to tissues by diffusion down pressure gradients.
   a. oxygen diffuses across lung membranes and is picked up by hemoglobin
      i. four iron-containing heme groups on each hemoglobin molecule bind oxygen reversibly
         a. binding to the first heme unit increases the affinity of the other three for oxygen
         b. this process is called cooperative binding, and the extent of complexation can be measured by looking at percent saturation of hemoglobin
         c oxygen binding is pH dependent
      ii. factors that affect oxygen binding to hemoglobin
         a. the Bohr effect shows that increased pH causes in hemoglobin to bind more oxygen and that reduced pH causes oxygenated hemoglobin to unload its oxygen (Figure 26.3)
         b. decreased oxygen partial pressure stimulates the synthesis of erythrocytes, thus increasing the oxygen-carrying capacity of the blood
         c. increased carbon dioxide partial pressure causes additional unloading of oxygen at the tissue level
         d. the presence of 2,3-bisphosphoglycerate which lowers the affinity of hemoglobin for oxygen, which makes it easier to unload oxygen at decreased partial pressures of oxygen

D. Transport of carbon dioxide
   1. Carbon dioxide generated as waste in the cells diffuses into the interstitial space and then into the blood owing to a concentration gradient.

carbon dioxide partial pressure in cells
carbon dioxide partial pressure in interstitial space
carbon dioxide partial pressure in blood

   2. In the blood, a small percentage of carbon dioxide is converted into bicarbonate ion, HCO_3^-, under the influence of carbonic anhydrase which uses zinc as a cofactor
      a. $CO_2\text{ (aq)} + OH^-\text{ (aq)} + H^+\text{ (aq)} \rightleftharpoons HCO_3^-\text{ (aq)} + H^+\text{ (aq)}$
      b. packaging carbon dioxide in this way allows a greater amount to be transported in the blood
      c. HCO_3^- is exchanged for Cl^- when it diffuses out of cells in what is called the chloride shift
      d. H^+ generated in the reaction enhances the unloading of oxygen in the lungs by combining with hemoglobin
   3. About 20% of the carbon dioxide in the blood is carried as carbamate compounds that form by reaction of CO_2 and amino groups in hemoglobin; this reaction causes hemoglobin to unload oxygen.
   4. Combining the activity of carbon dioxide in the tissue and in the lungs produces the following equations (Figure 26.7):
      a. $CO_2 + H_2O \rightleftharpoons H^+ + HCO_3^-$ (tissue)
      b. $H^+ + HCO_3^- \rightleftharpoons CO_2 + H_2O$ (lungs)
      c. carbonic anhydrase catalyzes both reactions
   5. Respiratory and metabolic imbalances can produce the following conditions:
      a. respiratory acidosis occurs when CO_2 is not removed from the lungs rapidly enough to prevent its buildup; H^+ is not removed by reaction with HCO_3^- and the pH of the blood falls
      b. metabolic acidosis occurs when other physiological events cause the blood pH to fall
      c. respiratory alkalosis occurs when CO_2 is removed too rapidly from lungs, using H^+ and increasing blood pH
      d. metabolic alkalosis when other physiological events such as excessive vomiting cause the blood pH to increase

Key Terms Bohr effect, carbamate, chloride shift, chylomicron, chylomicron remnant, cooperative binding, high-density lipoprotein, interstitial space, lipoprotein particle, low-density lipoprotein, metabolic acidosis, metabolic alkalosis, P_{50}, respiratory acidosis, respiratory alkalosis, very low-density lipoprotein.

Key Figures and Tables

Figure 26.2	Structure of low-density lipoprotein
Figure 26.3	Oxygen-saturation curves for hemoglobin at two pHs
Figure 26.4	Reactions responsible for transport of oxygen and carbon dioxide in blood
Table 26.9	Components of human blood plasma
Table 26.10	Principal protein fractions of blood plasma
Table 26.11	Normal concentrations of organic substances in blood plasma
Table 26.12	Density and composition of lipoproteins
Table 26.13	Lipid levels in blood

26.4 Metabolic Characteristics of the Major Organs and Tissues

A. Different cells have different metabolic requirements and may use glucose, fatty acids, and amino acids in different proportions, preferentially, or equally.

B. The heart contains virtually no stored energy.
   1. it can use glucose, fatty acids, lactate, and ketone bodies as fuel
   2. most of its energy is derived from the oxidation of fatty acids
   3. heart muscle cannot function anaerobically, and lack of oxygen results in cell death; heart attacks (myocardial infarctions) occur when oxygenated blood is prevented from reaching the heart
C. Muscle cells choose fuel sources on the basis of the level of activity in which they take part.
   1. resting muscle uses primarily fatty acids
   2. activated muscle uses glucose from glycogen stores
   3. for short periods of time in extreme exertion, creatine phosphate is a high-energy compound that is also available as an energy source
D. Adipose tissue is composed of adipocytes that store triglycerides.
   1. chylomicrons transport fatty acids to adipocytes, where they are either absorbed for storage or complexed with serum albumin for transport
   2. glucagon in adipocytes triggers the breakdown of triglycerides, and insulin reverses it
      a. when glucagon levels increase, triglyceride breakdown exceeds synthesis and fatty acids are released from the cell
      b. when glucose levels rise, triglyceride synthesis exceeds breakdown and fatty acids are packaged for storage
E. High concentrations of sodium ions outside of kidney tubules maintain a strong osmotic gradient that withdraws water and concentrates urine.
   1. most of the oxygen consumed by the kidney is used to generate ATP for this process
   2. kidneys work aerobically and can use glucose, fatty acids, ketone bodies, and amino acids as fuels
   3. the buffering ability of the kidneys is an important part of the body's pH control system
      a. ammonia can be combined with excess H^+ to form ammonium ion that is excreted
      b. deamination of amino acids in the kidney can be reduced, thus conserving H^+
F. The liver acts as a clearinghouse, processing all absorbed nutrients, except triglycerides, and detoxifying foreign substances.
   1. glucose levels rise dramatically after a meal, and the enzyme hexokinase, which phosphorylates glucose, is overwhelmed (Fig. 26.5)
      a. glucokinase, a phosphorylating enzyme found only in the liver, converts all incoming glucose into glucose-6-phosphate
      b. the liver can convert glucose-6-phosphate into glucose when it is needed
      c. the liver can convert glucose-6-phosphate into glycogen when it is not needed
      d. glucose-6-phosphate in excess of these two needs is converted into a fatty acid and cholesterol intermediate
         i. the fatty acids are used to synthesize triglycerides and phospholipids
         ii. the cholesterol is converted in part into bile salts and stored in the gall bladder
   2. amino acids that enter the liver can be metabolized in a variety of pathways (Fig. 26.6)
      a. they can be used to replenish the liver's proteins
      b. they can be used to synthesize other tissue proteins
      c. additional amino acids can be deaminated and degraded to acetyl-*S*-CoA
         i. acetyl-*S*-CoA can be used to synthesize fatty acids
         ii. acetyl-*S*-CoA can be metabolized in the citric acid cycle to produce ATP
         iii. citric acid intermediates can be used to synthesize glucose by gluconeogenesis
         iv. citric acid intermediates can be used to synthesize glycogen by glycogenesis
      d. the liver participates in the glucose-alanine shuttle
      e. amino acids are used to build nitrogen-containing molecules such as heme, peptide hormones, and nucleotides (Fig. 26.6)
   3. fatty acids are the main oxidative fuel of the liver (Fig. 26.7)
      a. fatty acids are converted into acetyl-*S*-CoA, which produces ATP in the citric acid cycle
      b. excess acetyl-*S*-CoA is converted into ketone bodies, which supply energy through the citric acid cycle

c. fatty acids are incorporated into plasma lipoproteins synthesized in the liver; lipoproteins transport dietary lipids to adipose tissue for storage
d. free fatty acids form complexes with serum albumin and are transported to other tissues for use as oxidative fuel

G. Detoxification in the liver is done by oxidative enzyme systems that are specific for foreign substances such as drugs, food additives, and organic molecules; most products have hydroxyl groups attached in order to make them more water soluble and easier to excrete

Key Terms adipocyte, creatine phosphate, glucokinase, myocardial infarction

Key Figures and Box
Figure 26.5 Metabolic pathways for glucose in the liver
Figure 26.6 Metabolic pathways for amino acids in the liver
Figure 26.7 Metabolic pathways for fatty acids in the liver
Box 26.1 Nerve Anatomy

26.5 Cellular Communication

A. Hormones are chemical messengers.
1. they are secreted by specific glands in response to a neural or chemical signal; regulation may be in the brain or in the target organs themselves
2. they are transported by the blood to target tissues
3. they trigger a chemical response in target tissue
4. Hormones can be:
   a. low-molecular-mass molecules
   b. peptides
   c. steroids
   d. proteins
5. Target-tissue response to hormonal signals:
   a. hormone may interact with a receptor molecule on the cell surface
      i. this interaction triggers the production of a second-messenger molecule inside the cell
      ii. the second-messenger molecule modifies the activity of an enzyme or enzyme system
      iii. these hormones are rapidly deactivated and their results are short-lived; for example, epinephrine
   b. hormone may enter the cell and form a complex with a specific receptor protein
      i. the hormone-receptor complex interacts with the cell's DNA
      ii. the interaction modifies the production of a specific enzyme coded by the DNA
      iii. these hormones produce long-lasting results; for example, sex hormones

B. Brain and nerve tissue:
1. the brain does not store energy and needs a constant supply of glucose from circulating blood
2. the amount of glucose used by the brain is about 60% of the total resting human glucose consumption
3. glucose use in the brain is not mediated by insulin
4. the brain generates large electrical potentials across its cell membranes and transmits electrical signals between cells
   a. sodium-potassium concentrations are maintained against a gradient by the active transport enzyme system, sodium potassium ATPase
   b. sodium is continually pumped out of the cell and potassium is continually pumped in, with energy supplied by ATP
   c. the inside of the cell membrane is negatively charged with respect to the outside of the membrane

5. membrane depolarization by an incoming signal sets up a chain-reaction depolarization down an axon membrane (Box 26.1); this is called an action potential
6. neurotransmitters are released into the synaptic cleft when the action potential reaches a presynaptic terminal
   a. neurotransmitters diffuse to a postsynaptic terminal and bind to receptor sites
   b. to detect the next incoming signal, enzymes deactivate the neurotransmitter or it is reabsorbed into the cell
7. sensory neurons are specialized receptor cells that detect specific environmental changes in temperature, touch, or particular chemicals
   a. the intensity of environmental change can be correlated with the extent of depolarization of the cells, which is translated into a change in frequency of the action potential; the nerve discharge is the frequency of the action potential
   b. effector cells and organs are designed to carry out functions appropriate to the incoming signal
8. the hypothalamus in the brain produces many hormones and hormone-releasing substances in response to signals from the central nervous system
   a. these substances may affect systems directly or may trigger secretion of hormones by other glands
   b. signals received by the hypothalamus are relayed to the pituitary gland and from there to a target gland that releases the final hormone that causes the events
   c. the original signal becomes amplified as it travels down the pathway as larger and larger amounts of subsequent hormones or factors are released; this amplification is called a hormonal cascade

Key Term action potential, hormone, hypothalmic releasing hormone, nerve discharge, receptor cells, sensory neurons, sodium potassium ATPase

Key Figure and Table

Figure 26.8 Signal pathway for hypothalmic releasing hormones in higher animals

Table 26.14 Some hypothalmic releasing hormones

26.6 Metabolic Responses to Physiological Stress

A. Homeostasis is the normal steady state of the body.
   1. deviations from homeostatic levels of temperature, pH, chemical concentrations, and so forth, evoke physiological responses
   2. the responses are designed to restore the body to homeostatic conditions

B. The absorptive state describes the conditions in the body when nutrients are entering the blood.
   1. anabolic processes outweigh catabolic processes
   2. the pancreas is triggered to release the hormone insulin, which:
      a. enhances the entry of glucose into muscle and adipose tissue
      b. activates glycolysis in the liver
      c. increases the rate of fatty acid and triglyceride synthesis in the liver and adipose tissue
      d. inhibits gluconeogenesis in the liver
      e. increases glycogenesis in the liver and muscle
      f. increases amino acid uptake in muscle, leading to protein synthesis
      g. inhibits muscle protein degradation
   3. glucose is the principal fuel used by all cells in the absorptive state; excess is converted into glycogen and triglycerides for storage
   4. amino acids not needed for protein synthesis are converted into lipids for storage

C. The postabsorptive state describes the conditions in the body when the gastrointestinal tract is empty.
   1. metabolism shifts from anabolic to catabolic processes
   2. the presence of the hormone glucagon reverses the effects of insulin

   a. blood-glucose levels are kept constant by enhanced glycogenolysis in the liver
   b. glucose is diverted to brain cells as insulin is reduced and cells shift to other fuel sources
   c. glucagon increases lipase activity in adipose tissue and increases the release of fatty acids for oxidation

D. Metabolic shifts from anabolic to catabolic processes can also be triggered by the neurotransmitter epinephrine in response to anger or fear; this is a short-lived response.

E. Starvation is a continuation of the postabsorptive state that results in significant loss of protein mass and the presence of ketone bodies in the blood and urine.
   1. glucose supplies to the brain must be kept constant
      a. liver shifts to glucose-alanine shuttle to supply gluconeogenesis
      b. this catabolizes protein, especially in muscle
   2. body mass is lost, along with water and vitamins
   3. the oxidation of fats produces ketone bodies that the brain uses; adaptation of the brain to the use of ketone bodies reduces catabolism of protein

F. Diabetes mellitus is an insulin-related disease that is found in two forms:
   1. insulin-dependent diabetes is due to insulin not being produced and requires the injection of insulin for the body to use glucose; this is also called type-1 or juvenile-onset diabetes
   2. non-insulin-dependent diabetes, in which insulin is produced but is not effective in mediating the transport of glucose across cell membranes; this is also called type-2 or adult-onset diabetes
   3. both types are characterized by an inability of cells, other than brain cells and erythrocytes, to use glucose
   4. as in starvation, glucagon mobilizes adipose tissues to produce fatty acids for fuel
      a. ketone bodies and organic acids are produced; ketone bodies produce acetone smell on breath
      b. the normal blood pH is decreased; decreased pH can produce coma
   5. blood-glucose levels soar and the kidneys cannot absorb all the glucose, which then appears in the urine

Key Terms absorptive state, epinephrine, postabsorptive state

ARE YOU ABLE TO...

Objectives Section 26.1 Digestive Processes

Are you able to…

- explain the role of digestive processes (Exercises 26.1–26.5)
- outline the digestive process for proteins, carbohydrates, and lipids (Exercises 26.7, 26.8)
- identify the major enzymes and hormones that regulate digestion and absorption (Exercises 26.9, 26.10)
- identify the products of digestion of proteins, carbohydrates, and lipids (Exercise 26.7, 26.8)
- describe the transport of digestive products

Objectives Section 26.2 Nutrition

Are you able to…

- identify the classes of nutrients needed by the body (Table 26.2)
- explain the term basal metabolic rate (Table 26.4)
- explain the energy relation between nutrients (Table 26.3)
- identify the factors that affect basal metabolic rate
- explain the nutritional roles of carbohydrates, proteins, fatty acids, vitamins, and minerals (Tables 26.5, 26.6, 26.7, 26.8; Exercises 26.13, 26.14)
- discuss the biological value of proteins (Table 26.5, Exercises 26.11, 26.12)
- explain nitrogen balance (Exercise 26.89)
- give examples of dietary-deficiency diseases (Exercises 26.15, 26.16)

Objectives Section 26.3 Nutrient Transport

Are you able to...
- explain the role of blood
- describe the composition of blood (Tables 26.9, 26.10, 26.11, Exercises 26.17, 26.18)
- explain the process by which lipids are transported in the blood
- describe the structure of lipoproteins (Fig. 26.2)
- identify the classes of lipoproteins (Exercises 26.19, 26.20)
- explain the roles of each of the classes of lipoproteins (Exercises 26.21, 26.22, Expand Your Knowledge 26.97)
- compare the compositions of each of the classes of lipoproteins (Table 26.12)
- explain how oxygen is transported from the lungs to cells (Fig. 26.4)
- explain the role of hemoglobin in oxygen transport (Fig. 26.6, Exercise 26.24)
- identify the factors that can affect oxygen binding to hemoglobin and explain their effects (Fig. 26.3)
- explain cooperative binding (Fig. 26.6)
- explain the Bohr effect
- describe the process by which carbon dioxide is transported from cells to the lungs (Fig. 26.4, Exercise 26.27)
- explain the role of bicarbonate ion in carbon dioxide transport (Fig. 26.4, Exercise 26.29)
- explain the relation between hemoglobin and carbon dioxide (Fig. 26.4)
- identify factors that affect the transport of carbon dioxide and explain their effects
- explain the chloride shift (Exercise 26.28, Expand Knowledge 26.98)
- describe carbamates and their role in carbon dioxide transport (Exercise 26.30)
- discuss respiratory acidosis and alkalosis (Exercises 26.51, 26.32)
- discuss metabolic acidosis and alkalosis

Objectives Section 26.4 Metabolic Characteristics of the Major Organs and Tissues

Are you able to...
- discuss and compare the metabolic requirements of different cells and organs
- discuss the relation between various metabolic cycles (Exercise 26.38)
- identify the principal fuel source for brain, heart, muscle, adipose tissue, kidney, and liver cells (Exercises 26.33, 26.34, 26.35, 26.36, 26.37, 26.38, 26.39, 26.40)
- discuss the roles of each of the organs listed above (Exercises 26.47, 26.48, 26.49, 26.50, 26.51, 26.52, 26.53, 26.54, 26.57)
- explain the role of hormones in each of the metabolic processes listed above (Tables 26.5, 26.6, 26.7)
- explain how hormone cascades are generated (Exercise 26.68)
- explain how brain cells generate action potentials (Box 26.1; Exercise 26.73)
- explain the role of neurotransmitters in the transmission of nerve signals
- explain what is meant by sensory neuron
- discuss the role of chylomicrons in lipid transport (Exercises 26.41, 26.42)
- explain how the liver acts as a detoxifying agent (Exercise 26.54)
- explain the metabolic choices available in the liver for amino acids, monosaccharides, and fatty acids (Figs. 26.5, 26.6, 26.7)

Objectives Section 26.5 Cellular Communication

Are you able to...
- define hormone
- explain the role of hormones
- explain how hormones trigger reactions

Objectives Section 26.6 Metabolic Responses to Physiological Stress

Are you able to...
- define homeostasis
- describe factors that cause physiological stress
- describe the physiology of the absorptive state (Exercises 26.75, 26.77, 26.78)

- describe the physiology of the postabsorptive state (Exercises 26.76, 26.79, 26.80, 26.81)
- describe the physiology of starvation (Exercises 26.81, 26.82)
- describe the physiology of diabetes mellitus (Exercises 26.83, 26.84, 26.85, 26.86, 26.87, 26.88)
- compare the normal absorptive and postabsorptive states
- compare the normal states with starvation and diabetes

MAKING CONNECTIONS

Draw a concept map or write a paragraph describing the relation between the members of each of the following groups:

(a) hormone, neurotransmitter, electrical signal, axon
(b) metabolic process, hormone, enzyme
(c) protein, biological value, essential amino acids
(d) vitamins, minerals, fuel, enzymes
(e) glucose, amino acid, protein, energy value
(f) glucose, amino acid, protein, fuel
(g) glucose, amino acid, protein, absorption
(h) protein, carbohydrate, lipid, absorption
(i) stomach, digestion, pepsin, pepsinogen
(j) liver, glucose, amino acid, fatty acid
(k) liver, toxic material, hydroxyl group
(l) absorptive state, postabsorptive state, diabetes
(m) absorptive state, postabsorptive state, starvation
(n) carbon dioxide, oxygen, hemoglobin
(o) hemoglobin, pH, oxygen partial pressure, carbon dioxide partial pressure
(p) carbon dioxide, bicarbonate ion, pH, buffer
(q) acidosis, alkalosis, pH
(r) respiratory alkalosis, respiratory acidosis, lungs
(s) lipoprotein, chylomicron, cholesterol
(t) lipoprotein, hydrophilic, hydrophobic, cell receptors, blood
(u) VLDL, LDL, HDL

FILL-INS

Chemical messengers, such as _____ and _____ accomplish communication in a multicellular organism. _____ control the coordinated operation of organs. Metabolism includes _____ and _____ processes in cells. Foods are _____ to provide energy for _____ processes and _____ for biological molecules. Food is first digested by _____ reactions that break it into smaller molecules. The smaller molecules are further broken down into _____. The cells in the intestine can absorb only _____. Different cells have different metabolic requirements. The brain uses only _____ for its energy needs. Glucose use in the brain is not _____ dependent and, therefore, is not affected by _____. The brain is able to _____ and _____ electrical signals between cells. _____ in the brain carry chemical information between brain cells in response to _____. The brain is also the site of synthesis for many _____ and _____. Heart muscle uses _____ as its primary energy source and is able to function only in the presence of _____. Even a short period of _____ deprivation can lead to death of heart tissue, as is seen in _____. The heart does not store _____ and must be kept supplied continually. Muscle cells other than the heart are able to use _____. Energy is stored in adipose tissue as _____, which are broken down under the influence of _____. _____ reverses the stimulation of epinephrine. Kidneys use _____ as energy sources. Most of the kidney's work is done in maintaining an _____. The kidneys also help to control pH by controlling the excretion of _____, which carry H^+, and _____, which can combine with H^+. The liver receives all nutrients absorbed in the intestine with the exception of _____. The metabolic processing of amino acids, fatty acids, and glucose in the liver is determined by the _____ body. The _____ refers to the physiological conditions of the body when a meal is eaten. _____ presence and the dominance of _____ processes characterize this state. Several hours after a meal, the body is in a _____ state and must use _____ energy for its needs. _____ and the dominance of _____ processes characterize this state. The transport of carbon dioxide and oxygen in the blood is accomplished by _____ down their respective _____. _____carries oxygen from the lungs to the cells and is responsive to changes in _____. Carbon dioxide waste is carried in the blood as _____ and is enzymatically converted into _____ for excretion by the lungs. Lipids are transported in the body in _____ carriers. The _____ parts of these molecules allow the _____ core to move through the aqueous environments of blood and lymph fluids. Lipoproteins are distinguished from each other on the basis of the _____ in the molecule. More protein produces _____ molecules. _____ have the lowest protein content and are the _____ lipoprotein. Very low density lipids become low-density lipids as they _____. High-density lipoproteins are synthesized in the liver and remove _____ from degraded cells.

Answers

Chemical messengers, such as hormones and neurotransmitters, and electrical signals accomplish communication in a multicellular organism. Hormones control the coordinated operation of organs. Metabolism includes anabolic and catabolic processes in cells. Foods are catabolized to provide energy for anabolic processes and building materials for biological molecules. Food is first digested by enzymatically catalyzed reactions that break it into smaller molecules. The smaller molecules are further broken down into mono- and disaccharides, amino acids, fatty acids, and glycerol. The cells in the intestine can absorb only small molecules. Different cells have different metabolic requirements. The brain uses only glucose for its energy needs. Glucose use in the brain is not insulin dependent and, therefore, is not affected by diabetes. The brain is able to generate and transmit electrical signals between cells. Neurotransmitters in the brain carry chemical information between brain cells in response to electrical impulses. The brain is also the site of synthesis for many hormones and hormone-releasing agents. Heart muscle uses fatty acid as its primary energy source and is able to function only in the presence of oxygen. Even a short period of oxygen deprivation can lead to death of heart tissue, as is seen in myocardial infarctions. The heart does not store energy and must be kept supplied continually. Muscle cells other than the heart are able to use fatty acids, glucose from stored glycogen, and creatine phosphate. Energy is stored in adipose tissue as triglycerides, which are broken down under the influence of epinephrine. Insulin reverses the stimulation of epinephrine. Kidneys use glucose, fatty acids, ketone bodies, and amino acids as energy sources. Most of the kidney's work is done in maintaining an osmotic gradient to concentrate urine. The kidneys also help to control pH by controlling the excretion of ammonium ions, which carry H^+, and bicarbonate, which can combine with H^+. The liver receives all nutrients absorbed in the intestine with the exception of triglycerides. The metabolic processing of amino acids, fatty acids, and glucose in the liver is determined by the needs of the body. The absorptive state refers to the physiological conditions of the body when a meal is eaten. Insulin presence and the dominance of anabolic processes characterize this state. Several hours after a meal, the body is in a postabsorptive state and must use stored energy for its needs. Glucagon and the dominance of catabolic processes characterize this state. The transport of carbon dioxide and oxygen in the blood is accomplished by diffusion down their respective concentration gradients. Hemoglobin carries oxygen from the lungs to the cells and is responsive to changes in oxygen partial pressure and pH changes. Carbon dioxide waste is carried in the blood as bicarbonate ion and is enzymatically converted into carbon dioxide for excretion by the lungs. Lipids are transported in the body in lipoprotein carriers. The hydrophilic parts of these molecules allow the hydrophobic core to move through the aqueous environments of blood and lymph fluids. Lipoproteins are distinguished from each other on the basis of the relative proportions of lipid and protein in the molecule. More protein produces denser molecules. Chylomicrons have the lowest protein content and are the least-dense lipoprotein. Very low density lipids become low-density lipids as they unload their cargo of lipids. High-density lipoproteins are synthesized in the liver and remove cholesterol from degraded cells.

TEST YOURSELF

1. For each of the following organs, identify the fuel sources that it can use:
   (a) brain (b) heart (c) muscle (d) liver (e) kidney
2. Identify the part of the digestion system in which the following enzymes are found:
   (a) amylase (b) pepsin (c) colipase (d) carboxypeptidase (e) disaccharidase

3. Identify the action catalyzed by each of the enzymes in Question 2.
4. Match the hormones in column A with the functions in column B.

Column A	Column B
(a) Glucagon	(1) Triggers the release of bicarbonate ion from the pancreas
(b) Gastrin	(2) Responds to signals from nervous system and facilitates fight or flight response
(c) Insulin	(3) Stimulates the stomach to make acid
(d) Secretin	(4) Acts on the liver to promote the breakdown of glycogen to glucose
(e) Epinephrine	(5) acts on cells to increase glucose uptake

5. Match the lipoprotein or lipid in column A with the description in column B.

Column A	Column B
(a) Chylomicron	(1) Component of cell membranes
(b) VLDL	(2) Composed primarily of cholesterol; they circulate through body and make their contents available to cells
(c) LDL	(3) Transports cholesterol back to the liver from peripheral cells; composed primarily of protein
(d) HDL	(4) Synthesized in the liver to deliver triglycerides to cells
(e) Cholesterol	(5) Transports diet-derived lipids from the intestine to the rest of the body

6. Match each part (a–e) in column A to a description in column B.

Column A	Column B
(a) Carbon dioxide	(1) Carrier of oxygen in the blood
(b) Carbamate	(2) Binding of first oxygen to hemoglobin increases affinity of other heme groups for oxygen
(c) Hemoglobin	(3) Describes the relation between pH and hemoglobin's oxygen-carrying capacity
(d) Cooperative binding	(4) Hemoglobin–carbon dioxide complex
(e) Bohr effect	(5) Waste product of cellular metabolism

Answers

1. (a) glucose; (b) fatty acids, primarily, but can use glucose, lactate, and ketone bodies; (c) fatty acids, glucose, creatine phosphate; (d) fatty acids; (e) glucose, fatty acids, ketone bodies, amino acids.
2. (a) mouth; (b) stomach; (c) liver; (d) intestine; (e) within intestinal cells.
3. (a) amylase: hydrolysis of starch to small polysaccharides; (b) pepsin: cleavage of proteins to small polypeptides; (c) colipase: protein that along with bile salts enables lipases to hydrolyze triglycerides; (d) carboxypeptidase: cleavage of amino acids from the amino ends of small polypeptides; (e) disaccharidase: hydrolysis of disaccharides into monosaccharides.
4. (a) 4; (b) 3; (c) 5; (d) 1; (e) 2.
5. (a) 5; (b) 4; (c) 2; (d) 3; (e) 1.
6. (a) 5; (b) 4; (c) 1; (d) 2; (e) 3.